Just Code

Just Code

Power, Inequality, and the Political Economy of IT

Edited by

JEFFREY R. YOST and
GERARDO CON DÍAZ

Johns Hopkins University Press   Baltimore

Johns Hopkins University Press
2715 North Charles Street
Baltimore, Maryland 21218
www.press.jhu.edu

Library of Congress Cataloging-in-Publication Data is
available.
A catalog record for this book is available from the British
Library.
ISBN 978-1-4214-5211-1 (paperback)
ISBN 978-1-4214-5213-5 (ebook)
ISBN 978-1-4214-5214-2 (ebook open)

*Special discounts are available for bulk purchases of this book.
For more information, please contact Special Sales at
specialsales@jh.edu.*

EU GPSR Authorized Representative
LOGOS EUROPE, 9 rue Nicolas Poussin, 17000, La Rochelle,
France
E-mail: Contact@logoseurope.eu

Contents

Acknowledgments   vii

Introduction: Encoding an Analytic   1
Gerardo Con Díaz and Jeffrey R. Yost

PART I. HOW DOES CODE BECOME BOTH A SUBJECT AND A MEANS OF GOVERNANCE?

1. Delivering Solidarity: Platform Architecture and Collective Contention in China's Platform Economy   23
Ya-Wen Lei

2. Consent Code and Default Dramas   64
Meg Leta Jones

3. A Mirror, Not a Glass Door: Legal Code and Software Code in Practice   87
Justin Petelka, Megan Finn, Janaki Srinivasan, Elisa Oreglia, and A. P. Janani

4. Algorithmic Collusion, Modern Monopolies, and Their Market Power   109
Hamid R. Ekbia

5. Reopening the Politics of Openness in the Age of Cloud Computing: Reflections on Recent FOSS Relicensing   127
Shun-Ling Chen

6. The Great E-book Conspiracy   148
Gerardo Con Díaz

PART II. HOW DOES CODE BECOME INFUSED WITH SOCIAL VALUES, ASSUMPTIONS, AND BIASES?

7. The Standard Head   171
Stephanie Dick

8. Spanning Space and Time Barriers: Computerized Conferencing, Disability, and Citizenship   193
Elizabeth Petrick

9. Pushing Fintech: Testing Mobile Money, Financial Inclusion, and "Rural Women" in Peru 212
Mariel García Llorens

10. Corporate Culture Made Material: Ephemera and In/equity at Control Data Corporation, 1957–1975 236
Elizabeth Semler

11. Reassessing the Iconic and Unbundling the Ironic: IBM System Engineering, Gender, and Antitrust 255
Jeffrey R. Yost

12. Y2K and the Politics of Labor 274
Dylan Mulvin

PART III. WHAT DOES IT MEAN TO GRAPPLE WITH CODE?

13. From Programming to Platform Expertise: Technical Reformers and the Reinvention of Institutions 295
Shreeharsh Kelkar

14. Computers as Colonizers: British Computing Companies and Indian Technological Resistance, 1955–1975 325
Mar Hicks

15. The Mask of Humanity: Manipulation and Psychopathy at the Human-Computer Interface 344
Jennifer Alexander

16. Cryptography Goes Public: Contesting the Meaning of a New Field in the 1970s United States 372
Gili Vidan

17. Nodes and Codes: Iterating with the State in México 388
Héctor Beltrán

Epilogue: Artificial Intelligence—Braiding Irony, Paradox, and Possibility 411
Jeffrey R. Yost and Gerardo Con Díaz

Contributors 433
Index 435

Acknowledgments

As an event and a volume, *Just Code* was made possible by the generosity of the Alfred P. Sloan Foundation (grant G-2020-12686, *Just Code*) and by our University of Minnesota cosponsors: College of Science and Engineering; History of Science, Technology, and Medicine; Computer Science and Engineering; China Center; Minnesota Center for the Philosophy of Science; Writing Studies; Anthropology; Hubbard School of Journalism; Electrical and Computer Engineering; Human Rights Center; Institute for Advanced Study; School for Social Work; Center for Human Resources and Labor Studies; Center for Women, Gender, and Public Policy; Communication Studies; Geography; Environment and Policy Institute; Institute of Linguistics; Interdisciplinary Center for the Study of Global Change; and Sociology. The Department of Science and Technology Studies at the University of California, Davis, also supported this project by allowing Con Díaz the time he needed to work on it.

Yost and Con Díaz are grateful to Joshua Greenberg at the Sloan Foundation; Matthew McAdam at Johns Hopkins University Press; and the tremendous scholars who participated in the event and contributed to the volume. We would also like to highlight the contributions of Melissa Dargay (Charles Babbage Institute [CBI] Administrator and Communication Specialist), Amanda Wick (CBI Curator and Archivist), Lauren Ruhrold and Katherine Buse (Graduate Assistants during the event), and Nico Wen-Li Ong (Con Díaz's star Research Assistant, and our volume's Editorial Assistant). Con Díaz thanks Marisol de la Cadena for a very useful discussion of this volume's code analytic. We are also very grateful to the two anonymous referees recruited by Johns Hopkins University Press.

Chapter 1 was originally published as "Delivering Solidarity: Platform Architecture and Collective Contention in China's Platform Economy" in the *American Sociological Review* 86, no. 2 (2021). It was edited to match the style of other chapters in this volume and is republished here with permission.

Just Code

Introduction

Encoding an Analytic

Gerardo Con Díaz and Jeffrey R. Yost

Code is everywhere around us, and it can take countless forms. Computer code structures the software and platforms that sustain our digital lives, and legal codes can govern everything from the meaning of a signature to the availability and flow of online content. There are also informal codes that can vary immensely from one region to the next: consider an organization's dress code or codes of professionalism that dictate how we write emails. Children can speak in code, making up their own languages to send covert messages. Armed forces use cryptographic codes to confound enemy intelligence. And, of course, the secret code behind public-facing artificial intelligence (AI) systems such as ChatGPT is transforming what it means to know the world and participate in it.

This book is about code—a term we interpret broadly to include anything from software routines and bodies of law to rules of personal and professional behavior. Our goal is to explore how code structures and reinforces power relations around the world. The contributors draw from across the humanities and qualitative social sciences, but they all share an interest in using the study of code to explore how information technologies interact with the broader power imbalances that are built into and through them. Our shared goal is to go beyond showing that codes can amplify power imbalances. We want to understand *how* this amplification occurs by tracing information technologies across their broader technological, commercial, and sociocultural contexts.

To understand the challenges that code poses today, we need to acknowledge that the problems we are facing predate and exceed the codes behind ongoing media frenzies and political outrage. Codes are infused with, and have the capacity to perpetuate, the ways of knowing and being in the world that shape their design, adoption, or use. AI systems today are bringing these issues to the forefront of public discourse about technology, but the scope of the problem is much broader, and older, than what centering the analysis on AI would allow us to appreciate. Creating codes involves creating systems of governance, whether—to draw examples from this volume—the governance concerns how different electronic components behave and communicate with one another or with their users, how discussants express their thoughts to one another, or how managers and employees relate to one another and with their competitors.

Code has allowed the world's information to be more structured and more easily accessed than ever before, but the troubling reality is that it also has the capacity to amplify social biases and enforce grave injustices.[1] Labor and behavioral codes allowed the British government to push women away from the computing professions in the mid-20th century,[2] and the computer code built into Google's image search engine exhibits deep gender and racial biases against women and people of color.[3] Data encoded in New Zealand were infused with biases against the nation's Indigenous populations and allowed a child welfare agency in Pittsburgh, Pennsylvania, to place poor children in a never-ending cycle of being labeled as future victims of abuse.[4] There are just too many examples to name them all, and new ones seem to hit the news cycle every week.

It is essential to reflect on code and its governance functions by working across academic disciplines instead of within them. The scholarly wisdom on which we can draw is vast. Historians have known about code's ability to amplify power imbalances for quite some time; it comes as no surprise to learn that racial and gender biases have informed the formulation and interpretation of laws for centuries. Anthropologists and sociologists have also explored, for decades, how behavioral codes structure social interactions and regional change. And a new wave of scholarship in political science is foregrounding inequality in its analysis of politics and economics. Scholars in critical race studies, legal studies, and gender studies cut across these traditional disciplinary boundaries to uncover how biases become institutionalized and magnified across several axes of identity and belonging.

Critical code and software studies, both relatively new fields, have taken major steps in this direction. They analyze software and computer code using the immense scholarly tool kit available for interpreting texts and technologies and situate them in their broader sociocultural contexts and imaginaries. What we propose, however, is to follow ways of knowing, power relationships, and instructions and procedures as they move across *modes* of code—including but not limited to computer code. This brings us into close contact with some of the major themes in critical code and software studies (especially the politics of implementation, deployment, and maintenance). It also underscores a much bigger problem: understanding interrelationships among ways of representing the world, governing a system, and grappling with power imbalances.

A universal mapping of how movement across modes of code takes place would be a multigenerational task, assuming it were possible. What we offer is a departure point in this long-term project by aiming at one of the most powerful forces at play: *political economy*, by which we mean quite simply the interplay between political and economic systems. The connections between the two are difficult to miss nowadays: Seven of the eight largest companies in the world—Apple, Microsoft, Google, Amazon, Facebook, Alibaba, and Tencent—are information technology (IT) behemoths in fields ranging from computer and smartphone hardware to software development and digital services. In the modern era, there has never been a market concentration in one industrial field like that in IT today, nor has any single industry ever had as much power and influence on everyday lives. Those seven companies, along with lesser-known, yet sizable IT firms around the world, are increasing global wealth inequality as they rapidly grow revenues and earnings for their shareholders.

In 2020 we, coeditors Yost and Con Díaz, grappled with this new and richer scholarly landscape by organizing the symposium titled "Just Code," where a global and diverse group of speakers (many of whom are contributors to this volume) uncovered the historically, geographically, and socioeconomically contingent ways in which inequalities emerge and become embedded in everyday lives and technologies. The symposium demonstrated, decisively, that conceptualizing code broadly was a generative way of grappling with one of the most difficult problems in making sense of the modern world: understanding information technology through and across globally disparate lived experiences, industrial structures, and systems of governance.

This volume grew out of that effort. It proposes a new way of using code as an analytic and mobilizes it to interrogate the relationships among information technology, power, and political economy. We challenge the impulse to treat any one modality of code in isolation. We would object, for example, to frameworks that treat a nation's legal and regulatory codes as separate from the computer codes that enable their implementation. We posit, instead, that modalities of code develop jointly and that the mutual construction among them can transform the broader economic and political environments in which they were developed and deployed. Our overarching argument is that the interplay between multiple modalities of code and political economies perpetuates the broader sociotechnical power structures that enable their operation and development.

This approach allows us to avoid unnecessary reductionism in our efforts to explain how codes and power relate to each other. For example, rather than repeating legal scholar Lawrence Lessig's refrain "Code is law,"[5] we seek to understand more dynamic problems: how codes play a controlling and regulatory role, how they can amplify the injustices and inequities embedded in the text and interpretations of laws, and how these phenomena transform the broader socioeconomic dynamics that made them possible in the first place. Similarly, we do not seek to encapsulate this subject as variations on the popular idiom of coproduction—that ways of knowing the world and ways of governing it transform each other.[6] We do not necessarily oppose this idiom, which has been transformative across various fields and for which this volume provides some examples. However, we consider it insufficient to articulate our overarching goal of studying codes and their political economies in a way that blurs the analytical boundaries that distinguish knowing, governing, and coding.

The rest of this introduction presents our framework in three parts: First, we posit the deployment of code as an analytic in an effort to engage with interplays among representation, communication, and interaction. We embrace (and move beyond) earlier conceptions of the *political economy of code* because we allow for code to emerge and act simultaneously across its various modalities, without restricting the scope of its operation to any field of technological development and use. Second, we propose a refinement of our code analytic that allows us to focus on the relationships between code and political economy. The key to this refinement is recognizing that engagements with code occur in varying degrees of permanence—a phenomenon we can distill further by reflecting on the informational asymmetries and power relations involved. The final section introduces our contributors' work by reflecting on three major questions that our analytic allows us to tackle directly.

A note on method: We coeditors developed this analytic *after* closely studying the contributors' work. We introduced our authors to *Just Code* as a scholarly project. However, rather than imposing a more specific conceptual framework on them, we allowed one to emerge organically as we reflected on the enormous body of scholarly literature that preceded us and synthesized their chapters' collective wisdom. This method allowed us to craft an analytic that foregrounds the opportunities for revision, expansion, and synthesis that the authors' contributions highlight. It also means that our backgrounds as researchers in the United States working with Anglophone sources have infused the framework with a specific positionality. We had an extraordinary diversity of speakers at the "Just Code" conference and authors in this volume, so we have been mindful to call out limitations and opportunities in our approach. However, we hope that this work will inspire challenges and reinforcements from authors with access to literature and intellectual traditions that we do not know.

Code as Analytic

Two major scholarly transformations motivate our approach. First, in recent years, the social studies of information technology have undergone a complete overhaul. Research avenues that used to unfold in parallel across academic fields have come together and redefined one another. Hard boundaries between fields remain, especially at the institutional level, but it has become essential for scholars at all stages to participate in communities that are interdisciplinary not just by design but by necessity. Think about how difficult it would be, for example, to examine race and gender in the computing professions without acknowledging the broader politics of employment and education, or to investigate electronics manufacturing in the Global South without grappling with the colonial legacies that built local markets and economic possibilities.

Along the way, a new kind of fragmentation has crystallized: we are becoming increasingly aware of the stunning variability in the ways people and information technologies have related to each other across time, space, and socioeconomic setting. One long-term effort that has made this very apparent is the Why We Post series, published by UCL Press. Each volume in that series is an anthropological examination of social media use in a specific location: from an industrial town in China to a small seaside village in Colombia. These books have allowed the series to document local variations on global patterns: how social media does not necessarily make us more individualistic; how it can *be* education; and how it allows folks to both challenge and reinforce gendered, racialized, economic, and colonial legacies.

In the second major transformation, the upheavals the world has been experiencing since 2020 have prompted scholars to propose new synthetic frameworks for the study of political economy. For example, in 2022, four political scientists (Jacob Hacker, Alexander Hertel-Fernandez, Paul Pierson, and Kathleen Teen) coedited the book *The American Political Economy*. Grounded in comparative political economy, the work collected in that volume proposed a new thread of research called American political economy (APE). This field is "centrally concerned with the ways in which *institutional configurations* shape *coalitional politics* to produce long-term *developmental processes*" (emphasis in original).[7] The four editors proposed this tripartite method as a way of approaching what they considered to be the three most important features of US political economy: fragmentation of political authority, the constitutive role played by organized interests, and deep racial inequality.

Of course, the political economies that concerned Hacker and his colleagues were all grounded in the United States and its global histories. As much as their framework highlights the local variability of political power, it hinges on an enormous historiography that (though far from unified) can still come together under the banner of US history. In contrast, the full enormous breadth of the social studies of information technology and governance prevent us from distilling an organizing logic or analytical grammar from any one nation's history. Several major themes cut across the two and their allied fields (labor, gender, race, capitalism, and so on), but transforming one of them into *the* organizing principle for the analytic risks attributing to that theme disproportionate explanatory power.

These two transformations yield an opportunity to revise and expand how we envision the relationships between political economy and code. About 20 years ago, during the formative years of software studies, the phrase "political economy of code" was a key term. Scholars used it to encapsulate the socioeconomic factors that shape computer code as a textual and technological entity.[8] It allowed researchers to treat source code as a literary object that could not be separated from its social context, highlighting the shifts in labor, capital, and discourse that accompanied techno-scientific development. While productive, this understanding of code remained close to the machine. As one philosopher put it back then, "Code is labour crystallised in a software form that is highly flexible and which when captured may be executed indefinitely."[9]

Our approach encompasses (and exceeds) that earlier effort. We resist the phrase "political economy of code" not because we reject it as a concept but because the multiple meanings of "code" blur the phrase's meaning in ways that exceed the scope of early software studies. In our framework, the classic political economy of code becomes one of the many phenomena we could examine while tending to the broader interplay among representation, communication, and interaction. It is software studies' version of political-economic interventions that occurred in the early 2000s across several disciplines. It is neither more nor less important than its counterparts—as none of them is designed to hold the multiplicity of code in a single analytic.

We offer code as a central concept for investigating phenomena ranging from personal and institutional idiosyncrasies to transnational systems of governance and trade. Of course, code can be a specific object: at the level of an individual computer, "code" may refer to the texts that programmers write or to the sequences of ones and zeroes that a device can process. If a human being is operating this device, they may be following a dress code or a behavioral code. Some behavioral codes are institutional (e.g., a company's rules), but others are social (such as expectations of gender expression, accent, mannerism, and physical contact). Whatever building the person is sitting in will be subject to local building codes; and if the computer is connected to a network, it will be dealing with network status codes and processing encryption codes. To top it all off, this person's actions are governed by a country's legal codes, which are sometimes aligned with international legal codes like treatises.

It is the multiplicity of code as object that allows us to think "code" as concept. This can open and structure our inquiry because relations like the ones described above are inseparable from broader relationships between political systems and economic forces. The texts that programmers write at a company are subject to both that company's internal code libraries and intellectual property restrictions set by both state bureaucracies and competitors' surveillance.[10] For decades, lawyers have been finding ways of coding assets, blurring legal categories, and creating entirely new ones in ways that favor the creation of private wealth over legal consistency and social welfare.[11] Corporate expectations can punish deviance from hegemonic behavior, and laws that determine flows of capital can reinforce the racial logics on which they are grounded.[12] At the macroscopic level, global demand for a specific kind of innovator can shape state policy and perpetuate social hierarchies in the name of economic development.[13]

Following code without narrowing its meaning allows us to cut across all these different settings and political-economic phenomena. It allows us to consider the possibility that a specific snippet of source code could be the subject of a multimillion-dollar battle over legal code, as was recently the case in the blockbuster battle between Google and Oracle. It enables us to think about gendered codes of behavior and access as determinants of the kinds of problems that programmers seek to solve. And it lets us examine how intersections of identities and their associated behavioral codes affect how a person engages with a technology, how that behavior is regulated, and how the technology itself can encode these behaviors into widespread, black box technologies.

To deploy code as an analytic is to pay sustained attention to three interrelated core registers: (1) representation, (2) communication, and (3) interaction. We posit that these registers cannot be fully separated from one another and that there is a collection of phenomena—all major themes in the social studies of both information technology and governance—that allow us to move across them. These phenomena include processes such as abstraction, institutionalization, embodiment, algorithmization, and standardization. They can become connective tissues among the registers, challenging the impulse to assume that any engagement with code can unfold exclusively within any of the registers.

The code analytic asks us to move through and across these registers alongside the situations we study, not to determine which mix and relative weight of registers is the "right" one for analyzing a situation. Two examples can clarify what we mean: First, a snippet of source code is designed to cause a computer to behave in a certain way. It is written in a standard manner so that it can be communicated to someone or something. Depending on our approach to this situation, we may begin thinking in the interaction register (what a programmer wishes a computer to do), travel via algorithmization to the representation register (the creation of a computer program), and move toward communication by following the standardized language in which the programmer writes and shares this code.

In the second example, a court opinion can establish behavioral expectations by controlling a party's actions. Once again, depending on our aims, we may begin in the representation register by analyzing how each of the parties involved represented themselves, their interests, and the conflict. Then we could follow them into a legal structure (courts and their associated bureaucracies) with the power to control their interactions during and after the proceedings are completed. The outcome of this process (the court's opinion) will set behavioral expectations, and the way the court communicates it (the judges' reasoning) may determine whether the loser attempts to appeal or work around this result.

Full deployment of the analytic, however, involves embracing the excesses of the three registers in our investigation of code. As tools the registers are both descriptive and conceptual: movement through and across them, especially as they change themselves and exceed the meanings we assign to them, can help us structure descriptions of how and why people interact with codes in each setting. We may take a cue from anthropology in embracing the simultaneous generalities and specificities that code and its registers can hold together. We are thinking here of Peter Skafish, who wrote that concepts can hold multiple meanings precisely because engaging with them focuses our attention on the "divergence and interconnections" that allow them to do their work.[14] That insight certainly applies here: we seek out and follow the three registers in whatever way they express themselves, regardless of whether the terms themselves or the underlying concepts are explicitly invoked by the people and institutions we study or the analytical machinery we have mobilized so far in an analysis.

Our task is to develop the registers in a way that accommodates the disciplinary, geographical, and analytical diversity of scholarship with which we can approach code. The key to doing this task is noticing that all three registers—communication, representation, and interaction—involve relations between actors, some of them other than human. For example, Safiya Noble's work on Google Images demonstrated how search engines can perpetuate racial stereotypes by literally amplifying the biases embedded in images' metadata. All three registers are at work here, each one operating in multiple ways. Sometimes they involve humans, and other times they don't encompass anything that could be said to have a physical presence.

Recognizing the registers of code as relational enables reflection on why and how each of them can hold so many meanings, and why it is analytically essential to recognize and embrace this diversity. Marilyn Strathern once reflected that the term "relation" is "an attractor: a term that engages other terms, a concept in a field of concepts, an idea that draws in values and disseminates feelings, a substantive from which adjectives (relational) and abstractions (relationality) can be made exactly as though everyone knew what they meant."[15] (This is one key moment when our positionality becomes especially apparent. By thinking alongside Marilyn Strathern, we are borrowing from Anglophone anthropological traditions and infusing our work with a distinctly European American line of conceptual thought.)

Putting all these ideas together, we arrive here: To use code as an analytic is to tend to relations that emerge from, and can transform, what it means to communicate, represent, and interact. This attention invites inquiry into the objects we would call code (the bodies of law, the snippets of ones and zeroes, and so on), but it also pushes us to recognize that each instance is part of a broader bricolage of relations that unfold with, through, and beyond them. For example, when Virginia Eubanks followed imported child welfare software in Philadelphia, she was not concerned *only* with the software. She was thinking about how welfare law is communicated and implemented, how neoliberal markets allowed a New Zealand firm to provide software to Philadelphia, and how computer programming can perpetuate biases against Indigenous populations.

Temporality and Political Economy

How can we make this code analytic tractable, given that the three registers' relational natures make them as malleable and multiple as code itself? Our answer, perhaps because we are both trained as historians, is to structure relations by focusing on their temporality. By referring to temporality we do not mean the contextualization in time and space required for traditional historical thinking (though, of course, we welcome that). Instead, we mean to say that relations have varying degrees of permanence. Some relations are ephemeral—one-time interactions between two actors who will never meet again. Think, for example, of a person today using a credit card, or a traveling merchant in the 19th century passing through a town. Others are extremely long-term or even permanent. Relations between nations or fully subservient relationships such as enslavement could qualify in this category.

This approach is born explicitly from our own backgrounds as researchers, and we call on scholars to leverage their own positionalities in service of adding dimension and structure to the code analytic in new ways. This volume aims to study political economy, so we engage with the temporality of relations following the lead of business and economic historians Naomi Lamoreaux, Peter Temin, and Daniel Raff. Their work describes how economic actors relate to one another despite their informational asymmetry (uneven access to information relevant to their relation). They call the actors' efforts to resolve these problems "coordination mechanisms" and arrange them in a linear spectrum that we are mobilizing for our own purposes. On one end, there is pure market exchange (one-off transactions). On the other, there is pure hierarchy, permanent or (at least) extremely long-lived. In the middle are relationships where actors participate voluntarily for extended periods of time, such as repeat business at a local coffee shop.[16]

Power imbalances and informational asymmetry are tightly interwoven phenomena, regardless of the period and region we study. Caitlin Rosenthal has shown how enslavers in the American South and the Caribbean in the 18th century developed accounting practices to control and assess the economic value of enslaved people and their forced labor. Enslaved people were not privy to these metrics, but their everyday lives were inseparable from them.[17] A century later, according to James Poskett and Roger Cooter, phrenologists around the world linked scalp morphology to personalities in ways that magnified and perpetuated the values of industrial capitalism. A person could not know what their own head shapes said about them, and yet a few measurements were enough to determine their potential productivity and trustworthiness. Phrenology even became a political language that enabled reformers around the world to infuse assumptions about race and human character into their political and social projects and ambitions.[18]

The following example illustrates that the linear-relational approach is flexible enough to be used in phenomena that occur around the world, despite its roots in US business and economic history. Consider a person on a road trip and a merchant who accepts their credit card to make a purchase. This one-time interaction is made possible by contractual relations between the merchant and the credit card company's local office; a potentially long-term relation between the cardholder and the issuer; the cardholder's own short-term relation with the plastic card (which has an expiration date); and the merchant's infrastructural relation with their telecommunications provider and its hardware and software. If the merchant and the buyer live in different countries, treaties and national branches may be in play. Many of these relations are bound by contracts that stipulate specific conditions for their continuance, and all of them are governed by local and national frameworks on consumer rights, commerce, electronic transactions, data, manufacturing, and waste disposal.

Code is involved in all these relations. The code analytic pushes us to investigate this commercial exchange without creating a hierarchy of codes and without necessarily identifying every object we could call code. Our focus expands to encompass the broader dynamics among registers, using each of them to connect the transaction with broader economic relations of varying degrees of permanence. This perspective helps us reflect on how this seemingly simple scenario—a person bought a thing—could play out differently in other circumstances. Different consumer rights laws might have made the entire transaction impossible by granting the merchant the right to turn away customers based on religious faith. Other credit card fee structures might have pushed the merchant to request that the buyer use cash or buy additional items to compensate for the bank fees. The buyer might not have had a card in the first place if a shift in interest rates made it unwise to open an account, or maybe they were using a card only because they forgot their cell phone at home and couldn't use Apple Pay.

Political and economic circumstances do not necessarily shape code directly, but they can certainly influence the broader interplay of registers that code governs or from which code emerges. Generations of business and legal history have shown us how technology, governance, and business affect one another not necessarily in pairs but by having each one reconfigure the interplay between the other two. Our analytic enriches this framework by recognizing two major categories of catalysts of these reconfigurations: changes in how economic and political actors interact with each other; and changes in how (and if) they communicate, represent, and implement their interests and strategies.

To appreciate how the code analytic enables movement across modes, consider Google's relationships with US copyright law, based on Con Díaz's book *Everyone Breaks These Laws*.[19] One of the pillars of Google's ability to offer search services is its successful argument (in US courts) that making a thumbnail of an image and adding it to a database is allowable under copyright law because doing so *transforms* the nature of the work. What was once a creative work, according to Google, becomes a useful thing—an item in a database—and the company's actions become legal. Google's search engine is at the heart of this story, but no one code is necessarily the most important one. The story brings together codes in technology, business, legal doctrine, user behavior and expectations, judicial procedure, and the dissemination of creative works.

The Google situation illustrates the three registers and the different coordinating mechanisms and temporalities of the political and economic relations. Google's zeal to organize the world's information was manifesting through powerful lobbying, litigation, and public relations efforts. The company routinely fended off copyright complaints, especially from litigious pornographic companies that were seeing their print and online profits crash and estimated that a single win against Google could generate enough money to make up for repeated losses. Judges and lawmakers often conceded, sometimes despite great public outcry, that it would be unwise to impede progress in the development of internet technologies.

In the courts, as expected in a common law system, lawyers based their arguments on older court opinions. Google's lawyers were bent on mobilizing one US Supreme Court opinion, on music and free speech, to the company's benefit. When assessing the legality of a rap group's parody of a classic song, the Court had established that the use of a creative work is "transformative" if it alters the original "with new expression, meaning, or message." This was, according to Google's lawyers, what the search engine was doing through the unauthorized creation and display of thumbnails. The black box engine was transforming an image to the extent that it stopped being a creative expression and became an object within a technical system.

Over time, this legal development enabled the replication of a power imbalance among economic actors new and old. At face value, Google's reasoning established that regardless of what an image's creator wishes to do with their work, Google has the right to make a thumbnail and display it. However, this argument is available for anyone to use. It is a crucial component of the legal doctrines that the developers of AI image generators can cite to fend off charges of copyright infringement. Like a search engine, the logic goes, systems like DALL-E transform creative works into something useful, for example, an item in a training database for an AI system. Courts may eventually reject this argument, but as of this writing, there is no reason to believe they will. If courts do eventually reject this line of thought, the legality of Google Images could come into question once again.

And that is the crux of our approach: the code analytic allows us to connect and exceed specific instances of code as object. It is partly a way of "decentering" code (to borrow Eden Medina's term), in the sense that it reduces the explanatory burden that we demand from any specific instance of code as object.[20] But in addition to decentering, we think of retracing: we embrace the excesses of the three registers and inquire into their temporality to follow relations across modes of code and into the broader political and economic environments that frame their development. This works the other way, too, as we can trace political and economic relations through their implementations and across the various modes of code that can both enable and transform them.

Arrangement of the Book

With this framework in mind, we think of the chapters in this volume as opportunities to reflect on, and develop new ways of inquiring into, three major questions in the social studies of code. How does code become both a subject and a means of governance? How does code become infused with social values, assumptions, and biases? And what does it mean to grapple with code? We offer these ambitious questions as opportunities for reflection and refinement of foundational methodologies across academic disciplines. The remarks that follow extrapolate answers from our authors' works, offering new departure points in order to inspire expansion and challenge from future scholars. This introduction reflects on the long historical arc that connects these kinds of issues with the problems and possibilities that AI is posing today.

*Question one. How does code become both a subject
and a means of governance?*

No mode of code can govern in isolation from the others. Employee surveillance and student antipiracy software popular in the United States during the pandemic are recent examples, and there is an extensive literature on algorithmic control.[21] However, the control that these technologies enable is inseparable from the broader range of codes that frame their operation. This is a phenomenon that Ya-Wen Lei illustrates in chapter 1,

"Delivering Solidarity," where she studies collective actions among workers in China who rely on apps to sell their services. Lei documents the interplay among the technological, legal, and organizational aspects of control and management—what she calls "platform architecture." She shows how platform architectures vary according to the kinds of labor that a platform enables and can restrict the sorts of collective action in which workers can participate.

It follows that code can govern, and can become governed, by restricting and enforcing modes of interaction. These restrictions and enforcements do not necessarily have an overarching principle or rationale, as they are inseparable from the power struggles and individual or local idiosyncrasies that surround them. In chapter 2, "Consent Code and Default Dramas," Meg Leta Jones uncovers the breadth of this variability by examining something quite common: the windows and buttons that appear whenever we open a new website or app, asking us whether we consent to the use of cookies or allow an app to track our activity across other apps. These pop-ups and buttons, Jones shows us, are by-products of recurring and dramatic power struggles among codemakers in legislatures, courts, government agencies, corporations, and standard-setting bodies. When a website's default privacy settings change, it is a sign that the ongoing negotiations among stakeholders have yielded new assumptions about website visitors' and publishers' rights.

Code can govern across national boundaries, spreading one jurisdiction's legal philosophies and expectations to regions where they would not necessarily be present otherwise. Justin Petelka and his collaborators explore this phenomenon in chapter 3, "A Mirror, Not a Glass Door," on data subject access rights in India, the United Kingdom, and the United States. The chapter focuses on the European Union's General Data Protection Regulation (GDPR), which went into effect in 2018. It documents what the authors call "trickledown GDPR," or the ways multinational companies and infrastructures extend GDPR rights to people in countries with entirely different legal, cultural, and political conceptions of privacy rights. Code governance trickles outward from the EU, not on its own but through companies' efforts to establish and sustain a multinational presence.

The restrictions of interaction at the heart of code's governance functions can span the full linear spectrum temporality that we use to connect the code analytic with the study of political economy. However, since no form of code governs in isolation, much of the work of governing code (and governing through it) involves coordinating the work of multiple restrictions on interaction. This is the issue that Hamid Ekbia raises in chapter 4, "Algorithmic Collusion, Modern Monopolies, and Their Market Power," which examines what happens when different parties coordinate prices through automatic systems. This coordination can happen in markets for insurance and travel and even online stores like Amazon. Federal antitrust law in the United States (designed to prevent anticompetitive behavior) includes provisions against collusion. However, these laws were written in the 19th century. Ekbia shows us how they are fundamentally incompatible with the forms of collusion that algorithmic pricing systems enable nowadays. In short, one-time purchases can unfold without direct human intervention in contexts regulated by century-old laws that, according to Ekbia, may not be enough to prevent outrageous pricing and unfair competition.

Reflecting on what it takes to govern price involves considering what it means for something to be free. However, creating a lasting and enforceable bundle of rights over a product distributed for free can be difficult. Several works on the relationships between software and ownership have documented the challenges involved.[22] In "Reopening the Politics of Openness in the Age of Cloud Computing" (chapter 5), Shun-Ling Chen expands the scope of analysis in this field by attending to one of the most important documents in a developer's arsenal: the license. Free and open source software (FOSS) does not begin and end by declaring a program to be free or open source. Instead, releasing (and re-releasing) a program following FOSS principles requires clear communication of expectations regarding what others can do with the program. The exact wording involved is a patchwork of widely available templates, language developed in-house, and historical texts and philosophies accumulated since the 1970s.

These discussions of pricing illustrate that code's governance dimensions can involve simultaneous relationships that span multiple temporalities—from the legacy of 19th-century antitrust statutes to the one-time interaction it takes to buy an insurance product or download a computer program. This complex situation prompts us to study how (or if) stakeholders communicate with one another, as they are the ones tasked with developing coordinating mechanisms that allow them to function despite the informational asymmetries and power imbalances that they may be facing.

This takes us to the communication register of the code analytic—to investigate what it takes for stakeholders to create alliances and rivalries, negotiate outcomes, and articulate their goals and needs both to themselves and to others. This is what Con Díaz does in chapter 6, "The Great eBook Conspiracy," on Apple's negotiations with book publishers on the eve of the iPad's launch. The closed-door negotiations that Apple's managers led to counter Amazon's pricing strategies show how a handful of individuals can conspire to restrict business strategies for an entire industry in hopes of countering a powerful new entrant. Short-term and highly guarded communications—in the form of a conspiracy—yielded a transformative and lasting new structure for a young industry.

Question two. How does code become infused with social values, assumptions, and biases?

An enormous body of literature has shown that socio-technical systems can become infused with, and perpetuate, the worldviews of their creators and their users. A well-known example is Safiya Noble's *Algorithms of Oppression,* which showed that the results pages of Google Image searches have racial politics of their own. Black people were disproportionately represented in results for terms such as "unprofessional hair" or "mugshot," whereas white folks were predominant in results pages for "professionalism" and "friendliness." This is not to say Google Images is racist—that would be an oversimplification of the matter—but that the way images are cataloged, the keywords people choose for them, and users' selection and query patterns all contribute to the perpetuation of racially charged stereotypes.

The representation register of the code analytic helps us understand how this infusion and perpetuation occur. After all, human beings are the ones choosing keywords for images in commercial and public databases. They are also the ones programming search engine bots and calibrating the metrics that determine which pages rise to the top of a search query result. But the challenge in understanding this phenomenon on a purely technical level is that much of this work occurs behind closed doors. Google's search engine codes are trade secrets, so it is impossible for anyone other than the company's most trusted members to open the black box of computer code that allows this selection process to happen. When the company adjusts the code to comply with public outcry (as it did soon after Noble's book was published), end users and researchers are able to verify that public concerns have been addressed only by trying the system and pondering whether they see a change.

These impenetrable archival silences do not stop scholars from developing a rich and nuanced portfolio of approaches to what "representation" means and how it shapes social and technical landscapes. In "The Standard Head" (chapter 7), Stephanie Dick examines automatic facial recognition to illustrate the infusion of unrealistic assumptions and generalizations into technical systems in the hope of simplifying computational problems. She places a system developed in the 1960s in the context of century-old efforts to police human bodies, showing how the "standard" head used in criminal identification systems was modeled after only seven men. In other words, a computational problem was simplified by the assumption that the heads of seven men, when averaged, was somehow neutral. This means that the representational politics of technical systems can be based on universalizing assumptions that allow for a small set of people to stand in for much larger target demographics. We see this practice nowadays with AI systems such as Alexa and Siri, which have trouble understanding people who do not speak with "standard" accents for a given language.

The core tension here is to ensure or improve computational efficiency while also accounting for the intractable range of human diversity. This tension may lead even the best-intentioned developers to infuse their own biases and conceptions of the characteristics and technological needs of specific communities. This is what Elizabeth Petrick documents in chapter 8, "Spanning Space and Time Barriers," in which the history of computerized conferencing offers a glimpse into the relationships between disability and technology. Petrick shows that software developers thought of conferencing as an equalizing technology, in that it hid individual differences and allowed users to present themselves however they wished. Believing that disability prevented social participation, they thought that one way to overcome this barrier was to create a technology that allowed people to hide their identities behind a screen.

More generally, the tensions that emerge between efficiency and representation enable the development of systems designed for how developers *imagine* specific communities, as opposed to systems created *with* or *for* those communities. Chapter 9, "Pushing Fintech," by Mariel García Llorens, illustrates this distinction. Llorens examines a Peruvian financial platform to show how impoverished women were treated as testing subjects and talking points to propel the platform, as opposed to users with their own agency. It shows how the task of representing a demographic becomes so great that designers create something for themselves and consider their target demographic as a piece of the broader infrastructural and testing puzzles that they need to solve.

However, discussions of bias in code are only half of the matter, as the making of that code is also subject to social and cultural pressures of various degrees of permanence. Here we can rely on a wealth of work on the history of computing that documents how working with information technology became masculine-coded labor, and on newer work that traces worldwide efforts to dissolve this association. To quote an earlier work by one of our contributors, bias has been "a feature, not a bug," of information technology industries from the beginning.[23]

This is a complex story involving several moving pieces, ranging from the changing labor politics of the 1950s to structural issues in education and employment that resulted in the chronic underrepresentation of women and people of color in computing and engineering. In "Corporate Culture Made Material" (chapter 10), Elizabeth Semler uses the history of Control Data Corporation (CDC) as a case study in the perpetuation of this bias. She finds that the company's internal documents depicted its own employees as embodiments of CDC's own corporate culture. This depiction meant that white men, many of whom were in charge, became the standard visual representation of who could work at the company.

IT cultures can perpetuate long-standing power imbalances, but it would be unwise to depict the communities with less power as necessarily passive actors or victims, or as navigating historically unprecedented phenomena. Jeff Yost addresses this in chapter 11, "Reassessing the Iconic and Unbundling the Ironic," which examines gender and labor at International Business Machines (IBM). Yost shows how work performed by women was considered "nonelite" even though it was deeply technical and essential to the maintenance of client relations. Its importance to the company becomes evident if one embraces a broad scope of inquiry in business history, to include not just IBM's corporate offices and landmark projects but also "the field"—the clients' offices and laboratories, which women would visit to solve engineering problems and foster positive relations with the company.

More generally, what it means to *work* with code is historically contingent, as people and institutions in specific situations create and perpetuate (sometimes inadvertently) conceptualizations of labor that allow them to navigate their circumstances. Dylan Mulvin shows this in "Y2K and the Politics of Labor" (chapter 12), which traces the cultural coding of workers during the Y2K crisis. Some of them were "diamonds" and others were "raw material," and their political and cultural status was closely tied to the descriptors that applied to them. Once again, this pattern becomes evident only if we take a broad view of what we consider coding labor to include not just research, development, and implementation but also maintenance and repair.

Question three. What does it mean to grapple with code?

A clean separation between users and makers of code is not possible, because a person may be governed by the code they help create. We therefore propose the phrase "grappling with code" to encompass the range of possible relationships between people and code. This term is not meant as a replacement of other idioms but as one broad enough to comprehend the many forms and degrees of permanence that those relationships can take—from making law and designing networks to following a dress code and scrolling through a newsfeed. To grapple with code is to participate in interrelated practices of representation, communication, and interaction. It does not require direct and conscious engagement, as codes can act at a distance. Take, for example, how a change in privacy law can affect the handling of our online browsing data, regardless of whether we're aware of it.

Using this terminology and drawing on the social studies of information and data, the discussion of the previous questions underscores that simultaneous competing social and historical circumstances shape how a community grapples with code. Based only on the chapters mentioned so far, we can name circumstances such as short-term social, commercial, and political pressures; ways of knowing and structuring the world; and the broader histories and power imbalances that shape a community's relationship with the state. These circumstances can play a role explicitly or implicitly—by shaping, for example, the ultimate goals that guide decision making or the worldviews that a community takes for granted. They carry social biases and assumptions and respond to political-economic forces of varying degrees of permanence.

So, what does it mean to grapple with code? To answer this question, we must acknowledge that what code *is* depends on the community grappling with it, even if the object called code remains the same. Consider this example: A Google engineer's conception of what the search engine code is will be very different from that of a casual user. The code analytic can help us structure inquiry into how these conceptions differ from each other, inviting us to connect these conceptions of code to the circumstances that shape what it means to grapple with the search engine code. How do different ways of knowing and structuring the world shape what each of these two people would type into the search bar? To what extent might awareness of commercial pressures (like advertising) shape how each person interacts with the search results page?

What it means to grapple with code depends not only on the characteristics of the code itself but also on the affordances and worldviews of the people involved. This is what Shreeharsh Kelkar documents in "From Programming to Platform Expertise" (chapter 13). Kelkar contrasts the celebration of "information technology" in the mid-20th century with that of "Big Data" in the 21st to argue that the difference between the two cannot be reduced to the technological affordances of the respective times. Instead, he argues that each one of these periods celebrated a distinct form of expertise, and that the shift from one to the other was responsive to broader shifts in social formations, ways of knowing, and conceptions of reform and expertise.

The circumstances that shape how people grapple with code can be so broad that they encompass national projects of identity building and economic development. In chapter 14, "Computers as Colonizers," Mar Hicks shows how computing professionals in India developed a native computing industry in the mid-20th century to push back against foreign technological and economic influence and challenge assumptions about India's underdevelopment. Working at the Indian government's behest, they embraced—but eventually transcended—British technocratic ideals. Slowly they built a distinctly Indian industry in which foreign manufacturers' role as suppliers did not necessarily extend to determining the country's industrial dynamics.

Grappling with code can also be a deeply personal matter. It can prompt us to reckon with the anxieties and deep vulnerabilities that we and the people around us face every day. In chapter 15, "The Mask of Humanity," Jennifer Alexander reflects on how computer interfaces obscure human connection and hinder assessments of personal authenticity. Inspired by her son's and students' experiences, as well as her own, Alexander shows us how interfaces can be sites of pain, despair, and dread. She explores how interfaces can threaten common bonds in order to investigate what it means for a person to become vulnerable and exposed to pain.

In the public sphere, grappling with code can enable (or require) large groups of people to come to terms with their own hopes and anxieties. Gili Vidan illustrates this in "Cryptography Goes Public" (chapter 16), on the emergence of "public cryptography" as a field of study by the 1970s. Vidan shows how, over the course of a decade, cryptography transformed from a mathematical field associated with secrecy and military operations to a major problem in the public debates about American democracy. It became a matter of public concern—a shield against the real and imagined vulnerabilities born of living in an increasingly interconnected society.

Code and the circumstances that shape what it means to grapple with it can develop jointly. The notion of code work, which Héctor Beltrán introduces in our final chapter, "Nodes and Codes," is particularly useful to understand how this conjoined development happens. Based on ethnographic work at a 2015 Mexican hackathon, Beltran uses "code work" to describe how people use logics drawn from coding and design to think about the institutions and elements of state-driven infrastructures. He proposes that internationally prominent approaches to grappling with computer code (including the hacker ethic) yield ways of thinking about the Mexican state's infrastructures useful in alleviating economic inequality.

But we can also ask what it means to grapple with code by looking at ourselves, as scholars and practitioners. We invite scholars to investigate the circumstances that shape our communities of scholarship and praxis—not just as intellectual and professional domains but as deeply human domains. Where do our own personal interests in a particular code or method of analysis come from? How do our ways of grappling with code emerge as distinct professions or fields of study? In short, to indulge more intense self-reflexivity, what codes shape how we work with code? We hope that the pages that follow will encourage you to ask these questions often, especially after you are finished reading this book and examine your own life and work.

Notes

1. Ruha Benjamin, *Race after Technology: Abolitionist Tools for the New Jim Crow* (Cambridge: Polity Press, 2019).

2. Mar Hicks, *Programmed Inequality: How Britain Discarded Women Technologists and Lost Its Edge in Computing* (Cambridge, MA: MIT Press, 2018).

3. Safiya Noble, *Algorithms of Oppression: How Search Engines Reinforce Racism* (New York: New York University Press, 2018).

4. Virginia Eubanks, *Automating Inequality: How High-Tech Tools Profile, Police, and Punish the Poor* (New York: St. Martin's Press, 2019).

5. Lawrence Lessig, *Code, Version 2.0* (New York: Basic Books, 2006).

6. Sheila Jasanoff, *States of Knowledge: The Co-production of Science and the Social Order* (New York: Routledge, 2006).

7. Jacob Hacker, Alexander Hertel-Fernandez, Paul Pierson, and Kathleen Teen, eds., *The American Political Economy* (Cambridge: Cambridge University Press, 2021), 7.

8. Berry, *The Philosophy of Software*; Brown, *Regulating Code*. David Berry, *The Philosophy of Software: Code and Mediation in the Digital Age* (Basingstoke, UK: Palgrave Macmillan, 2011); Ian Brown and Christopher Marsden, *Regulating Code: Good Governance and Better Regulation in the Information Age* (Cambridge, MA: MIT, 2013).

9. Berry, *The Philosophy of Software*, 39.

10. Gerardo Con Díaz, *Software Rights: How Patent Law Transformed Software Development in America* (New Haven, CT: Yale University Press, 2019).

11. Katharina Pistor, *The Code of Capital: How the Law Creates Wealth and Inequality* (Princeton, NJ: Princeton University Press, 2019).

12. Sareeta Amrute, *Encoding Race, Encoding Class: Indian IT Workers in Berlin* (Durham, NC: Duke University Press, 2016), Anjali Vats, *The Color of Creatorship: Intellectual Property, Race, and the Making of Americans* (Redwood Hills: Stanford University Press, 2020).

13. Lilly Irani, *Chasing Innovation: Making Entrepreneurial Citizens in Modern India* (Princeton, NJ: Princeton University Press, 2019).

14. Peter Skafish, "Introduction," in Eduardo Viveiros de Castro, *Cannibal Metaphysics*, ed. and trans. Peter Skafish (Minneapolis: University of Minnesota Press, 2014), 16.

15. Marilyn Strathen, *Relations: An Anthropological Account* (Durham, NC: Duke University Press, 2020), 2.

16. Naomi Lamoreaux, Daniel Raff, and Peter Temin, "Beyond Markets and Hierarchies: Toward a New Synthesis of American Business History," *American Historical Review* 108, no. 2 (April 2003): 404–33.

17. Caitlin Rosenthal, *Accounting for Slavery: Masters and Management* (Cambridge, MA: Harvard University Press, 2019).

18. James Poskett, *Materials of the Mind: Phrenology, Race, and the Global History of Science, 1815–1920* (Chicago: University of Chicago Press, 2019).

19. New Haven, CT: Yale University Press, 2025.

20. Eden Medina, "Forensic Identification in the Aftermath of Human Rights Crimes in Chile: A Decentered Computer History," *Technology and Culture* 59, no. S4 (October 2018): S100–S133.

21. Kathleeen Griesbach, Adam Reich, Luke Elliott-Negri, and Ruth Milkman, "Algorithmic Control in Platform Food Delivery Work," *Socius* 5 (2019), https://doi.org/10.1177/2378023119870041.

22. Con Díaz, *Software Rights*; Christopher Kelty, *Two Bits: The Cultural Significance of Free Software* (Durham, NC: Duke University Press, 2008).

23. Mar Hicks, "Sexism Is a Feature, Not a Bug," in *Your Computer Is on Fire*, ed. Thomas Mullaney, Benjamin Peters, Mar Hicks, and Kavita Phillips (Cambridge, MA: MIT Press, 2021).

Part I

How does code become both a subject and a means of governance?

1

Delivering Solidarity

Platform Architecture and Collective Contention in China's Platform Economy

Ya-Wen Lei

The platform economy is on the rise globally. Following Kenney and Zysman, I use the term "platform economy" to refer to economic activities facilitated and enabled by digital platforms—multisided digital frameworks that shape the terms by which various actors interact based on the application of algorithms.[1] With the rise of the platform economy, concerns about its socioeconomic consequences have arisen as well.[2]

Of particular scholarly interest are the platform economy's modes of labor control and management, especially algorithmic control and management—a diverse set of technological tools and techniques that rely on data collection and algorithms to manage a workforce.[3] Scholars use the term "digital Taylorism" to emphasize how control has been reconfigured by the digital transformation of work. Unlike the earlier digital Taylorization of white-collar work, however, algorithmic control and management in the platform economy involves not only monitoring and evaluating work performance but also automatically assigning tasks based on performance metrics.

As labor control and management continue to evolve in the platform economy, long-standing questions arise about the relationship between labor control and resistance. Commenting on evolving modes of labor control, Hyman noted: "The emergent pattern of labour control contains its own emergent contradiction. The new disciplines imposed on workers can be expected to provoke unpredictable and disruptive forms of revolt."[4] Nonetheless, as Burawoy's work suggests, capitalists might still be able to manufacture consent or at least compromise.[5] Considering the indeterminate relationship between labor control and resistance, Hyman wrote that whether labor itself can move from "reactive tactics to a concerted strategy" is a critical question.[6]

Scholars have begun to analyze the relationship between labor control and resistance in the platform economy. On one hand, research suggests collective resistance is unlikely to occur, because work is often atomized.[7] In addition, platform companies have endeavored to obfuscate the precarious and exploitive nature of the work, partly by integrating dimensions of "gamification"—the use of game design in nongame contexts—in their platforms.[8] On the other hand, research shows emerging labor activism and solidarity in the platform economy. Tassinari and Maccarrone demonstrate that algorithmic control has led to shared grievances among couriers and collective action against platforms in Europe.[9] And yet scholars also suggest research on labor control and management should better attend to its multidimensionality. Veen, Barratt, and Goods argue that labor control is multifaceted and irreducible to algorithmic management; similarly, Moore and Joyce suggest managerialism should be viewed as sets of ideational and institutional forms that are neither static nor homogeneous.[10]

Building on the above insights, this study asks how and when labor control and management lead to collective resistance in the platform economy. I develop the concept of *platform architecture* to examine the multidimensional—technological, legal, and organizational—aspects of control and management in the labor process and the variable relationships between them. To be sure, factors beyond the labor process, such as state intervention, can influence labor control, management, and contention.[11] Nevertheless, rather than simultaneously analyzing all of the factors and processes, this study focuses on variable configurations of labor control and management and their relationship with collective contention.

Empirically, I examine varying patterns of labor contention in China's food delivery platform economy. China's online food delivery market—the largest in the world—emerged in 2009 and has expanded greatly since around 2015. In 2017, the RMB297 billion (US$46.5 billion) market had more than 305 million consumers.[12] Consistent with what economists have observed about the platform economy, China's online food delivery market is highly concentrated, dominated by two powerful platform companies, Meituan and Ele.me.[13] Together, these two companies had a combined 5.7 million registered couriers in 2018. With the expansion of the food delivery platform economy, strikes and protests organized by couriers in numerous Chinese cities have occurred despite various barriers to collective action (see figure 1.1).

The platform economy has taken off even while the country's economic growth rates have slowed. The state views the platform economy as a new engine of growth and a reservoir for absorbing surplus labor. Thus far, regulations and state intervention have focused on food safety, consumer protection, and traffic safety. As strikes and protests began to emerge, the All-China Federation of Trade Unions (ACFTU), an organ of and subor-

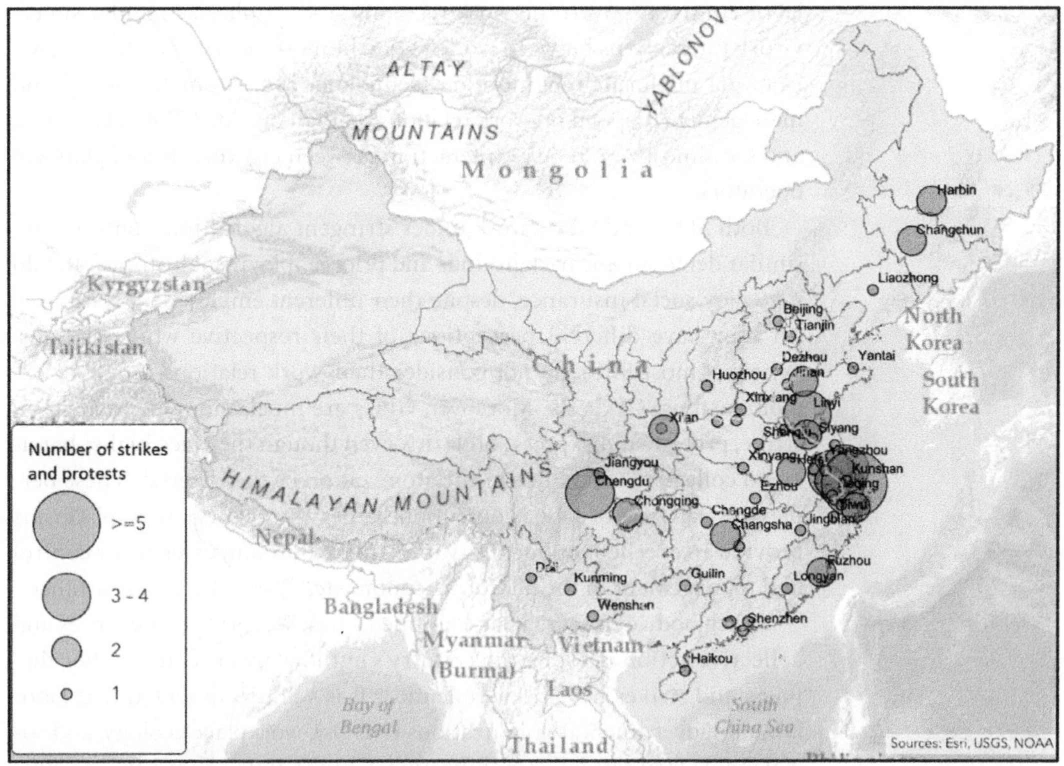

Figure 1.1. Geographic distribution of strikes and protests in China, 2017–2018.

dinate to the Communist Party–controlled state, announced it would accelerate establishment of trade unions for platform workers—yet, to date, this plan has barely been implemented. In short, the Chinese state has not tackled the labor or market competition problems that emerged with the platform economy.

Previous literature suggests employers want workers to be both dependable and disposable.[14] Indeed, Meituan and Ele.me operate two types of platforms, service and gig, to balance security and flexibility of the workforce. Service platform couriers (SPCs) and gig platform couriers (GPCs) both pick up food from a restaurant and deliver it to a customer. However, SPCs are full-time employees who work for a service station, that is, an actual physical station that coordinates orders, manages operations, and assigns couriers within a locality. In contrast, GPCs can decide when they want to work and do not share a workplace.

Essentially, the service platform provides the two companies with a stable workforce, and the gig platform provides them with a flexible workforce. These different purposes explain service platforms' and gig platforms' distinct architectures and how they exercise technological, legal, and organizational control and management differently. Although economic activities under both service and gig platforms are digitally mediated, service platforms rely less on technological control and management, have labor contracts governing work relations, and emphasize regular supervisory interactions between workers and human supervisors. In comparison, gig platforms rely more on technologies to automate control and management, depend on work relations that fall outside China's labor law, and see almost zero regular interaction between gig workers and platform operators.

Both SPCs and GPCs work under stringent algorithmic control, have similar demographic backgrounds and prior employment histories, and do *not* enjoy social insurance, despite their different employment statuses—yet they have different perceptions of their respective work relations. Whereas most SPCs do not consider their work relations unfair or exploitive, most GPCs do. Moreover, GPCs are much more likely to go on strike, protest, and express solidarity, even though they face higher barriers to collective action due to the atomization of their work.[15] These perplexing patterns of labor contention across the two types of platforms provide an excellent opportunity to examine how and when labor control and management in the platform economy leads to collective resistance.

A rich body of research has examined Chinese workers' grievances and collective action. Despite the country's authoritarian regime, labor disputes and strikes have been endemic.[16] It is well documented that labor law, nongovernmental organizations (NGOs), workplace ecology, and social media facilitate labor mobilization. China's 1994 Labor Law and 2007 Labor Contract Law provide the basic legal infrastructure for regulating labor relations, including the provision of social insurance. Prior to the intensification of state repression of civil society in 2013, labor NGOs and media helped disseminate knowledge about labor law. As political scientist Mary Gallagher argues, China's labor law regime is characterized by high standards but low levels of enforcement.[17] Employers commonly fail to meet their legal obligation to provide social insurance. In most labor contentions, aggrieved workers respond to violations of the Labor Contract Law by defending their legal rights and demanding state intervention and settlement of labor contracts.[18] The unique ecology of the workplace, particularly in factories and dormitories, and social media use further facilitate collective action.[19]

To explain labor disputes in China's food delivery platform economy, I draw on literature from sociolegal studies, collective action, labor studies, and platform economy scholarship to analyze how platform architecture influences labor's collective contention. Analyzing 68 in-depth interviews, ethnographic data, and 87 cases of strikes and protests, I compare the architecture of service platforms and gig platforms and examine their relationship to labor contention. I argue that specific differences in platform architecture diffuse or heighten collective contention. Within the service platform, technological control and management generate work dissatisfaction, but the legal and organizational dimensions of the platform architecture contain grievances and reduce the appeal of, and spaces for, collective contention. Conversely, within the gig platform, all three dimensions of platform architecture reinforce one another, escalating grievances, enhancing the appeal of collective contention, and providing spaces for mobilizing solidarity and collective action. As a result, gig platform couriers are more likely to consider their work relations exploitive and to mobilize contention even though they face higher barriers to collective action due to the atomization of their work. This study contributes to the intersecting fields of labor, collective action, organizations and work, economic sociology, sociolegal studies, and the sociology of information technologies.

How Platform Architecture Influences Collective Contention
Key mechanisms for collective contention

Escalating grievances and perceived injustice. Socio-legal and social movement scholars point out the importance of analyzing the interpretive process of constructing grievances and injustice frames in order to understand the emergence and escalation of contention.[20] They use the term "injustice frames" to refer to interpretations of experiences or conditions that show moral principles have been violated and ought to be redressed rather than tolerated.[21] Research shows that developing and sharing injustice frames is a necessary, albeit insufficient, condition for collective mobilization.

Building on social movement literature, scholars of industrial relations argue that labor's collective action requires constructing a sense of injustice from grievances. A work situation must be understood as wrong or illegitimate rather than simply dissatisfying.[22] The focus on constructing a sense of injustice resonates with labor research that analyzes the interplay between the labor process, morality, and solidarity. Perceived injustice often manifests in collective action frames, each of which comprises a set of interconnected ideas about the existence of problems, causes of and responses to the problems, and rationales for labor collectivism.[23] Although some research analyzes labor collectivism from the perspective of rational analysis, labor scholars find collective action frames go beyond the calculation of benefits and costs, and they highlight the moral imperatives for collective action.[24]

Critically, labor scholars point to the need to study moments of escalation, as they often become moments of collectivism and generate "cultures of solidarity," which Fantasia defined as emergent oppositional practices and meanings. At unusual moments of escalation, workers develop new repertoires of action, network ties, and moral valuations and suspend customary practices of daily life.[25]

Enhancing the appeal of collective contention. Even when grievances have escalated and collective action frames have been constructed, workers still have various options regarding whether and how to respond to perceived injustice. The literature suggests a collective action frame will not prevail if workers see problems as inevitable or soluble only by individual exit, not collective voice.[26] Indeed, research on low-paid workers in the United Kingdom finds that, although problems at work are widespread, satisfactory resolutions are rare because workers tend to discuss issues only with managers or just find another job.[27]

Socio-legal scholarship offers the concept of the "dispute tree" to capture the wide range of options a grievant can pursue, from formal legal disputes in court to institutionalized noncourt alternatives, individual informal legal mobilization, and extrainstitutional collective action.[28] Indeed, workers can choose one or a few options from the dispute tree. When workers develop a strong sense of injustice and consider institutionalized channels or individual informal legal mobilization ineffective, extrainstitutional collective action may become a more appealing option.

Making available free spaces for collective contention. Scholars also argue that collective action requires certain free spaces in which to develop. Free spaces, whether physical or virtual, are "small-scale settings within a community or movement that are removed from the direct control of dominant groups, are voluntarily participated in, and generate the cultural challenge that precedes or accompanies political mobilization."[29] Such critical spaces provide a relatively autonomous environment where community members can develop networks, skills, and solidarity that assist collective action. In the context of industrial relations, free spaces get relatively less oversight from management and are less influenced by fear of managerial retribution, normative control, and surveillance.[30] Free spaces allow workers to articulate grievances and a sense of injustice, develop collective action frames, and evaluate various dispute resolution options. Labor's collective contention does not necessarily require formal organization, but it does require rudimentary organizing efforts—efforts made possible in such free spaces.[31]

Platform architecture and how it matters

The multidimensionality of platform architecture. Most of the literature on labor control in platform economies emphasizes algorithmic control, but recent research has examined the multidimensionality of labor control and management.[32] Indeed, legal scholarship theorizes the coexistence of multiple models of control and regulation in society—law, technology, and social or organizational norms.[33] These insights can be applied at the level of labor process. For this reason, I analyze technological, legal, and organizational dimensions of control and management, as well as the relations among these dimensions. I use the term "platform architecture" to capture these multidimensional relations of control and management holistically.

"Technological control and management" refers to control and management that uses artifacts or "non-physical, systematic methods of making or doing things."[34] Prior studies show platform companies use algorithms not only to "match" consumers with service providers but also to manage and control workers' behavior, converting workers' general capacity for work into a specific and quantified amount of labor.[35] For example, Rosenblat and Stark detail how Uber uses algorithmic control, characterized by information asymmetries and opacity of technological design, to shape drivers' behavior. Griesbach and colleagues analyze how food delivery platforms use algorithms to calculate prices, influence workers' choices, and control worker activities.[36]

Platform economies also require legal arrangements that govern work and delimit the rights and obligations of the various parties involved. Most studies of platform economies concentrate on the contractual classification of workers: platform companies often seek to classify workers as independent contractors rather than employees.[37] This classification is critical, for it shapes the basic rights and obligations of both parties and the degree of work precarity and worker autonomy.

Building on Suchman's analysis of contracts as social artifacts,[38] I examine contractual design and processes in the platform economy rather than focusing only on classification. Legal scholars note the problems and inherent unjustness of contractual design in asymmetric contracts—that is, contract relationships between a dominant business and another market player with unequal bargaining power.[39] Because platforms tend to have enormous market power and resources to make legal arrangements, they can use asymmetric contracts as a means for labor control and management. To the extent that a contract is, as legal scholars theorize, a contingent relational process rather than a static transaction, problems can result from contractual design long after the initial consent.[40]

"Organizational dimension of control and management" refers to the extent to which management personnel can exercise normative influence and surveillance on platform workers. In traditional workplaces, management personnel tend to have a close supervisory relationship with workers and influence how workers respond to grievances. Platform workers, however, such as Uber drivers, often do not have a human supervisor.[41] Rather than hiring management personnel to supervise workers, some platform companies invest in professionals who develop technologies and analyze data to automate the supervision of workers. In such scenarios, management personnel manage professionals rather than workers.

How platform architecture influences collective contention. The technological, legal, and organizational dimensions of control and management can influence key mechanisms for collective action in at least three ways. First, studies of the platform economy show that the opaque, disciplinary, and arbitrary characteristics of algorithmic control and management can generate considerable grievances.[42] When platforms exercise stringent algorithmic control of workers' time and activities, workers not only feel dissatisfied with the work but consider the platforms unfair.[43] Extant literature thus suggests certain characteristics of technological control and management can generate grievances and a sense of injustice.

Second, the legal design of platform architecture can affect how workers construct grievances and injustice frames and, in turn, the perceived appeal of collective contention. Certain contractual designs can deviate from workers' moral norms. Legal scholars consider asymmetric contracts unfair.[44] Workers can share similar moral judgments when they experience enormously unequal power relations. Previous research on labor contention in China finds that workers developed a shared injustice frame that challenges the gap between "law on the book" and "law in action," and they invoked labor laws to demand resolution of their grievances.[45] I expect that when a platform's legal design is not based on labor law, workers are more likely to construct injustice frames based on a more abstract moral principle. Furthermore, when injustice frames are based on an abstract moral principle rather than a specific statutory clause, grievants are more likely to choose collective contention because it is difficult for them to turn injustice frames into a concrete legal claim in the institutionalized legal process.[46] In contrast, when injustice frames concern the gap between "law on the books" and "law in action," grievants can, relatively easily, make a legal claim through the institutionalized legal process or individual negotiations.

Third, the organizational design of platforms can influence the construction of grievances, the appeal of collective contention, and the availability of free spaces. Previous research on low-paid workers shows that communication between managers and workers often prevent grievances from escalating, and regular interactions can provide management opportunities to exercise normative influence on workers.[47] Recent research on the service sector further shows that when supervisors perform the emotional labor of managing workers' feelings and expressions, it can improve workers' emotional labor and increase their work satisfaction.[48] Accordingly, lack of regular interaction between management and platform workers under certain platform architectures could make grievances more likely to develop and escalate. The perceived implausibility of grievance resolution through management could also increase the appeal of collective contention. Finally, the absence of supervisory relationships between management personnel and workers can make certain free spaces available for mobilizing solidarity and collective action.[49]

Overlap of control and management. The three dimensions of labor control are analytically distinct, but they can operate simultaneously under a platform. Such overlap does not negate the need to distinguish the three dimensions analytically, as the ways the dimensions interact can vary across platform architectures and have different implications for collective contention.

The overlap of technological and legal control and management often results from the mutually enabling relationship between technologies and contracts under certain platform architectures. Technologies can mediate the formation and implementation of contracts. Platform companies often design a "clickwrap agreement," a digital prompt that allows individuals to accept or decline a digitally mediated agreement. Users cannot access a platform app unless they accept the clickwrap agreement. Because clickwrap agreements take the form of a nonnegotiated take-it-or-leave-it choice, they tend to give the party designing the agreement tremendous power and are often asymmetric contracts.[50]

Contracts can also enable and facilitate technological control and management. Algorithms operate according to a set of rules that can be changed to achieve certain goals, for example, to improve their optimization. Certain contractual designs, such as the right to change contractual terms unilaterally in a clickwrap agreement, allow companies to easily change contractual terms and work policies. This enables management to enact legal and technological updates to improve platforms' performance. Existing literature emphasizes that technological control and management occur in black boxes and remain opaque to platform workers.[51] However, certain updates are required by law to be overt to take effect and be followed by platform workers; when changes in technological control and management intersect with such legal changes, workers can clearly see the changes. As platforms' power becomes more visible, moments of simultaneous legal and technological change can become moments of grievance escalation.[52]

Technological and organizational dimensions of control and management can also overlap because platform companies use technologies to monitor workers, whether or not workers have a human supervisor. Technological and organizational control can interact in various ways. When a manager uses technologies to supervise workers, the two dimensions of control can have opposite effects on collective contention. Certain methods of technological control and management can generate grievances, but supervisors' mediation may reduce those grievances. When supervision is completely automated through technology, workers' problems with technological control might be exacerbated by the absence of managers to respond to the problems. Ultimately, automated supervision remains limited in its ability to exercise normative influence and perform emotional labor.[53]

In summary, my theoretical framework analyzes how platform architecture influences collective contention, which envisions workers developing a sense of grievance, finding collective contention appealing, and having access to free spaces for collective action. The technological dimension of platform architecture can generate considerable grievances. The legal dimension affects grievance construction and the appeal of collective contention. And platforms' organizational design can influence the construction of grievances, the appeal of collective contention, and the availability of free spaces. When different dimensions of control and management reinforce one another in ways that escalate grievances and perceived injustice, enhance the appeal of collective contention, and make free spaces available for collective contention, collective contention is more likely to arise.

Case Selection, Data, and Methods

I study the food delivery platform economy in China for two reasons. First, it has received little scholarly attention, despite its growing size in China's broader platform economy. Second, a growing literature focuses on the labor process in food delivery platform economies. Analyzing the Chinese case can build on insights in, and contribute to, this literature.

The study design is a comparative analysis of work relations under two different types of food delivery platforms—the service platform and the gig platform, each of which has a distinct platform architecture. A comparative design allows me to explain how platform architecture influences patterns of labor contention. Following the tradition of the comparative method, I understand and compare the cases as wholes rather than seeing different dimensions of platform architecture as independent variables.[54]

I draw on five qualitative methods to collect and analyze data. First, I conducted in-depth interviews with 30 SPCs and 30 GPCs in seven cities between 2018 and 2019. I recruited people who were couriers at the time of the interview or had been in the preceding 12 months. Table 1.1 shows the distribution of these couriers. At the beginning of the study, I attempted to diversify the interviewee pool according to location, given the possibility of variation across cities in terms of how platforms operate and how couriers work. When it became that clear there was no significant difference across cities, I did most of my interviews in Chongqing, a municipality in southwest China, given my access to research support there.

Interview questions included basic demographic information, employment history, work situation, interactions with other workers and supervisors, and experiences of strikes and protests. Many couriers shared their photos, videos, and social media communications with me, which helped me reconstruct collective action. Existing literature tends to neglect management's role, so I interviewed six people with management positions (four at service platforms, two at gig platforms).[55] I also interviewed a systems development engineer and an in-house lawyer to understand the technological and legal aspects of control and management. My assistants and I recorded and transcribed all 68 interviews.

Table 1.1. Distribution of interviewed couriers

Demographic variable	Service platform (*n* = 30)	Gig platform (*n* = 30)
Gender		
Female	3	2
Male	27	28
Age		
18 to 25	14	12
26 to 35	13	14
Over 35	3	4
Educational attainment		
Less than high school	10	11
High school	16	14
Some college	4	5
Full-time or part-time		
Full-time	30	21
Part-time	0	9
Migrant status		
Yes	24	24
No	6	6
City		
Chongqing	18	18
Fuzhou	2	2
Harbin	2	2
Linyi	2	2
Nanjing	2	2
Shanghai	2	2
Shenzhen	2	2
Employment history		
Prior experience in the manufacturing or construction sector	18	17
Prior experience as a service platform courier	N/a	3
Prior experience as a gig platform courier	1	N/a
Presence at protests or strikes organized by couriers		
Yes	3	15
No	27	15
Working as a courier at the time of the interview		
Yes	26	26
No	4	4

Notes: I collected income information from interviewees, but it is difficult to obtain precise information as neither SPCs nor GPCs have fixed earnings. In general, SPCs in my sample made from RMB4,000 to RMB5,000 (around US$564 to US$705) per month. Full-time GPCs in my sample made from RMB6,000 to RMB7,000 (around US$846 to US$987) per month, but they tended to work more hours per day than SPCs did.

Second, I conducted online ethnography between 2018 and 2019. I joined four nationwide discussion boards used by food delivery couriers in Baidu Tieba, one of the largest online communities in China. I also joined 10 social media groups formed by couriers. I examined the Weibo content of certain couriers who organized or participated in strikes or protests. Online ethnography enabled me to reach out to participants in collective action. I read discussions, listened to couriers' voice messages, and took notes and screenshots.

Third, I conducted on-the-ground ethnography in Chongqing. I began to conduct online ethnography with couriers in Chongqing two months before GPCs there organized a series of strikes and protests in May 2018. As figure 1.1 shows, Chongqing is one of the many cities in which platform couriers staged collective action. I went to Chongqing in June 2018 to conduct fieldwork. When I arrived in the city, many GPCs still had fresh memories of the collective action. I stayed in a plaza where gig couriers hung out and joined their conversations. One service platform station allowed me to observe their couriers' and supervisors' routines, and a gig platform manager allowed me to visit his office. After I left Chongqing, I continued to observe the situation through follow-up interviews.

I analyzed and collected interview and ethnographic data iteratively, moving between data and theories. Initially, I designed my interview questions based on literature on algorithmic control. After conducting the first 20 interviews and some online ethnography, I began to analyze the data. As a number of the emerging themes and codes did not appear in the literature on algorithmic control, my preliminary empirical analysis prompted me to develop a broader theoretical framework. I then revised my interview protocols based on the newly included literature. I did supplemental interviews with the first 20 interviewees and then conducted 48 more interviews.

Fourth, I compiled a dataset that contains 87 cases of strikes or protests organized by food delivery couriers between 2017 and 2018 in China. By "strike" I mean a period of time during which people refuse to work; "protest" refers to the public expression of opposition to something. All of the cases were both strikes and protests, as couriers not only refused to work but expressed their opposition to employers or platforms on banners or signs. I compiled the list from a professional news database (WiseNews), search engines, online communities, Weibo, WeChat, and the China Labour Bulletin. China Labour Bulletin is a labor NGO (nongovernmental organization) in Hong Kong that provides the most comprehensive data about labor disputes in China. The data I collected include texts describing strikes and protests and photos and videos. I documented slogans written on banners and signs. For each collective action, I coded the data in terms of type of platform, platform company, location, content of the complaint, and couriers' actions. The dataset is limited in terms of its exhaustiveness, so I use these data to complement my interviews and ethnography data, rather than trying to subject them to rigorous quantitative analysis.

Finally, I collected the texts of contracts between different actors involved in food delivery. I also compiled the texts of rules enacted by the two platform companies and their franchisees and local offices. In addition, I read 60 court decisions made by the Basic People's Courts or Intermediate People's Courts that listed either Meituan or Ele.me as one of the parties in the case. Most of the decisions concerned compensation following the death or injury of a platform couriers or other party in a traffic incident. The disputes in these cases are not central to collective contention, but the court decisions provide detailed information about the legal and organizational design of the service and gig platforms. My legal education and six years of work experience as a legal professional facilitated my analysis of legal texts.

Findings
Platform couriers

According to surveys conducted by Meituan and Ele.me, the companies' couriers are predominantly male (92%); 77% have rural household registration status; and 70% are migrants. Most couriers are in their 20s, with a middle or high school education. The most common prior occupation is factory worker. Among Meituan couriers, 40% reported working as a delivery courier because of the freedom it affords.[56]

Prior work experiences explain why the discourse of freedom advanced by the platforms had a strong purchase on couriers and how they thought about jobs, social insurance, and labor contention in general. Among my interviewees, 58% had worked previously in the manufacturing sector. Many were haunted by experiences or stories of harsh factory conditions. They said they would never want to work in a factory or go back to a factory, citing negative conditions such as the closed factory environment. Couriers desired freedom and sought to escape from manufacturing jobs. Strikes and protests were not foreign to the couriers who had worked in factories or construction sites, as many had heard of, if not personally participated in, such collective action. None of my interviewees wanted to work as couriers for the long run, but most full-time couriers had no concrete career plans.

Both SPCs and GPCs tended not to care much about social insurance, due in part to the weak enforcement of labor law and China's fragmented social insurance system.[57] In China, labor law requires employers to buy social insurance for employees. However, couriers understood that only more established factories, not smaller ones or employers in the service sector, would fulfill this obligation. Furthermore, labor scholars find young migrant workers often give up social insurance in exchange for higher wages because they do not have faith in the social insurance system and they are nonresidents of their cities.[58] Indeed, many couriers in my study thought that, due to their migrant status, social insurance protection would be minuscule even if employers followed the law.

Couriers themselves decide whether to work as an SPC or a GPC. The application processes for SPCs and GPCs are separate. Usually, couriers ask friends or couriers in social media groups how to choose between types of platforms and companies. For people who want a part-time job, being a GPC is the only choice because there is no part-time option for

SPCs. People who want a full-time job can choose between the two types of platforms. Some interviewees said they just followed their friends. Those who preferred to have a relatively structured life and a stable income tended to choose an SPC job. Some thought they could make more money as a GPC.[59] Others considered how familiar they were with the streets of the city district where they lived, as familiarity influences the on-time rate. For people who lack such familiarity, being a GPC might be a good option because supervisors can help. Prior to my contact with them, 3 of the 30 GPCs in my sample switched from a service platform to a gig platform, and only 1 of the 30 SPCs had switched from a gig to a service platform.

As table 1.1 shows, SPCs and GPCs shared similar demographic characteristics and employment histories. I did not find evidence suggesting the two groups had different levels of human capital, skills, or disposition. The only difference between the two groups was that all SPCs, but only 70% of GPCs, worked as full-time couriers. Full-time SPCs and GPCs wholly depended on platforms for their livelihood, whereas part-time GPCs used delivery work to supplement their primary source of income.[60]

Platform architecture

Urban dwellers who order food using a mobile app are familiar with the two major platform companies, but most consumers are unaware of the different types of platforms, let alone which type they are using for a given transaction. Patrons choose the restaurant and the food. In each transaction, the restaurant selects the type of delivery platform. In general, service platforms offer delivery within a three-kilometer radius, whereas gig platforms do not have such a constraint. When the distance is within three kilometers, restaurants can choose between the two platforms. For restaurants, the gig platform is the cheaper option, but it is also less reliable because GPCs are not monitored by supervisors. The pros and cons of the two types of platforms for restaurants are associated with their respective platform architecture.

Architecture of the service platform. The architecture of the service platform aims to establish a reliable workforce. As Liu, an in-house lawyer at Meituan, explained, building a dependable workforce requires a high level of supervisory power over couriers, but such power carries with it the legal responsibilities associated with conventional employment relations.

Both Meituan and Ele.me franchise their business to other companies to create a reliable workforce. Franchisees operate service stations in a local jurisdiction, and they employ and supervise district managers, service-station supervisors, and delivery couriers. Meituan and Ele.me provide their franchisees with technology—a platform system for managing delivery orders and couriers. The two franchisors oversee their franchisees through a set of key performance indicators (KPIs). These indicators are aggregated from the performance of individual couriers through the platform system. Cai, a district manager, described how KPIs work: "Platform companies set many KPIs, such as the on-time rate, customer review scores, the number of complaints, and so forth. If our KPIs are not good, Meituan can terminate our contract or give us a fine. Therefore, we ask everyone to have good KPIs."

Meituan and Ele.me's franchisees use the technology provided by the two franchisors—specifically, the platform system—to assist in managing and controlling the labor process. Technology is thus a tool used by management personnel, rather than something that replaces them. On one hand, franchisees and station supervisors are constrained by the platform system. Station supervisors mostly rely on the platform system and the algorithms embedded in it to dispatch couriers, but they can also manually dispatch couriers. Dispatch algorithms consider couriers' performance metrics when assigning tasks. The platform system incorporates various disciplinary rules, such as fines, and it sets the maximum time for completing each delivery. Franchisees can change certain parameters in the system, such as the amount of fines imposed on couriers.

Meituan and Ele.me's technological control and management also includes dimensions of "gamification."[61] The mobile apps are designed to induce couriers' participation in the "game," with a lively graphic design reminiscent of video games. Similar to popular video games, Meituan and Ele.me's apps classify couriers according to their performance metrics. Higher-level couriers earn certain privileges, such as priority in taking an order. The classification is dynamic, so couriers have to maintain their scores to avoid a downgrade. One SPC said, "I was amazed by the fact that working can be like video gaming. I competed with other couriers and tried to raise my level."

Legally, the franchisees, rather than Meituan or Ele.me themselves, form a labor contract with SPCs that classifies couriers as employees. How franchisees establish a labor contract with couriers varies across service stations, but such contracts must establish a salary structure, both parties' obligations, and the supervisory relationship. All of the labor contracts calculate salaries based on piece rates and the number of delivery tasks completed. Some contracts include a minimum monthly wage, but others do not. In most service stations, supervisors display disciplinary rules on the wall.

The organizational design of the service platform emphasizes the role of management personnel in supervising couriers. Management personnel aim to maximize KPIs through managing couriers. Huang, a service platform supervisor with a high school education, explained: "The platform system is a useful tool because it spares me time from assigning tasks manually, but ultimately my job is to manage people, who have emotion." Cai, a district manager, told me that when hiring supervisors, he considers people with prior experience as couriers to be desirable candidates who can work with SPCs to ensure excellent KPIs.

Station supervisors and couriers have regular face-to-face interactions. The daily interaction begins with a morning meeting. Yang, an Ele.me station supervisor, described the routines: "We begin our morning meetings at 10 a.m. Couriers sanitize their delivery boxes. Then we go over statistics. Our company has three service stations in this district. We announce how many orders, good reviews, and bad reviews each station received yesterday. I teach couriers how to behave in a civilized way. We practice how to apologize to customers. I have to do ideology work to better educate couriers, so they will have excellent KPIs. Many customers are crazy, so I often have to lift couriers up." Note that "ideology work" is the term for the Chinese Party–state's propaganda system for indoctrinating people. This statement shows station supervisors' effort to influence how couriers behave, deal with emotion, and generate high performance metrics through normative control and emotional labor.[62]

Architecture of the gig platform. The technical, legal, and organizational design of the gig platform aims to establish a flexible and disposable workforce. According to Liu, a lawyer at Meituan, gig platforms achieve this goal by surrendering most of the supervisory power of management personnel over couriers: "If management personnel are visibly involved in supervising couriers, the court will consider work relations as employment relations. That means platform companies do not have flexibility to adjust workforces. Also, platforms will have to take more legal responsibility."

Consequently, gig platforms automate most management and supervision tasks. As with technological control and management under the service platform, Meituan and Ele.me rely on technology to discipline and dispatch GPCs. Disciplinary rules, including fines and deactivation of an app account, are built into and implemented by the system. Dispatch algorithms incorporate rules that decide which couriers are given priority in choosing assignments. Wong, a systems development engineer at Meituan, explained the company's algorithm: "When a consumer makes an order, the system calculates the distances between the consumer and available gig couriers and the distances between the restaurant and available gig couriers. The algorithm also considers couriers' performance statistics. The system gives priority to gig couriers with better statistics." The two gig platform systems classify couriers into different levels. In addition, the platforms use points to measure a courier's credibility—how reliable the courier is—based on their behavior, such as their records of accepting or rejecting dispatch assignments. Similar to the service platform apps, the gig platform apps integrate elements and dynamics of gamification to induce couriers' participation in the game.[63]

The two gig platforms also use algorithms to set remuneration. Under the service platform, piece rates are generally fixed and spelled out in the contract by Meituan and Ele.me's franchisees. In comparison, Meituan and Ele.me's gig platforms use algorithms that factor in several factors (e.g., distance, condition of the destination, weather) to calculate piece rates. The pricing model is similar to that of Uber Eats.[64] Local managers who work for gig platforms can change some parameters in the system that influence remuneration—for instance, the basic rate per kilometer.

Both Meituan and Ele.me use technology to confirm the identity of gig couriers and monitor their compliance with rules. Because gig couriers do not have a human supervisor, both platforms require couriers to scan their face before they can take an order and to take a picture of themselves during their work. As Xue, a GPC, described the protocol, "Ele.me often ask me to upload a selfie when I'm on the road. They want to see whether I wear the company's helmets and vests and carry the company's delivery box." Failure to comply with such rules has disciplinary outcomes.

Meituan and Ele.me meticulously design their contracts to maximize the gig platforms' flexibility. Liu, a lawyer at Meituan, explained the legal work the two platforms have done:

> The legal relationship is complicated as there are many parties involved: platforms, couriers, restaurants, consumers, and people on the street. Our task is to minimize the company's liabilities and maximize flexibility. We avoid having an employment relationship with gig couriers. We don't want to be responsible for couriers' traffic negligence. We've left some space for the company to adjust contractual terms, policies, and, importantly, technologies.

Indeed, Meituan's clickwrap agreement stipulates that China's Labor Contract Law does *not* apply to the contractual relationship between Meituan or its subcontractors and GPCs. This legal classification is similar to counterparts in gig economies outside China.[65] Meituan's clickwrap agreement also stipulates that GPCs must accept every clause in the agreement and any rules published on Meituan's platform, including disciplinary measures. Meituan and Ele.me are similar to conventional companies that employ contingent workers and attempt to avoid liabilities associated with having employees. But, unlike conventional companies, Meituan and Ele.me retain for themselves the power to change contractual terms unilaterally through a nonnegotiated clickwrap agreement.[66] Asked whether he considered Meituan and Ele.me's contracts asymmetric, Liu—likely influenced by his awareness of my legal training—replied frankly: "The power of platforms and that of couriers are enormously asymmetric. We only have two major platforms in China. Tencent is the largest stockholder of Meituan, while Alibaba independently owns Ele.me. The two platforms have a lot of capital, political connections, and lawyers. What do couriers have?"

The organizational design of gig platform architecture minimizes the role of management personnel in managing and supervising couriers. Although Meituan's subcontractors and Ele.me set up offices in cities where the two companies' gig platforms operate, Peng, a manager who worked for Meituan's gig platform, explained that his office's main focus is monitoring and managing data and the platform system itself, not overseeing couriers. When I visited Peng's office, I noticed that most of the employees were either analyzing data or identifying irregular patterns on their computer screens. Much like Wong, the systems development engineer whom I interviewed, manager Peng saw the operation of the delivery platform as a mathematical problem of "multiple goal optimization," that is, a problem of "how to maximize revenues and customer satisfaction through dynamically optimizing the matching of orders and couriers with the best metrics." Couriers' experience was not something either Peng or Wong sought to optimize.

The number of gig platform offices in a city is much smaller than the number of service stations, as the former do not have the regular interactions with couriers that service stations do. To activate an app account, both Meituan and Ele.me require couriers to complete online training. They further require offline training, which is often outsourced, as a prerequisite to receive more than a certain number of delivery tasks or to reactivate an app account that was blocked when GPCs violated platform rules. I attended one such offline training, where an instructor talked about the platform company's rules, including fines and other forms of punishment, and ended with an exam that tested whether couriers could memorize the rules. Table 1.2 summarizes the platform architectures of the service and gig platforms.

Patterns of collective contention

Although SPCs and GPCs share similar demographic backgrounds and prior work experiences, I found that the frequency and injustice frames of collective action differed across the service and gig platforms. According to the management personnel I interviewed, the majority of delivery work was completed by SPCs, but GPCs organized much more collective actions than did SPCs. Among the 87 collective actions, 39% were organized by SPCs, and 61% were organized by GPCs. This finding is consistent with my interviews with management and couriers, which show most of the strikes and protests they knew of were organized by GPCs.

SPCs' collective actions typically targeted their employers—Meituan's or Ele.me's franchisee. Couriers communicated online and offline to prepare for collective action. They gathered outside their service station, raising banners and chanting their demands. There were typically fewer than 30 participants, as these actions included only SPCs at the same service station. In contrast, GPCs' collective actions usually targeted Meituan or Ele.me. GPCs organized action through social media and offline meetings.

Table 1.2. Platform architecture

Feature	Service Platform	Gig Platform
Guiding principle	The purpose is to create a reliable workforce. The design requires a higher level of supervisory power over couriers, at the expense of having legal responsibility based on employment relations.	The purpose is to create a flexible workforce. The design avoids legal responsibility based on employment relations, at the expense of having less supervisory power over couriers.
Technological dimension	Supplementing the tasks of management personnel —Using algorithms that consider performance metrics to dispatch couriers —Using algorithms to monitor and discipline couriers —Incorporating gamification	Automating management and supervision —Using algorithms that consider performance metrics to dispatch couriers —Using algorithms to monitor and discipline couriers —Incorporating gamification —Using algorithms to set and adjust remuneration —Using technologies to confirm couriers' identity and monitor whether couriers wear the company's helmets and vests
Legal dimension	Establishing legal relations based on a conventional employment contract —Based on a labor contract that classifies couriers as employees —Having a stable salary structure —Regulated by labor law	Avoiding legal relations based on a conventional employment contract and maximizing flexibility —Based on clickwrap agreements, which are asymmetric contracts —Not classifying couriers as employees —Falling outside labor law —Endowing the platform with the power to change contractual terms and platform rules unilaterally
Organizational Dimension	Management personnel in charge of managing and supervising couriers —Emphasizing management of couriers —Human supervisors can override certain algorithmic decisions. —Calculating the KPIs of supervisors and service stations from SPCs' KPIs —Imposing normative influences and surveillance on and performing emotional labor vis-à-vis couriers	Management personnel generally not in charge of managing and supervising couriers —Emphasizing management and monitoring of data and the platform system —Testing couriers on whether they can memorize important platform and disciplinary rules in online and limited offline training sessions

Because GPCs did not have a conventional workplace, they stood or sat in public space, such as plazas or the sidewalk outside restaurants. They attached signs to their scooters and held banners on the sidewalk so the public could see their grievances. GPCs also organized so-called flying pickets, in which they honked and rode scooters with strike or protest signs while patrolling outside restaurants or on the street to prevent non-participating GPCs from delivering food. The number of participants in a collective action organized by GPCs tended to be larger—often with more than 50 participants—than those organized by SPCs.

Table 1.3. Complaints in collective action

Content of complaint	Service platform (%)	Gig platform (%)
Decreasing piece rate, incentive, and subsidy	20.51	85.42
Disciplinary measures	35.29	14.68
Nonpayment of salary	37.93	0
Reorganization of franchisees	17.65	0
Work safety	17.65	6.25
Compliance with the Labor Contract Law	14.71	0
Requesting a written labor contract	17.65	0
Unilateral change of contract terms or platform rules	2.94	75.00

Note: The numbers denote the percentage of each kind of complaint among the collective actions organized by SPCs or GPCs. For instance, 35.29% of the collective actions organized by SPCs concerned disciplinary measures.

I analyze how SPCs and GPCs framed their complaints in the 87 cases. Couriers often expressed multiple complaints in a collective action. As shown in table 1.3, SPCs and GPCs had different concerns. For SPCs, the most salient concerns were nonpayment of their salary and disciplinary measures. Common phrases used by SPCs on banners included "Pay us salaries according to the contract and the law" and "The employer maliciously withholds our salaries. We need to eat!" For GPCs, the most salient concerns were reductions in piece rates and unilateral changes in contract terms or platform rules. Common slogans used by GPCs included "Ele.me reduces the rate maliciously, shortens delivery time, deceives couriers, and ignores lives"; "Meituan reduces the rate crazily. Brothers, stop taking orders. Resist Meituan!"; "Resist monopoly!"; "Despotic clauses!"; "Who can make the most despotic contracts? Ele.me!"; and "Ele.me exploits couriers; we need to feed our families; we want justice."

In summary, GPCs organized more collective actions than SPCs, and SPCs and GPCs had different injustice frames. The injustice frames used by SPCs expressed criticism of their employers for not paying salaries according to contracts and labor law. Their language did not question the legitimacy of contracts. In comparison, the injustice frames used by GPCs challenged the legitimacy of contracts, asymmetric power relations, the exploitation of capital-labor relations, and the morality of platforms. Table 1.4 summarizes the different patterns of collective contention across the two types of platforms.

Construction of grievances and collective contention under service platforms

Technological control and management. SPCs did complain about their work and named their injurious experiences, especially the pressure to maximize KPIs and the disciplinary measures built into the platform system. Several SPCs expressed that one complaint can cost RMB150 (around

Table 1.4. Varying patterns of collective contention

Characteristic	Service platform	Gig platform
Relative frequency	Less frequent	Much more frequent
Place of collective action	Outside a service station	In plazas, outside restaurants, or on the street
Who was target	Employers (franchisees)	Meituan or Ele.me
Number of participants	Usually fewer than 30	Often more than 50
Injustice frames	Violation of contracts and labor law but not challenging the legitimacy of contracts	Challenging the legitimacy of contracts, asymmetric power relations, exploitive work relations, and morality of platform companies

US\$22.30). On average, SPCs made RMB5 (around US\$.74) for each delivery, and the most productive couriers completed 50 to 60 delivery tasks each day. Thus, a fine could be a considerable proportion of an SPC's daily earnings. Many SPCs complained about the stress resulting from disciplinary measures built into the app. They talked about speeding in heavy traffic and severe weather conditions to deliver food on time. Similar to platform workers in the United States who are stressed by performing emotional labor, SPCs described feeling immense pressure to get excellent reviews.[67]

SPCs also complained about dispatch algorithms. Chen, a 26-year-old SPC, explained why he thought Meituan's algorithm was unfair: "It's clear that I am closer to the restaurant, but the system does not assign the order to me. Instead, the system assigns the order to a courier with higher KPIs. Top-five couriers keep getting orders, but others do not get many orders. This is unfair." Wang, a 27-year-old SPC, had a similar complaint: "The algorithm is stupid. When I am delivering an order, it always assigns me another with short delivery time. That decreases my on-time rate." Dissatisfaction with algorithmic control and management was prevalent among interviewees.

Legal control and management. Despite SPCs' complaints about technological control and management, I found their dissatisfaction seldom evolved into a grievance or a sense of injustice, in part because of their understanding of the legal relationship. SPCs believed their working conditions were expected based on their labor contracts. Several SPCs mentioned they knew the salary structure and disciplinary rules from the very beginning, and the rates and rules remained the same. As Dong put it: "You know how much you are going to get and what the service station expects you to do. If you are unhappy, don't take the job. You can leave the job anytime." Similarly, another SPC remarked: "The disciplinary rules are harsh, but the service station needs to ensure the quality of service.

And we know that from the contract and rules on the wall." These explanations show that the legal design of the platform architecture—specifically, the concrete and stable terms in a labor contract—helps stabilize labor relations.

In addition, SPCs distinguished dissatisfying from unacceptable aspects of work. They saw salary as the most important component of their contracts. As long as their employers paid their salaries, they did not have serious issues. Wang's view was typical among SPCs: "I'm dissatisfied with the dispatching algorithm, but it's not the most important issue. As long as my boss pays me the salary, I'm fine. I got my contract down in black and white. If my employer does not pay me, I'll have evidence to prove his violation of the contract and labor law. I can talk to supervisors. If they don't respond, I will go to the court." Asked whether going to the labor bureau or the court was difficult, Wang replied, "Couriers on WeChat say there are free templates available. We only need to fill out a form to file a complaint. You don't need a lawyer. Some couriers got money back successfully." When I asked Wang whether he would consider a strike or a protest if he did not receive his salary, he said he would not do so unless none of the other methods worked.

Wang's answer, which was common among SPCs, shows the relationship between injustice frames and the appeal of collective contention. Because SPCs' injustice frames were based on a well-established legal claim regarding violation of the labor contract and labor law, they believed they could solve problems through individual informal legal mobilization and institutionalized dispute resolution channels. They considered collective contention only when other branches of the dispute tree did not work.[68]

SPCs' focus on one contractual element—salary—was also reflected in how they thought about social insurance. Few SPCs were dismayed by the fact that service stations did not entirely follow the Labor Contract Law. Due to the often weak enforcement of labor law generally, SPCs had low expectations regarding social protection.[69] In fact, only one SPC said his employer bought social insurance for him. However, the SPCs and supervisors I interviewed said that because the turnover rate of SPCs was high, it was difficult for service stations to provide SPCs with social insurance. My interviewees told me that not even station supervisors receive social insurance. None of the SPCs I interviewed considered social insurance a critical component of their labor relations. Fang, a 26-year-old SPC, said: "Almost no employers in the service sector would buy social insurance for employees. Also, we don't think social insurance useful, so we don't ask the station to buy it for us. I bought social insurance when I worked in an automobile factory. I don't feel I'm in a better situation with social insurance because of my household registration status." SPCs cared mainly about wages. In most cases, it was only when a grievance about salary emerged that other issues, such as social insurance and disciplinary measures, were incorporated into the discussion.

Organizational control and management. Most SPCs told me that they communicate with their supervisors when problems arise and that this makes them feel more informed about the situation even when their complaints are not resolved. Many couriers understood that supervisors themselves are constrained by the platform system and Meituan or Ele.me. For example, Chen and Wang told me that although they complained about dispatch algorithms, it would be unreasonable to blame their supervisors because their supervisors are obligated to use the algorithms. As Chen explained: "I told my supervisor the platform system assigns orders unfairly. I received fewer orders than people with good KPIs. He told me he couldn't change the system, but he manually assigned me some orders from time to time after I talked to him. That made me feel better." This statement also shows that even when a complaint occurs, such as Chen's dissatisfaction with algorithmic control, supervisors' intervention could prevent problems from worsening.

Wu, a 30-year-old SPC, told me about a dispute in his station that could have led to a strike. When Wu and his fellow couriers did not receive their salaries on time, they discussed the issue on WeChat and talked to their supervisors, who knew SPCs could initiate a strike. Under pressure, Wu's supervisor asked the district manager to explain the situation to the couriers. The manager said his company had problems with Ele.me and would distribute salaries once they received payment from Ele.me. In the end, the SPCs received their salaries. "Because the management solved our problems," Wu said, "we didn't take further action like strikes." This instance shows effective management can reduce the appeal of collective action.

Most SPCs told me they had no serious problems with their supervisors, though two complained about supervisors' unfair manual assignment of tasks. I wondered why the supervisory relationship was not more ridden with tension. Chen explained that his supervisor's KPIs and the station's KPIs are the aggregation of SPCs' KPIs. He and his colleagues can retaliate against their supervisor by lowering their KPIs and then leaving their job if their supervisor is mean to them. Chen added that it is easy for SPCs to transfer their work to another service station if they have conflicts with supervisors.

In interviews with management, I also found organizational constraints on supervisors' behavior. Huang, a station supervisor, said: "It's better to educate and communicate with couriers nicely. We are in the service industry, not manufacturing. I need their smile, politeness, and good attitudes to make a living." Similarly, Cai, a district manager, told me that the key is "managing the hearts and minds of both customers and couriers" to get good KPIs. Cai pointed out that according to the franchise contract, the platform could terminate his company's right to operate the delivery business if a strike or protest occurred; franchisees also have obligations to oversee SPCs' behavior on social media. Essentially, the organizational constraints through KPIs and franchise contracts motivate supervisors to perform emotional labor and exercise normative influence—and surveillance—on SPCs.

Indeed, supervisor Huang said it is not difficult for management to discern what is going on among SPCs as he sees them every day and is in their social media groups. "You know who is doing what. . . . I do extra ideological work on unruly couriers." One aspect of ideological work, according to Huang, is telling SPCs that protests or strikes are illegal. Huang's explanation corresponds with the way Wu, an SPC, described his supervisor's surveillance: "He knew what we were discussing. It's quite difficult for us to discuss things. Also, couriers with a closer relationship with the supervisor might betray us." Huang's and Wu's statements reveal that couriers' workspace is not free but monitored by management, posing an established challenge to collective action.[70]

Collective contention. Despite organizational control and management, collective contention still arises on rare occasions. Three SPCs who reported participating in collective action provided similar accounts of why collective action happened. Ming described one situation: "We were supposed to receive our salaries. Our supervisors didn't know why the company [franchisee] hadn't paid us, but he agreed the company violated its obligation. Since some couriers said the company would go bankrupt soon, we decided to take action [a strike]. We didn't have time for the court." Ming's account reveals that the injustice frame for SPCs was violation of contractual terms. After the strike, Ming's company told SPCs the company had financial problems. SPCs eventually received around 80% of the salaries due them.

I also interviewed station supervisors and a district manager about collective action. Yang, a service station supervisor, commented: "Couriers rarely initiate a strike or protest as supervisors already solve most of the problems. Strikes occur only when franchisees have problems with Meituan or Ele.me or when there is an internal dispute among investors. These problems create financial difficulty for franchisees and cannot be solved by supervisors. Under such circumstances, franchisees try to pay less. Couriers and franchisees have more or less similar understandings about what is right and what is wrong." Yang's observation corresponds to something district manager Cai told me, that it would be difficult for franchisees to deny their obligation to pay salaries. The focus of negotiation from the perspective of franchisees, Cai explained, is to "let couriers know the company's difficulty and accept what the company can afford." Together, my interviews with SPCs and management suggest that, even when a collective action arose, couriers and employers did not have fundamentally different views on contractual terms and law. Rather than challenging the legitimacy of contracts, SPCs were demanding that employers honor the contracts.

Within the service platform, technological control and management can cause dissatisfaction among SPCs, but the legal and organizational design of the platform architecture often serves to contain grievances and make collective action less appealing. Collective contention seems less appealing to SPCs because their injustice frames (i.e., violation of contracts and labor law) are based on well-established legal claims and they can communicate with management personnel. Finally, the platform architecture's organizational design, particularly the constant presence of supervisors, also restricts the amount of free space for organizing collective contention.

Construction of grievances and collective contention under gig platforms

Technological control and management. Similar to SPCs, GPCs were critical of dispatch algorithms and disciplinary rules built into the platform system, but they also had problems with the calculation of remuneration, cuts in remuneration, and changes in dispatch and disciplinary rules. Several GPCs accused the gig platforms of "stealing money" from them. Qiang, a 29-year-old GPC, asserted: "Meituan is scamming us. They know the real delivery distance but use the linear distance between two points to calculate remuneration. We are not flying an airplane!"

Like gig workers in Europe, GPCs overwhelmingly criticized how platforms have lowered remuneration.[71] They believed platforms change algorithms to reduce remuneration, though the two platform companies have never openly acknowledged that they lower the piece rate. Much as Rosenblat and Stark's study of Uber in the United States showed, gig platforms and couriers in China have asymmetric information about the platform system.[72] The actual mechanism of algorithmic pricing remains a "black box," hidden from workers' view, so GPCs base their complaint about the falling rates on their observations. Although gig platforms use algorithms that consider many factors in calculating piece rates, couriers said it is evident that piece rates under similar circumstances have decreased over time.

GPCs often take screenshots of their tasks and share them online, so they believe they have evidence of the decreasing piece rates. For example, a GPC said that in the past, he received RMB5 (around US$.74) for a delivery task within 1.5 kilometers, but the rate went down to RMB3.5. There was consensus among my interviewees that Meituan and Ele.me had changed their algorithms and lowered the piece rate *stealthily*, hoping couriers would not notice. GPCs complained that decreasing piece rates had a significant effect on their lives. According to Shi, a 22-year-old GPC, "I was able to make 9,000 to 10,000 yuan [roughly US$1,278 to US$1,420] per month. Now I make around 6,000 yuan [around US$852] a month." Many GPCs reported having to work two more hours every day to maintain the same level of income.

GPC interviewees emphasized that along with decreasing piece rates, the platforms have also been changing the system's dispatch and disciplinary rules so as to restrict GPCs' freedom to decline a dispatch assignment. Similarly, Rosenblat and Stark found that Uber constrained drivers' freedom to reject a request.[73] When a GPC declines a dispatch assignment, Meituan's platform system deducts points from the courier's credit account. Couriers with low point totals are less likely to be assigned orders with higher remuneration. In some cities, Meituan has even made new dispatch rules that require offline training and restrict a courier's eligibility to take an order after they decline a certain number of dispatch assignments.

I saw a lot of criticism in online communities about these rules. For example: "Given that couriers do not have freedom, Meituan should stop using freedom in its advertisement to recruit gig couriers"; "The platforms promise to give us freedom, but now they treat us like their slaves. We are forced to take orders with low pay." These comments critiqued the contradiction between gig platforms' promise of freedom at work and the ways dispatch rules and algorithms negate such freedom. When GPCs realized gig platforms increase technological control to reduce the promised freedom, they saw the platform-courier relationship as a master-slave relationship and felt betrayed.

Legal control and management. Certain updates of dispatch and disciplinary rules in the gig platform system require corresponding legal arrangements, thus leading to the overlap of technological and legal control and management. Unlike changes in pricing practice, changes to dispatch and disciplinary rules are often transparent. Before implementing new dispatch and disciplinary rules in the platform system, local offices notify GPCs of the impending changes and justify them by asserting that their purpose is to "better regulate delivery behavior."

Local offices further affirm that they have the "ultimate legal right" to change platform rules according to contracts between the platform and couriers by pointing to the clickwrap agreement that couriers accept when installing the delivery app. However, none of my gig platform interviewees reported reading the legal texts when they installed the app and accepted the clickwrap agreement, and none knew that platforms had the right to change the content of the agreement and platform rules unilaterally. They all expressed anger about such clauses and other rules enacted by platforms unilaterally. GPCs called such contractual terms and platform rules "despotic clauses" (*bawang tiaokuan*) and pointed to them as evidence of exploitation.

I was curious to find out why the companies were so secretive when it came to changing pricing practices, yet open about changes in dispatch and discipline rules. Manager Peng explained that the companies have to announce the latter to shape GPCs' behavior: "We use the carrot and stick approach. If the new rules remain unknown to couriers, couriers will not change their behavior even after we incorporate the rules into the platform system." He emphasized the need to adjust platform rules dynamically:

> The purpose is to enhance the efficiency of dispatch assignments and customers' experience. Under the service platform, couriers must accept every order assigned by their station. Under the gig platform, couriers don't have such obligation, but the company wants every order to be taken by an excellent courier. Achieving this goal needs continuing improvement of dispatch and disciplinary rules. Now our platform system is powered by artificial intelligence and machine learning. It learns from the big data it collects and gives us feedback. These tools help us improve the system's efficiency by changing the rules dynamically.

Manager Peng's remark reminded me of my observations of offline training sessions, in which GPCs were required to memorize platform rules and tested on their content. Gig platforms, I realized, seek to turn couriers into a part of the platform system that automatically implements the rules. When companies change the rules, new rules have to be installed not only in the platform system but also in the minds of couriers themselves. When this is done successfully, human workers effectively become part of the platform technology.

Yet, unlike SPCs, GPCs do not internalize the rules through normative influence or emotional labor. Gig platforms just announce the changes in platform rules and corresponding disciplinary measures, assert their "ultimate legal rights," and test GPCs on the rules. In these moments, technological and legal control overlap, and the platform's power becomes salient. Deng described such a moment: "When I saw the platform's notice about the new dispatch and disciplinary rules in the system [the platform system] and the company's legal rights, I felt furious. Then I understood why couriers complained about despotic rules on WeChat. . . . Couriers were saying everything—from platform rules to legal clauses—is determined by the platform but not us." Deng's explanation shows that the revelation of platforms' unbridled legal and technological power turns moments of change into moments of grievance escalation.[74] Previous literature has focused on the legal classification of workers as employees or individual contractors,[75] but GPCs are most concerned about platform companies' unilateral and unconstrained legal and technological power. In the interviews, many GPCs emphasized that the platforms' exploitation is *blatant*. For example, one GPC remarked: "Now the platforms implement all kinds of despotic clauses. If you don't follow their rules or terms, they deduct money from you. They force you to accept their dispatch assignments with low pay. If you don't take their order, they deactivate your account. It is unfair. Platforms *blatantly* exploit couriers."

In general, gig couriers believed platforms used unfair legal arrangements, making the inputs and return for labor totally disproportional. Although the two platforms might not technically violate any existing law, they felt there were unfair loopholes in the legal system. Here, we can see how the legal dimension of the gig platform architecture generates grievances and heightens perceived injustice.

Asked whether they can address their grievances through the labor bureau or the court, most GPCs said no because they simply do not know which law platform companies are violating. Importantly, they understood that their legal situation differed from conventional employment relations, but they could not use legal terms to describe their status. As Zeng described their situation: "We don't have a contract. I mean a labor contract. I don't know what kind of status we have. I only know we're not their employees. I don't think we can go to the court or the labor bureau. That's for people with a labor contract. We don't know what kind of law the platforms have violated. We just feel they are wrong and things are unfair." In social media group discussions, most GPCs shared the same legal understanding as Zeng. As one GPC shared with a social media group composed of other GPCs, when he went to a labor bureau for legal advice, he was told the bureau could not help because it could only assist people with a labor contract. Consequently, GPCs see a disjuncture between their injustice frames that challenge the legitimacy of contracts and claims that can be accepted by dispute resolution institutions. Put another way, the institutionalized resolution "branches" in the dispute trees do not seem viable to GPCs.[76] For that reason, the legal dimension of the gig platform architecture makes collective action more appealing than institutionalized resolution channels.

Organizational control and management. The organizational design of the gig platform further intensifies grievances and increases the appeal of collective contention. As the gig platform largely automates control and management, GPCs do not develop social relationships with management personnel. Instead, they interact with the "cold" technological interface. GPCs use the apps to interact with platform companies. Similar to Uber drivers in the United States, they encounter difficulties trying to speak to platform representatives.[77] As Shi told me: "I tried many times, but they never responded to me. Last time when they changed platform rules, I contacted them to complain. I was not the only one. In the end, couriers became angrier and angrier. We had to do something else to get our voices heard." Indeed, other GPCs shared similar frustrations about not being able to solve their problems through apps or local offices. Frustration and anger escalated grievances. In most of the collective actions I analyzed, GPCs contacted platform companies but did not get any meaningful responses; they soon realized this approach would not address their grievances.

The disengagement of platform companies also means GPCs have relatively free spaces to mobilize solidarity and organize collective action, mostly online but also offline.[78] GPCs expressed their anger and sense of injustice in social media groups. I observed GPCs rename their social media groups as labor unions even though they could not form formal labor unions according to Chinese law. Many couriers discussed the importance of solidarity with such online statements as "We must have solidarity to defend our interests, brothers. Otherwise, we'll continue to be exploited by the platform." Some couriers began to organize collective action offline. To be sure, the platforms seek to undermine online organizing. In some cases, after local offices found GPCs were organizing a protest via social media groups, they deactivated organizers' app accounts and asked social media companies to block those groups. Nonetheless, GPCs were still able to form new groups.

Collective contention. Collective action occurred mostly when GPCs believed gig platforms had drastically decreased remuneration by changing algorithms or when platform offices notified GPCs of new platform rules and simply cited the platform's contractual right to do so. Usually, a core group of GPCs—those with existing interpersonal relationships—initiated a strike or protest. They did so in the absence of assistance from grassroots labor NGOs or government-organized trade unions. Consistent with findings in previous sociolegal studies, experience and knowledge mattered.[79] GPCs with experience working in the manufacturing sector played a crucial role in organizing such collective actions. Core organizers tended to be those who completed a huge amount of orders every day. Decreasing rates and changing rules cost such workers hundreds of RMB daily. Core organizers' attempts to improve their work situation was understandable, given that so many did not want to return to the manufacturing sector and did not have a feasible alternative career plan.

Here, I use a series of strikes and protests in Chongqing in 2018 to illustrate a typical mobilization process in online and offline free spaces. The core organizers belonged to a close-knit group. Tang, a 23-year-old GPC, recounted how his group organized a strike after Meituan announced a change in dispatch and disciplinary rules:

> We only had four people in the beginning. We became friends
> through a WeChat group. Four of us hang out after work two to

> three times a week. When we saw couriers got angry in social media groups, we discussed in a BBQ place and decided to organize a strike. . . . I saw how people in my factory went on strikes. They got what they wanted. [Author: Why didn't you talk to Meituan or go to the court or the labor bureau?] A lot of couriers already expressed their discontent to Meituan, but Meituan didn't respond. No one knows how to file a legal complaint under such circumstance. What we write would not be accepted by the court.

Tang's quote is consistent with the foregoing analysis about how the legal and organizational dimensions of control influence GPCs' perception of different branches in the dispute tree.

Tang and his friends made banners and signs featuring messages such as "Despotic clauses," "Resist exploitation," and "Stop taking orders." They chose these terms because other GPCs had used them in WeChat groups, where Tang and his friends recruited participants. They then visited several restaurants, because restaurant owners also suffer from asymmetric contractual relations with Meituan, especially increases in commission fees. Three restaurant owners lent their support, not only by hanging banners saying "Resist monopoly, resist Meituan" and "Resist despotic clauses," but also by providing GPCs with free meals to show solidarity. The core group sent out a message warning couriers that if nonstriking gig couriers were seen delivering food during a strike, their tires would be slashed. One core member organized a 10-person patrol team to this end.

After the strike and protest began in a plaza, GPCs uploaded screenshots from their delivery app to social media groups to further mobilize participation. They proclaimed their own offline status with comments like "Hundreds of brother couriers are offline. Don't be a bad apple. Our patrolling team is sanctioning bad apples on the street" and "We've determined not to be exploited by evil capitalists. Long live the working people!" GPCs often join multiple social media groups, so information spread quickly online and among GPCs who worked for different companies. Some nonparticipants encountered violence for disregarding warnings. Faced with potential sanctions, several GPCs decided not to work and instead went to protest sites as bystanders. Tang was surprised by what happened: "Initially, we had a strike in our district and drew about 100 couriers, but it soon spread to five other districts and lasted for a week. In each of the subsequent strikes, I saw 50 to 80 couriers. We initiated a strike against Meituan, but the strikes also attracted Ele.me couriers." Most participants were full-time GPCs, who were more dependent on platforms. One part-time GPC told me, "[The protest] isn't my business because I have a real job." He also said he was not as angry as full-time couriers as his stake in outcome was low. This finding resonates with literature that shows how economic dependency influences work experiences.[80]

Eventually, GPCs besieged one of Meituan's local offices in Chongqing for several hours. The core members announced to Meituan's local office specific demands concerning platform rules and remuneration. Zhang, a 20-year-old GPC with a high school education, recalled that moment:

> Finally, we saw those people [the managers] who've never responded to us. Four to five people came out from the office. They are all in their 20s, just like us. You can tell they are *college graduates*. The manager is a *woman*. They were so scared and cannot say anything to couriers. Although they did not tell us, we found they increased the rate. They also sent out a notice to withdraw the new platform rules. . . . After the protest, Meituan did not decrease the rate for eight months, but they deactivated a few participants' accounts. (emphasis added)

Zhang's recollection also reveals class and gender boundaries between management and couriers and the awkward absence of interaction when GPCs eventually encountered management.

Their singular focus on optimization of the platform system rendered gig platforms' local offices ill-equipped to address grievances and resistance. This tendency was evidenced by Meituan's attempt, in September 2019, to install more restrictive dispatching rules in Chongqing, despite the strikes in 2018. When many GPCs threatened another strike, Meituan finally gave up trying to install the new rules. GPCs were happy to see Meituan change its plan under such pressure, but they also recognized their inability to oversee the platform or to actualize such threats regularly.

I was interested to know why GPCs joined a collective action, particularly in the face of potential police suppression, rather than simply finding another job. Several GPCs told me they joined the collective action because the platform companies are so despotic and their rules are so unfair. Anger and a strong sense of unfairness, rather than an instrumental calculation about the potential pros and cons of collective action, prompted some GPCs who were not core organizers to join a collective action. Fang told me how such sentiment drew him into a protest: "I went there because the relationship is so unfair. I didn't know if we can change the situation, but I had to go because I was so angry." In fact, several GPCs were puzzled when I asked why they protested instead of finding another job. They said they would probably find a better job sooner or later because they did not plan to be a GPC for a long time, but they still wanted to teach platform companies a lesson. My finding thus resonates with research that shows collective action frames can go beyond material calculations and are often based on moral imperatives.[81]

A few couriers told me they were just bystanders. One GPC remarked: "When other couriers and I learned from social media groups that some gig couriers went on a strike in a plaza, many of us immediately dashed to the site. We were curious and wanted to see what happened." Interviewees who described themselves as bystanders shared the same sentiment as other GPCs, but they emphasized their need to make money to sustain their lives. Nonetheless, the presence of bystanders did escalate collective action. Not surprisingly, GPCs who decided not to join a collective action talked about needing to make money and thought such collective action would not work.

Asked why they were not afraid of police suppression, many GPCs said they were just bystanders or pretended to be bystanders; they did not think the police would arrest them as there were a lot of people there. Participants who joined the protests in Chongqing in 2018 said the police took away only a few participants and told them organizing a strike or protest was illegal. Most interviewees did not seem intimidated by the police.

Under the gig platform, technological control and management lead to grievances and perceived injustice. This dimension of control and management at times overlaps with, and is reinforced by, legal and organizational control and management, generating moments of escalation. GPCs' injustice frames (i.e., the illegitimacy of contracts) and the organizational design of the platform architecture make institutionalized dispute resolutions and individual informal legal mobilization seem implausible to GPCs. The platform companies' focus on managing the platform system rather than the couriers creates free spaces in which GPCs can turn moments of escalation into moments of collective action and solidarity.

Discussion and Conclusions

This chapter is among the first attempts to study collective contention in the food delivery platform economy in China. It joins recent efforts to analyze labor processes and contention in the platform economy.[82] Building on literature that suggests researchers should incorporate data about management's perspective in analyzing labor processes in the platform economy,[83] I examined a variety of data, including interviews with workers, management personnel, a lawyer, and an engineer. This article thus provides a multiperspective account of how platform companies and their franchisees control and manage the labor process. Specifically, I demonstrate how platform architecture—technological, legal, and organizational dimensions of control and management—influences labor process and critical mechanisms for collective contention. I also show that when different dimensions of control reinforce each other in escalating grievances and perceived injustice, and increase the appeal of and spaces for collective action, collective action and solidarity are more likely to emerge.

Before discussing the contribution of this study, I first address two analytic issues. The first concerns the 87 cases of collective action. As mentioned previously, the data are limited in their exhaustiveness. To strengthen

the reliability of my analysis, I examined the 87 cases along with interviews with couriers and management personnel. Results from the three
data sources are consistent. Data triangulation thus strengthens the reliability of my findings. The second issue is a partial alternative explanation. As mentioned, some GPCs worked first as SPCs and switched to
the gig platform once they became more familiar with their local area.
This suggests SPCs and GPCs might have different skill levels or characteristics due to the self-selection process. As shown in table 1.1, SPC and
GPC interviewees had similar demographic characteristics and prior employment history. Furthermore, after careful examination of the data, I
did not find SPCs and GPCs had different capabilities to participate in
collective contention. In short, evidence does not support the alternative
explanation.

This study makes three major contributions. First, it proposes the theoretical concept of *platform architecture* to capture not only the technological but also the legal and organizational aspects of labor control and
management, as well as the relationship among these different dimensions
of control. As such, the study deepens extant scholarship that focuses on
algorithmic control and management and sheds light on the various ways
these distinct dimensions of control and management operate, interact,
and shape the labor process jointly.[84]

Second, the study demonstrates and explains emergent collective action and solidarity in an unlikely setting. Recent scholarship suggests work
relations under platform economies can be analyzed through the lens
of labor process theory.[85] Although this theory is insightful for analyzing
control and management, it does not provide a comprehensive framework
for analyzing workers' responses, especially those revealing how workers
construct and react to grievances. This study contributes to the literature
by integrating labor process theory with sociolegal studies and literature
on collective action. This theoretical framework can thus help analyze
how the technical, legal, and organizational dimensions of platform architecture shape collective resistance by influencing the construction of
grievances, the appeal of collective contention vis-à-vis other responses to
grievances, and access to free spaces for collective action. Although work
is atomized and workers share neither workspace nor employers within
the gig platform, I show how gig workers mobilize to contest gig platforms. Following literature that suggests we focus on moments of escalation, I demonstrate that when different dimensions of control reinforce
one another, moments of escalation are more likely to evolve into collective action and solidarity.[86]

My findings contribute to a more nuanced understanding of the emergence and characteristics of solidarity in the platform economy. Much
like Tassinari and Maccarrone, I find expressions of solidarity—mostly
as couriers discuss similar grievances against platform companies in rel-

atively free spaces and subsequently participate in collective action. But I also examine the cross-cutting, uneven, and contested characteristics of such solidarity, which reveal both the potential for and the limitations of solidarity. On one hand, gig couriers working for one platform company can moblize support and solidarity from restaurant owners and couriers working for the other company. Such cross-cutting solidarity has not previously been documented by studies of collective action in the platform economy. I further provide evidence that some solidarity can prompt compromise from gig platform companies. On the other hand, my findings also reveal that solidarity is uneven and contested.[87] I uncover variation among gig workers in terms of their solidarity with other workers and involvement in collective action. Indeed, core actors in collective action share strong solidarity and support. Beyond core actors, however, many GPCs are united only by their shared animosity to the platforms. As a result of shared sentiments, potential sanctions, and curiosity, many GPCs showed up at strikes or protest sites, but only as bystanders. Some nonparticipating gig workers even encountered threats and violence.

Third, my finding about the legal design of platforms goes beyond the conventional concern with the employee-versus–independent contractor classification.[88] This study suggests the need to attend to how the enormous difference in contractual power between platforms and couriers intensifies couriers' sense of injustice and exploitation. As such, it contributes to literature on the role of contractual relationships in the market economy.[89]

The question may arise why employment classification is not a major concern for couriers in China, considering it is often an important issue in other contexts.[90] In China, one major difference between workers with and those without employee status is that the former are entitled to social insurance protection according to labor law, but the latter are not. Given the weak enforcement of labor law in China,[91] both SPCs and GPCs understood that employers in the service sector rarely purchase social insurance for workers. Therefore, SPCs did not expect to have social insurance, despite their employee status. Given no real difference between SPCs and GPCs in terms of social insurance, GPCs were not focused on employment classification.

What mattered more in grievance construction was how couriers perceived the predictability and fairness of their legal relationship. Under the service platform, labor contracts help stabilize labor relations. SPCs have clear expectations and believe their employers are constrained by contractual terms and labor law, at least regarding salary. Therefore, SPCs' injustice and collective action frames reveal only a legal contradiction—the violation of contractual terms or labor law. No one challenges the legitimacy of the legal order based on contracts and labor law. In contrast, under the gig platform, the contractual design enables intense algorithmic control and management by giving platforms unbridled legal and technological power. Legal scholars have long been concerned with the problems associated with clickwrap agreements and asymmetric contracts.[92] My analysis shows that when gig platforms exercise their unbridled contractual power to change platform rules, workers not only discern this power but also consider it and the rules made by such power to be "despotic." Importantly, GPCs can often mobilize restaurant owners and GPCs who work for the other gig platform to join collective action because they all suffer alike from asymmetric contractual power and share similar injustice frames. This finding is relevant beyond the Chinese context. As Stewart and Stanford point out, Uber has a similar contractual design with couriers in Australia.[93] In addition, Uber Eats and Grubhub in the United States have leveraged their market and contractual power to charge restaurants increasing commission fees.[94] The Chinese case shows that injustice frames based on asymmetric contractual power can contribute to cross-cutting mobilization and solidarity.

Finally, regarding the generalizability and limitations of this study, this chapter focuses on the labor process. Many of my findings about the relationship between platform architecture and collective resistance can be generalized to other contexts. As discussed, the ways gig platforms exercise technological, legal, and organizational control and management bear many similarities across contexts—in China, the United States, Australia, and Europe. Nonetheless, processes and factors beyond the labor process—from state regulation to political regime, trade union activity, differentiation of labor forces, and labor markets—can influence labor contention.[95]

To help evaluate the generalizability and limitation of my findings, I specify the following contextual characteristics of the Chinese case. First, the Chinese state has made little effort to restrain monopoly power or initiate antitrust investigations.[96] In contexts where the state does not fare well in overseeing market power, platform companies can have more unbridled contractual power in strengthening legal and technological control, which could generate a stronger sense of injustice. Second, platform workers do not have high expectations regarding social insurance protection in China. Platform workers in countries where people have higher expectations of job security and social welfare might have different injustice frames. Third, the Chinese case is characterized by its weak civil society, absence of independent labor unions, and authoritarian context—all of which make organizing collective action more difficult. Gig workers in China largely confine their actions to strikes and protests. In contexts where workers have more freedom and assistance from civil society actors and labor unions, collective action could be easier and manifest in various forms. For example, lawyers and civil society actors can help bring institutional changes through filing lawsuits in the court and influencing regulatory agencies.[97] Finally, platform workers in China share similar demographic characteristics. In contexts with a large internal division among workers, it might be more difficult for platform workers to organize collective action.[98] The limitations of this study also point to the need for future research. Efforts should be made to examine how factors beyond the labor process influence platform architecture and its relationship with labor contention.

Acknowledgments

I thank Kim Greenwell, Greta Krippner, Fangsheng Zhu, four anonymous reviewers, and the *American Sociological Review* editors for their insightful and constructive comments on my manuscript. Special thanks go to reviewer 2 for their heroic 2,596-word report, which tremendously improved the quality of the article. In addition, I thank Julia Adams, Deborah Davis, Mary Gallagher, Adam Reich, Diane Vaughan, Josh Whitford, and Jonathan Wyrtzen for their feedback and suggestions on earlier iterations of these ideas. The article also benefited from presentations at the Sociology Colloquium at Columbia University, the Comparative Research Workshop in Yale's Sociology Department, the 2020 annual meeting of the American Sociological Association, and the China law conference at the University of Michigan. A previous version of the paper was rejected by a special issue of *Socio-Economic Review*, but I sincerely thank the editor and two anonymous reviewers for their excellent suggestions and encouragement, which helped me see the potential of the project. My thanks also go to the informants, interviewees, and assistants who made my research possible. Finally, I thank the Dean's Competitive Fund and the Weatherhead Center for International Affairs at Harvard University for their generous funding support.

Notes

1. Martin Kenney and John Zysman, "The Rise of the Platform Economy," *Issues in Science and Technology* 32, no. 4 (2016): 701–9.

2. Kenney and Zysman, "Rise of the Platform Economy"; Juliet B. Schor and William Attwood-Charles, "The 'Sharing' Economy: Labor, Inequality, and Social Connection on For-Profit Platforms," *Sociology Compass* 11, no. 8 (2017): 1–16; Steven P. Vallas, "Platform Capitalism: What's at Stake for Workers?" *New Labor Forum* 28, no. 1 (2017): 48–59.

3. Min Kyung Lee, Daniel Kusbit, Evan Metsky, and Laura Dabbish, "Working with Machines: The Impact of Algorithmic and Data-Driven Management on Human Workers," in *Proceedings of the 33rd Annual ACM Conference on Human Factors in Computing Systems* (Seoul: ACM), 1603–12; Alex Rosenblat and Luke Stark, "Algorithmic Labor and Information Asymmetries: A Case Study of Uber's Drivers," *International Journal of Communication* 10 (2016): 3758–84.

4. Richard Hyman, "Strategy or Structure? Capital, Labour, and Control," *Work, Employment, and Society* 1, no. 1 (1987): 52.

5. Michael Burawoy, *Manufacturing Consent: Changes in the Labor Process under Monopoly Capitalism* (Chicago: University of Chicago Press, 1982).

6. Hyman, "Strategy or Structure?" 52.

7. Ruth Berins Collier, Veena Dubal, and Christopher Carter, "The Regulation of Labor Platforms: The Politics of the Uber Economy" (working paper, Berkeley Roundtable on the International Economy, March 1, 2017), https://brie.berkeley.edu/sites/default/files/reg-of-labor-platforms.pdf; Juliet Webster, "Microworkers of the Gig Economy: Separate and Precarious," *New Labor Forum* 25, no. 3 (2016): 56–64.

8. Rosenblat and Stark, "Algorithmic Labor and Information Asymmetries"; Sebastian Deterding, "Gamification in Management: Between Choice Architecture and Humanistic Design," *Journal of Management Inquiry* 28, no. 2 (2019): 131–36.

9. Arianna Tassinari and Vincenzo Maccarrone, "The Mobilisation of Gig Economy Couriers in Italy: Some Lessons for the Trade Union Movement," *Transfer* 23, no. 3 (2017): 353–57; Arianna Tassinari and Vincenzo Maccarrone, "Riders on the Storm: Workplace Solidarity among Gig Economy Couriers in Italy and the UK," *Work, Employment, and Society* 34, no. 1 (2020): 35–54.

10. Alex Veen, Tom Barratt, and Caleb Goods, "Platform-Capital's 'App-Etite' for Control: A Labour Process Analysis of Food-Delivery Work in Australia," *Work, Employment, and Society* 34, no. 3 (2019): 388–406; Phoebe V. Moore and Simon Joyce, "Black Box or Hidden Abode? The Expansion of Platform Work Managerialism," *Review of International Political Economy* 25 (2019): 1–23.

11. Burawoy, *Manufacturing Consent*; Collier, Dubal, and Carter, "Regulation of Labor Platforms."

12. "2018 Food Delivery Worker Employment Report," 199it, accessed July 27, 2020, http://www.199it.com/archives/823693.html.

13. Jean Tirole, *Economics for the Common Good* (Princeton, NJ: Princeton University Press, 2017).

14. Hyman, "Strategy or Structure?"

15. Webster, "Microworkers of the Gig Economy."

16. Mary E. Gallagher, "China's Workers Movement and the End of the Rapid-Growth Era," *Daedalus* 143, no. 2 (2014): 81–95; Lu Zhang, "Worker Protests and State Response in Present-Day China: Trends, Characteristics, and New Developments, 2011–2016," in *Handbook of Protest and Resistance in China*, ed. T. Wright (Northampton, MA: Edward Elgar, 2019).

17. Mary E. Gallagher, *Authoritarian Legality in China: Law, Workers, and the State* (New York: Cambridge University Press, 2017).

18. Gallagher, "China's Workers Movement"; Zhang, "Worker Protests and State Response."

19. Ya-Wen Lei, *The Contentious Public Sphere: Law, Media, and Authoritarian Rule in China* (Princeton, NJ: Princeton University Press, 2017); Ngai Pun et al., "Apple, Foxconn, and Chinese Workers' Struggles from a Global Labor Perspective," *Inter-Asia Cultural Studies* 17, no. 2 (2016): 166–85.

20. William L. F. Felstiner, Richard L. Abel, and Austin Sarat, "The Emergence and Transformation of Disputes: Naming, Blaming, Claiming," *Law and Society Review* 15, nos. 3–4 (1980): 631–54; William A. Gamson, "Constructing Social Protest," in *Social Movements and Culture*, ed. H. Johnston and B. Klandermans (Minneapolis: University of Minnesota Press, 1995), 85–106; Sandra R. Levitsky, "'What Rights?' The Construction of Political Claims to American Health Care Entitlements," *Law and Society Review* 42, no. 3 (2008): 551–90.

21. Gamson, "Constructing Social Protest."

22. Vanessa Beck and Paul Brook, "Solidarities in and through Work in an Age of Extremes," *Work, Employment, and Society* 34, no. 1 (2020): 5–16; John

Kelly, *Rethinking Industrial Relations: Mobilization, Collectivism, and Long Waves* (London: Routledge, 1998).

23. Beck and Brook, "Solidarities in and through Work"; Peter Gahan and Andreas Pekarek, "Social Movement Theory, Collection Action Frames, and Union Theory: A Critique and Extension," *British Journal of Industrial Relations* 51, no. 4 (2013): 754–76.

24. Glenn Morgan and Valeria Pulignano, "Solidarity at Work: Concepts, Levels, and Challenges," *Work, Employment, and Society* 34, no. 1 (2020): 18–34.

25. Rick Fantasia, *Cultures of Solidarity: Consciousness, Action, and Contemporary American Workers* (Berkeley: University of California Press, 1988), 17; Maurizio Atzeni, "Searching for Injustice and Finding Solidarity? A Contribution to Mobilisation Theory," *Industrial Relations Journal* 40, no. 1 (2009): 5–16.

26. John Kelly, "*Rethinking Industrial Relations* Revisited," *Economic and Industrial Democracy* 39, no. 4 (2018): 701–9.

27. Anna Pollert and Andy Charlwood, "The Vulnerable Worker in Britain and Problems at Work," *Work, Employment, and Society* 23, no. 2 (2009): 343–62.

28. Catherine R. Albiston, Lauren B. Edelman, and Joy Milligan, "The Dispute Tree and the Legal Forest," *Annual Review of Law and Social Science* 10, no. 1 (2014): 105–31.

29. Francesca Polletta and Kelsy Kretschmer, "Free Spaces," in *The Wiley-Blackwell Encyclopedia of Social and Political Movements*, ed. D. A. Snow, D. Della Porta, B. Klandermans, and D. McAdam (New York: Wiley Online Library, 2013), https://doi.org/10.1002/9780470674871.wbespm094.

30. Paul Thompson and Diane van Den Broek, "Managerial Control and Workplace Regimes: An Introduction," *Work, Employment, and Society* 24, no. 3 (2010): 1–12.

31. Kelly, "*Rethinking Industrial Relations* Revisited."

32. Kathleen Griesbach, Adam Reich, Luke Elliott-Negri, and Ruth Milkman, "Algorithmic Control in Platform Food Delivery Work," *Socius* 5, no. 1 (2019): 1–13; Lee et al., "Working with Machines"; Rosenblat and Stark, "Algorithmic Labor and Information Asymmetries"; Moore and Joyce, "Black Box or Hidden Abode?"; Veen, Barratt, and Goods, "Platform-Capital's 'App-Etite' for Control."

33. Lawrence Lessig, *Code and Other Laws of Cyberspace* (New York: Basic Books, 1999).

34. Gabrielle Hecht, *The Radiance of France: Nuclear Power and National Identity after World War II* (Cambridge, MA: MIT Press, 1998), 11.

35. Caleb Goods, Alex Veen, and Tom Barratt, "'Is Your Gig Any Good?': Analysing Job Quality in the Australian Platform-Based Food-Delivery Sector," *Journal of Industrial Relations* 61, no. 4 (2019): 502–27; Griesbach et al., "Algorithmic Control in Platform Food Delivery Work"; Moore and Joyce, "Black Box or Hidden Abode?"; Alex Rosenblat, *Uberland: How Algorithms Are Rewriting the Rules of Work* (Oakland: University of California Press, 2018); Rosenblat and Stark, "Algorithmic Labor and Information Asymmetries"; Alex J. Wood et al., "Good Gig, Bad Gig: Autonomy and Algorithmic Control in the Global Gig Economy," *Work, Employment, and Society* 33, no. 1 (2019): 56–75.

36. Rosenblat and Stark, "Algorithmic Labor and Information Asymmetries"; Griesbach et al., "Algorithmic Control in Platform Food Delivery Work."

37. Kenney and Zysman, "Rise of the Platform Economy"; Vallas, "Platform Capitalism"; Veen, Barratt, and Goods, "Platform-Capital's 'App-Etite' for Control."

38. Mark C. Suchmam, "The Contract as Social Artifact," *Law and Society Review* 37, no. 1 (2003): 91–142.

39. Vincenzo Roppo, "From Consumer Contracts to Asymmetric Contracts: A Trend in European Contract Law?" *European Review of Contract Law* 5, no. 3 (2009): 304–49.

40. Collier, Dubal, and Carter, "Regulation of Labor Platforms"; Ian R. Macneil, "Relational Contract Theory: Challenges and Queries," *Northwestern University Law Review* 94, no. 3 (1999): 877–907.

41. Pollert and Charlwood, "Vulnerable Worker in Britain and Problems at Work"; Rosenblat and Stark, "Algorithmic Labor and Information Asymmetries."

42. Rosenblat and Stark, "Algorithmic Labor and Information Asymmetries"; Tassinari and Maccarrone, "Mobilisation of Gig Economy Couriers in Italy"; Tassinari and Maccarrone, "Riders on the Storm."

43. Griesbach et al., "Algorithmic Control in Platform Food Delivery Work."

44. Roppo, "From Consumer Contracts to Asymmetric Contracts."

45. Gallagher, *Authoritarian Legality in China.*

46. Erik D. Fritsvold, "Under the Law: Legal Consciousness and Radical Environmental Activism," *Law and Social Inquiry* 34, no. 4 (2009): 799–824.

47. Pollert and Charlwood, "Vulnerable Worker in Britain and Problems at Work"; Thompson and van Den Broek, "Managerial Control and Workplace Regimes."

48. Barbara DiCicco-Bloom and Benjamin DiCicco-Bloom, "Secondary Emotional Labor: Supervisors Withholding Support and Guidance in Interdisciplinary Group Meetings in a Community Hospice Program," *Work and Occupations* 46, no. 3 (2019): 339–68.

49. Tassinari and Maccarrone, "Riders on the Storm."

50. Jessica L. Hubley, "Online Consent and the On-Demand Economy: An Approach for the Millennial Circumstance," *Hastings Science and Technology Law Journal* 8, no. 1 (2016): 3–40; Hilary Smith, "The Federal Trade Commission and Online Consumer Contracts," *Columbia Business Law Review* 2016, no. 2 (2016): 512–43; Roppo, "From Consumer Contracts to Asymmetric Contracts."

51. Lessig, *Code and Other Laws of Cyberspace*; Moore and Joyce, "Black Box or Hidden Abode?"

52. Atzeni, "Searching for Injustice and Finding Solidarity?"; Fantasia, *Cultures of Solidarity.*

53. Arlie Hochschild, *The Managed Heart* (Berkeley: University of California Press, 1983).

54. Charles C. Ragin, *The Comparative Method: Moving beyond Qualitative and Quantitative Strategies* (Berkeley: University of California Press, 1987).

55. Veen, Barratt, and Goods, "Platform-Capital's 'App-Etite' for Control."

56. "2018 Food Delivery Worker Employment Report."

57. Gallagher, *Authoritarian Legality in China.*

58. Mary E. Gallagher, John Giles, Albert Park, and Meiyan Wang, "China's 2008 Labor Contract Law: Implementation and Implications for China's Workers," *Human Relations* 68, no. 2 (2014): 197–235.

59. See the note in table 1 about SPCs' and GPCs' income.

60. Juliet B. Schor, William Attwood-Charles, Mehmet Cansoy, Isak Ladegaard, and Robert Wengronowitz, "Dependence and Precarity in the Platform Economy," *Theory and Society* 49 (2020) 833–61, https://doi.org/10.1007/s11186-020-09408-y.

61. Deterding, "Gamification in Management."

62. Hochschild, *Managed Heart.*

63. Deterding, "Gamification in Management."

64. Veen, Barratt, and Goods, "Platform-Capital's 'App-Etite' for Control."

65. Vallas, "Platform Capitalism"; Veen, Barratt, and Goods, "Platform-Capital's 'App-Etite' for Control."

66. Kalleberg, "Precarious Work, Insecure Workers"; Andrew Stewart and Jim Stanford, "Regulating Work in the Gig Economy: What Are the Options?" *Economic and Labour Relations Review* 28, no. 3 (2017): 420–37.

67. Noopur Raval and Paul Dourish, "Standing Out from the Crowd: Emotional Labor, Body Labor, and Temporal Labor in Ridesharing," in *Proceedings of the 19th ACM Conference on Computer-Supported Cooperative Work and Social Computing* (New York: ACM, 2016), 97–107; Rosenblat and Stark, "Algorithmic Labor and Information Asymmetries."

68. Albiston, Edelman, and Milligan, "Dispute Tree and the Legal Forest."

69. Compare Vallas, "Platform Capitalism."

70. Pollert and Charlwood, "Vulnerable Worker in Britain and Problems at Work."

71. Tassinari and Maccarrone, "Mobilisation of Gig Economy Couriers in Italy"; Tassinari and Maccarrone, "Riders on the Storm."

72. Rosenblat and Stark, "Algorithmic Labor and Information Asymmetries."

73. Rosenblat and Stark.

74. Atzeni, "Searching for Injustice and Finding Solidarity?"; Fantasia, *Cultures of Solidarity.*

75. Vallas, "Platform Capitalism"; Veen, Barratt, and Goods, "Platform-Capital's 'App-Etite' for Control."

76. Albiston, Edelman, and Milligan, "Dispute Tree and the Legal Forest."

77. Rosenblat and Stark, "Algorithmic Labor and Information Asymmetries."

78. Polletta and Kretschmer, "Free Spaces"; Tassinari and Maccarrone, "Riders on the Storm."

79. Levitsky, "What Rights?"

80. Schor and Attwood-Charles, "The 'Sharing' Economy"; Schor et al., "Dependence and Precarity in the Sharing Economy."

81. Morgan and Pulignano, "Solidarity at Work."

82. See, for example, Tassinari and Maccarrone, "The Mobilisation of Gig Economy Couriers in Italy"; Tassinari and Maccarrone, "Riders on the Storm."

83. Veen, Barratt, and Goods, "Platform-Capital's 'App-Etite' for Control."

84. Goods, Veen, and Barratt, "Is Your Gig Any Good?"; Griesbach, "Algorithmic Control in Platform Food Delivery Work"; Rosenblat, *Uberland*; Rosenblat and Stark, "Algorithmic Labor and Information Asymmetries"; Wood et al., "Good Gig, Bad Gig."

85. Veen, Barratt, and Goods, "Platform-Capital's 'App-Etite' for Control."

86. Webster, "Microworkers of the Gig Economy"; Atzeni, "Searching for Injustice and Finding Solidarity?"; Fantasia, *Cultures of Solidarity.*

87. Beck and Brook, "Solidarities in and through Work in an Age of Extremes"; compare Tassinari and Maccarrone, "Riders on the Storm."

88. Kenney and Zysman, "Rise of the Platform Economy"; Vallas, "Platform Capitalism"; Veen, Barratt, and Goods, "Platform-Capital's 'App-Etite' for Control."

89. Suchman, "Contract as Social Artifact."

90. Kenney and Zysman, "Rise of the Platform Economy."

91. Gallagher, *Authoritarian Legality in China.*

92. Hubley, "Online Consent and the On-Demand Economy"; Smith,

"Federal Trade Commission and Online Consumer Contracts"; Roppo, "From Consumer Contracts to Asymmetric Contracts."

93. Stewart and Stanford, "Regulating Work in the Gig Economy."

94. Christopher Gavin, "Grubhub Says a Cap on Delivery Fees in Boston Would Be an 'Overstep,'" *Boston.com*, May 19, 2020, https://www.boston.com/news/restaurants/2020/05/19/delivery-fee-cap-boston-grubub/.

95. Michael Burawoy, *The Politics of Production: Factory Regimes under Capitalism and Socialism* (London: Verso, 1985); Collier, Dubal, and Carter, "Regulation of Labor Platforms."

96. Yi Huang, "Monopoly and Anti-monopoly in China Today," *American Journal of Economics and Sociology* 78, no. 5 (2019): 1101–34.

97. Collier, Dubal, and Carter, "Regulation of Labor Platforms."

98. Tassinari and Maccarrone, "Riders on the Storm."

2

Consent Code and Default Dramas

Meg Leta Jones

"We're standing up to Apple for small businesses everywhere," read a full-page ad in the December 16, 2020, editions of the *New York Times, Wall Street Journal*, and *Washington Post*.[1] It went on, "Apple's forced software update [would] limit businesses' ability to run personalized ads and reach their customers effectively." The update would, according to the ad, decrease sales by 60% for every dollar spent. The Dutch version (printed under a photo of a bearded 30-something man pushing a stroller while scrolling on his phone) declared, "Nieuwe klanten werf je via Facebook" (You recruit new customers via Facebook). The next day, ads ran with the headline "Apple vs. the free internet," claiming that Apple's alteration would "change the internet as we know it—for the worse."[2] Apple's forced update would mean content providers (like "cooking sites and sports blogs") would have to start charging subscription fees and requiring in-app purchases, reducing the quality of free content.

"We hear your concerns, and we stand with you," the ad ends.

Who was the hero who mounted this campaign? The company took to the offensive when Apple changed the data-sharing default on an operating system update from opt-out to opt-in. The opt-in method is part of the App Tracking Transparency program that prompts the user to select between two options in a pop-up (figure 2.1). But by changing the consent code, Apple went after Facebook's bottom line.

Facebook gets almost 30% of all United States advertising dollars spent by arguing that it better delivers ads through better targeting with better data.[3] Like Google, which swallows up another 25% of advertising dollars, Facebook's business model revolves around having extraordinary access to people's activities, networks, and eyeballs. Both companies track users through partnerships with other sites and services by integrating their tools (e.g., the Facebook "Like" button[4] or Google Analytics), in addition

Figure 2.1. Sample pop-up requesting permission to track data.

to tracking users' activity while they are signed into Facebook or Google accounts or on Facebook or Google apps (e.g., Instagram or YouTube). However, Google and Apple control the most basic hardware (e.g., mobile phones and computers) and software (e.g., operating systems and browsers). They operate at lower layers in a network infrastructure that Facebook precariously sits atop. By changing the default, Apple revealed immense power over other companies, user data, and the practice of privacy. This chapter investigates the powers at play in default dramas and the political and economic tensions surrounding consent code.

Legal and Technical Defaults

"Code is law." The adage is a foundational concept of tech law. Building on Joel Reidenberg's pivotal notion that information policy rules were and could form through technology, Lawrence Lessig popularized concerns about using technology to regulate conduct.[5] Scholars have further probed the relationship between law and technology as a means of restricting behavior, as Megan Finn and colleagues do in chapter 3 of this volume. For example, Tarleton Gillespie critiqued the use of digital rights management technology to exercise and exceed copyright exclusions, and Karen

Levy critiqued the way block chain smart contracts misrepresent the important social workings and goals of contract law.[6]

Defaults are neither code nor law but a state that exists in both. As Ian Kerr distinguishes technical from legal defaults, technological defaults for computer software and hardware are simply preexisting values that a user can configure to reflect their preferences. Kerr reminds us of factory presets that are meant to be altered by consumers, "many of whom never knew that these presets could be changed, nor had any reason to change them."[7] Law includes defaults as well. Without express arrangements between parties, the law provides a default in many areas, including common law contracts. As "master of the offer," the offeror can specify the terms of acceptance, but if they do not, the offer is considered accepted when the acceptance is communicated by the offeree (not received by the offeror). This is known as the mailbox rule, but there are many other.[8] The default is the state before the user chooses.

While defaults can be changed depending on user preferences, defaults can also constrain behavior. The "default effect" refers to the changes in the likelihood a person will choose a particular option when it is or is not preset.[9] Sometimes default settings are buried and require a number of unapparent steps to alter. These may be designed to prevent users from changing them, reinforcing a desired outcome of the original default setting. Today we may even consider manipulative default settings as "dark patterns," trade practices that are potentially actionable as unfair or deceptive.[10] When the manipulation is toward a socially desirable end, it's called a nudge. Systems known as choice architectures are designed to increase the likelihood of a particular decision through nudges, carefully avoiding actual coercion. Using opt-out defaults, nudges have been used, for instance, to increase organ donation rates.[11] Nudges, choice architectures, and defaults inhibit autonomous choice on purpose.

Defaults and choice architecture are particularly relevant to consent for data sharing, because consent has been central to privacy and data protection laws around the world. Study after study shows how ill-equipped people are to consent to these data arrangements. In 1997, when only 36% of adults went online at all, 72% of Americans surveyed agreed new laws were needed to protect privacy on the internet.[12] By 2005, 80% of Americans knew they were being tracked across sites, but 85% did not agree to accept the everyday collection and sharing scenarios when presented with such a tracking request.[13] Surveys between 2003 and 2015 consistently showed half to three quarters of people still believed the existence of a privacy policy means that the site cannot share information without permission.[14] A 2008 study estimated it would take 76 workdays a year to read all of the privacy policies one was supposed to agree to.[15] Even if one were able to find and willing to spend that much time, the language of privacy policies is complex, serving legal ends not informational ends.[16] Hamid Ekbia highlights similar issues of complexity surrounding collusion claims in chapter 4. And, if one were to come to understand all these policies, there would be no way to negotiate or change the terms and often no alternative site, service, or app to choose instead.[17] The conditions define the "privacy self-management problem" and result in the "consent dilemma."[18]

The consent dilemma has led to two camps, one trying to improve consent and choice for the sake of control, and the other invested in defaults that promote a particular social outcome. In terms of the latter, Salomé Viljoen identifies what she calls "data relations."[19] Viljoen argues that existing critiques miss the collective operations of datafication and injustices beyond individual control or economic fairness. Viljoen showcases these shortcomings by asking what, if anything, is wrong with the US military purchasing location data from a Muslim prayer app. Neither individual control nor economic unfairness accounts for the "significance of the data flow between Muslim Pro and the US military . . . [that] drafts users—faithful Muslims—into the project of their fellow Muslims' oppression." Default debates allow us to argue over collective relations, systems, and realities in a way that choice alone does not. Defaults allow us to ask, as Viljoen asks, "What projects are worth being drafted into?"

Seemingly in the former camp, Apple Inc. CEO Tim Cook, in an interview with Kara Swisher, said his company has added privacy features every year and that "if you were designing such a system from scratch today, of course it should be your decision for what happens to your data."[20] But the timing of Apple's iOS change is no coincidence. Cook's answer ignores the legal pressure put on tech companies in recent years,[21] including what Petelka and colleagues call Trickledown GDPR (chapter 3). Companies, for the most part, are at the mercy of legal systems, which have directed technical code. Cook's answer also ignores pressure from standards organizations, participants in which would prefer that Apple slow down and join the crowd of companies, developers, and advocates trying to build new and improved interoperable systems.

Consent code—the defaults, the toggles, the cookie pop-ups—can be the product of or produce formal technical standards or legal rules or the whims of a company leader. Consent code hides intense power struggles between the ever-present forces of codemakers in law, corporations, and standards bodies. Like the feud between Apple and Facebook, interplay between and among these codemakers can be dramatic (another platform content battle is detailed by Con Díaz in chapter 6, on Apple, Amazon, and e-book publishers). This chapter looks at three consent default dramas: HTTP cookies, Do Not Track, and the ongoing Global Privacy Control. It analyzes company communications and code, public mailing list and meeting minute archives from standards organizations, and legal documents, including agency press releases, proposed bills, and legislative language in order to understand what kinds of power have been exercised and how.

By analyzing these materials for each drama, the chapter observes a movement toward user choice, with bursts of political action in the form of defaults. The shift occurs even with growing criticism of individual consent as a policy tool for achieving the goals of privacy and data protection law. Despite the tedious and repetitive nature of the opt-in–opt-out debate, default dramas are continual sites of remaking (albeit rehashing) socio-technical arrangements and reveal a great deal about the values and players in power at the time.

HTTP Cookies

When developers were trying to transform the web into a commercial network, they ran into a "memory problem."[22] Tim Berners-Lee designed the World Wide Web protocols to be stateless, that is, memory-less. One internet-connected computer requested a file dressed up in web protocols from another internet-connected computer. If all addresses and protocols were in place, the computer would send the requested file. Web developers, as they came to be called, wanted to do a lot more with the commercial web and began trying to maintain state.

On the WWW-Talk mailing list in spring 1995, Brian Behlendorf put the memory problem before the group: "Many content providers would find it useful to be able to detect common patterns or the effectiveness of various user interfaces. The problem is, of course, that HTTP is stateless, and beyond the hostname offers very little in the way of identification of unique 'trips' through the content site."[23] Behlendorf was building *HotWired.com*, the first commercial online magazine, and wanted to know more about how readers used the site. He proposed discussion of a new HTTP header called session-ID. Others were interested in functionality that required memory and shared their experiences with demands for metrics on users.

Lou Montulli, employee 9 at Netscape, replied with a detailed scheme for placing a small file known as a cookie on the visitors' hard drive.[24] When a user requested an HTTP object like a website homepage file, a server could send a piece of state information with the returned HTTP file. A cookie is simply a string of text that is placed in the memory of the browser. It's stored on a user's hard drive and associated with a value in a database available to the web server. Future HTTP requests sent by the user to the server will include the most recent value of the cookie pulled from the client's browser storage and sent back to the server.

Impressed, Behlendorf replied, "A very mature proposal—now we need some test implementations:) Any volunteers?" While the practicalities were debated, concerns and criticisms surfaced about how users could or would engage with the memory system. Behlendorf asserted, "The browser should allow the user to NULL their session ID at any time (or turn off the functionality, even if by default it is on)."[25] Koen Holtman, writing in from the Netherlands, provided the markup for a user control interface with two selections: one for remembering the user in order to provide session functionality and the other for "statistics-enhancing" session IDs. In the markup, Holtman set the default to "ask once for every sit, use reply in later session" for the first, and "no" to the second. Another developer expressed misgivings about burying this data creation and collection in the settings: "A while back someone asked what was wrong with the Netscape Cookie proposal. A long delayed response: it doesn't tell the user what it's doing."[26]

When state management became a subgroup within the Internet Engineering Task Force (IETF) HTTP Working Group at the end of 1995, what the user knew was central to the standard proposed. The IETF is a standards organization similar to the World Wide Web Consortium (W3C), set up by Berners-Lee in 1994.[27] Both are organized around working groups, delineated by subject matter and made up of volunteers without membership requirements, that maintain open mailing lists and are led by appointed chairpersons but make decisions through rough consensus. Led by Montulli and Dave Kristol, the state management subgroup of the IETF adopted Netscape's existing mechanism as a starting point. Other members included Holtman; Amazon's first employee, Shel Kaphan; Eric Sink of Netscape-competitor Spyglass Mosaic, and a small group of other developers. The first "Internet Draft" was published in February 1996 and distinguished between "verified" and "unverified" transactions—first-party versus third-party cookies.[28] Unverified transactions were flatly prohibited or could be allowed if a user could control them easily in the browser with a default set to reject them. After the draft was published by the IETF as Proposed Standard RFC 2109, titled "HTTP State Management Mechanism" in February 1997, the debate took off and participants fought to maintain the terms.[29]

One engineer clarified at the outset:

> The spec requires that you make sure that the same cookies are not used across domains; it allows you to do what you want *within a domain*. If you want to make sure that a dejanews user sees an advert three times and then gets moved to a different ad, you are welcome to use cookies to do so. If you wish to use cookies to make sure that a user sees an advert three times *at any site* before moving on to a different ad, sorry. Because the use of the same cookie across sites allows for the creation of very invasive clicktrail analysis, that has been ruled out. there is something which says that if you are using cookies, you need to manage them within your domain.[30]

A cofounder of DoubleClick, the largest online ad network company at the time, warned:

> In its current form, this section of the spec has potentially huge ramifications for Internet advertising networks and remote ad delivery services. . . . If the new RFC [request for comments] is adopted, these capabilities will be lost for networks, but still available and effective for the largest individual sites which accept advertising. I am unsure if ad networks, which provide an important economic model for smaller web sites, will be able to compete effectively in the long term if this happens.[31]

One version imagines protecting the privacy of individuals as they move across the web at the expense of ad networks, and the other imagines protecting advertisers supporting smaller sites that otherwise won't see marketing dollars.

The advocacy organization the Electronic Privacy Information Center (EPIC) joined the debate in April 1997, sending and posting letters to IETF, copying Microsoft's president, Bill Gates, and chief technology officer (CTO), Nathan Mhyrvold, and Netscape's president, Jim Barksdale, and CTO, Marc Andreesen. These were the early days of the "browser wars," and Microsoft remained under significant scrutiny from antitrust enforcers. Representing a group of organizations, including the Computer Professionals for Social Responsibility, Ralph Nader's Consumer Project on Technology, and the National Association of Elementary School Principals, EPIC wrote to express support of RFC 2109. It concluded, "We believe that 'transparency'—the ability of users to see and exercise control over the disclosure of personally identifiable information—is a critical guideline for the development of sensible privacy practices on the Internet. The alternative would be the surreptitious collection of data without the ability to exercise any control."[32]

Advertising organizations responded: "What concerns us is the tone of the proposal, which is that advertising is not good for us, so we want to avoid it. That begs the question, how is the Web going to be funded?"[33]

In the end, the standard didn't matter. Internet standards are voluntary, and the 24-year-old Montulli decided not to follow the one he had coauthored. Netscape and Microsoft set their increasingly competitive browsers to accept all cookies, which went against 2109. Both provided users the ability to change their preferences to alert them before a cookie was placed by each web server and to reject it, but both set the default to create and share cookies without any notification. Montulli reasoned:

> Any company that had the ability to track users across a large section of the web would need to be a large publicly visible company. Cookies could be seen by users so a tracking company can't hide from the

public. In this way the public has a natural feedback mechanism to constrain those that would seek to track them. . . . Governments have an ability to regulate the collection of data by large visible companies and has [sic] shown a willingness to do so. The public has a responsibility to keep pressure on both the companies that have the ability to track users and governments to enact reasonable privacy regulations and enforce them.[34]

Cookies were attractive to Montulli and the other developers because they supported their vision of a decentralized and commercial web. Cookies would give non-AOL sites a fighting chance at advertising dollars. If any of these Davids were somehow to transform into a Goliath, then it would be easy enough to see. The default was set to accept cookies, not because it reflected people's choices but because it served a potential business model that could be, if ever necessary, managed with legislation. Montulli's reflection essentially reads, "If you don't like it, change the legal default."

Violating the standard did not violate the law. DoubleClick came under investigation by the Federal Trade Commission (FTC) and 10 state attorneys general after the company acquired a fairly large offline direct marketing firm. The result was an order to do what DoubleClick had already done—inform users of online data collection and sharing practices and provide a means to opt out. As a leading member of the newly formed Network Advertising Initiative (NAI), DoubleClick provided users a page for opting out of their tracking, as well as that of the others in the organization.

The only relevant laws in the United States were found in the criminal code. A class action lawsuit was brought against DoubleClick in 2000 claiming violations of the Electronic Communications Privacy Act, the Computer Fraud and Abuse Act, and the Wiretap Act. These laws all prohibit certain forms of access to personal information and provide specific procedures for gaining access. All prohibitions may be altered by consent, however. A federal district court determined that sites, not users, affiliated with DoubleClick were authorized through their relationship to collect data.[35]

In Europe, the legal default is different. The European Union Data Protection Directive of 1995 required that member states pass laws prohibiting data collection unless one of six legal bases existed. One legal basis was and is consent of the data subject. The directive required consent to be "unambiguously given," defined as "any freely given specific and informed indication of his [the user's] wishes by which the data subject signifies his agreement to personal data relating to him being processed."[36] This definition of consent is used in the EU ePrivacy Directive, which is the policy that specifically governs cookies in Europe.

The initial proposal for the ePrivacy Directive did not include language about cookies, but edits were made by a Dutch member of the EU Parliament that would have prohibited cookies specifically, unless prior explicit, well-informed, and freely given consent of the users concerned has been obtained.[37] The change prompted the United Kingdom's Internet Advertising Bureau (IAB) to mobilize a "Save the Cookies" campaign.[38] Negotiations between the Parliament and the European Council took place under the influence of targeted lobbying. The language of ePrivacy Article 5 was changed to "the user concerned is provided with clear and comprehensive information in accordance with the [Data Protection] Directive . . . and is offered a right to refuse such processing by the data controller." Recital 25 includes a justification for the change from "prior notice" to "has access to clear information," stating that "cookies enhance surfing experience and provide for effective web services. Clear and comprehensive information will enable consumers to make an informed choice. In addition, the means to accept and/or reject cookies already exist in most browser software."[39] Consent was transformed into choice, and the legal default was changed to collection. The technological default was already set by US companies, and the legal default simply confirmed the compliance of this choice before choosing.

The ePrivacy Directive was adopted in 2002.[40] Advertisers and tech companies expressed their relief at dodging the bullet of opt-in consent. As *Media Week* explained, "[Opt-in] would have been catastrophic for everything from e-commerce to the set-up of websites, as cookies are used to identify computers and track either users' behavior online."[41] *Computer Weekly* observed, "A ban on cookies would have spelled disaster for Web-based industry. It would have led to the collapse of possibly hundreds of e-commerce companies, hastened the demise of the online advertising industry and stopped companies being able to study usability of a site or improve services for its users by looking at their online behaviour."[42] With little mention of consent or control, these debates were largely about data relations and the collective support that data provides to the development of the web. Both the technical and legal defaults supported the project of a global commercial computer network.

Do Not Track

The European Parliament tried again to change the legal default in 2007 when the ePrivacy Directive was being amended for concerns about digital rights management. With Article 5 up for discussion again, Parliamentary members seized the opportunity to renegotiate the terms regarding cookies. It proposed an amendment that would prohibit cookie tracking unless the user gave prior consent, which could be communicated through the browser setting. The Commission rejected the amendment and Parliament tried again, this time moving the bit about the browser setting to recital 66. The new Article 5 required that the "user concerned has given his/her consent, having been provided with clear and comprehensive information." The changes were accepted, but before the final signing, members of the Council commented in an addendum: "As indicated in recital 66, amended Article 5(3) is not intended to alter the existing requirement that such consent be exercised as a right to refuse the use of cookies or similar technologies used for legitimate purposes."[43] This odd translation, once again, interpreted "consent" as a right to refuse, a choice. The Article 29 Data Protection Working Party (A29WP), an advisory board made up of representatives from each member states' data protection authority, interpreted the changes differently. The A29WP saw two distinct obligations: to obtain consent and to provide clear and comprehensive information. The different interpretations provided an opportunity for confusion.

Meanwhile in the United States, the World Privacy Forum led a group of privacy advocates to propose their own response to increased data collection. In 2007, the coalition put forth the idea of a "Do Not Track" (DNT) list. Inspired by the popular Do Not Call List, the DNT list would have also been maintained by the Federal Trade Commission and populated with cookie-placing domains submitted by ad networks. The networks would have been charged with updating the list, and the FTC would be charged with promoting it. Then browsers could download the list and offer to block tracking.

There was little interest in 2007, but in 2009 Google offered a browser add-on for DoubleClick, which it bought in 2008.[44] Recall users could already opt out of cookies set by members of the Network Advertising Initiative, but this system used opt-out cookies that could be delivered when a user went to each member's website or from the NAI website. The opt-out cookies were deleted every time a user deleted all cookies. Google's add-on stabilized the DoubleClick opt-out cookie, making it possible to delete all cookies without eliminating the DoubleClick opt-out cookie. In discussing the flaws of opt-out cookies, privacy researchers Chris Soghoian and Dan Kaminsky surfaced the idea of a browser header.[45] Kaminsky argued that a browser header was a clear expression of the user's preference, and when Soghoian joined a Future of Privacy Forum meeting, he argued the same. He told the room he planned to add the header to his browser add-on and argue that internet advertisers were ignoring hundreds of thousands of clear signals.

Instead, when Soghoian began work at the Federal Trade Commission in 2010, Chair Jon Leibowitz announced interest in supporting a DNT mechanism at a hearing before the Senate Committee on Commerce, Science, and Transportation.[46] He specifically noted that this setup would make it easier for consumers to opt-out, instead of having to make choices site by site. By the end of 2010, an FTC preliminary staff report was published that suggested the implementation of a "persistent setting on consumers' browsers."[47] Although the report promoted the adoption of "privacy by design," it also recommended "a simple, easy to use choice mechanism for consumers to *opt out* of the collection of information about their Internet behavior for targeted ads."[48]

DNT headers and opt-out cookies were not the only option on the table. Microsoft Internet Explorer 9 allowed third parties to create blocking lists that users could add to the browser. Tracking protection lists carried two main benefits: they weren't created by Microsoft, keeping the company away from antitrust issues, and they were enforceable—content from the domains on the list simply would not load. A high-profile story that broke in 2008 revealed that browser developers at Microsoft had been building Internet Explorer 8 to block tracking techniques by default—at the time, only Apple's Safari browser blocked third-party cookies by default.[49] The developers lost an internal battle against advertising executives, but the battle didn't stay internal for long and was followed by the embarrassing realization that their privacy lists could be easily thwarted. TRUSTe, a data protection and privacy compliance company, distributed one of the first lists, which included 4,000 domains on an allow list and zero on the block list.

When the W3C took up the tracking issue, Microsoft and advocacy groups liked lists because headers were only requests that could be ignored while advertising groups like Google's opt-out cookies.[50] Soghoian had collaborated with Mozilla engineer Sid Stamm to build a prototype Firefox add-on with two HTTP requests, one that signaled opting out of behavioral ads and one opting out of tracking (the latter applying more broadly). Stamm had advanced the DNT header in the Firefox browser, with a simple header and interface, by early 2011. Shortly thereafter, Stamm, along with privacy researchers Arvind Narayanan and Jonathan Mayer, submitted a DNT header standard proposal to the Internet Engineering Task Force, but the organization declined. The W3C, on the other hand, launched the Privacy Interest Group (PING) to consider privacy across all web specifications. Participants rallied for a lengthy workshop and began meeting as the Tracking Protection Working Group (TPWG).[51]

The TPWG and DNT saga are infamous among those that work in privacy. Participants from tech companies, privacy and tech nonprofits, and advertising associations would need to determine how a server responded to a DNT header request. it got ugly. Did the header mean no information could be collected? Some information is always collected in order to send a response and included in server logs. What about cookies that are used to make sure the same ad is not seen multiple times or free visits have not been exceeded or items remain in shopping carts? In her recounting of the DNT standards process, Aleecia McDonald (who co-chaired the TPWG for DNT) shares highlights of "less than good faith" actions and contentious nature of the group. Like much of the W3C standards communication, TPWG interactions happened via email, which quickly ignited "into flame wars."[52] Participants disparaged others in the press, "including personal attacks and name calling." They complained about procedure and used stall tactics, "giving speeches" and "grandstanding." These moves were minor and common compared to the "derailment of careers."[53] Even with the ire and pandemonium, the working group managed to continue on for years.

Nicholas Doty's analysis of the DNT standard uses subheadings like "toxicity and personal attacks," "difficult people," "animosity," and "bad faith." Doty's interviews with participants reveal that DNT failed in part because it was too widely used and therefore too disruptive. When Microsoft turned DNT on by default in 2012, the number of browsers sending DNT signals was far more than many participants had bargained for. As one participant stated, "What is your profit margin? If your profit margin is smaller than the DNT adoption rate, then you must ignore all DNT signals or you'll go out of business."[54]

Microsoft changing all its browsers to opt-in might sound reasonable, but McDonald refers to it as "Microsoft's chaos monkey moment." The group had agreed on two points. First, users on a general browser like Explorer would have to choose by selecting DNT or something like "tracking is fine" (much like Apple's pop-up on the iPhone update). There would be no default for general-purpose browsers like Explorer. Although a selection would be made, there were doubts that users were really getting what they were trying to choose. McDonald worried that DNT "would inherently be a deceptive practice, since despite a catchy name it seemed unlikely to actually stop tracking."[55] Indeed, research showed that users expected that if they selected Do Not Track, they would not be tracked, but the signal wouldn't prevent all data collection, so the terms of what would still be collected had to be negotiated.[56]

Second, older browsers with no DNT settings would send a default, but they would be different between the United States and European Union. For US-based browsers that sent no signal, users had not made an active choice, and so tracking continued. For EU-based browsers, no signal meant users had not consented and so could not be tracked. This is particularly interesting because in 2012, the European Commission sent a letter to the W3C expressing little fuss over an opt-out default and instead insisting that users should be prompted to make a choice when they download or update their browser. "It is not the Commission's understanding that user agents' factory or default setting necessarily determine or distort owner choice." But the TPWG took the Article 29 Working Party guidance on the changes to the ePrivacy Directive seriously.[57] The position defied protests from the Internet Advertising Bureau. The organization's vice president expressed a number of objections: "I have serious problems with the way this group works and operates. I do not believe that we need to delve into (European) legal discussions . . . [Y]ou are pushing so hard for the acceptance of Article 29 WP opinion as the word of God on data protection issues . . . and I don't understand what you are trying to achieve with this. We may like what Article 29 WP says or not, but FACT is that it is JUST an opinion. It is not the law. And, frankly the UK, one of the most engaged EU Member States, is not following the supposed 'baseline.' "[58] The group was worried that if the specification didn't meet European legal standards, it would be ignored and didn't understand how "prior consent" could be interpreted other than an opt-in default.[59]

McDonald explained that some participants' involvement in the W3C effort was motivated by fear of increased legislation. Legal pressure in the United States mounted but fizzled much the same way DNT did in W3C. On the state level, California's Senator Lowenthal proposed drafting statewide regulations for DNT in early 2011.[60] California Assembly Bill 370 finally passed in 2013,[61] but was a largely ignored transparency law that merely required California-based companies to tell users how they respond to DNT requests. Over the course of 2011, members of the US Congress proposed the Do Not Track Me Online Act, Commercial Privacy Bill of Rights Act, Consumer Privacy Protection Act, Do Not Track Online Act, and Do Not Track Kids Act.[62] Privacy scholars and researchers overwhelmingly agreed that none of them were enough.[63] The efforts at the federal level failed entirely. Without clear legal support, the DNT standard was largely ignored.

The working group disbanded in 2019, citing insufficient deployment and lack of support.[64] In 2021, if you run a Google search in a Chrome browser for how to turn on the DNT signal in Chrome, the Google help page explains, "Most websites and web services, including Google's, don't change their behavior when they receive a Do Not Track request. Chrome doesn't provide details of which websites and web services respect Do Not Track requests and how websites interpret them."[65] Even today a user can turn on the Google Chrome browser's DNT setting, but Google sites will not respect the preference. DNT represents a unique political insight into the popularity of individual choice online when faced with radical reform through defaults.

Global Privacy Signal

Although corporate whims trivialized the consensus work at the W3C and IETF in the first two dramas, standards organizations attracted participation again after Google announced it would retire third-party cookies in 2022.[66] Google's announcement (as well as Apple's iOS update) followed adjustments by competitors in the face of legal pressure that the companies had long tried to resist. The General Data Protect Regulation (GDPR) passed in 2016, updating the 1995 Data Protection Directive and extending the same data-protecting laws to govern all member states, shifted the power dynamic. The GDPR speaks directly to defaults in two ways. First it requires affirmative consent, or opt-in, for those sites and services that choose to use consent as their legal basis (there are five other options) for collecting or processing data. The Article 29 Working Party released guidance on consent in the GDPR in November 2017. As the A29WP explained, consent works as a lawful basis for processing only "if a data subject is offered control and . . . a genuine choice with regard to accepting or declining the terms offered or declining them without detriment."[67] Consent must be freely given—that is, cannot be part of non-negotiable terms and cannot determine the functionality of site or service. Few scholars or commenters, however, take account of the ePrivacy Directive, which remains in its 2009 amended state and has yet to be updated into a regulation to match the GDPR.

California added legal pressure in 2018 as well. The strategy for getting the bill passed reads like a trick play—and is similarly irreplicable. Wealthy Bay Area real estate developer Alastair Mactaggart rubbed elbows at a cocktail party with a Google engineer who told him he'd be horrified if he knew the amount of data collected on users. Unknown in the state or federal political or tech scenes, Mactaggart simultaneously was educating himself on ballot initiatives and had "literally a shower thought" to put data privacy up for a vote by the people of California.[68] With the help of another parent from his kid's school but bankrolling the effort entirely alone, Mactaggart began working on a data privacy ballot initiative.[69] He enlisted the help of Ashkan Soltani, well known in tech politics as a savvy privacy expert. As the former chief technologist of the FTC and an active participant in the W3C's DNT efforts, Soltani knew from experience how hard Silicon Valley companies would push policy makers to prevent meaningful change to their operations.[70] By November the team presented Californians with a ballot initiative vote to slowly end third-party tracking, as well as a private right of action (meaning consumers could sue companies directly for violating the law as opposed to filing a complaint with a state agency). Tech companies scrambled to pressure state legislators to instead pass a less stringent law. Lawmakers hastily passed the California Consumer Privacy Act (CCPA) in 2018. The CCPA went into effect in 2020 but was amended that year by another ballot initiative called the California Privacy Rights Act (CPRA). Soltani coauthored both laws.

The CCPA required a "Do Not Sell My Personal Information" link on business homepages. Do Not Sell has been interpreted broadly, and the California attorney general specified that users may communicate their preference to opt out through an automated signal.[71] So, while companies have individual opt-out links on their sites, Californians should also be able to automatically send a signal to all sites or a group of selected sites. To make this option available, a group spearheaded by Soltani built the Global Privacy Control as a proposed specification. The GPC can currently be sent by browser extensions and is respected by a number of site partners, including the *New York Times* and *Washington Post*, but the group submitted the standard to the W3C in hopes that it would be built into all major browsers. Many correctly call this another go at DNT,[72] but the California attorney general didn't want to use DNT as its mechanism, because "the majority of businesses disclose that they do not comply with those signals . . . [and] that businesses will very likely similarly ignore or reject a global privacy control if the regulation permits discretionary compliance."[73] GPC is essentially DNT with legal teeth, but it is intended to be a communication of the user's privacy rights as opposed to a preference about tracking. According to the CPRA, the signal "must represent a consumer's intent and be free of defaults constraining or presupposing that intent."[74] But some of the browser extensions and apps, like DuckDuckGo, already enable GPC by default.[75]

James Rosewell, the CEO of a data analytics company, called the whole thing a Pandora's box and asked, "Should web browsers really become implementation mechanisms of specific government regulation?" Like the Internet Advertising Bureau in past DNT debates, Rosewell attempted to keep big companies and the law separate from technical standards. He and others found working within the W3C less than productive. "I've been—I don't know what the right word is—somewhere between upset and shocked," Rosewell said, "at just how much of a sort of vigilante group the W3C truly is."[76] An ad-tech company commented, "When you're going up against powerful companies that are very entrenched in the W3C, and you're saying something they don't want said, it can feel as though you're being gaslit, given contradictory information on rules that aren't applied later."[77]

Google is one such powerful company. Following its initial announcement about the retirement of cookies, Google published a second that pushed the date back to the end of 2023.[78] Why the delay? To replace third-party cookies, Google's announcement proposed a "Privacy Sandbox," a collection of proposals meant to get at different tasks previously taken up by third-party cookies.[79] Under international antitrust scrutiny, Google submitted the collection to the W3C's Improving Web Advertising Business Group, established in 2017 to embrace a " 'Web advertising by design' approach that includes the needs of all parties involved, that acknowledges the reality today of the advertising-funded Web sites, and that places privacy, security, accessibility, and user experience all in the forefront."[80] Business groups in the W3C say that putting the Sandbox in the W3C allows Google to act in collaboration with its competitors. The timeline for consensus is slower.[81] It gives related industries and interested parties more time to hammer out the details of a postcookie internet. Nonetheless, Google's behavior within the W3C drew the attention of the United Kingdom's Competition and Markets Authority, which was concerned that changes would "undermine the ability of publishers to generate revenue and undermine competition in digital advertising, entrenching Google's market power."[82]

Others, too, criticized Google's participation in the W3C. Google was the only member of the Privacy Interest Group to vote "no" on an update to its charter in 2019.[83] The change, which required a unanimous favorable vote, would have expanded the power of the PING to block new specifications that threaten web privacy. Narayanan and Mayer accused Google of "privacy gaslighting" for arguing that ending third-party cookies could be bad for privacy and that the company was trailblazing privacy-preserving advertising:

> Apple and Mozilla have tracking protection enabled, by default, today. And Apple is already testing privacy-preserving ad measurement. Meanwhile, Google is talking about a multi-year process for a watered-down form of privacy protection. And even that is uncertain—advertising platforms dragged out the Do Not Track standardization process for over six years, without any meaningful output. If history is any indication, launching a standards process is an effective way for Google to appear to be doing something on web privacy, but without actually delivering.[84]

Google has defenders, though. Facebook's Ben Savage, who actively participates on W3C efforts, accused Apple of "egregious behavior." Suggesting Apple wanted to drive apps on the iPhone away from ad-based models and toward in-app purchases (of which Apple takes a cut), Savage countered criticism of Google. He argued the company operates "in the open; announcing their intended privacy goals, proposing new APIs publicly, soliciting input, publicly responding to feedback, allowing people to test," whereas Apple "decided to just blow up everything first . . . and only start thinking about what to replace it with later."[85] Simply ending third-party tracking works for the Facebooks, Googles, and Apples of the world, Savage and Rosewell agree, but what about other apps and sites? The task for the group, according to Savage, is to "figure out **some** kind of new, highly-private approach to personalization that works for all those other apps and websites or else we have a future internet where only a handful of mammoth companies can succeed in offering free, ads-funded content and services."[86]

Because decisions are made by consensus at the W3C, causing trouble can conveniently appear like a lack of consensus. Soltani, who claimed that those defending the old guard were essentially attempting denial-of-service attacks on the group proceedings, expressed concern that new proposals from those whose businesses depend on the current system enshrine invasive tracking. "Fortunately in a forum like the W3C," he argued, "folks are smart enough to get the distinction. Unfortunately, policy-makers won't."[87]

Both the GDPR (in combination with the ePrivacy Directive) and the CPRA require a choice to change the legal and technical defaults—one set to collect and the other set to do not collect—that promote different projects users are drafted into. These arrangements are distinct from the DNT browser selection and the iPhone's App Tracking Transparency consent, where the goal is choice alone.

Conclusion

Facebook was not wrong about the impact of Apple's iOS update. Under the new App Tracking Transparency policy, 96% of iPhone users "ask app not to track," meaning only 4% opt to allow apps to track their activity across other companies' apps and websites.[88] Perhaps borrowing Facebook's old motto to "move fast and break things," Apple did not wait for companies or standards to shift but was also not motivated to act until it felt significant legal pressure to change its system. What if, instead of trying to get the public to turn on Apple in an attempt to drive the company to change its planned iOS update, Facebook had launched a campaign to get users to opt in? What if only a small portion of users actually opposed sharing their data or found surveillance capitalism troubling? Through these what-ifs we can see Viljoen's data relations.[89] How do we express support for the projects worth being drafted into?

During the spring of 2021, Facebook's ad campaign against Apple pivoted to television. When COVID-19 swept across the world in 2020, many broken infrastructures were revealed and forced into focus. Like many tech companies, Facebook appeared immune to the economic impact of COVID-19. At the end of 2020 its revenue was up 22%, pulling in almost $86 billion, and profits were up 58% at $29 billion. Small businesses are responsible for 75% of ad revenue. The company's new ads played a haunting rendition of "I Will Survive" while nostalgic images of businesses threatened by the lock-down health policies related to the pandemic called the viewer to action. "In the next three months, almost half of those small businesses, they could close if people don't do something. We have to keep our communities together. That's how we get through this." These ads played on the sentiment expressed by Facebook CEO Mark Zuckerberg at the company's earnings call in the summer of 2020 when he explained, "That's why I am often troubled by the calls to go after internet advertising, especially during a time of such economic turmoil like we face today with Covid. . . . The much bigger cost of such a move would be to reduce the effectiveness of ads and opportunities for small businesses to grow. This would reduce opportunities for small businesses so much that it would probably be felt at a macro-economic level. Is that really what policymakers want in the middle of a pandemic and recession?"[90]

Zuckerberg's arguments are certainly outdated in some ways (and increasingly criticized as unfounded[91]), but he's arguing about data relations.

Meanwhile Tim Cook appears a trailblazer, taking swings at surveillance capitalism and the attention economy but doing so using a now arcane notion in privacy: consent. Apple's revived and successful application of consent for its billion users worldwide will likely have a significant impact on the shape of tensions between consent and default in the future. Whether those impacts are progressive or regressive will depend on how codemakers in other companies, standards bodies, and legal settings incorporate the change into their next default drama.

Notes

1. Jack Nicas and Mike Isaac, "Facebook Takes the Gloves Off in Feud with Apple," *New York Times*, updated April 26, 2021, https://www.nytimes.com/2020/12/16/technology/facebook-takes-the-gloves-off-in-feud-with-apple.html.

2. Luke Dormehl, "Full-Page Facebook Ad Accuses Apple of Changing the Internet for the Worse [Updated]," *Cult of Mac*, December 17, 2020, https://www.cultofmac.com/730598/full-page-facebook-ad-accuses-apple-of-changing-the-internet-for-the-worse/.

3. Alexandra Bruell, "Amazon Surpasses 10% of U.S. Digital Ad Market Share," *Wall Street Journal*, April 6, 2021, https://www.wsj.com/articles/amazon-surpasses-10-of-u-s-digital-ad-market-share-11617703200.

4. In addition, most retails sites include Facebook software that sends data to the social media company whenever a purchase is made so that Facebook can use that data to show the retailer's ads to similar profiles.

5. Joel Reidenberg, "Lex Informatica: The Formulation of Information Policy Rules through Technology," *Texas Law Review* 76 (1997): 553–94; Lawrence Lessig, *Code and Other Laws of Cyberspace* (New York: Basic Books, 2000).

6. Karen E. C. Levy, "Book-Smart, Not Street-Smart: Blockchain-Based Smart Contracts and the Social Workings of Law," *Engaging Science, Technology, and Society* 3 (2017): 1–15, https://estsjournal.org/index.php/ests/article/view/107; Tarleton Gillespie, *Wired Shut: Copyright and the Shape of Digital Culture* (Cambridge, MA: MIT Press, 2009).

7. Kerr's piece is a commentary on Mireille Hildebrandt, *Smart Technologies and the End(s) of Law: Novel Entanglements of Law and Technology* (Northampton, MA: Edward Elgar, 2015). He organizes Hildebrandt's discussion of defaults into four categories: natural defaults, technological defaults, legal defaults, and normative defaults. Ian Kerr, "The Devil Is in the Defaults" *Critical Analysis of Law* 4, no.1 (2017): 98.

8. Randy E. Barnett, "The Sound of Silence: Default Rules and Contractual Consent," *Virginia Law Review* 78 (1992): 821; Ian Ayres and Robert Gertner, "Filling Gaps in Incomplete Contracts: An Economic Theory of Default Rules," *Yale Law Journal* 99 (1989): 87.

9. Isaac Dinner, Eric J. Johnson, Daniel G. Goldstein, and Kaiya Liu, "Partitioning Default Effects: Why People Choose Not to Choose," *Journal of Experimental Psychology: Applied* 17, no. 4 (2011): 332; Eric J. Johnson and Daniel Goldstein, "Do Defaults Save Lives?" *Science* 302 (2003): 1338.

10. Arvind Narayanan, Arunesh Mathur, Marshini Chetty, and Mihir Kshirsagar, "Dark Patterns, Past, Present, and Future: The Evolution of Tricky User Interfaces," *Queue* 18, no. 2 (2020): 67–92.

11. Richard Thaler and Cass Sunstein, *Nudge: Improving Decisions about Health, Wealth, and Happiness* (New York: Penguin, 2008), 178–83.

12. Susannah Fox, "The Internet Circa 1998," Pew Research Center, June 21,

2007, https://www.pewresearch.org/internet/2007/06/21/the-internet-circa
-1998/; "GVU's 8th WWW User Survey," Graphic, Visualization, and Usability
Center, Georgia Tech, 1997, https://www.cc.gatech.edu/gvu/user_surveys
/survey-1997-10/.

13. Joseph Turow, "Americans and Marketplace Privacy: Seven Annenberg
National Surveys in Perspective," in *The Cambridge Handbook of Consumer
Privacy,* ed. Evan Sellinger, Jules Polonestsky, and Omer Tene (Cambridge:
Cambridge University Press, 2018), 160.

14. Turow, "Americans and Marketplace Privacy," 161–62.

15. Aleecia M. McDonald, and Lorrie Faith Cranor, "The Cost of Reading
Privacy Policies," *I/S: A Journal of Law and Policy for the Information Society* 4
(2008): 543–68.

16. Woodrow Hartzog, "User Agreements Are Betraying You," *OneZero,*
June 5, 2018, https://onezero.medium.com/user-agreements-are-betraying-you
-19db7135441f.

17. Frederik J. Zuiderveen Borgesius, Sanne Kruikemeier, Sophie C. Boer-
man, and Natali Helberger, "Tracking Walls, Take-It-or-Leave-It Choices, the
GDPR, and the ePrivacy Regulation," *European Data Protection Law Review* 3
(2017): 353; Margaret Jane Radin, *Boilerplate: The Fine Print, Vanishing Rights,
and the Rule of Law* (Princeton, NJ: Princeton University Press, 2012).

18. Daniel J. Solove, "Privacy Self-Management and the Consent Dilemma,"
Harvard Law Review 126 (2013): 1880.

19. Salomé Viljoen, "Data Relations," *Logic,* no. 13 (May 17, 2021), https://
logicmag.io/distribution/data-relations/.

20. Tim Cook, "Is Apple's Privacy Push Facebook's Existential Threat,"
interviewed by Kara Swisher, *Sway,* April 5, 2021.

21. Alex Hern, "Apple Faces Privacy Case in Europe over iPhone Tracking
ID," *The Guardian,* November 17, 2020, https://www.theguardian.com/technol
ogy/2020/nov/17/apple-faces-privacy-case-in-europe-over-iphone-tracking-id.

22. John Schwartz, "Giving Web a Memory Cost Its Users Privacy," *New
York Times,* September 4, 2001, https://www.nytimes.com/2001/09/04/business
/giving-web-a-memory-cost-its-users-privacy.html.

23. Brian Behlendorf to WWW-Talk electronic mailing list, "Session Tracking," W3C, April 17, 1995, https://lists.w3.org/Archives/Public/www-talk/1995
MarApr/0456.html.

24. Lou Montulli to WWW-Talk electronic mailing list, "Session Tracking,"
W3C, April 18, 1995, https://lists.w3.org/Archives/Public/www-talk/1995Mar
Apr/0462.html.

25. Brian Behlendorf to WWW-Talk electronic mailing list, "Session Tracking," W3C, April 18, 1995, https://lists.w3.org/Archives/Public/www-talk/1995
MarApr/0468.html.

26. Koen Holtman to WWW-Talk electronic mailing list, "Session-Id and
Privacy Mechanism," July 22, 1995, World Wide Web History Project, http://
1997.webhistory.org/www.lists/www-talk.1995q3/0158.html.

27. For an in-depth discussion of how the IETF and W3C compare to other
standards bodies and to each other, see JoAnne Yates and Craig N. Murphy,
Engineering Rules: Standard Setting since 1880 (Baltimore: Johns Hopkins University Press, 2019).

28. Dave Kristol, "HTTP Cookies: Standards, Privacy, and Politics," *ACM
Transactions on Internet Technology* 1, no. 2 (November 2001): 151–98.

29. Dave Kristol and Lou Montulli, "HTTP State Management Mechanism
(RFC 2109)," Internet Engineering Task Force, February 1997.

30. Ted Hardie to WWW-Talk electronic mailing list, "Unverifiable

Transactions/Cookie Draft," W3C, March 18, 1997, https://lists.w3.org/Archives
/Public/ietf-http-wg-old/1997JanApr/0472.html.

31. Dwight Merriman to WWW-Talk electronic mailing list, "Unverifiable
Transactions/Cookie Draft," March 13, 1997, https://lists.w3.org/Archives
/Public/ietf-http-wg-old/1997JanApr/0416.html.

32. "Net Users Urge Standards Group to Protect Privacy," press release,
EPIC, April 7, 1997, https://www.epic.org/privacy/internet/cookies/ietf_letter
.html.

33. Rick E. Bruner, " 'Cookie' Proposal Could Hinder Online Advertising:
Privacy Backers Push for More Data Controls," *Advertising Age*, March 31, 1997,
https://adage.com/article/news/cookie-proposal-hinder-online-advertising/359.

34. Lou Montulli, "The Reasoning behind Web Cookies," *The Irregular
Musings of Lou Montulli*, May 14, 2013, https://montulli.blogspot.com/2013/05
/the-reasoning-behind-web-cookies.html.

35. In re Doubleclick Inc. Privacy Litigation, 154 F. Supp.2d 497 (S.D.N.Y.,
March 28, 2001).

36. Council Directive No. 95/46/EC of 24 October 1995 on the Protection
of Individuals with Regard to the Processing of Personal Data and on the Free
Movement of Such Data, art. 2(c), OJ. L 281/31, at 38 (1995) (Article 2(h)).

37. European Parliament, First reading (co-decision procedure), Proposal
for a European Parliament and Council directive concerning the processing of
personal data and the protection of privacy in electronic communications sector
(COM(2000) 385 - C5-0439/2000 - 2000/0189(COD)), A5-0374/2001 [2001] OJ
C140E/121 13.06.2002.

38. "Why the EU should 'Save Our Cookie,' " *Precision Marketing* 11 (November 16, 2001).

39. Sylvia Mercado Kierkegaard, "How the Cookies (Almost) Crumbled:
Privacy and Lobbyism," *Computer Law and Security Review* 21, no. 4 (2005):
310–22.

40. Directive 2002/58/EC, of the European Parliament and of the Council of
12 July 2002 Concerning the Processing of Personal Data and the Protection of
Privacy in the Electronic Communications Sector, 2002 O.J. (L 201) 37.

41. James Bainbridge, "Digital Cookies Allowed to Stay," *Media Week* 14
(June 7, 2002).

42. Simon Roberts, "Users Are Still Wary of Cookies," *Computer Weekly* 24
(June 20, 2002).

43. Council of the European Union, Addendum to "I/A" note: Adoption of
the proposal for a Directive of the European Parliament and of the Council
amending the Directives 2002/21//EC on a common regulatory framework for
electronic communications networks and services, and 2002/20/EC on the au-
thorisation of electronic communications networks and services (LA + S) (third
reading)—statements, 15864/09 ADD 1 REV 1, Brussels, 18 November 2009.

44. "Google Analytics Opt-Out Browser Add-On," Google, accessed December 9, 2024, https://tools.google.com/dlpage/gaoptout.

45. Chris Soghoian, "The History of the Do Not Track Header," *Slight Para-
noia*, January 11, 2011, http://paranoia.dubfire.net/2011/01/history-of-do-not
-track-header.html.

46. US Senate Committee on Commerce, Science, and Transportation,
Consumer Online Privacy, July 27, 2010 (prepared statement of the Federal Trade
Commission), https://www.commerce.senate.gov/2010/7/consumer-online
-privacy.

47. Federal Trade Commission, "Protecting Consumer Privacy in an Era of
Rapid Change," Preliminary FTC Staff Report, December 2010.

48. Federal Trade Commission, "FTC Staff Issues Privacy Report, Offers Framework for Consumers, Businesses, and Policymakers," press release, December 1, 2010, https://www.ftc.gov/news-events/press-releases/2010/12/ftc -staff-issues-privacy-report-offers-framework-consumer (emphasis added).

49. Nick Wingfield, "Microsoft Quashed Effort to Boost Online Privacy," *Wall Street Journal*, August 2, 2010, https://www.wsj.com/articles/SB1000142 4052748703467304575383530439838568.

50. Aleecia McDonald, "Stakeholders and High Stakes: Divergent Standards for Do Not Track," in *The Cambridge Handbook of Consumer Privacy*, ed. Evan Sellinger, Jules Polonestsky, and Omer Tene (Cambridge: Cambridge University Press, 2018), 253.

51. World Wide Web Consortium, "W3C Workshop on Web Tracking and User Privacy: Workshop Report," W3C, March 2011, https://www.w3.org/2011 /track-privacy/report.html.

52. McDonald, "Stakeholders and High Stakes," 258–59.

53. Nicholas Doty, "Enacting Privacy in Internet Standards" (PhD diss., University of California, Berkeley, 2020), 113; McDonald, "Stakeholders and High Stakes," 258.

54. McDonald, "Stakeholders and High Stakes," 263–64.

55. McDonald, "Stakeholders and High Stakes," 263–64.

56. Chris Jay Hoofnagle, Jennifer M. Urban, and Su Li, "Privacy and Modern Advertising: Most US Internet Users Want 'Do Not Track' to Stop Collection of Data about their Online Activities," paper presented to Amsterdam Privacy Conference, 2012; Aleecia McDonald and Jon M. Peha, "Track Gap: Policy Implications of User Expectations for the 'Do Not Track' Internet Privacy Feature," paper presented at Telecommunications Policy Research Conference (TPRC), September 25, 2011.

57. Robert Madelin, European Commission, to World Wide Web Consortium Tracking Protection Working Group, Ref. Ares(2012)743354, W3C, June 21, 2012, https://lists.w3.org/Archives/Public/public-tracking/2012Jun/att-0604 /Letter_to_W3C_Tracking_Protection_Working_Group.210612.pdf; Rigo Wenning to TPWG W3C electronic mailing list, "Consent," W3C, June 6, 2012, https://lists.w3.org/Archives/Public/public-tracking/2012Jun/0098.html.

58. Kimon Zorbas to TPWG W3C electronic mailing list, "Examples of Successful Opt-In Implementations," W3C, June 14, 2012, https://lists.w3.org /Archives/Public/public-tracking/2012Jun/0413.html; Kimon Zorbas to TPWG W3C electronic mailing list, "DNT Concerns," W3C, December 5, 2012, https:// lists.w3.org/Archives/Public/public-tracking/2012Dec/0065.html#replies.

59. Rigo Wenning to TPWG W3C electronic mailing list, "ACTION-174: Write Up Implication of Origin/* Exceptions in EU Context," W3C, June 6, 2012, https://lists.w3.org/Archives/Public/public-tracking/2012Jun/0109.html #replies.

60. SB 761, 112th Cong. (2011).

61. A.B. 370, 113th Cong. (2013).

62. Do Not Track Me Online Act of 2011, H.R. 654, 112th Cong. (2011); Commercial Privacy Bills of Rights Act of 2011, S. 799, 112th Cong. (2011); Consumer Privacy Protection Act of 2011, H.R. 1528, 112th Cong. (2011); Do Not Track Online Act of 2011, S. 913, 112th Cong. (2011); Do Not Track Kids Act, H.R. 1895, 112th Cong. (2011).

63. Meg Leta Jones and Jenny Lee, "Comparing Consent to Cookies: A Case for Protecting Non-use," *Cornell International Law Journal* 53 (2020): 97–132.

64. Plehegar, "WG closed," W3C/dnt GitHub, January 18, 2019, https:// github.com/w3c/dnt/commit/5d85d6c3d116b5eb29fddc69352a77d87dfd2310.

65. "Turn 'Do Not Track' On or Off," Google Chrome Help, accessed December 9, 2024, https://support.google.com/chrome/answer/2790761?hl=en&co=GENIE.Platform%3DDesktop.

66. "Building a More Private Web: A Path towards Making Third Party Cookies Obsolete," Chromium Blog, January 14, 2020, https://blog.chromium.org/2020/01/building-more-private-web-path-towards.html.

67. Article 29 Data Protection Working Party, "Guidelines on Consent under Regulation 2016/679 17/EN WP 259 3" (2017), European Commission, accessed December 9, 2024, https://ec.europa.eu/newsroom/article29/item-detail.cfm?item_id=623051.

68. Brian Fung, "The Unlikely Activist behind the Nation's Toughest Privacy Law Isn't Done Yet," CNN, October 10, 2019, https://www.cnn.com/2019/10/10/tech/alastair-mactaggart/index.html.

69. Betsy Morris and Eliot Brown, "The Real-Estate Developer Who Took on the Tech Giants," *Wall Street Journal*, June 29, 2018, https://www.wsj.com/articles/the-real-estate-developer-who-took-on-the-tech-giants-1530308857.

70. Nicholas Confessore, "The Unlikely Activists Who Took On Silicon Valley—And Won," *New York Times*, April 14. 2018, https://www.nytimes.com/2018/08/14/magazine/facebook-google-privacy-data.html.

71. Text of Modified Regulations, California Consumer Privacy Act Regulations Proposed Text of Regulations, California Office of Attorney General, March 11, 2020, https://www.oag.ca.gov/sites/all/files/agweb/pdfs/privacy/ccpa-text-of-second-set-mod-031120.pdf?.

72. Gilad Edelman, "Do Not Track Is Back, and This Time It Might Work," *Wired*, October 7, 2020, https://www.wired.com/story/global-privacy-control-launches-do-not-track-is-back/.

73. Final Statement of Reasons, California Office of Attorney General, accessed December 9, 2024, https://oag.ca.gov/sites/all/files/agweb/pdfs/privacy/ccpa-fsor.pdf.

74. Cal. Civ. Code § 1798.185(a)(19).

75. "Global Privacy Control (GPC) Enabled by Default in DuckDuckGo Apps & Extensions," DuckDuckGo News, January 28, 2021, Spread Privacy, https://spreadprivacy.com/global-privacy-control-enabled-by-default/.

76. Issie Lapowsky, "Concern Trolls and Power Grabs: Inside Big Tech's Angry, Geeky, Often Petty War for Your Privacy," *Protocol*, July 13, 2021, https://www.protocol.com/policy/w3c-privacy-war.

77. Lapowsky, "Concern Trolls and Power Grabs."

78. Kate Kaye, "Cheat Sheet: Google Extends Cookie Execution Deadline until Late 2023, Will Pause FLoC Testing in July," *DigiDay*, June 24, 2021, https://digiday.com/marketing/cheat-sheet-google-extends-cookie-execution-deadline-until-late-2023-will-pause-floc-testing-in-july/.

79. Justin Schuh, "Building a More Private Web," Google The Keyword Blog, August 22, 2019, https://www.blog.google/products/chrome/building-a-more-private-web/.

80. Liam Quin, "Improving Web Advertising (by Changing the Web)," W3C, December 21, 2017, https://www.w3.org/community/web-adv/2017/12/21/improving-web-advertising-by-changing-the-web/#more-4.

81. Allison Schiff, "An Inside Look at the W3C with Strategy Lead Wendy Seltzer, As Debate Swirls around the Privacy Sandbox," *AdExchanger*, April 26, 2021, https://www.adexchanger.com/privacy/an-inside-look-at-the-w3c-with-strategy-lead-wendy-seltzer-as-debate-swirls-around-the-privacy-sandbox/.

82. "Notice of Intention to Accept Commitments Offered by Google in Relation to Its Privacy Sandbox Proposal," Competition and Market Authority

Case 50972, June 11, 2021, https://assets.publishing.service.gov.uk/media
/60c21e54d3bf7f4bcc0652cd/Notice_of_intention_to_accept_binding_com
mitments_offered_by_Google_publication.pdf.

83. "CMA to Investigate Google's 'Privacy Sandbox' Browser Changes,"
press release, CMA (Competition and Market Authority), Gov.UK, January 8,
2021, https://lists.w3.org/Archives/Public/public-privacy/2019JulSep/0056.html.

84. Jonathan Mayer and Arvind Narayanan, "Deconstructing Google's
Excuses on Tracking Protection," Freedom to Tinker, August 23, 2019, https://
freedom-to-tinker.com/2019/08/23/deconstructing-googles-excuses-on-tracking
-protection/.

85. Ben Savage (@btsavage), Twitter, July 9, 2021, 4:14 p.m., https://twitter
.com/BenSava78155446/status/1413411126105329672; Ben Savage (@btsavage),
Twitter, June 20, 2021, 4:45 p.m., https://twitter.com/BenSava78155446/status
/1413411617530916865.

86. Ben Savage (@btsavage), Twitter, June 20, 2021.

87. Soltani quoted in Lapowsky, "Concern Trolls and Power Grabs."

88. Samuel Axon, "96% of US Users Opt Out of App Tracking in iOS 14.5,
Analytics Find," *ArsTechnica*, May 7, 2021, https://arstechnica.com/gadgets
/2021/05/96-of-us-users-opt-out-of-app-tracking-in-ios-14-5-analytics-find/.

89. Viljoen, "Data Relations."

90. Jeff Beer, "This New Facebook Ad Targets the Only People Left Who
Actually Like the Brand," *Fact Company*, February 27, 2021, https://www.fast
company.com/90609058/this-new-facebook-ad-targets-the-only-people-left
-who-actually-like-the-brand.

91. Thomas Huang, *Subprime Attention Crisis: Advertising and the Time
Bomb at the Heart of the Internet* (New York: Farrar, Straus, and Giroux, 2020).

3

A Mirror, Not a Glass Door

Legal Code and Software Code in Practice

Justin Petelka, Megan Finn, Janaki Srinivasan, Elisa Oreglia, and A. P. Janani

What was supposed to be like a glass door that lets me see through what Google is doing with my data, instead of making this a glass door, they've just made it a mirror.
—Participant P15, India, talking about their Google data

In this chapter, we explore the uneven and shifting practices of data access rights through the experiences of people requesting their data from different companies in India, the United Kingdom, and the United States in 2019–21. This is the period after the General Data Protection Regulation (GDPR) of the European Union (EU) went into effect in 2018 and during a period of legal change, in the United States with the implementation of the California Consumer Privacy Act (CCPA) in June 2020, and in India with the drafting of the 2019 Digital Personal Data Protection Bill.[1] While the GDPR applies to data processed within the EU or from EU residents, we argue that GDPR's scope in this period extended beyond EU borders because multinational company infrastructures extended from the EU to countries with different regulations, compliance and enforcement protocols, and even cultural conceptions of data privacy. Building on analyses of *written laws,* we evaluate the influence of the GDPR beyond the EU (called the Brussels Effect) and the CCPA's influence within the United States (called the California Effect) by analyzing *data infrastructure* outputs.[2] We explored the byproducts of both legal code and software code by working with research participants in the three countries to request, and to examine the process of requesting and accessing, their personal data from the same multinational companies and regionally operating companies. Our research articulates what we call *trickle-down GDPR,* an important mapping of the ways corporations translate national laws into their technical infrastructure and then how people ultimately experience those laws in different regions.

Rather than focusing only on legal doctrine or technical infrastructure, we map individuals' experiences of both in a concrete, ground-up way in order to understand the connection between the making and the intended effect of laws, and the actual practices that emerge from these laws and technical infrastructures. We interviewed 19 participants in the United States, India, and the EU who made 38 unique data subject access requests (DSARs) to companies, and we documented their experience of this process. We show that trickledown GDPR does appear to be real: companies that fulfilled data access requests in the EU also extended those rights to data subjects in India and the United States. Companies that do not operate in the EU, however, mostly did not respond to data requests. There were exceptions: one US company that did not operate in the EU inconsistently responded to data requests, and participants in India were unable to access data from Amazon. We examine how the variability in companies' interpretation of GDPR legal code is incorporated in the software code that powers the infrastructures that process data.

The experience of people requesting access to their personal data is not reducible to just the legal code or the software code: any analysis of the efficacy of the GDPR and similar laws must consider the lived experiences of users too. While we show that data access rights exist for non-EU citizens and residents who are requesting data from companies that operate in the EU, we are also circumspect about how meaningful these rights are. Further, these experiences of users also introduce new questions such as, How do GDPR's data access rights address information asymmetries between civilians and tech corporations? Data requests are expected to provide a certain level of transparency and be a cornerstone of data protection, but most of our research participants felt that GDPR and data access rights did not allow them to catch a glimpse of the companies' inner workings as they expected—they were a mirror, not a glass door.

In the chapter epigraph, we invoke *An Engine, Not a Camera*, in which author Donald MacKenzie showed that financial economics was in fact "an active force transforming its environment, not a camera passively recording it."[3] Whereas many participants were expecting a glass door they could peer through to see some of the logic and modus operandi of a data controller, they instead found a mirror—a reflection of only *themselves* in their data. Furthermore, when people requested access to personal data, what data companies had about people remained unclear. Despite the performance of transparency (and compliance with transparency-focused laws like GDPR), no participant could read corporate privacy policies or see their data and come away with an idea of which data were being collected by the company, how they were kept, or what purpose the data serve for the company. The companies produced the problem of personal data access and, through control of the corporate infrastructure, the supposed solution: the process of data request and access.

This paper unfolds over two parts. The first describes the scholarly background that helped us map the relations between legal code and software code. The second part describes our efforts to understand how software code and legal code translate into user experiences. In the first part, we begin with personal data access in different legal regimes in the EU, United States, and India. We then discuss different conceptualizations of the application of local laws to transnational infrastructures. Following this review of legal regimes and their deployment in transnational contexts, we discuss the scholarship on GDPR compliance practices. In the second part, we describe our research process, starting with our research design, structured to understand whether and how different companies (both those that operate across and those that operate within regions of our study) incorporate the GDPR into the process of personal data requests.

Legal Code and Software Code

The next two subsections describe the different historical and political backgrounds of the EU's, United States', and India's approaches to data protection and data regulation. Rights of access, rectification, cancellation, and objection (ARCO) are now defined and protected by different laws and regulations in different jurisdictions, particularly the European Union's 2018 GDPR. The 1995 Data Protection Directive (DPD95) and the GDPR created requirements for organizations that collect and process "personal data," called "data controllers."[4] Among the requirements for data controllers is to recognize a series of informational rights granted to "data subjects." To comply with these obligations, EU organizations in every sector implemented procedures for rights bearers to access their personal data—reflecting a concept with a history reaching back a half century.[5]

European legislators began discussing the idea of data protection, including the right of access to personal data, more than 50 years ago. Increasing international data flows and widespread access to telephony, data storage, and databases created issues of jurisdiction; in response, some European regions passed data protection laws in the 1970s.[6] This legislation signaled the beginning of tensions around data flows that are still playing out today between multinational corporations and states. Corporations argued that information was immaterial—a special kind of good that cannot be depleted and needs other information to multiply. European states, however, saw information as another good, produced in specific places, whose flow and boundaries had to be controlled by the state itself or through international agreements. These two perspectives together provided scaffolding for the growing production of information as a commodity. The Organisation for Economic Co-operation and Development (OECD), the European Economic Community (EEC), and other entities, began to issue data protection laws and attempted to regulate international data flows.[7] Alongside these efforts, "developing countries" started to push back against the free flow of information, which they saw as asymmetrical.[8] In the late 1980s and early 1990s, the emergence of the World Wide Web, with its supranational standards and governance, the increasingly global presence of US tech companies, and the libertarian politics that underpinned the computing industry of this era slowed attempts at global regulation of transnational data flows.[9] Yet European countries continued to enact national data protection laws, which were harmonized into a single cohesive framework by the EU through the DPD95, followed by the GDPR.[10]

The right of data subjects to request the data held about them through a DSAR is a cornerstone of GDPR and the subject of this chapter.[11] Legal scholars Jones and Kaminski argue that ARCO rights such as DSARs are crucial to data protection: "Only through knowing about and participating in what information governments and companies hold on them can individuals check unchecked power, regain some (but not absolute!) control, and push back against manipulation in the information age."[12] Thus, DSARs can be understood as a form of information transparency like that emerging from the Freedom of Information Acts (FOIAs) in the United States and United Kingdom and the Right to Information Act (RTI) in India.[13] While individual data access is one of the fundamental building blocks in the GDPR's complex regulatory scheme, DSARs cannot alone ensure data protection, as Ausloos and Dewitte have argued: "The right of access both acts as a necessary first step enabling the exercise of most other data subject rights, and *as a strategic tool to assess compliance with data protection law more broadly*."[14] Thus, though data access has limited effectiveness in regulation and is focused on individual rights, "the right of access can therefore be a vehicle not just at an individual level but also as the catalyst to pursue legislative reform and policy considerations."[15] The potential of DSARs to check power asymmetries lies in collective action.[16]

Alternative Approaches to Data Regulation

By comparison, US approaches to ensuring privacy and protecting data from corporations are not based on natural rights or collective harms. The implicit assumption parroted by corporations is that consumers trade personal data for goods and services by consenting to different legal agreements. However, this transactional approach obscures the fact that users don't know what data are collected about them and that platforms share their data with other, undisclosed organizations. The current transactional approach to consumer data was not always the case, as illustrated by the National Data Center efforts in the 1960s, the Fair Credit Reporting Act of 1970, the Ware reports, the Department of Health, Education and Welfare 1973 report (which included a first description of Fair Information Practices [FIPs]), and the US Privacy Act of 1974.[17]

Today, US information privacy rights are enacted through a patchwork of sector-specific laws, often through regimes of notice-and-choice in which individuals are treated as consumers who theoretically can decide whether to participate in using a service based on transparent and accurate descriptions of data collection by different entities.[18] This individualized consumer model presumes rational deliberation and real choice—an approach to privacy that does not reflect how people actually work.[19] While the consumer model's appeal in the United States has deep ties to notions of progress, technology, and innovation, the gap it creates between the imagined citizen of US market discourse and a real citizen is a legal fiction according to which consumers are presumed to have read and understood a company's data practices when they accept the terms of a privacy notice.[20]

The US approach to privacy has resulted in data protection laws distinct from the EU tradition, such as the California Consumer Privacy Act (CCPA), which became effective January 1, 2020.[21] The CCPA and other state data laws such as Virginia's Consumer Data Protection Act oblige for-profit entities that collect consumer personal data to recognize the rights of state residents, such as the right to access personal data and other ARCO rights.[22]

The informational rights regime in India appears to draw on both the European and American values described above and aspects of the local legal landscape and politics. India's Digital Personal Data Protection Act was passed in 2023, though a data protection law had been discussed since at least 2017. The 2023 law came from a fresh draft of a bill that was tabled in 2022, after the 2019 version was scrapped. Many aspects of the 2023 law, as well as the preceding 2019 draft bill and the Srikrishna Committee report of 2018, appear to mirror the GDPR. These legal documents share the broad definition of "processing" of data, many forms of data are classified as "sensitive," and they include the requirements of data minimization, purpose limitation, and consent.[23]

While influenced by the GDPR, India's Data Protection Act and draft bills were also influenced by the reaffirmation of privacy as a fundamental constitutional right based on the 2017 Puttaswamy judgment of the Indian Supreme Court.[24] The Srikrishna Committee highlighted in its expert report that the "core of informational privacy is a right to autonomy and self-determination in respect of one's personal data. Undoubtedly, this must be the primary value that any data protection framework serves."[25] The Indian act (and earlier draft bills) differ significantly from the GDPR in the exceptions it makes for the state; legal scholar Graham Greenleaf has labeled a previous draft "GDPR-lite with Chinese characteristics."[26] Another important provision in the 2023 act is its amendment to the Right to Information Act that abolishes the obligation to provide a citizen any information seen as personal (thus providing an absolute rather than a conditional exemption).[27] As of this writing, the provisions of the Indian Personal Data Protection Act and the draft rules required to operationalize the act that were released for public consultation in January 2025 are subjects of fierce debate in academic, popular, and policy circles,[28] particularly against the backdrop of the country's recent pervasive use of digital ID systems,[29] the 2018 National Policy on AI, the national Automated Facial Recognition System, and, ironically, the act's amendments to the Right to Information Act that would reduce the ability of citizens to access information.[30]

Translating Legal Code into Technical Code

We recruited participants from three countries—the United States, the United Kingdom, and India—with different legal regimes in order to understand the interaction of transnational and local laws in 2019–21. The potentially internationalizing qualities of the web set up tensions between local laws and global infrastructures. On one hand, examining the laws of different regions often showcases variations in interpretation of words by class, work expertise, and geography. In *Trying Leviathan*, Burnett shows how a trial about whale oil regulation ultimately hinged on whose categorization of a whale would prevail—the naturalist's as a mammal or the layperson's as a fish. In the New York case of *Maurice v. Judd* (the subject of *Trying Leviathan*), the interpretation of locals held primacy over the interpretation of nonlocal "easterners."[31] On the other hand, a transnational organization must negotiate multiple, sometimes conflicting, national laws in order to maintain business activities in different countries, including development and deployment of technical infrastructure. IBM, one corporation with a global footprint, set a historical precedent by working within and shaping many legal and political regimes.[32]

Given the ambitions of personal data access and the nature of transnational corporations that must comply with the GDPR within EU borders, a closer look at how EU laws shape global practices is warranted. Legal scholar Anu Bradford has characterized the transatlantic impact of EU regulation as the "Brussels Effect," suggesting that the EU can unilaterally regulate global markets through market mechanisms.[33] Paul M. Schwartz suggests, however, that the expansion of EU law has more to do with the appeal of high standards for data protection and its portability to other legal contexts.[34] The diffusion of legal regimes for data protection into low-resource countries may also play into historical arcs of marginalization and colonization.[35] Studies of on-the-ground privacy practices and belief show variability in expectations and understanding of "privacy" across countries and within them, based on social position.[36] For instance, ideas about privacy vary within India based on people's social position and their varied encounters with the state.[37]

Our study suggests that in addition to EU influence on laws, the power of the GDPR lies in transforming other nations' digital infrastructures. As Jones and Kaminski argue, this transformative effect is deliberate: "Much of the work the GDPR aspires to do is beneath the surface, changing corporate infrastructure and processes and reprioritizing decision-making around data protection rights and values."[38] Thus, the impact of the GDPR regulation emerges materially in private internet infrastructures that extend beyond EU borders. GDPR requires that states and corporations work together to enact the regulation in an arrangement called "collaborative governance."[39] The ongoing collaborative governance between transnational corporations that operate infrastructure and the EU creates bordercrossing material infrastructure that sits in the gap between "law in books" and "law in action."[40] The results of data requests are the material representations of the law in action or the ways a corporation has enacted GDPR.[41]

Understanding the Impact of the GDPR

Theoretical and legal scholarship on rights in the EU analyze the legal and historical contexts of international data policy and interpret the requirements of law.[42] In contrast to law in books, scholarship on ARCO rights in practice (law in action), breaks apart the exercise of data rights into smaller subprocesses and focuses on evaluating company compliance. While it is possible to see how written laws compare with one another, it is much harder to see how different companies have interpreted and enacted the GDPR and how they have translated the legal code into software code that enables something like a DSAR. While there are provisions in the GDPR for auditing, the audits are not required to be public.[43] Scholars have thus adopted a range of approaches to analyzing GDPR's effects through automated measurements of GDPR compliance such as privacy policy statement length, topics covered in privacy policies, requirements of data portability requests, and security risks to data subjects and data controllers.[44]

In the US context, law and technology scholar Ari Waldman argues that rendering data privacy laws ineffective is not an accident but a product of one of the core goals of data privacy law, which, both in the United States and globally, is to make privacy "doable" for companies.[45] Waldman borrows the concept of *legal endogeneity* from socio-legal scholar Lauren Edelman to describe how data privacy law does not constrain, restrict, or guide regulated entities but is instead shaped by the regulated entities themselves.[46] Data privacy laws provide compliance professionals (e.g., lawyers, engineers, privacy professionals, and risk managers) a wide berth to interpret what "compliance" means in practice and to design processes by which to fulfill corporate goals of efficiency and profitability that act as symbols of following the law. These symbolic structures in turn shape the everyday practices of judges, regulators, data controllers, and data subjects who, over time, normalize these symbols as adequate data protection and privacy measures and even hail them as groundbreaking. Similarly Con Diaz's history of patent law explains how corporations' interpretation of the concept of software, and thus what is regulated, are bound to their interests.[47] In the area of DSARs, corporations have substantive leeway in interpreting the GDPR; this research suggests that corporations decide how to implement GDPR within their technical infrastructures so as to obey the letter but not the intention of the law.

Researchers have also assessed the "practical reality" of exercising ARCO rights in the EU by using various rubrics, such as success versus failure, compliant versus noncompliant, complete versus incomplete, easy versus difficult, and satisfactory versus unsatisfactory.[48] These approaches to analyzing the technical infrastructure of both automated methods and DSAR compliance metrics is important for ensuring GDPR compliance. However, they do not necessarily capture the gaps in human experience of engaging with such infrastructure, mediated by users' own technical infrastructure, skills, and expectations, or the meaning making that happens as people make sense of their experiences dealing with a corporation's technical infrastructures. Our DSAR-focused study seeks to add an additional layer of analysis to Norris and colleagues "law in action" by looking not only at results but also at processes in order to understand some of the practical accomplishments of GDPR in the EU and beyond.

Personal Data Requests and Data Access in Practice

To understand how DSARs are addressed in different legal regimes, we recruited 19 participants from India (7 participants), the United States (6), and the United Kingdom (6). Data collection was conducted in overlapping phases through surveys and interviews between September 2019 and March 2021. Crucially, all UK access requests were completed (and data retrieved) before the United Kingdom left the European Union on January 31, 2021, and all US requests were completed after the CCPA came into effect on January 1, 2020, by participants outside California and prior to the passage of India's 2023 Personal Data Protection Act. Participants were recruited from single states (Washington in the United States, Karnataka in India) or nations (England in the United Kingdom) through our personal networks. Our participants were residents of these states but were not necessarily citizens and tended to have high levels of education.

In addition to understanding how DSARs are experienced with transnational companies in different countries, we also wanted to find out how DSARs worked in companies that did not operate across all three regions, and whether *transnational* and *domestic* companies had different practices. Understanding whether transnational companies fulfilled DSAR requests outside the EU and whether companies that did not operate in the EU fulfilled DSAR requests helped us understand the reach of the GDPR. We selected several transnational companies that operate in all three regions—Google, Amazon, Facebook, and Spotify—and a number of domestic companies that operate in only one of the regions: Monzo and

Ryanair in the United Kingdom; Alaska Airlines and Venmo in the United States; and Reliance Jio, Paytm, and Flipkart in India.[49] In addition to region, we also selected companies based on economic sector, focusing on transnational media, technology, retail sales, and regionally operating aviation and financial companies.

Although we intended to conduct the same protocol in all three regions simultaneously, we ended up using two different protocols. Our initial protocol was inspired by Ausloos and Dewitte's 2018 study.[50] We created a series of questionnaires that walk respondents through the process of requesting personal data from two services and reporting their experiences; after the questionnaires were completed, we interviewed participants. We began recruiting participants in all three regions in September of 2019. While 7 of 10 participants in India completed the initial protocol, participants in the United States and United Kingdom struggled.

We then revised our protocol for US and UK participants, replacing questionnaires with two remote interviews via Zoom to first record people's experiences in making DSARs and then analyzing the data they received. We recruited participants from outside California in the United States (the only state with data privacy laws at the time of the study) and United Kingdom and asked them to send DSARs to two different services. During one- to two-hour interviews, participants were encouraged to think aloud while sharing their screen (only if the participant was reasonably confident a file did not contain any sensitive information).[51] The interviewer also encouraged participants to ask questions if they ran into obstacles—which they consistently did.

Table 3.1 summarizes what our participants received. Our work shows evidence of trickledown GDPR: 27 out of the 38 access requests made by our participants resulted in a participant receiving some kind of data. With the exception of Amazon DSARs from India, all of the transnational companies responded to DSARs with data regardless of the requestor's location. And most, though not all, US and Indian companies that did not operate in the EU did not fulfil data requests. We hypothesize that Venmo, the US company that does not operate in the EU yet complied with DSARs (albeit inconsistently, returning different file formats to different participants), might have been influenced by CCPA. But, since the timing of our work overlapped with CCPA implementation, more research is needed to understand the effects of CCPA and other data protection laws in the United States.[52]

The Experience of Codes

In this section, we discuss our findings, focusing on how access requests worked in different jurisdictions and for different companies. Our description focuses on the process and the results of DSARs because both played an important role in how participants interpreted their rights under data

Table 3.1. Company responses to DSAR requests

Service	Region	Channels used for DSAR	DSAR channel locations	Total data requests	Successful data requests	Total requests analyzed*	Time to complete	Returned data format	User chooses format?	Formats available	*Read me* file included in data returned?
Google	Multiple	Dedicated service	Account settings, help documentation, community help	6	6	6	Same day	Multiple	Yes	Products, delivery method, frequency, file type and size	No
Amazon	Multiple	Dedicated button	Help documentation	6	4	4	5–15 days	Multiple	Yes	Data categories	Sometimes
Facebook	Multiple	Dedicated button	Account settings, help documentation, contact form	7	7	6	Same day	Multiple	Yes	Categories, date range, format, media quality	No
Spotify	Multiple	Dedicated button, chat	Privacy settings, help documentation	5	5	5	2–8 days	JSON	No	N/a	yes
Flipkart	India	Phone, email	Privacy policy	3	0	0	N/a	N/a	N/a	N/a	N/a
Reliance Jio	India	Phone, email	Customer service	1	0	0	N/a	N/a**	N/a	N/a	N/a
Paytm	India	Phone, email, chat	Customer service	2	0	0	N/a	N/a	N/a	N/a	N/a
Venmo	US	Contact form, email, chat	Privacy policy, customer service	2	2	2	20–23 days	PDF, CSV	Sometimes	Data range for CSV	Sometimes
Alaska Air	US	OneTrust request form	Privacy policy	2	0	0	N/a	N/a	N/a	N/a	N/a
Monzo	UK	Contact form, email, chat	Privacy policy, customer service	2	1	1	18 days	PDF	No	N/a	Yes
Ryanair	UK	OneTrust request form	Privacy policy	2	2	1	2–3 days	DOC, web portal	No	N/a	No
Total				38	27	25					

*Number of requests for analysis. (Two participants who received data, P7/Ryanair and P11/Facebook, were unable to go over them at the time of our interview.)

**A Reliance Jio representative confirmed which categories of data were being held but emphasized that they could not disclose them to our participants.

protection laws. A complex data request process, followed by files that were not easy to interpret or even to open, influenced people's perception of how transparent the process really was. While past work has framed DSAR activities as successes or failures, the experiences of our participants defied clear metrics. Therefore, rather than framing our findings on the data access request in terms of success or failure, we describe them according to the ways requestors experienced DSARs in order to illustrate the range of ways legal codes were translated into software codes. Both the data that requestors did and did not receive, and the process of making the data request were essential to users' understandings of whether DSARs were actually helping users gain insight into what kinds of personal data corporations held.

Transnational companies (e.g., Facebook, Google, Spotify, and Amazon) had processes for providing data on request. Unlike the processes that domestic companies followed, those of the transnational companies were both integrated (that is, not operated by a third-party service) and at least partially automated. Facebook and Spotify provided "download my data" buttons in their settings, while Amazon placed its in the help documentation. Google provided a special service (Google Takeout) for requesting data from all its products separately or at once. Transnational companies usually returned data the same day but took no longer than a week (table 3.1). For those domestic companies that provided data at all, the data was returned within 20 to 23 days (with the exception of Ryanair, who returned data in 2 days).

To the extent that the DSAR and GDPR's informational regime hinges on access, individuals need to be able to request data. Participants often could not figure out how to conduct a DSAR, as we explain with examples from Amazon and Monzo. Of the transnational companies, Amazon posed notable challenges for our participants. In all three countries, participants struggled to identify where and how to download data from Amazon. Ultimately, both participants from India were unable to retrieve any data from Amazon, despite contacting customer service. Where a "download my data" button did not exist, participants faced other challenges. Participant P7 spent several hours trying to request and unlock data from the UK-based payment service Monzo by exchanging emails with the company asking for data and finally scheduling a call with a representative to obtain the password over the phone.

While policy readability serves a critical role in compliance, the approach adopted by our participants highlights that using privacy policies does not reflect the processes users undertake in trying to obtain data.[53] Participants looked at privacy policies less than a third of the time when they were making requests of transnational companies. On the other hand, during almost half of the request attempts, participants used Google with searches such as "access personal data amazon" (P9, UK). Thus, although data download processes were touted as streamlined and straightforward, they were not. Comments from our participants warrant future research about how personal data is transferred and combined with data from other firms.

If making a data request took these many forms, the steps involved in receiving the data were no less varied or complex. Since various actors and services were responsible for enabling access requests, the receipt of data could fail in many ways and at several stages. For instance, while requesting data from the UK-based airline Ryanair, P7 encountered OneTrust, a privacy management service. Two days after successfully requesting data, OneTrust sent an expiring link via email to P7, which the participant then used to successfully access their data on OneTrust's web platform. However, P7 was not aware that the link would expire and did not download the data before our interview. Similar to other participants who encountered this problem, we simply rescheduled the interview and P7 remade the request. When we reconvened, P7 noted that OneTrust had not yet sent the expiring link as expected. After some cooperative troubleshooting, we discovered that, coincidentally, Gmail was experiencing an outage and the participant was unable to receive emails at all. We made one more attempt the following day, but ultimately were unable to receive the verification email from OneTrust. This example underlines the complexities of underlying infrastructures that support data access in practice, and also shows that in some cases third parties have at least limited access to personal data.

The Black Box of Data Collection Remains a Black Box for Users

When participants received data, most across geographies described being concerned about what the companies really understood about them, how and why companies had classified them as certain types of users, or why they were given certain kinds of data. Our research participants were not privy to the reasons why companies chose to keep certain data about users and not others. But they were less concerned about this information asymmetry than about the imbalance of power between themselves and corporations.

Participants did not know what to make of the data that were provided to them. They speculated as to whether the data were complete and why certain information was included or excluded—deliberately or because of incompetence. This led to more guessing about the for-profit companies that held the data. Rather than increasing the sense of control, for some participants the process of requesting their data did the opposite. For example, users did not understand why they received data from services they did not use but that were owned by the companies from which they had requested data. A participant making Amazon requests was surprised to learn that there was data about them associated with an Amazon product (Alexa device) that the participant did not own. When asked whether they thought the data they received were all that the company held on them, one participant expressed confusion: "I obviously do not think that this is all they have. . . . I have no reason to think that they do [have more data] but, yeah, some part of me thinks that they for sure have more stuff. But, yeah, there's no basis for this. It's a gut feeling" (P5, United States, regarding Spotify and Amazon data requests). The participant couldn't know whether these were all the data that a corporation had about them.

The many formats of data added to the confusion and sense of being powerless experienced by participants: they felt upset at being analyzed by companies in ways that were not accessible to themselves. The fact that data came packaged inconsistently as PDF reports, JSON files, or CSV spreadsheets led to frustration. One participant expected "something more in text format, or something even in spreadsheet format. Something, a report, of how am I behaving or how am I interacting with such softwares or application. I wasn't expecting codes and riddles or anything. I have no idea what this is" (P8, UK).

Participants wondered about the link between their behavior on a specific platform. P5, based in the United States, wondered about being labeled an "alcohol buyer" on the basis of Spotify activity despite self-identifying as a practicing Muslim. The corporation's code gave participants access to their personal data, but this only led confused participants to speculate as to whether the data were accurate. Moreover, esoteric formats and categories made participants feel powerless: "It gives me the raw data, but it doesn't really give me any idea necessarily on the power of this data. And especially, and here's another thing, is like my data individually may not be worth a whole lot, but when it is in aggregate with every other customer's data, that's worth a lot. They could do anything they wanted with this information" (P4, United States, Amazon and Alaska Airlines data request). Ultimately, participants expressed a sense of disappointment on realizing that having control of their own data did not mean much in terms of seeing what happens to the data *within* companies.

Expectations about what the GDPR allowed was confusing and sometimes empowering for users both inside and outside the EU. Participants in the EU contrasted their impression that GDPR should enhance transparency with the opaqueness of the process in practice. As P10 (United Kingdom) pointed out, "The interesting thing is that GDPR makes it look like, 'Oh, here, I've got all this control,' but if you actually look at the control it's offering you, it doesn't actually explain that, okay, if I say this, what does it really mean for my browsing future?" (P10, concerning GDPR cookie notices).

Interviewees in India and the United States thought that the GDPR guarantees EU citizens a right to privacy—a popular yet incorrect understanding of data protection.[54] Comparison with European experiences came up in several interviews with non-EU participants. For example, P4 (United States) said regarding Amazon data: "I have no way of knowing like if I lived in another country, like if I lived in Europe, would they be obligated to send me more. I have no way of knowing for sure if there is more information on me. I just kind of have to trust that this is all … [t]his is all there is." Another participant imagined that, for companies that carry on business globally and must comply with GDPR and California privacy laws, it was easier to apply them to other jurisdictions rather than turning them on and off. In India, research participants who were aware of GDPR knew that it did not apply to their country; they wondered whether domestic companies intended to comply, especially when taken in conjunction with the Indian Supreme Court's 2017 Puttaswamy judgment, which reaffirmed privacy as a fundamental constitutional right in India. Participants were also confused about which data protection laws applied to them, and felt the burden was on them to understand applicability: "So if I can figure out which data protection laws apply to me, I can figure out what rights I've got" (P6, United States, on Alaska Air's privacy notice).

Even with the confusion about how laws might be relevant for people living in different places, the possibility that users could obtain their personal data from corporations could be empowering. After successfully requesting data from Google (but not Amazon), P15, from India, attempted to work through Flipkart's resistance to giving them data: "I pressed [the Flipkart representative] a little bit. I said, 'See, I tried it on Amazon and I got it. I tried it on Google and I got it.' At least Google by that time I'd already given the request, so I know it's possible. Amazon, I just simply told [Flipkart] that [Amazon gave me data]." Though unsuccessful, P15's experience with data controllers who complied with EU data laws shaped their expectations as to whether data controllers in India (Flipkart) should also comply with data access requests.

How to See the Nexus of Legal Code and Software Code

In this second part of the chapter, we have shown the nexus of legal and technical infrastructures that support rights of access in practice, which can only be seen through a bottom-up approach that focuses on the entire process by which law is implemented in practice. We have distilled much of this complexity in the simplified model of figure 3.1 to show how the law in action is a tangle of corporate and individual practices, business and personal motivations, and technical infrastructures. (The law in practice also includes how law is enforced in legal proceedings, in courts, and so forth and is not addressed in this chapter.)

On the "data supply" side, there is the company's technical infrastructure, wherein the legal framework is translated into actionable requests via software and code. On the "data demand" side is the personal technical

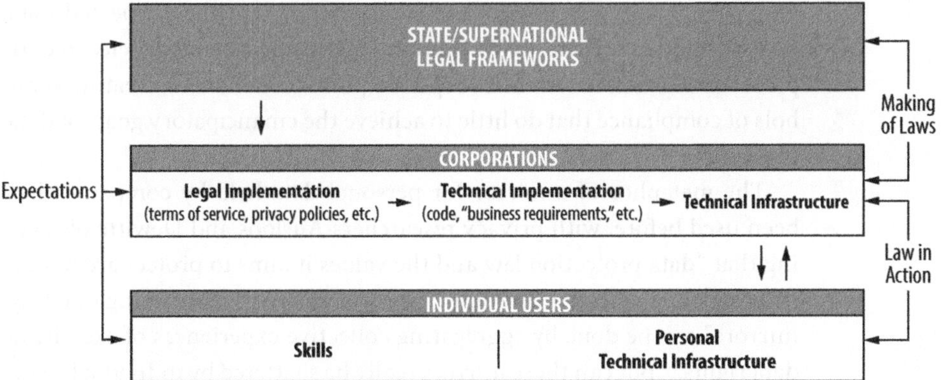

infrastructure of data requesters, often a complex combination of various devices, software, and connectivity. There is also a wide variation in the (technical) personal skills that might be necessary to pursue the data requests and to open and interpret the files received. DSARs are the connective tissues between these two technical infrastructures and the wider influence of laws affecting data and the expectations they might engender (even if unfounded in law).

Figure 3.1. Provisional model of law in action for individual users acting in a private capacity.

Conclusion

As we note in our opening to this chapter, in *An Engine, Not a Camera*, MacKenzie emphasized that the field actively transformed their object of study. As a tribute to this insight, we borrow a quote from one of our research participants for the chapter's title, "A Mirror, Not a Glass Door" and to show how companies, through their acquisition of personal data and control of data infrastructure, construct the conditions of data requests and data access:

> What I initially thought was a very powerful option, that GDPR . . .
> Oh my God. I know it's not like they can keep so much data about
> me. They have to show me what all they know about me. But, in fact,
> the way they're giving me data, clearly, it's like, they're holding a
> mirror to me. Yeah. So it's literally that right then what was supposed
> to be like a glass door that lets me see through what Google is doing
> with my data, instead of making this a glass door, they've just made
> it a mirror. They're like, oh, you want your data here? This is the data
> you gave me. They're just holding up a mirror to my face and saying,
> these are the data that you gave. This is not what this was. At least in
> my understanding, that's not what I expected. But I realized that
> that's what they're giving. So that is something that I find was an
> important thing. So there's still, whatever dump I've gotten leaves
> me no more intelligent than I was before I got this data. (P15, India,
> on Google data)

Our participants did not get the transparency that they expected data access would give them. As Waldman suggests, legal endogeneity in data protection laws can entrench corporate power over data by creating symbols of compliance that do little to achieve the emancipatory goals of data protection laws.

The metaphor of a mirror for personal data held by companies has been used before, with privacy researchers Ausloos and Dewitte observing that "data protection law and the values it aims to protect are but an illusion when assessed through a one-way mirror." "Shattering one way mirrors" can be done by aggregating collective experiences of exercising data rights.[55] But can these mirrors really be shattered by individuals who aim to find out how a company "sees" them through their data? Our work suggests that while aggregated assessment of data access requests could shatter mirrors, most people's experience remains that of looking in a one-way mirror.

Instead of a straightforward story about legal code regulating software code and producing particular results, the combination of software and legal code to enable personal data requests and access produced non-deterministic experiences for users because of variations in their personal computing infrastructure, individual expectations, and expertise (as indicated in figure 3.1). While the experience of our participants in attempting DSARs was mired in confusion and speculation, we also saw some glimpses of what Michael McCann calls "rights consciousness" as our participants interpreted their rights and asked for their data.[56]

Despite all of these confusions, our research provides evidence for trickledown GDPR: residents of India and the United States have personal data access rights for most instances when companies operate in the EU. Multinational corporations incorporate data privacy and protection laws into data infrastructures, extending the borders of legal regimes. A number of questions then emerge from our explorations of the variable experiences of trickledown GDPR: Are Indian and US residents experiencing a kind of European imperialism as EU rights are extended to contexts with different ideologies, cultures, and modes of data governance? Should we read trickledown GDPR as European values pushing back against the imperialism of dominant US technology firms? Have transnational companies empowered US and Indian residents with new rights? Does trickledown GDPR translate in other regions beyond those we examined? For example, China's new data protection laws were modeled on GDPR. Our experience shows that variation in the ways laws are implemented make it unlikely to be a universal experience, and the nuances are hard to predict with Chinese firms and require more research. As we have shown, moving from symbolic compliance into real protections cannot be done solely by examining legal and technical codes but must be grounded in people's actual experiences and goals.

Acknowledgments

We are grateful to Jef Ausloos for sharing research protocols and generous advice on designing this research project. Meg Leta Jones, Gerardo Con Díaz, and Jeff Yost provided important feedback on earlier drafts of this chapter. Meg Young helped pilot research protocols. This research was supported by the University of Washington Information School's Strategic Research Fund.

Notes

1. In 2023 India passed the Digital Personal Data Protection Act.

2. Anu Bradford, "The Brussels Effect," *Northwestern University Law Review* 107, no. 1 (Fall 2012): 1–67, https://scholarlycommons.law.northwestern.edu/nulr/vol107/iss1/1; Anupam Chander, Margot Kaminski, and William McGeveran, "Catalyzing Privacy Law," *Minnesota Law Review* 105 (2021): 1736–1802, https://minnesotalawreview.org/wp-content/uploads/2021/04/3-CKM_MLR.pdf.

3. Donald MacKenzie, *An Engine, Not a Camera: How Financial Models Shape Markets* (Cambridge, MA: MIT Press, 2006), 12,

4. General Data Protection Regulation (GDPR), Article 4(1).

5. Viktor Mayer-Schönberger, "Generational Development of Data Protection in Europe," in *Technology and Privacy: The New Landscape*, ed. *Philip E. Agre and Marc Rotenberg* (Cambridge: MA: MIT Press, 1998).

6. Philip L. Frana, "Telematics and the Early History of International Digital Information Flows," *IEEE Annals of the History of Computing* 40, no. 2 (April 2018): 32–47, https://doi.org/10.1109/MAHC.2018.022921442.

7. Frana, "Telematics," 41.

8. Sandra Braman, "Vulnerabilities of the State and the New World Information and Communication Order," *Media Development*, no. 3 (1991): 3; Miriyam Aouragh and Paula Chakravartty, "Infrastructures of Empire: Towards a Critical Geopolitics of Media and Information Studies," *Media, Culture, and Society* 38, no. 4 (May 1, 2016): 559–75, https://doi.org/10.1177/0163443716643007.

9. Frana, "Telematics"; Armand Mattelart, *The Information Society: An Introduction*, trans. Susan G. Taponier and James A. Cohen (London: Sage, 2003); Fred Turner, *From Counterculture to Cyberculture* (Chicago: University of Chicago Press, 2006).

10. The history of data protection legislation in the United Kingdom is both distinct and entwined with that in the EU. In the United Kingdom, the debate on privacy and data protection started early with a number of private members' bills introduced in the House of Commons and in the House of Lords from 1961 to 1972 without being passed into law. The 1972 report of the Committee on Privacy recommended against a general privacy law, but a section of the report dealt with the collection and handling by computers of personal information, and possible misuse. It recommended 10 principles for the use of personal data by computers, which included specifying the purpose of holding data, giving subject access to data, and keeping data accurate, up-to-date, and for limited periods. It took until 1984, however, for the government to enact the Data Protection Act, which focused on data automatically processed, rather than on the right to privacy. Adam Warren and James Dearnley, "Data Protection Legislation in the United Kingdom: From Development to Statute, 1969–84," *Information, Community, and Society* 8, no. 2 (2005): 238–63. The right to privacy became part of the 1998 Data Protection Act, which enacted the EU Data Protection Directive of 1995. The 1998 act was replaced by the 2018 Data Protection

Act, which enacted the EU GDPR in the United Kingdom. Although the United Kingdom left the EU in 2020, it retained the UK GDPR. See the "Overview—Data Protection and the EU," ICO: Information Commissioner's Office (UK), accessed February 3, 2025, https://ico.org.uk/for-organisations/data-protection-and-the-eu/overview-data-protection-and-the-eu/.

11. Mayer-Schönberger, "Generational Development of Data Protection in Europe."

12. Meg Leta Jones and Margot E. Kaminski, "An American's Guide to the GDPR," *Denver Law Review* 98, no. 1 (2021): 111, https://papers.ssrn.com/abstract=3620198.

13. Jones and Kaminski, "An American's Guide to the GDPR."

14. Jef Ausloos and Pierre Dewitte, "Shattering One-Way Mirrors: Data Subject Access Rights in Practice," *International Data Privacy Law* 8, no. 1 (February 1, 2018): 7 (emphasis ours), https://doi.org/10.1093/idpl/ipy001.

15. Clive Norris, Paul de Hert, Xavier L'Hoiry, and Antonella Goletta, eds., *The Unaccountable State of Surveillance: Exercising Access Rights in Europe*, Law, Governance and Technology Series (Cham, Switzerland: Springer International, 2017), 5, https://doi.org/10.1007/978-3-319-47573-8_1.

16. Jan Philipp Albrecht, *Hands Off Our Data!* (Strasbourg: Greens / European Free Alliance in the European Parliament, 2015; René Mahieu and Jef Ausloos, "Recognising and Enabling the Collective Dimension of the GDPR and the Right of Access," Law Archive, July 2, 2020, https://doi.org/10.31228/osf.io/b5dwm.

17. For extensive discussion of the shift away from projects such as the National Data Center, the Ware reports on the 1970s, and the 1974 Privacy Act toward the commercialization of the internet under the Clinton administration, see Meg Leta Jones, *The Character of Consent: The History of Cookies and the Future of Technology Policy* (Cambridge, MA: MIT Press, 2024); Dan Bouk, "The National Data Center and the Rise of the Data Double," *Historical Studies in the Natural Sciences* 48, no. 5 (November 1, 2018): 627–36, https://doi.org/10.1525/hsns.2018.48.5.627; Christopher Loughnane and William Aspray, "Rethinking the Call for a US National Data Center in the 1960s: Privacy, Social Science Research, and Data Fragmentation Viewed from the Perspective of Contemporary Archival Theory," *Information and Culture* 53, no. 2 (May 2018): 203–42, https://doi.org/10.7560/IC53204; Robert Waterman McChesney, *Digital Disconnect: How Capitalism Is Turning the Internet against Democracy* (New York: The New Press, 2013); Paul M. Schwartz and Karl-Nikolaus Peifer, "Transatlantic Data Privacy Law," *Georgetown Law Journal* 106 (2017): 137, https://works.bepress.com/paul_schwartz/71/; Willis H. Ware, "Security Controls for Computer Systems: Report of Defense Science Board Task Force on Computer Security" (Santa Monica, CA: RAND Corp., February 1970); Fred H. Cate, "The Failure of Fair Information Practice Principles," in *Consumer Protection in the Age of the Information Economy*, ed. Jane K. Winn, Markets and the Law (London: Routledge, 2006).

18. Daniel J. Solove, "Privacy Self-Management and the Consent Dilemma," *Harvard Law Review* 126 (2013): 1880–1903; Julie Brill, "Remarks by Commissioner Julie Brill—Privacy 3.0 Panel," presented at Conference of Western Attorneys General Annual Meeting, Santa Fe, NM, July 20, 2010, https://www.ftc.gov/public-statements/2010/07/presentation-commissioner-julie-brill.

19. Julie E. Cohen, "What Privacy Is For," in "Symposium: Privacy and Technology," *Harvard Law Review*, 126, no. 7 (2013): 1904–33, https://harvardlawreview.org/wp-content/uploads/2013/05/vol126_cohen.pdf; Alessandro Acquisti et al., "Nudges for Privacy and Security: Understanding and Assisting

Users' Choices Online," *ACM Computing Surveys* 50, no. 3 (August 8, 2017): 44:1–41, https://doi.org/10.1145/3054926.

20. Schwartz and Peifer, "Transatlantic Data Privacy Law," 150–55; Lon L. Fuller, *Legal Fictions*, Notations edition (Stanford, CA: Stanford University Press, 1967); Solove, "Privacy Self-Management and the Consent Dilemma."

21. Chander, Kaminski, and McGeveran, "Catalyzing Privacy Law."

22. Cal. Civ. Code §1798.115; Va. Code § 59.1-573(A).

23. Graham Greenleaf, "GDPR-Lite and Requiring Strengthening—Submission on the Draft Personal Data Protection Bill to the Ministry of Electronics and Information Technology (India)," UNSW (University of New South Wales) Law Research Paper No. 18-83, September 20, 2018, http://dx.doi.org/10.2139 /ssrn.3252286.

24. Justice K. S. Puttaswamy (Retd) v. Union of India on 26 September, 2018, Indian Kanoon, https://indiankanoon.org/doc/127517806/

25. Committee of Experts under the Chairmanship of Justice B. N. Srikrishna, *A Free and Fair Digital Economy Protecting Privacy, Empowering Indians* (Ministry of Electronics and Information Technology, Government of India, 2018), 10, https://www.meity.gov.in/writereaddata/files/Data_Protection_Committee _Report.pdf.

26. Greenleaf, "GDPR-Lite and Requiring Strengthening—Submission on the Draft Personal Data Protection Bill."

27. Section 8(1)(j) of the RTI Act, 2005, currently reads as follows: "8(1) Notwithstanding anything contained in this Act, there shall be no obligation to give any citizen: (j) information which relates to personal information the disclosure of which has no relationship to any public activity or interest, or which would cause unwarranted invasion of the privacy of the individual unless the Central Public Information Officer or the State Public Information Officer or the appellate authority, as the case may be, is satisfied that the larger public interest justifies the disclosure of such information." Section 44(3) of the Digital Personal Data Protection Act of 2023 replaces clause 8(1)(j) with: "information which relates to personal information."

28. For critiques of the 2019 draft by the drafters themselves of the Srikrishna Committee report, see Megha Mandavia, "Personal Data Protection Bill Can Turn India into 'Orwellian State': Justice BN Srikrishna," *Economic Times* (Mumbai, updated December 19, 2019, https://economictimes.indiatimes.com/news /economy/policy/personal-data-protection-bill-can-turnindia-into-orwellian -state-justice-bn-srikrishna/articleshow/72483355.cms; and Rahul Nair, "The Personal Data Protection Bill, 2019: A Constitutional Critique," Constitutional Law and Philosophy (blog), January 8, 2020, https://indconlawphil.wordpress .com/2020/01/08/the-personal-date-protection-bill-2019-a-constitutional-cri tique/. For critiques of the current (2023) act, see Rashmi Rajagopal, "New Data Protection Law Draws Criticism," *Deccan Herald* (Bengaluru, India), updated August 15, 2023, https://www.deccanherald.com/india/karnataka/bengaluru /new-data-protection-law-draws-criticism-2648820; and Gayatri Malhotra, "India's Data Protection Law Does Little for Privacy while Bolstering the State's Surveillance Powers," *Scroll*, September 26, 2023. For critiques of the draft rules published in January 2025, see Aaratrika Bhaumik, "Draft Data Protection Rules: What Are the Concerns Related to Localisation Norms and Government Exemptions?" *The Hindu*, January 16, 2025, https://www.thehindu.com/news/national /draft-data-protection-rules-what-are-the-concerns-related-to-localisation -norms-and-government-exemptions/article69094728.ece; and the Internet Freedom Foundation's statement of January 4, 2025, https://internetfreedom .in/statement-on-the-draft-dpdp-rules-2025/.

29. Subhashis Banerjee, "Short on Nuance: Draft Data Protection Bill Has Lost Sight of The Intricacies of Digital Identity," *Indian Express* (Mumbai), December 6, 2018, https://indianexpress.com/article/opinion/columns/aadhaar-data-protection-bill-privacy-cambridge-analytica-5480398/; Smriti Parsheera, "Adoption and Regulation of Facial Recognition Technologies in India: Why and Why Not?" Data Governance Network Working Paper 5, December 5, 2019, https://ssrn.com/abstract=3525324 or http://dx.doi.org/10.2139/ssrn.3525324; Vidushi Marda, "Every Move You Make," *India Today*, November 29, 2019, https://www.indiatoday.in/magazine/up-front/story/20191209-every-move-you-make-1623400-2019-11-29.

30. For a critique of the bill before its passage as an act that pertains to its effect on the RTI, see " 'Regressive Amendments to RTI Act': NCPRI Flags Concerns over Digital Personal Data Protection Bill," *The Wire*, August 4, 2023, https://thewire.in/rights/regressive-amendments-to-rti-act-ncpri-flags-concerns-over-digital-personal-data-protection-bill. For a discussion of the act after it was passed, see M. Sridhar Acharyulu, "How the 'Strict' Data Act Is Diluting RTI," *Down to Earth*, September 8, 2023.

31. D. Graham Burnett, *Trying Leviathan: The Nineteenth-Century New York Court Case That Put the Whale on Trial and Challenged the Order of Nature* (Princeton, NJ: Princeton University Press, 2007), 196.

32. Regarding IBM, see, for example, Corinna Schlombs, "Engineering International Expansion: IBM and Remington Rand in European Computer Markets," *IEEE Annals of the History of Computing* 30, no. 4 (October 2008): 42–58, https://doi.org/10.1109/MAHC.2008.66; Eden Medina, "Big Blue in the Bottomless Pit: The Early Years of IBM Chile," *IEEE Annals of the History of Computing* 30, no. 4 (October 2008): 26–41, https://doi.org/10.1109/MAHC.2008.62; Colette Perold, "IBM's World Citizens: Valentim Bouças and the Politics of IT Expansion in Authoritarian Brazil," *IEEE Annals of the History of Computing* 42, no. 3 (July 2020): 38–52, https://doi.org/10.1109/MAHC.2020.3010892.

33. Bradford, "Brussels Effect."

34. Paul M. Schwartz, "Global Data Privacy: The EU Way," New York University Law Review 94, no. 4 (October 2019), https://nyulawreview.org/issues/volume-94-number-4/global-data-privacy-the-eu-way/.

35. Nora McDonald and Andrea Forte, "The Politics of Privacy Theories: Moving from Norms to Vulnerabilities," in *CHI '20: Proceedings of the 2020 CHI Conference on Human Factors in Computing Systems, April 25–30, 2020, Honolulu, HI* (New York: Association for Computing Machinery, 2020), 1–14, https://doi.org/10.1145/3313831.3376167; Payal Arora, "Decolonizing Privacy Studies," *Television and New Media* 20, no. 4 (May 2019): 366–78, https://doi.org/10.1177/1527476418806092.

36. See, for example, Syed Ishtiaque Ahmed, Md. Romael Haque, Irtaza Haider, Jay Chen, and Nicola Lee Dell, " 'Everyone Has Some Personal Stuff': Designing to Support Digital Privacy with Shared Mobile Phone Use in Bangladesh," in *CHI '19: Proceedings of the 2019 CHI Conference on Human Factors in Computing Systems* (New York: Association for Computing Machinery, 2019), 1–13, https://doi.org/10.1145/3290605.3300410; Rishabh Bailey, Smriti Parsheera, Faiza Rahman, and Renuka Sane, "Disclosures in Privacy Policies: Does 'Notice and Consent' Work?," National Institute of Public Finance and Policy Working Paper 246, December 11, 2018; Seeta Peña Gangadharan, "The Downside of Digital Inclusion: Expectations and Experiences of Privacy and Surveillance among Marginal Internet Users," *New Media and Society* 19, no. 4 (April 2017): 597–615, https://doi.org/10.1177/1461444815614053; Margaret C. Jack, Pang Sovannaroth, and Nicola Dell, " 'Privacy Is Not a Concept, but a Way of Dealing

with Life': Localization of Transnational Technology Platforms and Liminal Privacy Practices in Cambodia," *Proceedings of the ACM on Human-Computer Interaction* 3, issue CSCW (November 7, 2019): 1–19, https://doi.org/10.1145/3359230.

37. Janaki Srinivasan et al., "The Poverty of Privacy: Understanding Privacy Trade-Offs from Identity Infrastructure Users in India," *International Journal of Communication* 12 (March 1, 2018): 20, https://ijoc.org/index.php/ijoc/article/view/7046.

38. Jones and Kaminski, "An American's Guide to the GDPR," 116.

39. Jones and Kaminski, "An American's Guide to the GDPR," 111; Margot E. Kaminski, "Binary Governance: Lessons from the GDPR's Approach to Algorithmic Accountability," *Southern California Law Review* 92, no. 6 (2019): 1529, https://doi.org/10.2139/ssrn.3351404.

40. Norris, *Unaccountable State of Surveillance*, 5. On border-crossing materials and policy, see Emilie Cloatre and Robert Dingwall, "'Embedded Regulation:' The Migration of Objects, Scripts, and Governance," *Regulation and Governance* 7, no. 3 (2013): 365–86, https://doi.org/10.1111/j.1748-5991.2012.01152.x.

41. For a discussion of material representations of patent law, see Alain Pottage and Brad Sherman, *Figures of Invention: A History of Modern Patent Law* (Oxford: Oxford University Press, 2010).

42. Schwartz, "Global Data Privacy"; Schwartz and Peifer, "Transatlantic Data Privacy Law"; Bradford, "Brussels Effect"; Paul De Hert, Vagelis Papakonstantinou, Gianclaudio Malgieri, Laurent Beslay, and Ignacio Sanchez, "The Right to Data Portability in the GDPR: Towards User-Centric Interoperability of Digital Services," *Computer Law and Security Review* 34, no. 2 (April 2018): 193–203, https://doi.org/10.1016/j.clsr.2017.10.003; Bert-Jaap Koops and Ronald Leenes, "Privacy Regulation Cannot Be Hardcoded: A Critical Comment on the 'Privacy by Design' Provision in Data-Protection Law," *International Review of Law, Computers, and Technology* 28, no. 2 (May 4, 2014): 159–71, https://doi.org/10.1080/13600869.2013.801589.

43. Margot E. Kaminski and Gianclaudio Malgieri, "Algorithmic Impact Assessments under the GDPR: Producing Multi-layered Explanations," *International Data Privacy Law* 11, no. 2 (April 2021): 125–44, https://doi.org/10.1093/idpl/ipaa020.

44. See, for example, De Hert et al., "The Right to Data Portability in the GDPR"; Thomas Linden, Rishabh Khandelwal, Hamza Harkous, and Kassem Fawaz, "The Privacy Policy Landscape after the GDPR," *Proceedings on Privacy Enhancing Technologies* 2020, no. 1 (January 7, 2020): 47–64, https://doi.org/10.2478/popets-2020-0004; Janis Wong and Tristan Henderson, "How Portable Is Portable? Exercising the GDPR's Right to Data Portability," in *UbiComp '18: Proceedings of the 2018 ACM International Joint Conference and 2018 International Symposium on Pervasive and Ubiquitous Computing and Wearable Computers* (Singapore: Association for Computing Machinery, 2018), 911–20, https://doi.org/10.1145/3267305.3274152; Luca Bufalieri, Massimo La Morgia, Alessandro Mei, and Julinda Stefa, "GDPR: When the Right to Access Personal Data Becomes a Threat," paper presented at 2020 IEEE International Conference on Web Services (ICWS), Beijing, October 19–23, IEEE Xplore, December 2020, https://doi.org/10.1109/ICWS49710.2020.00017; Jatinder Singh and Jennifer Cobbe, "The Security Implications of Data Subject Rights," *IEEE Security Privacy* 17, no. 6 (November 2019): 21–30, https://doi.org/10.1109/MSEC.2019.2914614.

45. Ari Ezra Waldman, *Industry Unbound: The Inside Story of Privacy, Data, and Corporate Power* (Cambridge: Cambridge University Press, 2021), 102.

46. Lauren B. Edelman, *Working Law: Courts, Corporations, and Symbolic Civil Rights*, Chicago Series in Law and Society (Chicago: University of Chicago Press, 2016), https://doi.org/10.7208/9780226400938.

47. Gerardo Con Díaz, *Software Rights: How Patent Law Transformed Software Development in America* (New Haven, CT: Yale University Press, 2019), https://yalebooks.yale.edu/book/9780300228397/software-rights.

48. Norris et al., *Unaccountable State of Surveillance*; Lokman Tsui and Stuart Hargreaves, "Who Decides What Is Personal Data? Testing the Access Principle with Telecommunication Companies and Internet Providers in Hong Kong," *International Journal of Communication* 13 (2019): 15; Tobias Urban et al., "Measuring the Impact of the GDPR on Data Sharing in Ad Networks," in *ASIA CCS '20: Proceedings of the 15th ACM Asia Conference on Computer and Communications Security* (New York: Association for Computing Machinery, 2020), 222–35, https://doi.org/10.1145/3320269.3372194.

49. Alaska Airlines did business outside the United States but had no direct flights to EU member states at the time this research was conducted.

50. Ausloos and Dewitte, "Shattering One-Way Mirrors."

51. Clayton Lewis, "Using the 'Thinking-Aloud' Method in Cognitive Interface Design," IBM Thomas J. Watson Research Center, Yorktown Heights, NY, February 17, 1982, https://dominoweb.draco.res.ibm.com/reports/RC9265.pdf; K. Anders Ericsson and Herbert A. Simon, "Verbal Reports as Data," *Psychological Review* 87, no. 3 (May 1980): 215–51, https://doi.org/10.1037/0033-295X.87.3.215.

52. Chander, Kaminski, and McGeveran, "Catalyzing Privacy Law."

53. Aleecia M. McDonald and Lorrie Faith Cranor, "The Cost of Reading Privacy Policies," *I/S: A Journal of Law and Policy for the Information Society* 4, no. 3 (2008): 543–68, https://heinonline.org/HOL/P?h=hein.journals/isjlpsoc4&i=563.

54. Jones and Kaminski, "An American's Guide to the GDPR."

55. Ausloos and Dewitte, "Shattering One-Way Mirrors."

56. Michael W. McCann, *Rights at Work: Pay Equity Reform and the Politics of Legal Mobilization*, Chicago Series in Law and Society (Chicago: University of Chicago Press, 1994), https://press.uchicago.edu/ucp/books/book/chicago/R/bo3630053.html. McCann defines "rights consciousness" as "the ongoing, dynamic process of constructing one's understanding of, and relationship to, the social world through use of legal conventions and discourses" (7).

4

Algorithmic Collusion, Modern Monopolies, and Their Market Power

Hamid R. Ekbia

In April 2011, Michael Eisen, a *Drosophila* biologist at the University of California, Berkeley, asked a postdoc assistant to purchase a copy of Peter Lawrence's *The Making of a Fly* for his lab. The 1992 book was out of print, so the assistant was pleased to see 15 used copies and 2 new ones available on Amazon. He was shocked, however, to see the prices for the two new copies from two credible sellers set at $1.7 and $2.2 million. These two price tags looked random, suggesting that they were set by a computer. But how did they get so out of whack? Eisen tells the rest of the story: "Amazingly, when I reloaded the page the next day, both prices had gone UP! Each was now nearly $2.8 million. And whereas previously the prices were $400,000 apart, they were now within $5,000 of each other. Now I was intrigued, and I started to follow the page incessantly. By the end of the day the higher priced copy had gone up again. This time to $3,536,675.57. And now a pattern was emerging."[1]

The pattern in question essentially reveals an apparent coordination between the two sellers whereby prices go up while also converging—that is, "bordeebook: was offering a price that was 1.270589 times the price offered by "profnath," while the latter set its price at 0.9983 times the for- mer (table 4.1). In other words, the sellers continued adjusting their prices with respect to each other, one upward and the other one downward, and that is how they converged toward each other. In this fashion, within 10 days, the price peaked around $23 million until the item was removed from Amazon's website.

The patent absurdity of this case brought it to an end without tangible damage, but it revealed an inherent vulnerability of current pricing sys- tems and the opportunity that they can create for all kinds of mischief— from price escalation and price discrimination to collusion and even the possibility of a retailer selling something they do not actually own. In fact,

Table 4.1. Changes in price for *The Making of a Fly* documented by Michael Eisen in 2011

Date	Profnath's price (US$)	Bordeebook's price (US$)	Profnath price over previous Bordeebook price	Bordeebook price over Profnath price
April 8	1,730,045.91	2,198,177.95		1.27059
April 9	2,194,443.04	2,788,233.00	0.99830	1.27059
April 10	2,783,493.00	3,536,675.57	0.99830	1.27059
April 11	3,530,663.65	4,486,021.69	0.99830	1.27059
April 12	4,478,395.76	5,690,199.43	0.99830	1.27059
April 13	5,680,526.66	7,217,612.38	0.99830	1.27059

in recent years dynamic and algorithmic pricing have become a common practice in different industries such as banking, insurance, travel, and retail, turning these issues into matters of practical concern. In oligopolist situations, for instance, firms can independently arrive at interdependent decisions.[2] In "captive traps" situations, this tendency can lead to price escalation, especially when the same pricing algorithm is used by competitors in the same industry. The Accenture software Partneo, for instance, originally designed for Renault to price auto parts, was also used by Nissan, PSA Group (formerly PSA Peugeot Citroën), Chrysler, and Jaguar Land Rover, leading to a total price escalation of €2.6 billion at the expense of consumers.[3]

Legal Challenges and Remedies: From Prohibition to Liability

These developments have led antitrust experts to reconsider the notion of collusion and the scenarios that might give rise to collusive behavior. Joseph Harrington, for instance, considers a fictitious scenario similar to the booksellers' story described above, with the proviso that the managers of bookselling firms I and II, unbeknownst to each other but dissatisfied with the existing profit margins, are experimenting with the adoption of a software program that "is not a pricing rule but rather a learning algorithm that experiments with different pricing rules in search of one that yields the highest profits." To that end, "this program uses all available information including firms' past prices (for firm II's prices are available online), firm I's past sales, the cost of each product, the time of the year, and so on."[4] In an ideal but practically feasible scenario, the two programs manage to raise profits for both firms, as seen in the case of Accenture above. The question, then, is whether or not this can be considered unlawful collusion.

Traditionally, the law has understood collusion as a "causal relationship" between the behaviors of two or more firms. According to that interpretation, collusion is *not* about a firm setting a supracompetitive price (that could be done through, for instance, lobbying the government) but about one firm inducing other firms to move their prices up, which implies some type of mutual understanding or, in the language of American law, a meeting of minds and, in the language of European law, "joint intention"[5] or "concurrence of wills."[6] An implicit reward-punishment scheme, in other words, should be in place for an arrangement to be legally considered collusion. That is why in dealing with illegal collusion, jurisprudence has to exclude the possibility of firms acting independently. To effect that exclusion, courts often seek evidence of an *overt* act of communication. The challenge for the law, therefore, is to find evidence of mutual agreement in situations where overt communication is *not* in evidence.

Section I of the Sherman Antitrust Act (1890), for instance, does not prohibit collusion; it prohibits communication among firms with the purpose of achieving agreement. With that standard, Harrington argues, firms that collude through the use of algorithms do not violate the Sherman Act.[7] In the bookseller scenario, the managers were not aware of the possibility of collusion, and hence there was no mutual understanding between them, let alone communication.

What about the software programs that they used? When it comes to computers setting prices, the legal challenge is multiplied because there are no easy answers to questions such as the following:

- Can algorithms *induce* other algorithms to behave in a certain way?
- Can two algorithms *coordinate* pricing?
- Can computers *communicate* in "covert" ways?
- Can computers *discriminate* on the basis of price?
- Can computers *identify* monopolistic behavior?
- Can computers *learn* to collude?
- Can computers *collude*?

These questions have fueled intense debate among scholars and practitioners of antitrust law in recent years. Harrington maintains that price-setting algorithms "can be transmitting data to each other and acting on that data so as to yield coordinated pricing, but that does not imply [that the algorithms] understand they are coordinating to restrain *competition*. And, without understanding, there cannot be mutual understanding. Given [that] no *agents*—human or artificial—in those firms have a "meeting of minds," the firms do not have an agreement and thus have not violated Section 1 of the Sherman Act."[8] The US Department of Justice seems to agree, following a similar logic: "Absent concerted action, independent adoption of the same or similar pricing algorithms is unlikely to lead to antitrust liability even if it makes interdependent pricing more likely. For example, if multiple competing firms unknowingly purchase the same software to set prices, and that software uses identical algorithms, this may effectively align the pricing strategies of all the market participants, even though they have reached no agreement".[9] "Even though it is collusion which is harmful," in other words, "jurisprudence has made the communication that facilitates it illegal."[10] Based on this thinking, the proposed remedy is the per se prohibition of certain types of algorithms—that is, liability would be determined by (1) examining the pricing algorithm's code to determine whether it is a prohibited pricing algorithm; or (2) entering data into the pricing algorithm and monitoring the output in terms of prices to determine whether the algorithm exhibits a prohibited property.[11]

Not Just Code: Algorithmic Inspection and the Limits of Transparency

The idea of per se prohibition is based on an implicit assumption about the possibility of reading the "mind" of a computer: "Courts cannot read the minds of those who are choosing prices, and use communication in its place. In contrast, we can, in principle, 'read the mind' of an autonomous artificial agent by reading its code or 'simulating its mind' by entering input and observing output."[12] The assumption here is that we can understand, control, and govern systems by opening their black box and inspecting the contents—that transparency is an effective mechanism for accountability, in other words. As a hallmark of the Enlightenment, this line of thinking has a long history in law, policy, philosophy, and elsewhere.[13] More recently, the thinking has found a new thrust in computing through ideas such as "code audits," "algorithmic transparency," "algorithmic inspection," and "explainable AI." Diakopolous, for instance, discusses some of the ways by which algorithmic transparency can be accomplished, among them, disclosure of specific human involvement; data quality, accuracy, and uncertainty; and the input features to the algorithm, their weights, and the rationale behind them.[14]

Despite its strong appeal, however, transparency is rather limited in its scope and effectiveness, sometimes even leading to more confusion, obfuscation, and opacity.[15] There are many reasons for these limitations, but for our purposes here, I focus on three key arguments based on the complex, dynamic, and distributed character of computing and software systems.

First, in terms of complexity, many software systems are too large and too convoluted for anyone to examine and to understand, let alone predict and control. This observation applies even to some of the "insiders" involved in the design and development of algorithms.[16] Examples abound. The Google search engine, for instance, is a multicomponent system designed by teams in different divisions, making it impossible for individuals to understand the flow and logic of the employed algorithms.[17] This unknowability is even more of an issue for machine-learning algorithms that change their logic as they process and "learn" from millions of heterogeneous data items.[18] That is why the idea of per se prohibition based on algorithmic audit of dynamic pricing systems has limited utility. The opacity of these systems is *not* in the algorithms, such as backpropagation, designed by human beings but in the *models* that they produce, especially in cases when the data itself is complex and the space of possible models is immense.[19] This is an "emergent" kind of opacity that arises from interactions between straightforward human-designed algorithms and large amounts of data.

Second, in terms of dynamism, software systems are not static objects that easily lend themselves to inspection in the same way that, for instance, a physical artifact might. They are, rather, "quasi-objects" that change and mutate in a continuous and contested manner.[20] One cannot examine and understand them by taking a snapshot of their operation. Even if the source code of a system is transparent, its meaning and "interpretation" changes at run time—a basic fact about software that is often lost in discussions of software semantics.[21] When it comes to pricing algorithms, this constant change means, among other things, that these systems have to be continually assessed and monitored, creating an unending race between auditors and proprietors and making it almost impossible for an automated system to be regulated by something that is not similarly automated.

Third, and most fundamentally, software systems are distributed and dispersed among many different components. An algorithmic system, as Ananny and Crawford emphasize, "is not just code and data but an *assemblage* of human and nonhuman actors."[22] To examine and understand such an assemblage, we need to look not just into but also *across* these systems as they get embedded in socioeconomic, political, and cultural environments. Even if we take the narrow view of software as just code and data, at least four different places in the "pipeline" would demand auditing and explanation: the input data, the algorithm, the model that emerges from the algorithm, and the decisions made by the model. Even if the data are simple and the algorithm is straightforward, there is no guarantee that the model is understandable. In fact, often the opposite is true: "Simple algorithms applied to simple datasets can nevertheless lead to inscrutable models."[23] It is a major misconception, in short, to think of software as a single system that we can peer into as we would with physical artifacts of yore.

In summary, for transparency to work effectively as a mechanism of governance and accountability, systems must be simple, static, and localized—attributes that hardly apply to systems of dynamic pricing that are currently in use. We can think of such attributes as "conditions of visibility" for algorithmic auditing.[24] Beyond these technical conditions, however, many other conditions need to be in place for law to be able to effectively deal with the issue of algorithmic collusion. The rest of this chapter expounds these conditions of possibility.

Expanding the Conversation: From Computer Collusion to Conditions of Possibility

If pricing algorithms are so opaque and our options for auditing them so narrow, what other options are available for us to harness them and to hold their proprietors accountable? To answer this question, we need to step back and ask a more basic question: *Whence do algorithms derive their power?* We need, in other words, to look beyond the code at the conditions that make the code possible and give it its efficacy. Rather than asking if computers can collude, we should ask, What does it take for computers to collude?—or, more accurately, What are the *conditions of possibility* for computers to collude?" In short, before we think about what can be done, we should figure out how we got here.

In what follows, I break up these conditions into parallel developments in business, economy, law, and regulation. Multiple lines of inquiry should be pursued in these areas. Here, I focus on the most immediate drivers, focusing on three parallel developments in the second half of the twentieth century: (1) the rise of platforms as vehicles of modern monopolies; (2) the shift in antitrust thinking from business competition to consumer welfare; and (3) the decline in the enforcement of antitrust law due to the multiplying effects of the above developments 1 and 2 and the challenges that these effects pose for regulatory bodies.

Market Power: From Open Systems to Modern Platforms

Before the rise of modern platforms, the use of market power to exclude other companies from competition was a major consideration of US courts in dealing with antitrust cases.[25] This was the core argument in the historical case of *U.S. v. AT&T* in 1981, in which the judge sustained the allegation that the company "monopolized the intercity services markets by frustrating the efforts of other companies to compete with it in the market on a fair and reasonable basis."[26] That reasoning can hardly be applied against current platform companies such as Amazon that appear to adopt an "inclusionary logic" toward their competitors,[27] treating them as "frenemies" who can openly market and sell their products in the Amazon Marketplace.[28] In such situations, the dominant position of Amazon would seem to be more similar to that of Kodak than AT&T in the 1970s, a position deriving "from growth or development as a consequence of a superior conduct, business acumen, or historic accident."[29]

The origins of this inclusionary logic can be traced back to "open-system standards"—the idea of building computers and devices using modular components with standard interfaces that allow system integration. Pioneered by IBM, this strategy enabled the vertical specialization of the microcomputer industry, with IBM as the dominant supplier of personal computers, and Intel and Microsoft the default producers of chips and operating systems, respectively.[30] It was partly through the adoption of open-system standards that IBM's sales grew by 11 times in a matter of two years (1982–84), exceeding the revenues of its top eight rivals, including Apple.[31] The business model that emerged from the adoption of open standards had a dual character. On one hand, it created opportunities for new firms and entrepreneurs to innovate and develop specialized niche products such as digital watches, calculators, and checkout scanners.[32] On the other hand, it erected high barriers for entry, giving rise to a quasi-monopolistic environment in essentially all areas of growth. In an economic sense, therefore, open-systems standards can be considered the Trojan horse that opened the way for the computerization of the economy and the rise of platforms in later decades. These developments, along with the rise of the "portfolio theory" of the firm in the 1980s, facilitated the emergence of platforms as we know them today.[33] Once again in modern history, technical competence and organizational innovation helped drive business and economic change.[34]

To understand the significance of these developments, one can compare modern computerized systems with earlier ones, including some adjacent industries with close historical ties with computing. One such industry, for instance, was electrical manufacturing, in which major firms engaged in anticompetitive behaviors with adverse effects on pricing from the 1950s through the 1970s. What was significant, however, was that these behaviors did not take the traditional form of collusion prohibited by law, but rather the form of "public signaling" that facilitated collusion without the direct violation of antitrust law—somewhat like two gas stations on opposite corners of an intersection openly displaying their prices to customers and to each other. This kind of behavior in the electrical manufacturing industry led antitrust authorities to require manufacturers to limit public signaling and the amount of information that they could share with each other—a strategy that proved successful in curbing collusion in the long run.[35] The question is, Why wouldn't a similar strategy work in the computerized environment of modern economies, especially when platforms with their open-system strategies constitute the dominant business model of our times?

The answer to this apparent paradox lies in the very "openness" of modern platforms. A platform, from the perspective of business enthusiasts, is a model "that connects two or more mutually dependent groups in a way that benefits all sides."[36] To make this model work, "platforms harness and create large, scalable networks of users and resources that can be accessed on demand."[37] The term "users" in this characterization is applied rather loosely, lumping together a heterogeneous set of actors— all the way from suppliers, software and app developers, distributors, and competitors to consumers and other targeted audiences. By the same token, however, that same loose characterization provides the key to the paradox. The networks organized by platforms are the "natural" outcome of the adoption of open-systems standards. By facilitating transactions among producers and consumers of commodities and services, these networks reduce transaction costs and enable value and information flows that were inconceivable with earlier business models built on a linear concept of a supply chain.[38] That free flow of information also accounts for the rise of dynamic pricing and the possibility of algorithmic collusion. Modern platforms come in different shapes and types—Amazon, eBay, and Uber are platforms, and so are Apple, Google, and Facebook. What is common among them, however, is their access to troves of information enabled by inclusionary logic that trumps a basic premise of current antitrust law—"exclusionary conduct" toward other firms.

While this logic seems an insurmountable barrier for the law, a deeper issue has to do with key assumptions in economic thought about how markets work, chief among them the assumption of "perfect competition" based on the idea of "information symmetry" among different market players—between consumers and producers, for instance, but also among different producers. Devout proponents of markets have long been aware of the unrealistic nature of these assumptions, but that has not shaken their faith in the supremacy of markets over alternatives such as central planning. Friedrich Hayek famously made the "local knowledge" argument against the feasibility of large-scale coordination of economic activity. In its place, he advocated the price system as the only viable machinery that "enables individual producers to watch merely the movement of a few pointers" in order to get the pulse of the economy.[39] Markets, according to this logic, solve the problem of local knowledge by aggregating information and distributing it in the form of prices.

In the current networked environment, Hayek's "pointers" can easily be monitored and manipulated by major platforms having asymmetric access to information that they freely collect about the market, without even "going out" to get them. This is an environment where, according to Alibaba's chief strategy officer, Ming Zeng, "data is collected automatically as a by-product of the business itself."[40] This observation exposes the twisted logic of Uber's CEO, who once said, "We are not setting the price, the market is setting the price . . . we have algorithms to determine what that market is."[41] According to this logic, the market is at once an autonomous mechanism and also the product of an algorithm.

In sum, the rise of platforms as the dominant business model of our times has created a complex economic environment that plays against some of the basic assumptions and premises of legal and economic thinking. While this economic world might be uncharted territory for current antitrust law, the practice of the law itself has also changed throughout the decades in ways that favor, rather than adjust, the dominance of platforms.

The Legal Environment: From Business Competition to Consumer Welfare

In 1890, when the US Congress passed the Sherman Act, the focus of the law was to curb price discrimination and predatory pricing practices of major corporations that intended "to destroy and make unprofitable the business of their competitors." This thinking was further entrenched by the Clayton Antitrust Act in 1914. It was during the same period that the Federal Trade Commission (FTC) was created by the Congress to put a leash on vertical integration and the related practices of leverage and foreclosure, whereby a company could use its dominance in one line of business to either acquire dominance in another line or to put its competitors at a disadvantage. A famous case of the application of the Sherman Act during this early period was the US Supreme Court ruling against Standard Oil, which tried to cross-subsidize its operations, cutting prices below production cost in regions where it had competition and charging monopoly prices in others. The Supreme Court ordered the company to be broken up in order to stop this behavior. During this time, in the era of industrial production, the main focus of antitrust legislation in the United States was protecting small businesses from the predatory behavior of large corporations. The economic structuralism behind this thinking considered concentrated market structures as the key drivers of anticompetitive behavior.

Economic structuralism continued to thrive into the 1960s, giving rise to landmark cases such *Brown Shoe Co. v. United States* (1962) and *United States v. Philadelphia National Bank* (1963). In the former case, the courts blocked the vertical merger of a shoe manufacturer and a leading retailer on the grounds that the merger would "foreclose competition" and "enhance Brown's competitive advantage over other producers, distributors and sellers of shoes." A similar decision was made in the second case.

This thinking started to change in the liberalized economic environment of the midcentury, when the image of a happy consumer reigned with equal force in the minds of people, politicians, and policy makers.[42] It was within this context that antitrust thinking started to shift as well. Under the sway of the Chicago School, which highlighted the administrative efficiencies and benefits of vertical mergers (as opposed to horizontal ones), legal thought shifted toward consumer welfare. Economists such as Richard Bork considered predatory pricing irrational, advocating price theory (as opposed to market regulation) as the right perspective for antitrust legislation. The implication was that the "rational" logic of competition would prevail over the irrational practice of predatory pricing. Many years later, Alan Greenspan, the chair of the Federal Reserve, described the shift: "In the 1970s and 1980s, there was a significant shift in emphasis from a relatively deterministic antitrust enforcement policy to one based on the belief (under the aegis of the so-called Chicago School) that those market imperfections that are not the result of government subsidies, quotas, or franchises would be assuaged by heightened competition. Antitrust initiatives were not seen as a generally successful remedy. More recently, limited avenues for antitrust policy are perceived by policymakers to enhance market efficiencies."[43]

The Regulatory Environment: Light-Touch Enforcement

The antitrust environment of late twentieth century was gradually influenced by the Chicago School of economic thought, with its fundamental belief in the self-correcting mechanism of the market through competition and its narrow view of consumer welfare as having to do only with prices. From this perspective, antitrust enforcement was not needed to create or maintain the conditions necessary for effective competition.[44] Many of the policies of earlier decades, including the 1950 law that restricted horizontal and vertical mergers, were either relaxed or reversed by the Reagan administration, and an environment of business takeover prevailed.[45] These shifts made it increasingly difficult for regulatory bodies to effectively tackle modern platform monopolies in the United States and elsewhere.

Throughout the ensuing years, the US government and the Congress strove to create a policy and regulatory environment favorable to the Silicon Valley venture capital. In 1975, the Security and Exchange Commission (SEC) facilitated early initial public offerings (IPOs) for not-yet-profitable ventures by making it less costly for investors to trade stocks. The US Department of Labor followed suit in 1979, allowing pension-fund managers to invest in speculative assets such as new ventures—a decree that cost millions of Americans their life savings during the Enron debacle in 2001.[46] In 1978 and then 1981, the US Congress, with intense lobbying from the American Electronics Association and the National Venture Capital Association, reduced the ceiling of the capital gains tax rate from roughly 40% to 28% and then 20%, providing stronger incentives for investment in high-tech firms. Finally, in 1993 the Clinton-Gore administration released to the private sector the publicly funded infrastructure developed by DARPA and expanded by the National Science Foundation that came to be known as the internet, catapulting a new era of capitalism.

A landmark court case reflecting this trend was *Matsushita Electric Industrial Company v. Zenith Radio Corporation*, in which the Supreme Court found the allegation of predatory pricing schemes implausible. Justice Lewis Powell, who wrote for the majority, concluded that "cutting prices in order to increase business often is the very essence of competition."[47] This logic devolved into the "light touch" antitrust policies of later years. In the landmark case of *Brooke Group Ltd. v. Brown & Williamson Tobacco Corp.*, for instance, the Supreme Court rejected below-cost pricing as insufficient "to permit an inference of probable recoupment and injury to competition," in the words of Justice Anthony Kennedy, who wrote the majority opinion.[48] In so ruling, the Court established a doctrine (the light-touch approach) that has continued to dominate antitrust thinking into our own times.

Throughout the following decades, this doctrine gave rise to colossal vertical mergers in the United States. Prominent cases of large mergers include DuPont-Conoco and BP–Standard Oil of Ohio during the Reagan administrations; Exxon-Mobil and Pfizer–Warner-Lambert during the Clinton administrations, AOL–Time Warner and AT&T–BellSouth during George W. Bush's tenure, Dow-DuPont and Comcast-NBC during the Obama administrations, AT&T and Time Warner during Trump's first term, and Microsoft and Blizzard under Biden. Altogether, a back-of-the-envelope calculation reveals a rough total of $5 trillion in inflation-adjusted dollars transacted in large mergers and acquisitions between 1980 to 2020. In the process a small number of global corporations managed to gain a strong hold of the economy of the globe.

The fallout from these policy shifts became more pronounced in the aftermath of the Great Recession of 2008–10, when the globe witnessed the repetition of the capitalist crisis cycle and the emergence of new arrangements that have cast some new players into economic and political prominence—an environment in which Apple and Google provide 90% of smartphone operating systems, Google controls roughly 70% of the world's online search activity, and Amazon controls roughly 50% of the entire e-commerce retail market's gross merchandise volume.[49] The situation was so grim that the US House of Representatives held multiple hearings with CEO's of four of the so-called Big 5 (Amazon, Apple, Facebook, Google, and Microsoft) on the basis of the 2020 report of the Judiciary Committee "Investigation of Competition in Digital Markets."[50] "Four companies have turned 'scrappy' start-ups," the report bemoaned, into "the kinds of monopolies we last saw in the era of oil barons and railroad tycoons."

Despite this diagnosis, however, attempts by enforcement bodies such as the SEC have thus far faced limited success in curbing the trend. The situation, in this respect, is not dissimilar to what prevailed at the turn of the twentieth century. In her study of the "great merger movement" in American business between 1895 and 1904, Naomi Lamoreaux notes the irony that the system of regulation weighed more heavily on small firms. During this period, more than 18,000 firms with substantial shares (between 40% and 70%) of their relevant markets disappeared into (horizontal) consolidations.[51] The shock of these rapid and intense developments generated public outcry, driving politicians at different levels of government into action. These efforts were largely thwarted, however, leading the economic historian Gabriel Kolko to characterize the countermovement as "the triumph of conservatism."[52] Lamoreaux elegantly described the situation: "Faced with growing competition, the [corporate] officers of consolidations had actively sought federal regulation as a means of stabilizing their markets and protecting their positions of dominance. And federal officials had obliged. Dazzled by the equation of bigness and efficiency, sharing business's distaste for 'wasteful' competition, they were never much interested in breaking up consolidations."[53]

Today, we seem to be facing a similar situation, where major corporations use a similar logic in their efforts against the breaking up of their monopolies. In his testimony to the House of Representatives, Facebook's CEO, for instance, warned against the possibility of "larger competitors like Google and some of our Chinese rivals" getting access to Facebook's (now Meta's) systems; and some legislators seem to agree, expressing concern over "heavy-handed regulation."[54] It is difficult to meet the standards of current antitrust law, even when there is a will to fix things among regulators. A federal judge, for instance, dismissed the Federal Trade Commission's antitrust lawsuit against Facebook, reasoning that the agency had failed to provide evidence of a monopoly beyond the "vague" claim that Facebook "maintained a dominant share of the US personal social networking market (in excess of 60 percent)."[55] In his ruling, the judge wrote that sites like Facebook "are free [for anyone] to use, and the exact metes and bounds of what even constitutes a [personal social network] service—i.e., which features of a company's mobile app or website are included in that definition and which are excluded—are hardly crystal clear."[56] This reasoning speaks to some of the challenges that law enforcement faces in dealing with current monopolies.

Possible Remedies

The narrow focus of current antitrust practice on high prices or low output limits the possibilities of intervention at the moment of incipiency; it postpones legal intervention to the point when major platforms become dominant enough to curb competition. Rather than countering this trend according to the terms set by corporations, law can stop them by thinking ahead, using some of the existing legal tools when available, and devising new tools when needed. What options are available for such a reversal?

Legally, antitrust law can use current tools to stop anticompetitive behavior through the concentration of capital before things get to the point of no return. These tools might include the Robinson-Patman Act in the United States, Article 102(c) of the Treaty on the Functioning of the European Union, and the Equality Act of 2010 in the United Kingdom. These laws might be able to curb anticompetitive behaviors "in their incipiency."[57] Historical precedent provides useful ideas here. In the 1980s when AT&T was broken up, concepts such as portability and interoperability were employed by the law, allowing customers of competing phone companies to freely call one another without restrictions. Similar ideas have been invoked in dealing with current monopolies. Imagine a world where a user of Dropbox can easily move their data to some other cloud service, where a company that uses Amazon Web Services would not have to incur hefty costs and fees in migrating to another platform, where Google wouldn't block competitors from accessing real-time user data, and where Facebook users could seamlessly communicate with users of rival platforms.[58]

Technically, regulatory bodies can also devise new ways to deal with the black box of proprietary algorithms such as those involved in dynamic pricing or those that seek to absorb consumer resources by nudging and persuasion. To that end, they can apply techniques such as "counterfactual exploration" to examine the response of pricing algorithms to various combinations of inputs.[59] As discussed earlier, algorithmic audit has serious limitations because of the opacity, complexity, and distributed character of these systems. But specific outputs can be examined by tweaking the inputs so as to yield useful insights about the system's inner workings.

The real remedy, however, would require drastic changes in the political and economic environment. Major platforms have enormous power, giving them great leverage in the current environment. Their power has material, social, cultural, and political dimensions. Materially, for the first time in human history, private corporations are involved in the creation of vast infrastructures, from satellite and communication networks (including thousands of server farms and millions of miles of optical cables) to self-driving cars, drone operation networks, and proprietary algorithms whose workings are kept as commercial secrets not only from rivals but also from regulatory bodies. The invisible character of these systems confounds current antitrust frameworks, making it difficult for governments to exert meaningful oversight. Amazon's CEO, for instance, asserted many years ago that the goal of this e-commerce platform was not "to build an online bookstore or an online retailer, but rather a 'utility' that would become essential to commerce."[60] Today, Amazon uses sophisticated schemes and algorithms of dynamic pricing, changing prices on items as frequently as 2.5 million times a day, while engaging in first-degree price discrimination. These invisible algorithms and structures are further compounded by the organizational complexity of current companies. The top 100 multinationals have an average of 20 holding companies each, many of them domiciled in low-tax jurisdictions, and more than 500 affiliates operating in more than 50 countries, which allows them extensive recoupment across markets. Their large cash reserves, often parked in offshore hideaways, has made it easy for them to buy startups and to lobby extensively for regulatory environments that are favorable to their interests.

Socially, through the Trojan horse of open systems, tech companies have positioned themselves as indispensable hubs and obligatory passage points in different lines of business. It is becoming increasingly difficult, if not impossible, these days to have a mobile phone without either a Google or an Apple operating system, to shop online or to use a cloud service without first going through Amazon or Microsoft, to develop an app or sell and buy music without going to Apple, to communicate with people without the interposition of Facebook or X (formerly Twitter), to even develop agricultural products without Monsanto hovering over, and so on. These corporations suppress or channel innovation because small start-ups, increasingly incapable of competing with them, often decide instead to sell themselves to large corporations, which can provide everything from legal services to quality control.

Culturally, through lock-in and incipiency standards, modern high-tech monopolies manage to hide the coercive character of their practices. The total effect of invisible infrastructures and inclusive platforms is to render a small number of corporations de facto providers of goods and services on a global scale. The process by which corporations attain this status is not, in fact, dissimilar to prospecting behavior among early American pioneers and gold rush explorers who claimed ownership of property by putting a stake in it. This is the kind of behavior that current high-tech corporations are manifesting, placing themselves in a situation of "prospective" monopoly through mechanisms of predatory pricing, platform extortion, and network exploitation. We do, indeed, live in a new gold rush era, but one that seems to favor only a minuscule number of prospectors.

Politically, major-platform practices have proven to be decisive in the workings of democratic institutions such as elections—most outrageously in the interventions of the world's richest man in the US presidential elections of 2024. Furthermore, the close ties between major corporations and governments have taken the revolving-door phenomenon to a new level, where former government officials turn into senior executives and lobbyists of major corporations, and vice versa. How much real change can one expect in a political environment where trillionaires dictate government in Washington, DC?

Current legal and policy remedies for correcting course may have little impact on the inner workings of this enormous power. Divesting Facebook as pursued by the FTC, for instance, might be a step in the right direction, but it is by no means adequate in addressing the underlying issues. In light of this fact, it makes sense to pursue in earnest current suggestions to turn platforms into public utilities owned and run by statutory organizations at different local, regional, national, and international levels.[61]

Looking Ahead: Not Just Code, Not Just Price, and Not Just Policy

The main thrust of this chapter is to change the conversation around the question of algorithmic collusion (and other similar questions), as well as the remedies proposed for dealing with them. Questions raised by antitrust scholars about the possibility of algorithmic collusion are interesting but narrow because the broader situation involves many moving parts other than technology. Certainly, computing technology is rapidly changing in terms of the speed and capacity by which it can integrate, process, and analyze large amounts of data, including dynamically changing market information, in real time. But it is not only technologies that are changing; so are the social, economic, political, and regulatory environments in which they operate. We cannot meaningfully answer these questions, therefore, without taking account of the embedding environment.

In fact, narrow questions might put our thinking on the wrong track because, as Stephanie Dick shows in this volume, reform and critique can sometimes strengthen these systems, buying as they do into some of the very underlying assumptions on which such systems are built.[62] By "systems," Dick has the New York State Identification and Intelligence System(NYSIIS,) in mind, but one can make a similar argument about any system built from an allegedly "objective" perspective—in our case, from the perspective of an algorithmic pricing system. Actually, the claim about these pricing systems goes even further, invoking themes of fairness and legality. While it might be fair and legal for your corner bakery to "dynamically" change prices at the end of the day, pricing algorithms pose a different set of issues that can be understood only through a more fundamental rethinking of current practices.[63]

We are at a moment of unknown novelties in the global economy, and we have to think innovatively in order to make this system work for the majority of human beings rather than for a small number of corporations. To achieve such innovation, we cannot go back to early-20th-century antitrust when the focus was only on small business, or to the mid-20th century when consumers were the primary focus of law. Rather than the narrow reductionist notion of "consumer welfare," having to do only with commodity prices, we should adopt a broad notion of "citizen welfare," consisting of wages; healthcare, childcare, education, retirement, and other social benefits; data protections and privacy; environmental protection; and so forth. The challenge for antitrust law in dealing with modern monopolies is not dissimilar to that of labor law in dealing with contractual labor: they both have to overcome the inclusionary logic of platforms. In this environment, tying a "policy knot" across these different domains might be just the first step in bringing about serious change.[64] Recent developments in Chinese government policy toward Big Tech corporations might be driven by political motivations, but at the same time, they echo the same concerns raised elsewhere—namely, "that digital markets tend towards monopolies and that tech firms hoard data, abuse suppliers, exploit workers and undermine public morality." In this light, China might in fact be a "policy laboratory" for experimenting with ideas such as the public ownership of modern utilities.[65]

Notes

1. Michael Eisen, "Amazon's $23,698,655.93 Book about Flies," *it is NOT junk* (blog), April 22, 2011, https://www.michaeleisen.org/blog/?p=358.

2. Michal S. Gal and Niva Elkin-Koren, "Algorithmic Consumers," *Harvard Journal of Law* 30, no. 2 (Spring 2017): 309–53.

3. Yann Philippin, "The Captive Trap: Global Consulting Firm Accenture Bought Software Showing How to Boost Prices of Captive Car Parts for Renault and Shared the Results with Top Competitors," *The Black Sea*, May 31, 2018, https://theblacksea.eu/stories/captive-trap/.

4. Joseph Harrington, "Developing Competition Law for Collusion by Autonomous Artificial Agents," paper presented at Dechert LLP and the New York State Bar Association's Antitrust Law Section "Antitrust in High-Speed: Colluding through Algorithms and Other Technologies," January 8, 2018, 331.

5. Sathya S. Gosselin, April Jones, and Annabel Martin, "Are Antitrust Laws Up to Task? A US/EU Perspective on Anti-competitive Algorithm Behavior," Hausfeld LLP, August 10, 2017, reposted to Lexology, accessed February 17, 2025, https://www.lexology.com/library/detail.aspx?g=8210db07-cff3-4e1d-85e2-28542b18ba0d.

6. Nobuyuki Hanaki, Rajiv Sethi, Ido Erev, and Alexander Peterhansl, "Learning Strategies," *Journal of Economic Behavior & Organization* 56 (2005): 523–42.

7. Harrington, "Developing Competition Law"; Joseph Harrington, "Developing Competition Law for Collusion by Autonomous Artificial Agents," *Journal of Competition Law and Economics* 14, no. 3 (2018): 331–63.

8. Harrington, "Developing Competition Law" (journal article), 343.

9. US Department of Justice, "Algorithms and Collusion—Note by the United States," written contribution submitted for the 127th Organisation for Economic Co-operation and Development Competition committee meeting on June 21–23, 2017, DAF/COMP/WD(2017)41, May 26, 2017, 6.

10. Harrington, "Developing Competition Law" (journal article), 344.

11. Harrington, "Developing Competition Law" (conference paper), 3.

12. Harrington, "Developing Competition Law" (journal article), 351.

13. Christopher Hood, "Transparency in Historical Perspective," in *Transparency: The Key to Better Governance?*, ed. Christopher Hood and David Heald, 1–23 (Oxford: Oxford University Press, 2006).

14. Nicholas Diakopolous, "Accountability in Algorithmic Decision Making," *Communications of the ACM* 59, no. 2 (February 2016): 56–62.

15. Mike Ananny and Kate Crawford, "Seeing without Knowing: Limitations of the Transparency Ideal and Its Application to Algorithmic Accountability," *New Media and Society* 20, no. 3 (2018): 973–89.

16. Christian Sandvig, Kevin Hamilton, Karrie Karahalios, and Cedric Langbort, "Auditing Algorithms: Research Methods for Detecting Discrimination on Internet Platforms," paper presented at "Data and Discrimination: Converting Critical Concerns into Productive Inquiry," preconference to 64th Annual Meeting of the International Communication Association, Seattle, May 22, 2014.

17. Nick Seaver, "Knowing Algorithms," paper presented at Media in Transition 8 conference, Cambridge, MA, 2014.

18. Jenna Burrell, "How the Machine 'Thinks': Understanding Opacity in Machine Learning Algorithms," *Big Data and Society* 3, no. 1 (2016).

19. Michael Kearns and Aaron Roth, *The Ethical Algorithm: The Science of Socially Aware Algorithm Design* (Oxford: Oxford University Press, 2020): 10.

20. Hamid R. Ekbia, "Digital Artifacts as Quasi-Objects: Qualification, Mediation, and Materiality," *Journal of American Society for Information Science and Technology* 60, no. 12 (2009): 2554–66.

21. Brian Cantwell Smith, *On the Origin of Objects* (Cambridge, MA: MIT Press, 1996).

22. Ananny and Crawford, "Seeing without Knowing," 983 (emphasis in original).

23. Kearns and Roth, *Ethical Algorithm*, 174.

24. Tarleton Gillespie, "The Relevance of Algorithms," in *Media Technologies*, ed. Tarleton Gillespie, Pablo Boczkowski, and Kirsten Foot, 167–94 (Cambridge, MA: MIT Press, 2014).

25. Ben M. Enis and E. Thomas Sullivan, "The AT&T Settlement: Legal Summary, Economic Analysis, and Marketing Implications," *Journal of Marketing* 49, no. 1 (Winter 1985): 130.

26. United States v. AT&T, 524 F. Supp. 1336, 1357 (D.C. September 11, 1981).

27. Hamid R. Ekbia, "Digital Inclusion and Social Exclusion: The Political Economy of Value in a Networked World," *Information Society* 32, no. 3 (2016): 165–75.

28. Ariel Ezrahi and Maurice Stucke, "Artificial Intelligence and Collusion: When Computers Inhibit Competition," *University of Illinois Law Review* 2017: 1775–1810.

29. Berkey Photo, Inc. v. Eastman Kodak Co., 603 F.2d 263, 274 (2nd Cir. 1979); cert. denied, 100 S. Ct. 1061.

30. William Lazonick, "The New Economy Business Model and the Crisis of US Capitalism," *Capitalism and Society* 4, no. 2 (2009).

31. Alfred D. Chandler Jr., *Inventing the Electronic Century: The Epic Story of Consumer Electronic and Computer Industries* (New York: Free Press, 2001): 119.

32. Peter J. Schuyten, "To Clone a Computer: Neophyte National Semiconductor Has Jolted the Industry," *New York Times*, February 4, 1979, https://www.nytimes.com/1979/02/04/archives/to-clone-a-computer-neophyte-national-semiconductor-has-jolted-the.html.

33. Greta R. Krippner, *Capitalizing on Crisis: The Political Origins of the Rise of Finance* (Cambridge, MA: Harvard University Press, 2011).

34. Steve W. Usselman, "IBM and Its Imitators: Organizational Capabilities and the Emergence of the International Computer Industry," *Business and Economic History* 22, no. 2 (Winter 1993).

35. David F. Lean, Jonathan D. Ogur, and Robert P. Rogers, "Does Collusion Pay . . . Does Antitrust Work?," *Southern Economic Journal* 51, no. 3 (January 1985): 828–41.

36. Alex Moazed and Nicholas L. Johnson, *Modern Monopolies: What It Takes to Dominate the 21st Century Economy* (New York: St. Martin's Press, 2016), 5.

37. Moazed and Johnson, *Modern Monopolies*, 5.

38. Michael Porter, *The Value Chain from Competitive Advantage* (Cambridge, MA: Harvard University Press, 1985), xxx.

39. F. A. Hayek, *The Road to Serfdom* (Chicago: The University of Chicago Press, 1944): xxx.

40. Ming Zeng quoted in Moazed and Johnson, *Modern Monopolies*, 71.

41. Marcus Wohlsen, "Uber Boss Says Surging Prices Rescue People from the Snow," *Wired*, December 17, 2013, https://www.wired.com/2013/12/uber-surge-pricing/.

42. Luc Boltanski and Eve Chiapello, *The New Spirit of Capitalism* (New York: Verso, 2007); Hamid R. Ekbia and Bonnie Nardi, *Heteromation and Other Stories of Computing and Capitalism* (Cambridge, MA: MIT, 2017)

43. Alan Greenspan, "The Effects of Mergers," testimony before the US Senate Committee on the Judiciary, June 16, 1998.

44. Ezrahi and Stucke 2017.

45. Krippner, *Capitalizing on Crisis*, 8.

46. Matsushita Electrical Industrial Company, Ltd. v. Zenith Radio Corp., 475 U.S. 574, 106 S. Ct. 1348, 89 L. Ed. 2d 538 (1986).

47. *Matsushita Elec. Industrial Co. v. Zenith Radio Corp.*

48. Brooke Group Ltd. v. Brown & Williamson Tobacco Corp., 509 U.S. 209, 226, 113 S. Ct. 2578, 125 L. Ed. 2d 168 (1993).

49. Stephanie Chevalier, "Projected Retail E-Commerce GMV Share of Amazon in the United States from 2016 to 2021," Statista, July 27, 2022, https://www.statista.com/statistics/788109/amazon-retail-market-share-usa/.

50. Cecilia Kang and David McCabe, "House Lawmakers Condemn Big Tech's 'Monopoly Power' and Urge Their Breakups," *New York Times*, October 6, 2020, https://www.nytimes.com/2020/10/06/technology/congress-big-tech-monopoly-power.html.

51. Naomi R. Lamoreaux, *The Great Merger Movement in American Business, 1895–1904* (New York: Cambridge University Press, 1985): 2.

52. Gabriel Kolko, *The Triumph of Conservatism: A Reinterpretation of American History, 1900–1916* (New York: Free Press of Glencoe, 1963).

53. Lamoreaux, *Great Merger Movement in American Business*, 160.

54. Jon Reid, "AT&T Antitrust Fight Gives Lawmakers Road Map to Rein in Big Tech," *Bloomberg Law*, March 24, 2021, https://news.bloomberglaw.com/tech-and-telecom-law/at-t-antitrust-fight-gives-lawmakers-road-map-to-rein-in-big-tech.

55. Gilad Edelman, "Lina Khan's Theory of the Facebook Antitrust Case Takes Shape," *Wired*, August 19, 2021, https://www.wired.com/story/lina-khan-theory-facebook-antitrust-case-takes-shape/.

56. FTC v. Facebook, Inc., 560 F. Supp. 3d 1, 4 (D.C. 2021).

57. Dan Mogin, Jonathan Rubin, and Jennifer M. Oliver, "Will the Antitrust Laws Get an Update?," *National Law Review*, February 18, 2021, https://www.natlawreview.com/article/will-antitrust-laws-get-update.

58. Reid, "AT&T Antitrust Fight Gives Lawmakers Road Map to Rein in Big Tech."

59. Kearn and Roth, *Ethical Algorithm*, 174.

60. Lina M. Khan, "Sources of Tech Platform Power," *Georgia Law Technology Review* 2 (2018): 325.

61. James Muldoon, "Don't Break Up Facebook—Make It a Public Utility," *Jacobin Magazine*, December 16, 2020, https://www.jacobinmag.com/2020/12/facebook-big-tech-antitrust-social-network-data.

62. See chapter 7, this volume.

63. Ezrahi and Stucke 2017.

64. Meg Leta Jones, "The Ironies of Automation Law: Tying Policy Knots with Fair Automation Practices Principles," *Vanderbilt Journal of Entertainment and Technology Law* 18, no. 1 (Fall 2015): 102.

65. "Unruly Response: Xi Jinping's Assault on Tech Will Change China's Trajectory," *The Economist*, August 14, 2021, https://www.economist.com/leaders/2021/08/14/xi-jinpings-assault-on-tech-will-change-chinas-trajectory.

5

Reopening the Politics of Openness in the Age of Cloud Computing

Reflections on Recent FOSS Relicensing

Shun-Ling Chen

On August 29, 2018, a clause was added to the MIT Open Source License of Lerna, widely used project management software, to ban a list of contractors of the US Immigration and Customs Enforcement (ICE), including Microsoft, Palantir, and Amazon. Jamie Kyle, a Lerna developer, explained that this was a response to ICE's action against immigrants, particularly the separation of children from their parents at the border. Kyle viewed this relicensing as a way to prevent him and other Lerna contributors from inadvertently aiding ICE's work.[1] The announcement immediately ignited heated debates. The next day, Daniel Stockman, a Lerna project maintainer with whom Kyle had initially consulted, revoked the new clause. Stockman expressed sympathy with Kyle's noble intentions, but apologized for making "the rash decision" to support the change.[2]

The short-lived clause challenged some fundamental principles of the free and open source software (FOSS) community. Begun in the early 1980s, the Free Software Movement (FSM) was a response by software developers to a growing legal and industrial structure that they perceived as a threat to their shared practices and culture. One cornerstone of the movement is the use of alternative software licenses, which serve both as communal pacts and as legal strategies. Instead of exercising full intellectual property control and treating source code as a trade secret, proponents of FSM choose to release their source code and use a diverse array of licenses to enable everyone to exercise what many developers call the four freedoms: to run, copy, and modify their programs and share subsequently modified versions.[3] Free software advocates emphasize that "free" in "free software" refers to the four freedoms that everyone should be able to enjoy, rather than implying the software is free of cost. "Open source software" is a closely related term, preferred by those who emphasize the efficiency of a software development model in which users are enabled to build on existing code. Despite these differences, the definitions of "free software" and "open source software" are more or less the same. The substantial similarity allows people from either camp to work under a larger umbrella, the FOSS, or FLOSS movement (*L* standing for the Spanish or French *libre*).[4] FOSS gained momentum in 1998, as several incidents demonstrated that it was more than just a radical practice shared by a small number of hardliners but had tangible impacts on the software industry.[5]

Adding a clause that bans certain users from using the software, as Kyle did, would make Lerna no longer a FOSS project. Kyle admitted that "it's not news . . . people can use open source for evil." But he found ICE was overreaching to the point that he must prioritize other values over what he had agreed as "part of the whole deal" when contributing to Lerna. The short-lived ban was an explicitly political act in a community often characterized as politically agnostic.[6] Kyle is not the first developer to attempt to block the use of software for malevolent purposes, and this incident is part of a reemerging debate about whether the FOSS community should limit itself to only the four freedoms. In an effort to avoid unintentionally contributing to various forms of injustice, activist-developers have crafted new licenses to address other pressing political and socioeconomic concerns.

Another recurring topic of discussion within the FOSS community revolves around its relationship with the business sector. For example, one of the longest-standing disagreements between free software and open source proponents involves their different perspectives on "copyleft," that is, a FOSS license featuring a clause that mandates derivative works be released under the same terms. Free software advocates see copyleft as a mechanism for ensuring the expansion of the commons. In contrast, open source advocates perceive copyleft as an unnecessary constraint on subsequent developers. As technology and business practices continue to evolve, FOSS developers have debated whether existing licenses adequately safeguard the basic freedoms. Since 2018, several FOSS companies have reignited these discussions by introducing a new family of licenses, allegedly designed to prevent mega cloud providers from taking advantage of the community.

By drawing attention to these two recent developments in FOSS licensing, this chapter underscores the evolving nature of both the FOSS community and the power structures that FOSS developers have attempted to address over the past few decades. It renews the debate about what it means to be "free" and "open." This chapter begins with a brief overview of FOSS history, tracing its origin and its rise as a prominent development model in the software industry. It then examines the two new waves of discussions within the FOSS community that began in 2018. These rekindled debates also suggest that political agnosticism within FOSS remains a point of contention, and this tension has (re)emerged as the porous community has interacted with its ever-evolving environment.

The FSM represents a recursive public—a community of developers offering their critique by building an ecosystem that sustains its own existence and facilitates self-reproduction.[7] FOSS licenses serve as the tools that advocates have devised to carve out space in the law that they found constricting. As a legal tool, the usefulness of licenses should diminish, if not disappear, when copyright is no longer the cornerstone of the contested power structure. However, as evidenced by emerging trends in relicensing and the related debates, licenses continue to play a curious role in the FOSS community. Yet the relationship of these new licenses with copyright law has shifted from restoring user freedoms to enabling a degree of authorial control, even when the objective is to safeguard other freedoms.

FOSS: Sociolegal Practice and Copyright Critique

In the early days of modern computing, software was not sold as a separate product but bundled with the hardware. Developers often had access to the source code of programs they ran, allowing them to modify the code as needed or share it, along with their modified versions, with others. IBM's unbundling in 1969, along with the rise of personal computers in the second half of the decade, transformed software into a distinct and valuable commodity.[8] As the software industry gradually enlisted intellectual property laws to protect their revenues, the day-to-day practices of developers posed potential infringements. Intellectual property and non-disclosure agreements gradually reshaped the culture of a community that had long cherished free exchange and communication.

In the 1980s, with the advent of personal computers marking yet another significant transformation in the software industry, Richard Stallman initiated the "GNU's not Unix" (GNU) project in 1983 and founded the Free Software Foundation (FSF) in 1985.[9] The GNU project aimed to build an ecosystem of free software that would offer a viable alternative to the dominant proprietary software world. To enable the four freedoms (run, copy, redistribute, and modify), users must have access to the source code.[10] In 1989, these freedoms were inscribed in the GNU General Public License (GPL), which included a copyleft clause. This clause mandates that subsequent derivative works follow the same license, ensuring downstream users can enjoy an equivalent level of freedom.

Some scholars have described the GPL as a "hack" of the copyright system, drawing on the concept of hacking in computing. Essentially, they argue, the GPL represents a nonconformist interpretation of copyright law.[11] GPL users first assume the position of authors under copyright law. Then, instead of using copyright to police and control access to their work, they subvert conventional licensing terms to facilitate sharing and collaboration. The GPL's copyleft clause mandates that derivative works must be released under the same terms. In other words, rather than relinquishing all rights as an individual author, the GPL empowers users to surpass conventional copyright practices and establish a community with shared values regarding the distribution of intellectual efforts.[12] In this sense, the GPL serves as a community pact in the form of a copyright license.

This community pact can also evolve to address changes in the environment. For example, in response to practices such as hardware manufacturers incorporating free software into devices without distributing the corresponding source code, or software companies' increasing reliance on patent law as an additional layer of legal protection for software, the Free Software Foundation initiated a revision process for the GPL in 2005 in order to address these issues. Since the GPL is one of the most widely adopted FOSS licenses, the revision process for what became GPL version 3 (GPLv3) provided an opportunity for FOSS developers to rejuvenate communal norms. This revision took 19 months, 4 drafts, and 2,635 public comments and featured dozens of events held in different continents. It can be likened to "non-governmental transnational lawmaking."[13] Similarly, in response to new challenges posed by cloud computing, another extensive community discussion gave birth to the GNU Affero General Public License (AGPL).[14]

The AGPL is nearly identical to the GPL, except for one clause (article 13) that mandates the availability of source code for download when a modified work is used as part of a cloud service. In contrast, the GPL requires the source code be made available only when the modified version is distributed as software. This distinction in requirements arose as a result of the emergence of cloud service as a significant way of delivering software, a loophole that did not exist in the earlier context. The AGPL was an early example showing how the FOSS community addressed the fresh challenges brought by the era of cloud computing. These challenges have gained greater prominence in recent years, which I further examine in the following section.

Free software permits commercial uses and sales. To make the concept more appealing to the mainstream business world, Eric Raymond introduced an alternative term, "open source software" (OSS), in 1998. This alternative label was meant to tone down the adversarial connotations of "free" and to shifting the focus to open source as an efficient and effective software development model.[15] The Open Source Initiative (OSI), established in 1998, provided a distinct definition of OSS and offered a certification program for open source licenses. While the free software and OSS camps may have differed in their politics, motivations, and goals, they largely agreed on what constitutes free or open source software. One common misunderstanding is that free software licenses are copyleft, whereas open source licenses are not. In fact, licenses that meet the OSI's open source definition generally satisfy the Free Software Foundation's definition of free software as well. Some differences still exist; for example, FSF may discourage the use of a license that meets the free software definition but contains clauses that raise other concerns. Nevertheless, with proprietary software companies as a common foe, the umbrella term "FOSS" brought the two camps together in the late 1990s.

In 1998, the FOSS community reached a significant milestone. Netscape, the company renowned for its eponymous web browser, was rapidly losing market share to Microsoft's Internet Explorer. To counter this trend, Netscape decided to make its browser open source. (Netscape later morphed into Mozilla Firefox, which remains a widely used browser.) In the same year, Eric Raymond released the Halloween Documents, which included leaked internal memoranda about Microsoft's strategies toward OSS (as phrased in the Documents). The Documents were accompanied by Raymond's own commentaries and responses. In contrast to Microsoft's public statements, which often discredited OSS projects, the Documents recognized them as credible. They acknowledged the potential of OSS projects to meet commercial quality standards, noted their complexity and scale, praised their rapid development process, and acknowledged their competitiveness. The Documents also revealed that Microsoft's FUD (fear, uncertainty, and doubt) tactics against OSS were ineffective or could even backfire. Instead, they suggested that Microsoft should consider incorporating the OSS model into its development process.[16]

The Halloween Documents incident illustrates another critique within the FOSS community of copyright law. In 1976, when Microsoft was still a fledgling company known as Micro-Soft, Bill Gates penned his now widely known "Open Letter to Hobbyists." This letter set the tone for Microsoft's decade-long campaign in favor of software copyright. The central premise was that professionals must be rewarded for their investment in producing quality products, and hobbyist users should pay for their use of these products, as failure to do so would impede progress benefitting society.

Gates's letter reiterated a rationale that generations of new-technology adopters had employed to advocate for the expansion of the copyright system to encompass the output of their respective technologies. This rationale hinged on the dichotomy between professionals and laypersons, based on the assumption that the former were more capable of producing quality products and driving societal progress. This perspective prioritized authors' rights by sidelining users in copyright law.[17]

Such rhetoric did not resonate well among hobbyists and FOSS supporters, for whom the line between professionals and hobbyists was more a matter of affiliation than capability. In addition, the proliferation of information and communication technologies significantly reduced the marginal cost of software reproduction and distribution. A strong copyright system could encourage rent-seeking behavior, and the potential threat of copyright violation could stifle user engagement in innovation—the promotion of which is, ironically, copyright's own legislative purpose. Unconvinced by proprietary companies' advocacy for software copyright, FOSS advocates began developing software in their own way as a critique of a strong copyright regime. The leaked Microsoft documents further validated FOSS as viable alternatives.

Since 1998, FOSS has continued to expand its influence in the new millennium and has even become mainstream in the software industry. For instance, in 2020, approximately 84% of all smartphones sold used the Android mobile operating system, which is based on a Linux kernel, which is free and open source. Moreover, around two-thirds of websites are powered by FOSS servers. WordPress, a FOSS content management system, boasts a market share of more than 60%.[18] The most popular browser, Chrome, with 70% market share, is built from Google's FOSS project Chromium. While Chrome itself is proprietary freeware, the majority of its code remains available as FOSS. In the database field, FOSS also commands around 70% of the market share.[19] Furthermore, tech giants such as Google, Facebook and Microsoft routinely release their artificial intelligence projects as open source, further cementing the influence of FOSS in the technology landscape.[20]

Many companies have integrated FOSS into their business model, and the code for numerous FOSS projects is today largely written by employees of for-profit enterprises.[21] Various reasons prompt software companies to embrace FOSS as a development strategy. For instance, consider Mozilla, which became open source because Netscape chose not to maintain it in the face of stiff competition from Microsoft when it bundled Internet Explorer with its popular operating system. Software companies may opt for the FOSS model so as to cultivate a large user and developer base, ultimately expanding their market share. A substantial user base also enables companies to benefit from external contributions, such as bug reports, fixes, and input for prioritizing certain features over others.

Microsoft's embrace of Linux reflects a significant shift in the software industry's engagement with FOSS. Historically, Microsoft had been one of FOSS's primary antagonists. Beginning with Bill Gates's aforementioned letter to hobbyists," Microsoft had actively advocated strong copyright enforcement and was a prominent player in the Business Software Alliance. In the late 1990s, the Halloween Documents brought to light Microsoft's adversarial relationship with Linux and FOSS in general.

In 2001, Craig Mundie, then vice president of Microsoft, labeled GPL as "viral" and a grave threat to intellectual property. Steve Ballmer, Microsoft's CEO at the time, referred to Linux as a "cancer" and the GPL as something that "attaches itself . . . to everything it touches."[22] As Microsoft gradually revised its business strategy, however, it began to foster a warmer relationship with the FOSS community.

By 2011, Microsoft had risen to become one of the top five corporate contributors to Linux.[23] In 2014, after Ballmer stepped down as CEO, his successor Satya Nadella declared, "Microsoft loves Linux," and the company even built its Azure Cloud Switch based on Linux.[24] In 2016, Microsoft joined the Linux Foundation and became the organization with the most contributions on GitHub, a popular FOSS project hosting platform. In 2018, Microsoft also became a member of the Open Innovation Network and began permitting others to use its extensive portfolio of over 60,000 patents for Linux-related projects.[25] Microsoft's acquisition of GitHub for US$7.5 billion in 2018 raised concerns among FOSS developers and prompted some projects to migrate to competing platforms. However, GitHub has remained an independent subsidiary of Microsoft and a vibrant hub for FOSS projects.[26]

Reversing it previously held animosity, Microsoft now plays the role of a good citizen by contributing code and supporting the FOSS community. These developments have led some to proclaim that Microsoft has "surrendered" in this epic fight and decided to join the winning side.[27] Here, it may be worth redrawing the distinction between "free software" and "open source software." As mentioned earlier, "open source" supporters devised the term to make it more appealing to the industry by focusing more on the efficiency of a software development model rather than user freedom. It might be fairer to say that open source has won, surpassing both the proprietary software development model and free software idealism. When Microsoft Research invited Stallman to give a talk in 2019, Azure's CTO (chief technology officer) tweeted the event as "OSS-related Microsoft news," leaving out the *F* in FOSS.[28]

For the FOSS community, licenses are instruments for private ordering, allowing the community to self-regulate under the licensing terms. For that reason, FOSS advocates, as private parties, develop specific license terms for navigating the industrial and legal environment and facilitating self-regulation. Significant changes to the technical, socioeconomic, and political environments over the past few decades have also largely transformed the licensing discussion. Nevertheless, the same old topic—the definition of freedom versus openness—remains at the center of the debate. This chapter now turns to two contexts in which this has been especially clear: open source companies' new licensing model as a strategy to compete with cloud oligarchs, and individual developers' attempts to introduce clauses or licenses to address social and economic injustices that were typically beyond the scope of FOSS licenses.

New Politics in FOSS: Two Directions
A new generation of open source businesses facing off against mega cloud services

The business model of earlier FOSS companies, such as Red Hat (one of the first commercial Linux distributors), relied on offering services and support for community-developed software.[29] Many newer FOSS companies offer cloud-based services and have much more control over their software, with as much as 90% of a project's code written by their employees, though individual FOSS contributors still play an important role in the software development.[30] These FOSS companies often employ a hybrid strategy. They may keep certain code proprietary or induce commercial licensing by choosing a more complicated FOSS license. The main challenge for these companies is to compete with mega cloud providers such as Amazon Web Services, which can offer FOSS as part of their services but holds back from contributing. A recent report from a market intelligence firm on the status of the open source market exemplifies a challenge faced by FOSS companies: it is wise for cloud providers to simply take popular open source software developed by a third party and, "better yet, replicate every line of source code from that third-party service, give it a creative name and make it proprietary."[31] In 2018, to prevent mega cloud service providers from free riding, a firm called Redis Labs restrategized by introducing what it later called a source-available or noncompete license.

Founded in 2012, Redis Labs built its business based on the Redis database management system, first developed by Salvatore Sanfilippo and released as open source in 2009.[32] Redis Labs offers both data management software and database-as-a-service solutions that support scalable and real-time data processing on major cloud providers and its own could service. With Twitter as one of its early use cases, its customers now include major credit card issuers like Mastercard and American Express and large retailers such as Staples. Building on the Redis foundation, Redis Labs has developed modules (add-ons to Redis) that were originally released under the Affero General Public License.

Redis Labs found that the AGPL was inadequate in preventing mega cloud service providers from abusing the open source companies and the community. In 2018, Redis Labs switched the license for their modules from AGPL to "Apache 2.0 modified with Commons Clause."[33] The Commons Clause is a template that can be added to any license of choice. Its primary purpose is to remove the right to sell the software that was included in the license grant. It states, "Without limiting other conditions in the License, the grant of rights under the License will not include, and the License does not grant to you, the right to Sell the Software."[34] The term "sell" is broadly defined to include "practicing any or all of rights . . .

to provide to third parties for a fee or other consideration . . . a product or service whose value derives, entirely or substantially, from the functionality of the Software."

Although the Apache license is a permissive open source license that allows users to "reproduce, prepare Derivative Works of, publicly display, publicly perform, sublicense, and distribute the Work and such Derivative Works in Source or Object form" (article 2), Redis Labs's modified Apache 2.0 version, with the addition of the Commons Clause, does not grant users the right to sell the software. It is important to note that, by definition, FOSS must allow anyone to use the software for any purpose. Redis Labs admits that the Commons Clause is not open source and that adding it to a license renders it no longer free or open, as it restricts a "field of use" (sell). Instead, it is a "new hybrid licensing model."[35]

Another company soon followed in Redis Labs's footsteps but brought new challenges to the open source licensing model. Founded in 2007, MongoDB Inc. is known for its eponymous cross-platform database system that supports web and mobile applications.[36] MongoDB is now a publicly traded company with a wide range of customers, including KPMG, Barclays, and Genentech. Like Redis Labs, MongoDB has its own cloud service but can also be deployed on major cloud providers. Similar to Redis Labs, MongoDB is also concerned that open source licenses would make it "too easy for large cloud vendors to capture all the value but contribute nothing back to the community."[37]

Instead of simply adding the Commons Clause to a FOSS license, which admittedly renders it non–open source, MongoDB attempted a more ambitious approach with its new license, the Server Side Public License (SSPL). Unlike the Commons Clause, which discriminates against a field of use (selling), the SSPL replaced AGPL's clause regarding cloud service (article 13) with a more demanding requirement. Both the AGPL and the SSPL are revisions of the GPL, but the SSPL stretched the copyleft requirement further to cloud service providers.

Whereas the AGPL asks source code of the licensed work and modified work be available for download only when a modified work is part of a cloud service, the SSPL's copyleft clause extends to cover all other programs that interact with it and make it functional as part of a cloud service. If the licensee makes "the functionality of a program or a modified version" available to third parties in a cloud service, the licensee "must make the Service Source Code available . . . under the terms of this License." Article 13 goes on to define "Service Source Code" to include the source code for not only "the Program or the modified version" but also for "all programs that you use to make the Program or modified version available as a service, including, without limitation, management software, user interfaces, application program interfaces, automation software, monitoring software, backup software, storage software and hosting software, all such that a user could run an instance of the service using the Service Source Code you make available." MongoDB states that article 13 does not apply to the whole stack of software and that the actual scope of what users should contribute back needs to be parsed out case by case.[38] However, the license language is undeniably broad, and there is no clear way to determine which code is covered and which not.

When announcing this shift, MongoDB asserted the close relationship between the SSPL and the GPLv3 and provided a detailed comparison between the two licenses. MongoDB claimed that the SSPL is still open source and that strong copyleft is appropriate and needed in the cloud computing environment. MongoDB further stirred the waters by asking the Open Source Initiative to certify the SSPL as an open source license.[39] Despite these postures, MongoDB's announcement on the OSI "license review" mailing list that the SSPL was seeking OSI approval immediately raised doubts. One substantial question was whether the broad copyleft clause violates the OSI definition, which requires that an open source license must not restrict other software distributed with the licensed work.[40] In addition, many respondents in this thread, including Bruce Perens of OSI, Bradley M. Kuhn of the Software Freedom Conservancy, and Red Hat developer Florian Weimer, were unconvinced that MongoDB was genuinely trying to find a new way to defend the FOSS community. They expressed concerns that SSPL's strong copyleft clause could be a proprietary strategy intended to compel potential users and competitors into a commercial license.

FOSS companies can, and often do, employ hybrid licensing models as part of their business strategy. Redis Labs' relicensing garnered attention but sparked only limited controversy because it affected only the modules developed in-house and because the company acknowledged from the outset that the Commons Clause was not open source. MongoDB's SSPL might have generated less controversy had it not been submitted to the OSI for open source certification. By doing so, MongoDB challenged the definition of open source by asking the OSI's endorsement of the SSPL, which could be part of the company's hybrid strategy. As a result, not only the license terms but also MongoDB's motivations came under scrutiny.

MongoDB may have genuinely believed that the cloud computing environment, including the oligarchic structure of the market, had changed so much that the AGPL was no longer adequate to protect the FOSS community, and that another moment for renewing the social contract of FOSS had arrived. If that had been the case, it would have been appropriate for MongoDB to call for the community to deliberate on how to update existing licenses or initiate a public review process. Instead, MongoDB drafted the SSPL behind closed doors and unilaterally decided to relicense.

MongoDB eventually withdrew the SSPL from the OSI review process but continued to use the license. Within months, Amazon followed a path similar to what the market intelligence firm advised, as mentioned earlier. Amazon rolled out its own DocumentDB in its cloud service as a MongoDB-compatible alternative. MongoDB may have believed the SSPL would help level the playing field by compelling mega cloud providers to contribute back essential elements in their cloud services, including other code in the stack that is not a part of or a modified form of MongoDB's software. With most of its code developed in-house, MongoDB may even have felt entitled to retain more control.

This approach is not entirely unprecedented. When Netscape turned its browser into a FOSS project, it adopted the Mozilla Public License and also released part of the code under the Netscape Public License (NPL). While the NPL may qualify as a FOSS license, it has an asymmetrical clause that permits Netscape to package code contributed by the community into its future proprietary version. Netscape, as the first major company to turn a prominent product into free and open source software, defended this licensing approach as a way to compensate itself for taking this step.[41] Even though the FSF counts NPL as a free software license, it discourages people from using it.[42] However, in 1998, Netscape held the moral high ground as the first major company to make such a transition. MongoDB may not enjoy the same moral authority to demand such an asymmetry. MongoDB's SSPL has been viewed as a proprietary business strategy disguised with open source–like language.[43] In addition, clause V of amendments v. 10 and v. 1.1 of the NPL limits its asymmetrical position to two years after the initial release of code, while the SSPL does not impose such a time limit.

Whether MongoDB's assertion that the SSPL was open source was genuine or not, this episode raises fundamental questions about FOSS that merit revisiting. First, what is the relationship between copyleft and FOSS? Copyleft has traditionally been associated with FOSS and other public licenses. However, can a copyleft clause be so extensive that it disqualifies a license as FOSS? OSI's stance on this matter is clear, as its definition requires that an open source license should not restrict other software. Nevertheless, this question is not explicitly addressed in the free software definition. Second, what is the scope of copyleft? If copyleft can extend beyond the original code and its derivative works, how far can it reach, and what criteria should be used to draw that line? Third, what effects will the changing industry landscape have on FOSS? The industry environment has evolved significantly since the inception of the Free Software Movement. The free software camp has championed users' ability to control software, with Stallman strongly opposing cloud service due to concerns about losing this control. Yet, in an environment where cloud computing has become dominant, should license maintainers like the FSF or other leaders in the FOSS community take a more proactive role in leading discussions about new definitions or norms? Is the issue of free riding by mega cloud services beyond the scope of what any FOSS license can address?

The changing demographic landscape and radicalized politics

While the new generation of FOSS businesses seeks to provide licensing fixes for the new techno-industrial environment, some FOSS developers have taken a different path by crafting licenses designed to tackle socio-economic injustices.

The Anti-996 License was initiated by Chinese developers on GitHub (996.icu) to address the widespread phenomenon of overworked developers in China. Long work schedules (9 a.m. to 9 p.m., 6 days a week; hence, 996) take a severe toll on software developers, sometimes resulting in hospitalization. Chinese companies, ranging from startups to tech giants, have formally or informally adopted this grueling work schedule.[44] As offshore software outsourcing can influence the working conditions of software developer elsewhere, the Anti-996 License requires licensees to abide by stricter labor protection standards. Similar to typical FOSS licenses, the Anti-996 License grants users "the right to use, reproduce, modify, prepare derivative works of, distribute, publish and sublicense the licensed work." However, it mandates that licensees must "strictly comply with all applicable laws, regulations, rules and standards of the jurisdiction relating to labor and employment where the individual is personally located, was born or naturalized, or where the legal entity is registered or operating (whichever is stricter)." In cases where the jurisdiction lacks such laws or regulations, the licensee is required to comply with "Core International Labor Standards." The Anti-996 license also asks the licensees not to "induce, suggest or force its employee(s), whether full-time or part-time, or its independent contractor(s), to agree . . . to weaken or relinquish his or her rights or remedies." With the aim of "creat[ing] an open source software license that advocates workers' rights," 996.icu quickly became one of the most starred (equivalent to "liked" on social media sites) projects on GitHub.[45] This substantial GitHub support notwithstanding, the Anti-996 License may not align with either the open source or free software definition, which requires a FOSS license to let anyone use the program.

Another emerging trend in the FOSS community revolves around the moral tension created by the fact that software can be used for harmful purposes, as illustrated by the story of Jamie Kyle mentioned earlier. Kyle was not the first person to insert a moral stance in a copyright license. In 2005, a project modified the General Public License by adding a clause to prohibit military use. When this modification gained media attention in 2006, the FSF soon asked its developers to reconsider the change. To avoid legal disputes and to remain on SourceForge, a major open source project hosting site, the developers removed the clause within days.[46]

Back in 2002, when President George W. Bush started the "war on terror" and pursued those his administration deemed "evildoers," computer programmer Douglas Crockford joked that he tried to do his part by writing the JSON license, which is literally the MIT license with an added clause stipulating that "the Software shall be used for Good, not Evil." While this clause could potentially have disqualified the JSON license as open source or free software, sarcasm within the FOSS community allowed its circulation for some time. Google hosted a project under the JSON License on its open source hosting site until it noticed this vague restriction in 2009.[47] As archives found on the Wayback Machine show, the FSF did not classify the JSON license as a nonfree license until April 2010. The Apache Foundation allowed the use of the JSON license in Apache products until 2016, reversing an earlier decision made in 2008 based on the assumption that the "no evil" clause was a joke.[48] These earlier incidents demonstrate that political agnosticism in FOSS has been well institutionalized—by the FSF and the OSI as maintainers of the definitions, other open source projects such as Apache, and by hosting websites that provide the infrastructure for collaborative software development.

Kyle was only one of those who began to reevaluate the inadvertent consequences of their commitment to open source in the context of recent radicalized politics.[49] In 2019, Caroline Ada Ehmke wrote the first version of the Hippocratic License "to start a conversation about ethics in open source and programmers' control over their work."[50] The Hippocratic License (v. 2.1) grants licensees broad permission as long as they use the software in a manner consistent with "human rights laws and human rights principles." The license broadly defines the former as "any applicable laws . . . that protect human, civil, labor, privacy, political, environmental, security, economic, due process, or similar rights" and bases the latter on the "United Nations Universal Declaration of Human Rights and United Nations Global Compacts."

Another example is the Anti-Capitalist Software License (ACSL). Admittedly not FOSS, it permits free use by "individuals and organizations that do not operate by capitalist principles" and explicitly excludes law enforcement and military entities, as well as their contractors. Everest Pipkin, one of the authors of ACSL, said that, while he wants to let others reuse his work, he does not want his labor to be used against his moral center.[51]

A common response to these activism-oriented ethical licenses is the concern that if individuals with diverse political agendas begin attaching their own clauses, it may erode the pragmatic consensus that has enabled contributors to collaborate effectively under the FOSS umbrella. Licenses with different agendas may not only be legally incompatible with one another but also hinder collaboration. The four examples of activism-oriented ethical licenses pose different challenges to FOSS. The Anti-996

License is part of a labor rights campaign with a rather narrow scope. It is the only one among these licenses that does not explicitly mention developers' regaining control of their code. While its claim to be an open source license was not without questions, it did not, to my knowledge, face significant challenge. The Anti-Capitalist Software License has a clear political agenda and offers a broader critique of labor conditions in software development than the Anti-996 License presents. Its backers appear to be dissatisfied with political agnosticism but does not directly challenge the definition of FOSS. The modified MIT License in Lerna had rather a limited scope, targeting a specific agency and its policy at a particular time. This ephemerality limited its potential impact. Kyle's initial announcement suggested an acknowledgment that the added clause would not meet the FOSS definitions, indicating that it was not aimed at challenging them.

The Hippocratic License represents an ambitious project with broadly generalizable principles. Unlike the JSON License, which uses rather vague language, the Hippocratic License seeks to establish a more widely accepted standard by referencing international human rights declarations. Dissatisfied with the political agnosticism within FOSS, Ehmke cofounded the Organization for Ethical Source and advocated for the community to transition from "open source" to "ethical source." The Hippocratic License claims to meet the open source definition, though disagreements are anticipated. Ehmke ran for election to the OSI board in 2020, seeking to "update the Open Source Definition to open the door for ethical open source licenses."[52] While her candidacy was unsuccessful, Ehmke did open up a new round of debate about the meaning and purpose of the FOSS movement.

Free or Open? The Purposes and Limits of Public Licenses

Despite some ideological differences among FOSS advocates, a core understanding of freedom and openness has traditionally revolved around the removal of restrictions on users' ability to use, copy, modify and distribute software. However, this consensus can be challenged when the technological, industrial, demographic, and sociopolitical environment undergoes significant transitions, as demonstrated by the two lines of licensing debates that have emerged since 2018.

First, as FOSS gradually became a prominent development model, the politics was no longer about the struggle between evil profit maximizers who strangle the public with copyright law, on one hand, and FOSS advocates who sought to liberate it, on the other. The FOSS community has periodically grappled with how to update licenses to adapt to the evolving technological and industrial environment. The emergence of source-available or noncompete licenses like the Commons Clause and the SSPL are among the latest developments in this ongoing debate. The debate culminated when MongoDB sought to obtain Open Source Initiative certification for the SSPL. MongoDB could be posing a legitimate question: Can the FOSS model remain sustainable now that oligarchic cloud providers, functioning as infrastructure providers, can effectively extract value from the community? MongoDB's approach, which did little to engage the community, left it vulnerable to accusations of being self-serving and hindered constructive discussions about extractivism in the cloud-computing environment.

Second, FOSS developers raised questions about the assumed political neutrality and shared values of the FOSS community, as reflected in the Lerna incident. The widespread adoption of FOSS as a development model also means its developers are no longer to be found only in North American and Europe. FOSS developers in other countries experience very different working conditions and cultures. The Anti-996 License, by drawing attention to the impacts of offshoring on employment in the Global North, successfully invoked a shared identity and garnered support from FOSS developers across continents. Although its terms may not align with the free software or open source definitions, it did not draw close scrutiny.

Driven by recent social justice activism, some developers revisited the assumed political neutrality of the FOSS movement, reopening debates about the meaning of freedom and openness and the movement's overarching purpose. The Organization for Ethical Source challenges the existing consensus, arguing that "software freedom must always be in service of human freedom."[53] Similarly, the Anti-Capitalist Software License denies "usage to those that exploit labor for profit" and to the military and law enforcement.[54] Advocates of these licenses do not target copyright law as the main institution for social transformation. Instead, these licenses serve as an organizing tool for developers' "producer activism," akin to consumer activism, and inevitably invite questions about their enforceability and effectiveness as a tactic. But then, why would these new ethical codes continue to be embodied in copyright licenses?

In the early days of the software industry, software developers who worked in academic institutions began to see their culture of sharing and collaboration encroached on by proprietary software practices that were protected by the institution of copyright. To rebuild a community in which sharing and collaboration is possible, they crafted and adopted alternative licenses using the power granted to them under copyright law. These alternative licenses represented the reification of shared values, or a code of conduct accepted by the community. The scope of license is limited to releasing source code from the author's exclusive control under copyright law. Yet, as software makes up an essential layer of information and communication technologies, alternative licenses enable FOSS proponents to be less dependent on the socioeconomic structure that early proprietary companies sought to entrench with copyright. By crafting and adopting these licenses, the FOSS community provided sociopolitical critiques that targeted both conventional copyright law and proprietary business practices. One critique is there is room for negotiation within the existing legal institution. Copyright law, like other property law, may appear as a product of possessive individualism.[55] Yet copyright licenses may be crafted that relinquish authorial control and thus remove this artificial boundary dictated by legal code. A second critique is that strong copyright enforcement and protection as championed by proprietary software companies may encourage them to invest in developing more software, but overprotection can offset copyright's legislative purposes, as it encourages rent-seeking behavior and obstructs technological advancement.

Through collective practices enabled by these alternative licenses, FOSS advocates not only rebuilt a community but validated both of these critiques, thereby exercising a transformative impact on the broader industrial and political environment. Viewed against this context, it may be easier to understand why copyright licenses have claimed a unique status in FOSS developers' socio-technical practices. Indeed, even though copyright law alone cannot address power concentration in the cloud-computing market or broad ethical and social justice questions, new generations of open source companies and activism-oriented developers still try to hitch new codes of conduct to copyright licenses.

These newer licenses have gone beyond challenging the resource distribution under copyright law to the extent that they are rather seen as attempts to define the relationship between contributors and the behavior of subsequent users beyond the mere distribution of resources. But no matter whether the purpose is to stop free riding and extractivism, to defy capitalist logic, or to prevent human rights violations, the motivation behind both source-available licenses and ethical licenses is to regain at least some of the authorial control relinquished by FOSS licenses. Thus, ironically, both types of licenses accept the superior role of authors granted by copyright, rather than seeking to transform it. This fact partly explains why source-available licenses were immediately met with skepticism from the FOSS community, seen as potentially masking proprietary business strategies. On the other hand, activism-oriented developers employing ethical licenses may embrace a more genuine criticism of the proprietary model, but they are willing to leverage the copyright system and exercise authorial control in pursuit of what they see as more pressing goals and more fundamental freedoms that fall beyond the realm of copyright law.

The recent surge in generative artificial intelligence (GenAI) and the potential for extractive activities it has unleashed have extended or intensified licensing debates. Individuals who have used FOSS or similar licenses (e.g., Creative Commons licenses) to release their works are increasingly concerned that these licenses might inadvertently facilitate corporate extractivism, allowing companies to easily exploit their work for training GenAI models that could ultimately replace human jobs.[56] There is also a new trend among AI companies to embrace new "responsible AI" licenses that include extensive ethical clauses as a response to growing public anxiety that new technologies may be used for harmful purposes.[57] Compared to the previous round of licensing debates outlined in this chapter, more individual developers and artists are now voicing concerns about corporate extractivism. Conversely, more corporations are embracing ethical licenses as a way to endorse "responsible AI" models. A full discussion of these recent developments is beyond the scope of this chapter. It suffices to say that as new technological advancements in AI continue to reshape power dynamics and introduce fresh concerns, reasserting authorial control with copyright licenses is gaining renewed support.

FOSS developers are not alone in wishing to avoid being seen as exploitable resources or inadvertently aiding malevolent purposes. In a world where mundane digital traces of daily activities can be exploited as data resources enabling tools for manipulation or surveillance, many have highlighted an increasing concentration of market power in information technologies and the growing socio-technical power in the hands of data cartels.[58] Many have championed stronger regulatory intervention such as antitrust enforcement.[59] Software developers may be better positioned than the general public not only to offer a forceful critique of the existing material reality but also to develop (elements of) alternative infrastructures. From this perspective, the relicensing episodes illustrated in this chapter are not merely peculiar practices of a subculture but part of a larger effort to shape the socio-technical future. It is essential to acknowledge, however, that copyright may not be a perfect tool for combating extractivism or to prohibit unethical behavior. Copyright law includes mechanisms, such as fair use, that seek to balance different interests and does not grant authors absolute power to prevent all unwanted subsequent uses. Even if we leave aside the question of whether copyright licenses can effectively achieve these goals, the real challenge lies in whether a community pact can reach sufficient critical mass to establish a new consensus in the fast-moving technological and economic landscape and lay the foundation for a new recursive public.

Notes

1. Jamie Kyle (@jamiebuilds) on Lerna issue tracker, "Add text to MIT License banning ICE collaborators," no. 1616, GitHub, August 29, 2018, https:// github.com/lerna/lerna/pull/1616. For more on the MIT license, see "Exploring the MIT Open Source License: A Comprehensive Guide," MIT Technology Licensing Office, accessed February 3, 2025, https://tlo.mit.edu/understand-ip /exploring-mit-open-source-license-comprehensive-guide.

2. Daniel Stockman (@evocateur) on Lerna issue tracker, "Restore Unmodified MIT License," no. 1633, GitHub, August 30, 2018, https://github.com/lerna /lerna/pull/1633.

3. "What Is Free Software?" Free Software Foundation, GNU Operation System, accessed July 20, 2021, https://www.gnu.org/philosophy/free-sw.en .html.

4. Except where the discussion refers to only one of the two camps, this paper uses FOSS, the umbrella term.

5. E. Gabriella Coleman, *Coding Freedom: The Ethics and Aesthetics of Hacking* (Princeton, NJ: Princeton University Press, 2012).

6. E. Gabriella Coleman, "The Political Agnosticism of Free and Open Source Software and the Inadvertent Politics of Contrast," *Anthropological Quarterly* 77, no.2 (Summer 2004): 507–19.

7. Christopher M. Kelty, *Two Bits: The Cultural Significance of Free Software* (Durham, NC: Duke University Press, 2008).

8. Gerardo Con Díaz, *Software Rights: How Patent Law Transformed Software Development in America* (New Haven, CT: Yale University Press, 2019), 91.

9. Christopher Tozzi, *For Fun and Profit: A History of the Free and Open Source Software Revolution* (Cambridge, MA: MIT Press, 2017), 64, 84.

10. Richard M. Stallman, "What Is the Free Software Foundation?" *GNU's Bulletin* 1, no. 1 (February 1986): 8–9, https://www.gnu.org/bulletins/bull1.txt.

11. Coleman, *Coding Freedom*, 69–70.

12. Shun-Ling Chen, "To Surpass or to Conform—What Are Public Licenses For?," *Journal of Law, Technology, and Policy* 2009, no. 1 (2009): 107–39.

13. Eben Moglen, "A History of the GPL v3 Revision Process," Directorate General for Internal Policies, Policy Department C: Citizens' Rights And Constitutional Affairs, European Parliament (2013), Software Freedom Law Center, accessed December 13, 2024, https://www.softwarefreedom.org/resources/2013 /A_History_of_the_GPLv3_Revision_Process.pdf.

14. Tozzi, *For Fun and Profit*, 263.

15. Eric S. Raymond, "Goodbye, 'Free Software'; Hello, 'Open Source,'" February 8, 1998, catb.org, http://www.catb.org/esr/open-source.html.

16. Eric S. Raymond, "The Halloween Documents," catb.org, accessed April 30, 2021, http://www.catb.org/~esr/halloween/index.html (see Halloween I, VII).

17. Shun-Ling Chen, "Exposing Professionalism in United States Copyright Law: The Disenfranchised Lay Public in a Semiotic Democracy," *University of San Francisco Law Review* 49, no.1 (2015): 57–121.

18. "Smartphone Market Share," IDC (International Data Corporation), updated April 28, 2021, https://www.idc.com/promo/smartphone-market-share /os; "Historical Trends in the Usage Statistics of Web Servers," W3Techs, accessed Aug 1, 2021, https://w3techs.com/technologies/history_overview/web _server; "Usage Statistics of Content Management Systems," W3Techs, accessed Aug 1, 2021, https://w3techs.com/technologies/history_overview/content _management.

19. Gregg Keizer, "Top Web Browsers 2020: Edge Makes Double Digits," *ComputerWorld*, November 2, 2020, https://www.computerworld.com/article /3199425/top-web-browsers-2020-chrome-takes-a-punch-firefox-stays-alive .html; George Anadiotis, "The State of Open Source Databases in 2019: Multiple Databases, Clouds, and Licenses," *ZDNet*, October 1, 2019, https://www.zdnet .com/article/the-state-of-open-source-databases-in-2019-multiple-databases -clouds-and-licenses/.

20. Cade Metz, "Google Just Opened Sourced TensorFlow, Its Artificial Intelligence Engine," *Wired*, November 9, 2015, https://www.wired.com/2015 /11/google-open-sources-its-artificial-intelligence-engine/; John Mannes, "Facebook Open Sources Caffe2, Its Flexible Deep Learning Framework of

Choice," *TechCrunch*, April 19, 2017, https://techcrunch.com/2017/04/18/face
book-open-sources-caffe2-its-flexible-deep-learning-framework-of-choice/;
Allison Linn, "Microsoft Releases CNTK, Its Open Source Deep Learning
Toolkit, On GitHub," *Microsoft: The AI Blog*, January 25, 2016, https://blogs
.microsoft.com/ai/microsoft-releases-cntk-its-open-source-deep-learning
-toolkit-on-github/.

21. Mike Volpi, "How Open-Source Software Took Over the World,"
TechCrunch, January 13, 2019, https://techcrunch.com/2019/01/12/how-open
-source-software-took-over-the-world/.

22. "Microsoft CEO Takes Launch Break with the *Sun-Times*," *Chicago
Sun-Times*, June 1, 2001, https://web.archive.org/web/20011211130654/http://
www.suntimes.com/output/tech/cst-fin-micro01.html.

23. Steven J. Vaughan-Nichols, "Top Five Linux Contributor: Microsoft,"
ZDNet, July 17, 2011, https://www.zdnet.com/article/top-five-linux-contributor
-microsoft/.

24. Tozzi, *For Fun and Profit*, 245.

25. Klint Finley, "Microsoft Calls a Truce in the Linux Patent Wars," *Wired*,
October 11, 2018, https://www.wired.com/story/microsoft-calls-truce-in-linux
-patent-wars/.

26. Phillip Tracy, "Microsoft's GitHub Purchase Sends Talent to Rival
GitLab," *Daily Dot*, June 4, 2018, https://www.dailydot.com/debug/microsoft
-github-gitlab/; Simon Bisson, "GitHub after Microsoft: How It Has Changed,"
InfoWorld, January 22, 2019, https://www.infoworld.com/article/3335256
/github-after-microsoft-how-it-has-changed.html. GitHub's Copilot, a cloud-
based artificial intelligence tool codeveloped with Open AI released in 2021, was
contested by FOSS developers for using software hosted on GitHub to train its
machine learning engine. A class action lawsuit against GitHub and Microsoft
was filed in 2022 for a variety of infringements, including the violation of FOSS
licenses. This controversy is beyond the scope of this chapter but will be briefly
addressed in its last section.

27. Steven J. Vaughan-Nichols, "Open Source Has Won, and Microsoft Has
Surrendered," *ComputerWorld*, November 28, 2016, https://www.computer
world.com/article/3144063/open-source-has-won-and-microsoft-has-surren
dered.html.

28. Aditya Tiwari, "Richard Stallman Went to Microsoft's Office—Here's
What Happened," *FossBytes*, September 6, 2019, https://fossbytes.com/richard
-stallman-microsofts-office/.

29. Tozzi, *For Fun and Profit*, 201–2.

30. Volpi, "How Open-Source Software Took Over the World."

31. "Open-Source Software Has Changed the Way Software Is Developed:
Here's Where the $33B Industry Is Headed," CB Insights, updated August 1,
2019, https://www.cbinsights.com/research/report/future-open-source/.

32. Dave Nielsen, "Redis Turns 10," *Redis Labs* (blog), February 26, 2019,
https://redislabs.com/blog/redis-turns-10/; W. Chen, "3rd Time Is a Charm!
Garantia Data Is Now Redis Labs," *Redis Labs* (blog), January 29, 2014, https://
redislabs.com/blog/3rd-time-is-a-charm-garantia-data-is-now-redis-labs/.

33. Heather Meeker, "Commons Clause: Understanding an Evolution in
Software Licensing," Redis Labs, accessed August 1, 2021, https://redislabs.com
/wp-content/uploads/2018/10/Commons-Clause-White-Paper.pdf. This document was prepared on behalf of Redis Labs and reflects its position.

34. "The Commons Clause," accessed August 1, 2021, https://commons
clause.com/.

35. Meeker, "Commons Clause." The Commons Clause approach caused

users some confusion. In 2018, Redis Labs dropped the approach and relicensed its modules under the Redis Source Available License. See Yiftach Shoolman, "Redis Labs' Modules License Change," *Redis Labs* (blog), February 21, 2019, https://redislabs.com/blog/redis-labs-modules-license-changes/.

36. MongoDB was developed by the software company 10gen, which changed its name to MongoDB Inc. in 2013.

37. "Server Side Public License FAQ," MongoDB, accessed August 1, 2021, https://www.mongodb.com/licensing/server-side-public-license/faq.

38. Eliot Horowitz to OSI license-review mailing list, "[License-review] Approval: Server Side Public License, Version 1 (SSPL v1)," October 25, 2018, http://lists.opensource.org/pipermail/license-review_lists.opensource.org/2018 -October/003789.html.

39. Eliot Horowitz to OSI license-review mailing list, "[License-review] Approval: Server Side Public License, Version 1 (SSPL v1)," October 16, 2018, http://lists.opensource.org/pipermail/license-review_lists.opensource.org/2018 -October/003603.html.

40. "The Open Source Definition," Open Source Initiative, ver. 1.9, last modified March 22, 2007, https://opensource.org/osd.

41. Steven Weber, *The Success of Open Source* (Cambridge, MA: Harvard University Press, 2004), 184.

42. Richard Stallman, "On the Netscape Public License (Original Version)," GNU Operating System, March 12, 1998, https://www.gnu.org/philosophy /netscape-npl-old.html.

43. Jonathan Corbet, "Who Wrote 3.0—From Two Points of View," *LWN. net*, July 13, 2011, https://lwn.net/Articles/451243/.

44. Klint Finley, "How GitHub Is Helping Overworked Chinese Programmers," *Wired*, April 4, 2019, https://www.wired.com/story/how-github-helping -overworked-chinese-programmers/.

45. "Readme.md," 996.ICU, updated October 20, 2019, https://github.com /996icu/996.ICU; Klint Finley, "How GitHub Is Helping Overworked Chinese Programmers."

46. M. J. Wolf, K. W. Miller, and F. S. Grodzinsky, "On the Meaning of Free Software," *Ethics and Information Technology* 11 (2009): 281–82.

47. Stephen Shankland, "'Don't-be-evil' Google Spurns No-Evil Software," *CNET*, December 28, 2009, https://www.cnet.com/news/dont-be-evil-google -spurns-no-evil-software/.

48. Jack Edge, "Apache and the JSON License," *LWN.net*, November 20, 2016, https://lwn.net/Articles/707510/.

49. Kyle, "Add text to MIT License banning ICE collaborators."

50. Klint Finley, "An Open Source License That Requires Users to Do No Harm," *Wired*, October 4, 2019, https://www.wired.com/story/open-source -license-requires-users-do-no-harm/.

51. Patrick Klepek, "The Path to Destroying Capitalism Might Go through a Software License," *Vice*, August 26, 2020, https://www.vice.com/en/article /5dzam3/the-path-to-destroying-capitalism-might-go-through-a-software -license.

52. Caroline Ada Ehmke (@CarolineAda), Twitter, February 18, 2020, https://x.com/CoralineAda/status/1229456857317683201.

53. "Ethical Source: Open Source, Evolved," OES: Organization for Ethical Source, accessed August 1, 2021, archived at Internet Archive Wayback Machine https://web.archive.org/web/20210727234614/https:/ethicalsource.dev/.

54. "The Anti-Capitalist Software License," accessed August 1, 2021, https:// anticapitalist.software/.

55. C. B. Macpherson, *The Political Theory of Possessive Individualism: Hobbes to Locke* (Oxford: Clarendon Press, 1962), Wendy J. Gorden, "A Property Right in Self-Expression: Equality and Individualism in the Natural Law of Copyright," *Yale Law Journal* 102, no. 7 (1993): 1503–1609.

56. See, for example, Eryk Salvaggio, "A New Contract for Artists in the Age of Generative AI," Tech Policy Press, August 25, 2023, https://techpolicy.press/a-new-contract-for-artists-in-the-age-of-generative-ai/.

57. See, for example, Carlos Muñoz Ferrandis, "OpenRAIL: Towards Open and Responsible AI Licensing Frameworks," *Hugginface* blog, August 31, 2022, https://huggingface.co/blog/open_rail.

58. See, for example, Nick Couldry and Ulises A. Mejias, *The Costs of Connection: How Data Is Colonizing Human Life and Appropriating It for Capitalism* (Stanford, CA: Stanford University Press, 2019), Sarah Lamdan, *Data Cartels: The Companies that Control and Monopolize our Information* (Stanford, CA: Stanford University Press, 2022).

59. See, for example, Tim Wu, *The Curse of Bigness: How Corporate Giants Came to Rule the World* (London: Atlantic Books, 2020), Jonathan B. Baker, *The Antitrust Paradigm: Restoring a Competitive Economy* (Cambridge, MA: Harvard University Press, 2019).

6

The Great E-book Conspiracy

Gerardo Con Díaz

On January 27, 2010, while announcing the launch of the iPad, Steve Jobs introduced the iBookstore.[1] He told an eager audience that Amazon had done a great job of pioneering e-book functionality with its Kindle, and that Apple is "going to stand on their shoulders and go a bit further." Jobs tapped on the iPad's screen, which displayed a digital bookshelf—an image of a wooden bookshelf with thumbnail-sized book covers arranged like magazines in a stand. Jobs then tapped on a button on the upper left corner, causing the shelf to rotate on its central vertical axis and reveal the fully integrated iBookstore. He described this as "kind of like a secret passageway" from the shelf to a full digital library. It was a seamless integration of reading, touching, and purchasing.

The digital bookshelf showed the covers for highly popular trade books, including Kathryn Stockett's *The Help*, Stephenie Meyer's *Eclipse* (from the *Twilight* series), Thomas Friedman's *Hot, Flat, and Crowded*, and Ted Kennedy's *True Compass* (which Jobs purchased as part of his demo). Jobs announced that books on *New York Times* best-seller lists would be available for purchase and that Apple "got five of the largest publishers in the world that are supporting us in this and are going to have all their books on the store." What he didn't tell the audience was that this announcement was the culmination of several months of aggressive negotiations among Apple, Amazon, and the publishers whose books Jobs was spotlighting. In fact, just a few days earlier, some of the contracts that allowed him to make these promises still had not been signed.

At the heart of these negotiations was one question: How much should an e-book cost? Amazon, whose Kindle had arguably launched an e-book revolution, was charging $9.99 for any e-book. Large publishers were deeply dissatisfied with this pricing model, in part because it represented a drastic reduction of their profit margins, especially for best-selling hardcovers and illustrated volumes such as cookbooks. This dispute created an opportunity for one of Apple's most celebrated negotiators, Eddy Cue, to enlist several major publishers in the company's iBookstore. Through Cue, Apple promised publishers a popular new major platform that centralized e-book purchases through Apple's latest sensation in hand-held devices, the iPad. Indeed, Cue was confident that reading a book on an iPad would be desirable enough for users to pay more than Amazon's $9.99.

This chapter investigates Apple's negotiations with publishers on the eve of the iPad's launch. It argues that the early development of e-books as consumer goods hinged on the outcome of an industry-wide battle between two pricing models: the traditional wholesale model that retailers used for physical books (and that Amazon had adapted for use with the Kindle) and the agency model, which priced e-books using the principles that guided Apple's pricing of apps and songs. Variations of this narrative are scattered across the literature on antitrust law (the field of law designed to preclude firms from engaging in anticompetitive behavior) because a US district court ruled in 2013 that Apple had conspired with publishers to raise the price of e-books. Judge Denise Cote, who oversaw the case, found "overwhelming evidence" of this conspiracy. Apple appealed all the way to the Supreme Court, which declined to review the case in 2016. Since then, legal commentators have analyzed Apple's dealings in great detail in the service of critiques (or celebration) of governmental interference in industry affairs.[2]

This argument and narrative show how the study of pricing can open inquiry into the emergence of new technologies as distinct commercial and legal entities. This analysis decenters strict economic readings of prices as functions of supply and demand and favors their study as boundary objects in industrial ecosystems—that is to say, prices can mean very different things to the many actors who encounter them. Certainly, customers may see them as an expenditure of capital required to access a good or service; and firms may see their own prices as the outcome of the coordinated efforts in their research, development, and marketing divisions. However, prices can be many other things: tools to structure a market for a new kind of product, coordinating mechanisms that create or challenge industry standards, and even legal evidence that courts can use to recalibrate the power dynamics of an entire market sector. To publishers negotiating with Apple, Amazon's $9.99 price was a rallying call to pitch Apple against the world's most powerful online retailer.

This chapter also illustrates how legal archives can help scholars bypass the obstacles they encounter when dealing with more recent firms in which written communications occurs through email. In the case of Apple, court records granted me access to the company's internal market research, email correspondence between Jobs and his executives, draft and annotated contracts, private exchanges between publishing executives negotiating with the company, and even handwritten notes created in haste during meetings. Many of these materials were introduced into the public record by the District Court of the Southern District of New York in the course of investigating the United States Department of Justice's claim that Apple was engaging in a conspiracy. Such archives are constructed through the legal battles from which they emerge, but their careful and legally informed use allows us to examine decision making and strategy in firms that would normally not make their records available for research.

The story that follows is arranged chronologically. It shows how Apple's executives developed specific visions for their company's entry into publishing, negotiated with publishing executives, and ultimately deployed a strategy to counter Amazon's newfound control of the electronic publishing industry. The aim of this story is not to weigh in on the legal merits of the case but to showcase how a small group of powerful people negotiated aggressively at lightning speed to rebalance market power in this emerging industry. The conclusion reflects on the opportunities and methodological subtleties involved in using legal archives to write business histories like this one.

Against Amazon

By 2009, Eddy Cue, a half-Cuban product developer, had become crucial to Apple's expansion into online marketplaces. Since joining the company in 1989, the computer scientist and economist had steadily climbed the corporate ladder by launching flagship projects like the Online Store (1998), the iTunes Store (2003), and the App Store (2008). He was a talented and aggressive negotiator, instrumental to Apple's deals with movie studios, television networks, record labels, and app developers. With the App Store up and running on iPhones and iPod Touch devices, Cue set out to develop a new online store for use with Apple's latest device: the iPad, which Jobs would announce in a few months, on January 27, 2010.[3]

Apple's executives hoped that the iPad would solidify the company's reputation as a maker of multipurpose portable devices. Other portable devices at the time (e.g., MP3 players and Kindle e-readers) were designed with a single purpose in mind. With the right apps, however, the iPad could replace all single-purpose devices, especially e-readers and music players. The iPad's most marketable features—large color touch screen, wireless and optional cellular connectivity, camera, and microphone— would transform the device into a one-stop shop for users' multimedia needs. The iPad would be both a continuation and an advancement of the features that Apple celebrated in the iPhone and iPod, and an opportunity for Apple to pursue new opportunities in online marketplaces.

Cue was especially interested in having the iPad replace the Kindle as the top e-reader in the market, but Jobs had objected to the development of an e-book store for several months on the grounds that Apple's product selection would not offer users a reading experience distinct enough to motivate users to adopt Apple as their e-book provider.[4] However, Cue and his team of product managers produced a confidential research report in early 2009 suggesting that Jobs's objection might be unfounded. The report noted that "consumers don't want multiple devices" and that readers were already purchasing books through Apple.[5] In fact, the team estimated that more than 3 million unique users had downloaded at least one e-reading app through iTunes (totaling 12 million e-book downloads).

The report had two important implications. First, Apple already had a large customer base for e-books. In fact, users were willing to sacrifice ease of access to their e-books in order to read on an Apple device. One of the Kindle's most celebrated features was that users could access hundreds of thousands of titles through a single interface. In contrast, iTunes allowed users to purchase books in two ways: through individual apps dedicated to one book each, or through e-reading apps that offered access to a relatively small collection of books. Users would then have to read their books either on a computer or on an iPhone's small screen. Second, app developers that served as intermediaries between readers and book publishers were taking a portion of the profit from e-book sales. These "unnecessary middlemen" were profiting from e-book sales not because they were authors or publishers but because they had created the apps that allowed users to access e-books. This meant that there was an opportunity for Apple to profit if it centralized all e-book purchasing through a single app developed in-house.[6]

Cue saw an opportunity for immense profit. His team estimated in February 2009 that the 2008 book-publishing market was worth $43 billion dollars. This was more than the music ($7 billion) and movie ($32 billion) industries combined.[7] Audiobooks and e-books constituted just over $1 billion of the publishing market. Cue believed that this amount was not higher for three reasons: consumers were unwilling to purchase more single-purpose electronic devices; the Kindle's price was too high and its design was not appealing enough; and the inventory of e-books available for purchase was still too small. If Apple entered the industry, however, it could potentially fix these three problems and cause yearly e-book revenues to skyrocket, as shown in figure 6.1.

Cue was certain that Apple users would flock to an e-book store, but such a store would succeed only if users could access a very large collection of books in whatever genre they desired. Meeting this demand would require Apple to sign deals with as many publishers as possible, especially the "Big Six" houses: Penguin, McMillan, HarperCollins, Simon & Schuster, Hachette, and Random House. The publishers Cue was talking about were the Big Six. The top five among them controlled more than 50% of the market (as shown in figure 6.2), and they expected that their digital

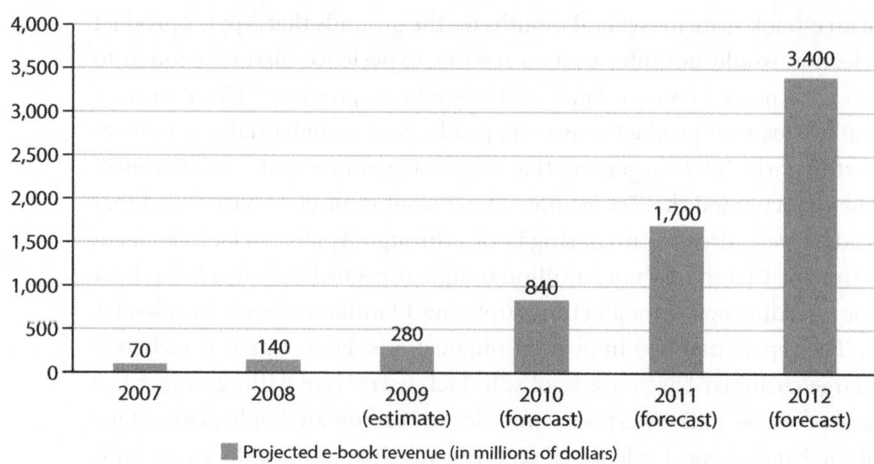

Figure 6.1. Growth projections for e-book revenues, according to Apple's confidential 2009 report, in millions of dollars. The 2009 estimate is based on data compiled by the Association of American Publishers. The 2010–2012 forecasts assume that Apple entered the industry.

sales would double in the coming year even without Apple's participation, from 2% of their total sales to 4.2%.[8]

While pursuing these deals, Apple would have the advantage of being a latecomer in the e-book market. Since the introduction of the Kindle in 2007 (when the first generation of devices sold out five and a half hours after their release), the online retailer had become the leading e-book seller in the world.[9] However, publishers were deeply dissatisfied with Amazon's pricing strategy for e-books, which was an adaptation of the retailers' extensive experience selling physical books. The company used what commentators would later call the wholesale model for sales: in exchange for a higher-volume purchase (as determined by physical sales projections and, later, digital sales records), Amazon would receive a lower price-per-unit from publishers. On top of this, Amazon capped e-book prices at $9.99, selling them at a loss in exchange for its dominance as the number one retailer. Other retailers, like Barnes & Noble, would match this price, which risked causing readers to expect to pay $9.99 for all e-books.

Publishers were looking for ways to stop Amazon from enforcing these price caps. At Penguin, a confidential three-year outlook plan predicted that "competition for the attention of readers will be most intense from digital companies whose objective may be to disintermediate traditional publishers altogether." The report warned: "This is not a new threat but we do appear to be on a collision course with Amazon, and possibly with Google as well. It will not be possible for any individual publisher to mount an effective response, because of both the resources necessary and the risk of retribution, so the industry needs to develop a common strategy."[10] These alliances were becoming appealing among the Big Six, especially now that Amazon was considering launching a Kindle 2 device. For example, Hachette's CEO, Arnaud Lagardere, emailed one of his executives to say that the world's major publishers were eager "to create an alternative

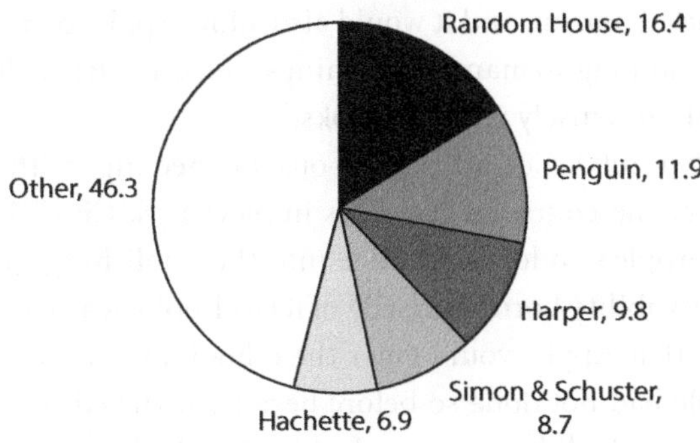

Figure 6.2. Market share (%) of the five biggest publishers, circa 2009, as reported to Eddy Cue during his initial meetings with publishing executives.

platform to Amazon for ebooks." Their strategy would not necessarily be to compete with Amazon but to "force it to accept a price level higher than 9.99, and also to favor the emergence of standards more open than the Kindle."[11] That "wretched $9.99 price point," Lagardere would later tell another publisher, is "becoming a de facto standard." Similarly, Simon & Schuster's CEO, Carolyn Reidy told Leslie Moonves (CEO of its parent company, CBS) that "unless other publishers follow us, there's no chance of success in getting Amazon to change its pricing practices."[12] She and Moonves agreed that "without a critical mass behind us, Amazon won't 'negotiate,'" but she did not know just how many publishers would join the fight. "We will keep thinking of how to attack the problem . . . of current e-book pricing," she wrote. "But clearly we need to gather more troops and ammunition first!"

While the publishers looked for ways to raise Amazon's prices, Cue was brainstorming possible strategies for competing with Amazon in the e-book market.[13] He told Jobs that Apple could treat Amazon "like any other similar developer" and cultivate a commercial relationship with the company. Perhaps Amazon could pay Apple some percentage of all sales through the Kindle app for Apple devices. As another option, Apple could develop an e-reader for the Kindle, charging an affiliate fee like the ones that Amazon used for products sold through ads placed on third-party websites. Or perhaps Apple and Amazon could become exclusive resellers for each other; Amazon could be the exclusive reseller for iTunes, and Apple could be Kindle's. This would not necessarily generate immediate monetary gains, but Cue told Jobs that if Amazon executives "really value books and want to own the category moving forward[,] maybe they would consider it."

Cue was confident that publishers' resentment of Amazon would make it easier for Apple to strike deals with them. In February 2009, he told Jobs that "it would be very easy for us to compete and I think trounce Amazon by opening up our own e-book store. The book publishers would do almost anything for us to get into the e-book business."[14] Getting all of them to sign with Apple would give each of them leverage to renegotiate their deals with Amazon, and it would also allow Apple to announce that the iPad was—among so many other things—an e-reader with access to a vast catalog of immensely popular books.

In December 2009, Cue set up one-on-one meetings with the Big Six publishers over the course of two days in New York City.[15] His strategy was to pitch Apple's sudden entrance into the publishing industry as a last-minute possibility born primarily of its technological achievements.[16] He told them that Apple would enter the e-book market within 90 days and that Apple had not done so before because it lacked the "machines" required to do so. He did not say much about the iPad, other than describing it generically as a web-enabled device with a color touchscreen. He also told the publishers that Apple wanted to have a low price point for e-books and that it would never sell at a loss.[17]

Cue learned from these meetings that publishers were eager to have their e-books treated like other forms of digital media in Apple's stores. They showed interest in the FairPlay rights system (which allowed users to preselect devices on which media could be played or displayed), the use of a single app for all e-books, preordering capabilities, users' ability to redownload their purchases, and Apple's offer to provide more thorough sales reports than Amazon. Publishers also expressed willingness to give Apple what a member of Cue's team called "good, rich metadata" that they would update weekly, a large collection in underserved categories in the digital market (including cookbooks, travel guides, and children's books), and entire first chapters as samples.[18]

The publishers were genuinely excited about using deals with Apple to counter Amazon's pricing strategies. The interest at Hachette was so strong that its CEO emailed Cue soon after the meeting to say he was "really just thrilled to know" about Apple's plans. "As I understand our business interests are very much aligned," he wrote, "I can tell you I will save no effort to help Apple turn the launch into a fantastic success!"[19] At Simon & Schuster, Reidy told Moonves, "Terrific news! . . . [Apple] cannot tolerate a market where the product is sold significantly more cheaply elsewhere, . . . i.e., they don't want Amazon's $9.95 to continue."[20] A HarperCollins executive told one of his colleagues, "It all seemed to go too well. I thought I was dreaming."[21] Harper's internal report on the meeting noted that Apple was "serious about making books cool for the iTunes generation and expect to make up a significant percent of our market share in a very short timeframe."[22]

Back at Apple, Cue felt optimistic about his strategy. After completing half the meetings, he told Jobs, "Nothing scared me or made me feel like we can't get these deals right away. Clearly the biggest issue is new release pricing and they want a proposal from us. Everyone was ecstatic to see Apple and what it could mean for their industry. . . . Amazon is definitely not liked much because of selling below cost for NYT Best Sellers [because it is] not being upfront on their deal with them."[23] Cue also noted that the publishers had expressed great dissatisfaction with Amazon's text-to-speech feature (which Amazon had introduced without notice, threatening the audiobook market) and book-lending programs through devices like Barnes & Noble's Nook.

To move forward, Apple would now have to figure out its own e-book pricing structure, along with the many other legal and commercial details of the deal. Two days after the meetings, Cue received an email from Keith Moerer, the managing director of Apple's Worldwide Podcasting, who would soon direct the company's iBooks program. Moerer reported that all publishers "agreed that digital prices should be less than physical—20% or 30% being the commonly expressed range." This meant that a $9 digital price for *New York Times* best sellers that sold for $12 in physical form "is both appropriate and should be acceptable for everyone." Perhaps, Moerer wondered, Apple should just "adopt an across-the-board 25% discount off physical wholesale for all books."[24]

A few days later, Cue told Jobs that Moerer's proposal was unlikely to work as a universal pricing strategy because it was less than publishers wanted, especially for expensive best sellers such as the Harry Potter series and Stephen King novels.[25] It was time to go back to the drawing board and develop a new pricing mechanism for iBooks. However, Moerer, Cue, and Jobs had another pricing model at their disposal: the agency model, which the app store used, wherein the maker of a product sets the price for the consumer and Apple takes a hefty royalty: 30%. It seemed likely that the publishers would strike a deal with Apple, but in the process they would need to examine just how much pricing control they were willing to give a newcomer to their industry. In the words of a Hachette executive, "The devil will be in the detail around terms."[26]

Agent Apple

On December 22, 2009, Cue contacted all the publishers to propose that they adopt the agency model in their iBook Sales. He called Reidy, for instance, to present the agency model as the "better way to do it" and propose the 30% royalty.[27] He added that e-book prices would generally range from $0.99 to $12.99, while specialized books or those with color in them could sell for more. According to Cue, this was the only way of dealing with the market's pricing "craziness." The lower end of that range was shockingly low, but Cue suggested that Apple would not want to offer discounts directly to consumers, as Amazon would. "You are doing the discounting," Cue told Reidy; "you are directly offering the consumer a bargain."

Soon after the new year, Cue emailed the publishers to further delineate what he called the "principal agency model."[28] Apple would sell publishers' products as an agent for their account. If publishers chose to dis-

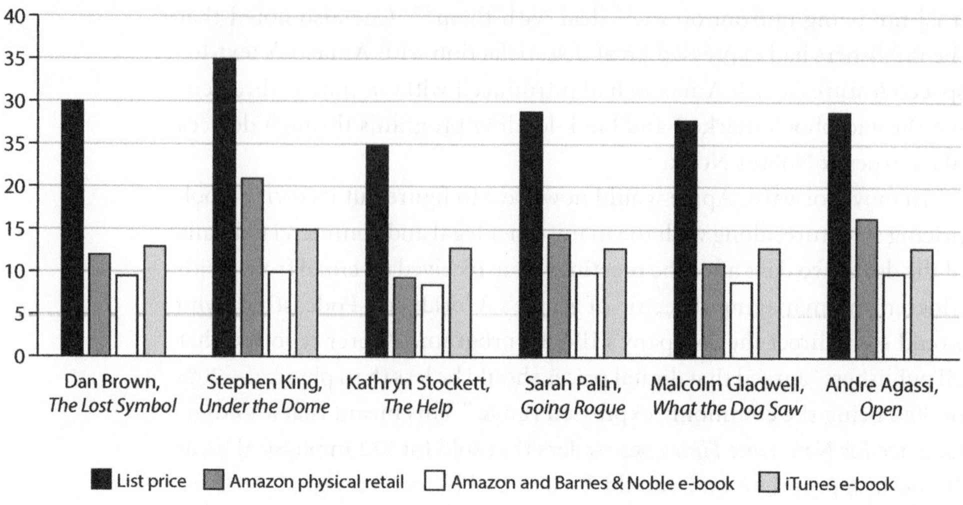

Figure 6.3. List prices (US dollars) for physical books, compared with their physical and e-book retail prices on Amazon and iTunes and Apple's proposed price for them.

tribute an e-book, then they would need to make it available on iTunes, though they had the freedom to offer it through other distributors as well. The $0.99 to $12.99 price range would apply to books sold in physical form for $35, in dollar increments as determined by the publisher. More expensive books would retail starting at $14.99 using $5 increments, up to 50% of the retail price for a physical book. In his email exchanges, Cue also clarified that Apple would want these terms to apply across the board with all of a publishers' e-book distributors, and he strongly opposed higher price points for hardcovers that retailed under $35.[29]

Figure 6.3, based on Apple's internal research records, shows the actual prices for a few *New York Times* (*NYT*) best sellers, alongside Apple's proposed price for them.[30] Five of the Big Six seemed somewhat willing to work with Apple's proposal even though they wished to see higher prices overall. At Random House, however, Madeline McIntosh (the publisher's number two executive) worried that Apple's terms would simply not generate enough profit.[31] McIntosh wanted Apple to use the agency model only for new releases (as opposed to their backlist of previously published books) and to create more price tiers. Cue was somewhat receptive to the first of these, but he was adamantly opposed to the second one. When Moerer asked him if he'd be willing to read McIntosh's proposal for the new tiers, Cue wrote, "No, I don't want a proposal. This is our offer. I am willing to add $14.99 for above $30. We need to start being very firm on price."

On January 11, 2010, Kevin Saul, an attorney in Apple's Legal Affairs Department, sent the formal e-book distribution agreement to the publishers.[32] For the most part, the agreement formalized the discussions that Cue had had with the publishers. It also featured a clause, which Saul had composed, guaranteeing that Apple would always have the lowest price for a new-release e-book.[33] The clause read: "If, for any particular New Release in hardcover format, the then-current Customer Price at any time is or becomes higher than a customer price offered by any other seller ("Other Customer Price"), then Publisher shall designate a new, lower Customer Price to meet such lower Other Customer Price."[34]

This is an example of what lawyers call a most favored nation (MFN) provision, a term that originated in international relations: members of the World Trade Organization cannot discriminate among their trading partners, so a country offering a special treatment to another must offer the same treatment to all countries.[35] In situations such as Apple's, lawyers can write an MFN clause that replace this equal-treatment-for-all approach with an agreement guaranteeing that their client will always be the most favored—that is, that they will receive conditions equal to or better than all other parties in similar positions.[36]

Saul believed that an MFN clause tying Apple's e-book pricing to physical books was the only way to ensure that Apple would remain price competitive in the long run. He worried that since Apple would not be the one setting prices, it would be very easy for publishers to set iBookstore prices higher than those found at other retailers. It would also be possible for physical-book sellers who purchased at wholesale prices to sell books for less than the iBookstore would sell the e-book versions of it. This was especially important when dealing with Amazon, which was willing to sell physical copies at a loss to remain the dominant retailer. Saul would later summarize his reasoning: "By ensuring that Apple would be able to match the lowest prices on the market for new release titles, Apple would be price competitive regardless of whether a publisher conducted business with other retailers on a wholesale or on an agency model."[37]

By the next day, Cue and Moerer recognized that they would need to be more flexible with their pricing tiers, especially because Apple was unwilling to take less than a 30% royalty.[38] Executives at Pearson and Penguin told Moerer that they had "agonized" over the proposal and, for two reasons, couldn't see a way to move forward.[39] First, best-selling authors would never agree to the dip in per-book royalties that Apple's pricing would generate, and it was unlikely that promises to sell even more copies than usual would appease them. Second, Apple's pricing research was based on *NYT* best sellers, so its conclusions did not apply to most books in publishers' catalogs, which would never make that list. These less popular books would necessarily have lower sales, which meant that their authors would be especially skeptical about the idea of reducing the per-copy royalties they received.

Cue sent an email to Jobs seeking approval for a higher pricing structure that would "push [the publishers] to the very edge and still have a credible offering in the market."[40] Cue proposed that, while publishers would be free to choose lower prices if they wished, books listed for $30 in the *NYT* best seller list would not be sold for more than $12.99 on iTunes.

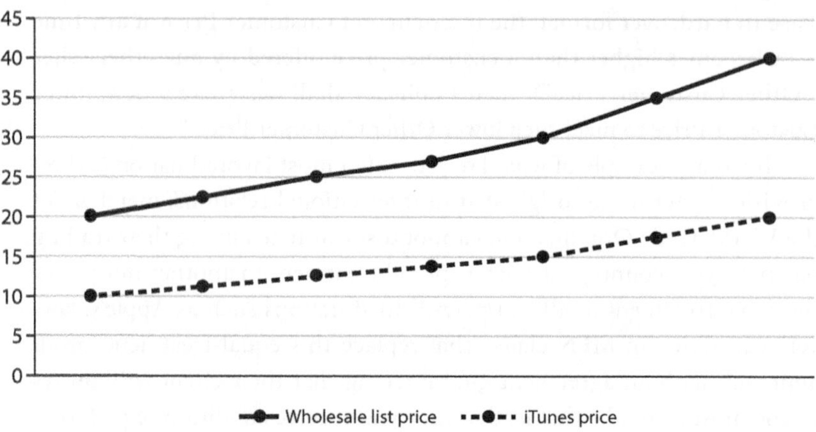

Figure 6.4 shows this new pricing structure. Jobs told Cue he could "live with this" as long as publishers "move Amazon to the agent model too for new releases for the first year." Otherwise, Jobs told Cue, "I'm not sure we can be competitive."

Cue presented this new pricing structure to the publishers while affirming Apple's unwavering stance on the 30% commission. "It is the margin we use with all music labels, tv studios, Hollywood studios, and app developers," he told each publisher. "We cannot compromise this without damaging those multi-billion dollar businesses."[41] Cue's email implied that the very existence of the iBookstore depended on whether Apple and the publishers could strike this deal within the week: "The most important goal for us is to build a book store that will be the biggest in the world (sales) as quickly as possible. Our time and resources are precious and we don't want to spend them unless we can build a real business (big revenues) for us as well as you. Our alternative is to let others build book store apps (like we already have from Amazon, Barnes & Noble, etc.) which gives our devices many book solutions. But we can do way better!"[42] This urgency, paired with Apple's unwillingness to compromise on its royalties, annoyed the publishers. At Simon & Schuster, for example, Reidy told Moonves that Apple was sticking to the 30% fee "with no give at all" and that Apple and Simon & Schuster "aren't that far apart, if they will listen/talk."[43]

For the next few days, publishers scrambled to decide whether to take Apple's deal. Macmillan's Brian Napack insisted that he would need time limits on Apple's ability to cap e-book prices, so that a book priced at $30 in wholesale would be offered for $14.99 just for the first three months. He also told Cue that Macmillan "would not agree to match other resellers prices"—that is, that it would not agree to the MFN—and that Apple would need to provide the publisher with "identity and transaction info of purchasers of our titles."[44] At Hachette, one executive told his colleagues that

Apple's deal "represents a significant loss in profit margin" and had the potential to further erode it by causing authors to demand higher royalties to make up for the drop in retail price.[45] Penguin executives were a little more receptive, though they worried that the new device was "an important unknown" about which they needed more information in making decisions.[46]

Ultimately, all these discussions of pricing were part of a much bigger project that all publishers were engaged in, even if they were working independently from one another: eliminating Amazon's $9.99 price cap. The publishers knew they shared this goal, but the records available for research suggest that firms had kept it to themselves during the negotiations with Apple. Their managers knew that contracts with Apple would give them leverage to negotiate with Amazon, especially if they could present their collective agreements as the first step in an industry-wide restructuring that Amazon would need to support to remain competitive. After all, Amazon could potentially retaliate against an individual publisher that demanded higher prices, but it would be harder for it to retaliate against all the biggest publishers at the same time. Publishers likely expected to begin these negotiations after signing with Apple, but a media leak made their time frame much shorter than they expected.

Amazon Strikes Back

On January 18, 2010, the *Wall Street Journal* published a short article titled "Publisher in Talks with Apple over Tablet."[47] Its author, a staff journalist named Jeffrey Trachtenberg, had heard from "people familiar with the situation" that HarperCollins and Apple were in negotiations "to make electronic books available for the introduction of a new tablet device." Trachtenberg did not have much more information. Still, he reported that these negotiations "represent a direct challenge to Amazon," which would face serious competition if Apple's "newest creation"—what he conjectured was a multimedia tablet—featured two capabilities the Kindle didn't have: color displays and video capabilities.

The next day, Michael Cader, the founder of *Publishers Lunch* (an industry news magazine) ran a story claiming that Thrachtenberg's article "is just a sliver of the story that is emerging, albeit changing all the time."[48] Cader reported that Apple's negotiations with the Big Six were confidential, far for complete, and unlikely to be wrapped up before the launch of its new device. Price negotiations aside, the publishers' goal was to change what it meant to sell an e-book: they sought to "change the basic selling terms of e-books with at least one major trading partner in a way that lets the publisher take back control of pricing and reassert their vision of the value of an electronic version of a book."

Amazon responded swiftly and forcefully by threatening to remove the publishers from the picture for lower-price books. On January 20, the company announced that authors who used the Kindle Digital Text Platform—Amazon's self-publishing tool—would receive a stunning 70% in royalties from the list price starting June 30, 2010.[49] This meant that an author selling a self-published e-book for $8.99 would earn $6.25 in royalties. According to Amazon's vice president of Kindle content, Russ Grandinetti, "Today, authors often receive royalties in the range of 7 to 15 percent of the list price that publishers set for their physical book, or 25 percent of the net that publishers receive from retailers for their digital books. . . . We're excited that the new 70 percent royalty option for the Kindle Digital Text Platform will help us pay authors higher royalties when readers choose their books."[50]

This new program applied only to e-books priced between $2.99 and $9.99, if their list price was at least 20% lower than the "lowest physical list price for the physical book" (therefore excluding Amazon's own price for the physical copy).[51] Public domain books were ineligible for the program, as were any sales made outside the United States and books that could not be adapted to Kindle's text-to-speech function.

This announcement caused publishers to consider Apple's proposed terms more seriously, if only to combat Amazon.[52] Apple's 30% royalty still seemed too high and its pricing tiers too low, but they were better than Amazon's new terms. The discussions at HarperCollins illustrate just how quickly publishers were willing to move despite the deal's sensitivity and potential high impact. Charlie Redmayne, the publisher's chief digital officer, told his colleagues, "I think we find ourselves in an invidious position here. Having Apple competing with Amazon is a good thing but we are being asked to take sides—and to take sides against our second largest global partner and to do so by the end of the week!! . . . Ultimately they need us to agree [to] this deal—their share price has rocketed on the back of a reading proposition—if they launch without a deal it will fall. This IS important to them and we need to leverage that." Redmayne didn't think that Apple's proposal was "a great deal" and questioned "whether it is a good enough deal to give us the stomach to take Amazon." Still, executives were looking for a way to justify accepting Apple's terms. For instance, Robert Zaffiris, a senior vice president in charge of corporate new-business development, was already pondering whether they should just "agree to the terms to move quickly (as they seem inflexible on terms)."[53]

After meeting with Amazon on January 20, HarperCollins CEO Brian Murray told senior executives that he was "inclined (barely) to bring Apple into the business [to] introduce the agency model" and that this model would have to operate "on terms that work for Amazon, B&N [Barnes & Noble], and consumers."[54] He added that the Big Six would be "better off with Apple in the business in a significant way than not in the business," and he implied that Amazon's move suggested that it was best to "get another major player in the business."[55] There were still several points that needed settling—especially commission caps, which would allow the publishers to offset the lower list prices that Apple was proposing.

The next day—less than a week before the iPad's announcement—Cue was confident that Simon & Schuster and Penguin would definitely come on board. Random House had definitively declined. Markus Dohle, Random House's CEO, told Cue that Apple's high commission, paired with the potential changes in pricing policies across all retailers that Apple's deal could cause, "would be detrimental to the growth of a healthy, long term digital model."[56] This meant that it was time to increase the pressure on the remaining publishers (Macmillan, HarperCollins, and Hachette). Cue told Jobs: I'm confident we have 2 even though not yet signed. I have given all others a drop dead of a 1–1 meeting tomorrow morning. They keep chickening out so I have to give them a real drop dead time or they won't make up their minds. If I get a no from them then I think you should call them to make a final attempt. In the end, they want us and see the opportunity we give them but they're scared to commit!" These three publishers' main concern continued to be the pricing. The Macmillan and Hachette CEOs had "legal concerns over the price matching," and HarperCollins's was so concerned about prices that he "just may not have the guts to do this."[57]

Over the next few days, the publishers slowly started to align with customized versions of Apple's deal. Cue and Apple attorneys exchanged annotated drafts of the agreements and quickly negotiated the finer points of the deal—the limits of the MFN, the time frame during which a recently published title would be deemed a new release, and the royalties (which several publishers wanted to cap at 10% for a limited time). Jobs himself wrote to James Murdoch at HarperCollins, where negotiations appeared to have screeched to a halt. After a back-and-forth that was not generating a deal, Jobs summarized the situation to Murdoch as having to choose one of three options:

1. Throw in with Apple and see if we can all make a go of this to create a real mainstream ebooks market at $12.99 and $14.99.
2. Keep going with Amazon at $9.99. You will make a bit more money in the short term, but in the medium term Amazon will tell you they will be paying you 70% of $9.99. They have shareholders too.
3. Hold back your books from Amazon. Without a way for customers to buy your ebooks, they will steal them. This will be the start of piracy and once started there will be no stopping it. Trust me, I've seen this happen with my own eyes.[58]

The publishers' continued resistance meant that meeting the January 27 deadline would require Apple to make a few concessions.[59] To begin with, Apple ultimately agreed to let HarperCollins take an author's books off iTunes if they object to having their books offered for sale there. The publisher also negotiated the option to create new business models such as subscriptions or book clubs and, most important, the creation of an MFN for HarperCollins that would ensure the publisher would be getting the best treatment possible.[60] By January 26, all five publishers had informally agreed to work with Apple, and Amazon was responding with announcements of Kindle-exclusive works from prominent and emerging authors alike.[61]

On the eve of the iPad's announcement, other publishers started preparing to participate in the coming industrial seismic shift. For example, Pearson's CEO, Marjorie Scardino, told her board to prepare for Apple's "long-awaited entry into the e-reader market," for major trade publishers' alliances with it, and for Pearson's upcoming opportunity "to change models for the future, starting with e-books." She explained that the wholesale and agency models could not "coexist very long," so Pearson would be telling its "re-selling middlemen" that they would be transitioning to the agency basis in the future. Scardino thought that fighting back against Amazon's terms by siding with Apple was important because Amazon had "been trying steadily to remove publishers and take over the whole e-book chain themselves," challenging Pearson's "physical book business" through aggressive discounts on digital content and "romancing agents and authors publishing physical books." This encroachment had generated an unsustainable situation for Pearson and had the potential to teach customers that all books are worth no more than $10 each. "This all may not cause even a flicker, or it may be a firestorm," Scardino wrote, "but I wanted you to be ready whatever happens."[62]

At the iPad's announcement in early 2010, Jobs seemed confident not just about publishers' desire to join but also about Amazon's prices rising. A journalist named Kara Swisher filmed him having the following exchange with journalist Walt Mossberg.

> *Mossberg:* Why should she buy a book for $14.99 on your device when she can buy one for $9.99 from Amazon or Barnes & Noble?
>
> *Jobs:* That won't be the case.
>
> *Mossberg:* You won't be $14.99 or they won't be $9.99?
>
> *Jobs:* The prices will be the same.
>
> [. . .]
>
> *Jobs:* Publishers are actually withholding their books from Amazon because they're not happy.[63]

Jobs's implication—that Amazon would be forced to raise prices—angered publishers in large part because their negotiations with Amazon had already gotten off to a rocky and unexpected start. "I cannot believe that Jobs made the statement," one Simon & Schuster executive wrote in an email. "Incredibly stupid."[64]

Despite the publishers' chagrin, Jobs's prediction helped to create unity among them. "Interesting that we did the Apple deal with no contact with other publishers," Macmillan's John Sargent would later write in an email, "yet when Jobs announced he had 5 on the agency plan things were clear."[65] Though brief, Jobs's presentation and his bold postevent commentary helped publishers see beyond their individual contracts with Apple and take aim at the broader industrial shift that they wanted. "The optics make it look like I stood alone," Sargent remembered, "but in the end I had no doubt that others would eventually follow.[66] Hachette's Nourry told Sargent, "I can ensure [*sic*] you that you are not going to find your company alone in battle," and a Penguin executive wrote to say, "Bravo."[67]

Publishers leveraged their contracts with Apple into an ultimatum for Amazon: either create a new agency relationship with higher prices, or at least five of the largest publishers in the world will delay the release of e-books with the retailer.[68] Macmillan became especially prominent at this stage because it was the first one to open formal negotiations with Amazon. The negotiations started from a difficult place: Amazon removed the "buy" buttons from the listings for several Macmillan titles and redirected some of them to a page that read: "We're sorry. The Web address you entered is not a functioning page on our site." Users reported that Macmillan titles from their wish lists had disappeared. According to one reader: "I asked about the missing 'Buy It Now' button and asked if Amazon was in some dispute with Macmillan. I was told that sometimes Amazon likes to promote the Amazon Marketplace."[69]

On January 31, Amazon accepted Macmillan's terms. As a press release addressed to its customers explained: "We want you to know that ultimately . . . we will have to capitulate and accept Macmillan's terms because Macmillan has monopoly over their own titles, and we will want to offer them to you even at prices we believe are needlessly high for e-books. Amazon customers will at that point decide for themselves whether they believe it's reasonable to pay $14.99 for a bestselling e-book."[70]

By March, Amazon had established agency relationships with the publishers. Its negotiators drove a hard bargain, but the publishers' coordinated actions backed the retailer into a corner. One of Amazon's negotiators would later testify that if he made a small concession to one of the publishers, the other ones would soon ask him for the same thing. This allowed each of the publishers to create an MFN-style relationship with the retailer and signal that Amazon's role was not so much negotiating individual deals as it was delineating its own rights in a new industrial environment. The retailer's negotiators were ruthless, of course, but their leverage quickly eroded—especially after industry analyses suggested that

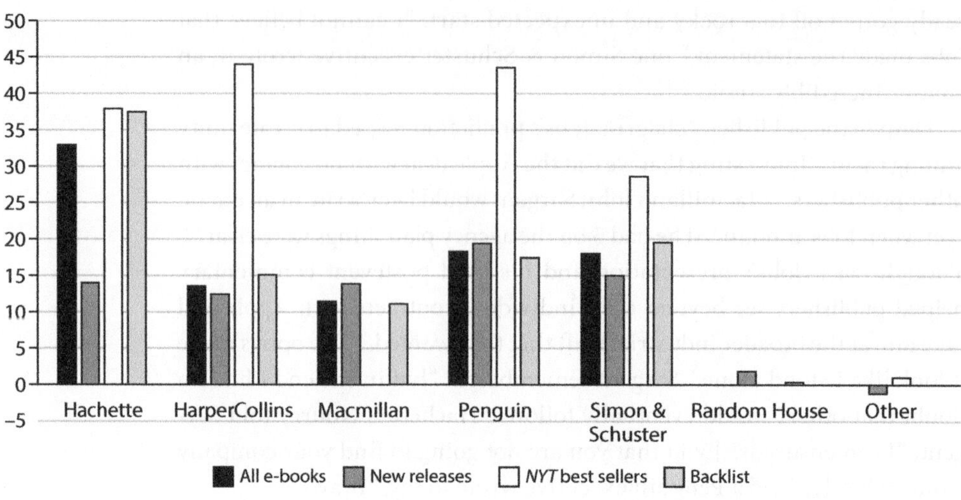

Figure 6.5. Weighted-average price increases of e-books on Amazon after signing the agency deal. Note that Random House (which did not sign an agency agreement in the period covered by this data) saw very little price increase and that all publishers other than the Big Six saw price drops.

readers could simply go to other online retailers (especially Barnes & Noble) if Amazon removed "buy" buttons in the course of negotiations.[71]

For the next few months, all publishers saw their e-book prices increase substantially. Figure 6.5 illustrate this situation, with weighted price increases of up to 44%. Publishers started signing agency deals with other publishers, and a new industry standard started to crystallize. By December 2010, the last major change required for this standard to take hold was having Random House adopt it: "When we get Random House," Cue told Jobs, "it will be over for everyone."[72] By the end of the year, however, as Random House explored the release of new, illustrated e-books, Cue's refusal to launch an app from Random House unless the publisher signed an agency agreement made this a reality. Jobs then took aim at the rest of the publishing industry: with books under Apple's belt, it was time to approach newspapers, magazines, and new subscription services that could help customers build long-term relationships with their content developers.[73]

Conclusion

Markets are not neutral arenas of exchange but contested spaces where competing visions of the future collide. Apple's entry into the e-book market, armed with the agency model and the most favored nation clause, was a strategic intervention in the rapidly changing dynamics of the publishing industry. This entry challenged Amazon's established dominance by introducing a framework that granted publishers more control over their products' prices. Apple's contractual arrangements were effective because they recalibrated the relationships among retailers, publishers, and consumers, enticing publishers to operate in ways that advanced Apple's vision for the industry's future.

Efforts to counter a dominant firm may require smaller ones to find a precarious balance between immediate gains and long-term risks. For publishers, the alliance with Apple was both an opportunity and a gamble. It allowed them to counter Amazon's $9.99 standard, a figure that had undermined their profitability and constrained their autonomy. But aligning with Apple meant accepting new constraints, including rigid pricing tiers and a revenue-sharing model that limited their flexibility. Amazon's response shows how dominant firms can protect their market dominance by bypassing traditional relationships and creating new pathways of influence. Its decision to offer higher royalties to self-published authors was a tactical move. It had the potential to grow a segment of the market while ensuring the company's control over it, further transforming the e-book market into a space where Amazon dictated the terms of participation. This would reinforce Amazon as the central hub for e-book distribution and challenge the hierarchies that had long governed publishing.

The negotiation between publishers, Apple, and Amazon underscores a crucial feature of markets undergoing technological change: they are constructed not through the inevitabilities of technological innovation but through deliberate contests over value, control, and participation. Apple's use of the agency model and most favored nation clause was a calculated effort to impose new norms on the industry, and Amazon's countermeasures sought to retain dominance by redefining its relationships with key stakeholders. Apple and Amazon thus forced publishers to reconsider their role in an increasingly platform-driven publishing market where the terms of engagement were shaped by the swift strategic interventions of powerful newcomers.

Notes

1. Steve Jobs, iPad announcement, January 27, 2010, YouTube, posted by pil.com, May 5, 2015, ttps://www.youtube.com/watch?v=zZtWlSDvb_k (49:20).

2. See, for example, Chris Sagers, *United States v. Apple: Competition in America* (Cambridge, MA: Harvard University Press, 2019).

3. Declaration of Eddy Cue, April 26, 2013, DX-714, in Deferred Appendix at A1755, United States v. Apple, Inc. et al., No. 13-3741 (2d Cir. July 1, 2014).

4. Declaration of Eddy Cue, in Deferred Appendix at A1764.

5. "eBooks Opportunity (Apple Confidential Report)," February 19, 2009, PX-0735, in Deferred Appendix at A268.

6. "eBooks Opportunity," PX-0735, at A268.

7. The report was based on data collected by trade associations in music and film, as well as the accounting firm PricewaterhouseCoopers. "e-books Opportunity," PX-0735, at A268.

8. Carolyn Reidy to Leslie Moonves, email, December 17, 2009, PX-0510, in Deferred Appendix at A348.

9. On the Kindle selling out, see Nilay Patel, "Kindle Sells Out in 5.5 Hours," *Engadget* (blog), November 21, 2007, archived at Wayback Machine, citing capture dated November 23, 2007, https://web.archive.org/web/20071123084048 /http://www.engadget.com/2007/11/21/kindle-sells-out-in-two-days/.

10. John Makinson, "Penguin Three Year Plan 2009," confidential communication, August 4, 2009, PX-0133, in Deferred Appendix at A286.

11. Arnaud Nourry to Arnaud Lagardere, translated email, July 29, 2009, PX-0392, in Deferred Appendix at A285.

12. Carolyn Reidy to Leslie Moonves, email, September 23, 2009, PX-0344, in Deferred Appendix at A290.

13. Eddy Cue to Steve Jobs and Scott Forstall, email, February 19, 2009, PX-0027, in Deferred Appendix at A282.

14. Cue, email to Jobs and Forstall, PX-0027, at A282.

15. Eddy Cue to Steve Jobs, email, December 15, 2009, PX-0050, in Deferred Appendix at A326.

16. Carolyn Reidy, handwritten notes, December 16, 2009, PX-0519, in Deferred Appendix at A336–A339; Elinor Hirschhorn, handwritten notes, December 16, 2009, PX-0359, in Deferred Appendix at A340–A343; Brian Murray to Charlie Redmayne, email, December 16, 2009, PX-0861, in Deferred Appendix at A333; Leslie Hulse to Charlie Redmayne et al., email, December 16, 2009, PX-0301, in Deferred Appendix at A334.

17. Reidy, email to Moonves, PX-0510 at A348.

18. Keith Moerer to Eddy Cue, email, December 15, 2009, PX-0693, in Deferred Appendix at A330.

19. Arnaud Nourry to Eddy Cue, email, December 15, 2009, PX-0298, in Deferred Appendix at A329.

20. Reidy, email to Moonves, PX-0510, at A348.

21. Murray, email to Redmayne, PX-0861, at A333.

22. Hulse, email to Redmayne et al., PX-0301, at A334.

23. Reidy, email to Moonves, PX-0510, at A348.

24. Keith Moerer to Eddy Cue, email, December 18, 2009, PX-0039, in Deferred Appendix at A353.

25. Eddy Cue to Steve Jobs, email, December 21, 2009, PX-0043, in Deferred Appendix at A354.

26. Murray, email to Redmayne, PX-0861, at A333.

27. Carolyn Reidy to Dinnis Elau et al., email, December 21, 2009, PX-0540, in Deferred Appendix at A355.

28. See, for example, Eddy Cue to Carolyn Reidy, January 4, 2010, PX-0021, in Deferred Appendix at A359.

29. Eddy Cue and John Sargent, email thread, December 8, 2009 to January 4, 2010, PX-0476, in Deferred Appendix at A361–A363.

30. Keith Moerer to Eddy Cue, email, January 9, 2010, PX-0023, in Deferred Appendix at A383.

31. Madeline McIntosh to Amanda Close et al., email, January 10, 2010, PX-0174, in Deferred Appendix at A386–A387.

32. Kevin Saul to John Sargent, email, January 11, 2010, PX-0248, in Deferred Appendix at A390.

33. "E-book Agency Distribution Agreement," in Saul, email to Sargent, PX-0248, at A391–A402.

34. "E-book Agency Distribution Agreement," in Saul, email to Sargent, PX-0248 at A394. See also Declaration of Kevin Saul, April 28, 2013, DX-550, in Deferred Appendix at A1671–A1677.

35. See "Principles of the Trading System," World Trade Organization, accessed December 13, 2024, https://www.wto.org/english/thewto_e/whatis_e/tif_e/fact2_e.htm.

36. Cite COMM LAW book.??

37. Declaration of Kevin Saul, DX-550, at A1674.

38. Eddy Cue to Steve Jobs, email, January 13, 2010, PX-0026, in Deferred Appendix at A498; Keith Moerer to Eddy Cue, email, January 12, 2010, PX-0894, in Deferred Appendix at A495–A497.

39. Moerer, email to Cue, PX-0894, at A495–A497.

40. Eddy Cue to Steve Jobs, email, January 14, 2010, PX-0055, in Deferred Appendix at A499.

41. Eddy Cue to David Young, email, January 16, 2010, PX-0059, in Deferred Appendix at A502; Eddy Cue to Carolyn Reidy, email, January 16, 2010, PX-0511, in Deferred Appendix at A504; Eddy Cue to John Sargent, email, January 16, 2010, PX-0512, in Deferred Appendix at A506; Eddy Cue to David Shanks, email, January 16, 2010, PX-0120, in Deferred Appendix at A508; Eddy Cue to Brian Murray, email, January 16, 2010, PX-0513, in Deferred Appendix at A510.

42. See, for example, Cue, email to Young, PX-0059, at A502.

43. Carolyn Reidy to Leslie Moonves, email, January 18, 2010, PX-0537, in Deferred Appendix at A512.

44. Brian Napack to Eddy Cue, email, January 19, 2010, forwarded in Keith Moerer to Kevin Saul, email, January 19, 2010, DX-187, in Deferred Appendix at A513, A515.

45. Maja Thomas to David Young, email, January 19, 2010, PX-0521, in Deferred Appendix at A516.

46. Tom McCall to David Shanks and Susan Kennedy, email, January 19, 2010, PX-0584, in Deferred Appendix at A518.

47. Jeffrey Trachtenberg, "Publisher in Talks with Apple over Tablet," *Wall Street Journal*, January 18, 2010, https://www.wsj.com/articles/SB1000142405274 8704541004575011092145509872.

48. Michael Cader, "Big Six Negotiate with Apple, Ready New Business Model for eBooks," *Publishers Lunch*, January 19, 2010, https://lunch.publishers marketplace.com/2010/01/big_six_negotiate_with_apple_re/.

49. "Amazon Announces New 70 Percent Royalty Option for Kindle Digital Text Platform, Enabling Authors and Publishers to Earn More Royalties from Every Kindle Book Sold," Amazon Press Release, January 20, 2010, DX-196, in Deferred Appendix at A534–A536.

50. "Amazon Announces New 70 Percent Royalty Option," DX-196, at A534.

51. "Amazon Announces New 70 Percent Royalty Option," DX-196, at A534.

52. Charlie Redmayne to Brian Murray, email, January 20, 2010, forwarded in Brian Murray to Undisclosed Recipients, email, January 20, 2010, PX-0307, in Deferred Appendix at A526–A527.

53. Robert Zaffiris to Brian Murray et al., January 20, 2010, in Brian Murray and Charlie Redmayne, email thread, PX-0303, in Deferred Appendix at A531.

54. Murray, email to Undisclosed Recipients, PX-0307, at A526–A527.

55. Brian Murray to Charlie Redmayne, email, January 20, 2010, PX-0303, in Deferred Appendix at A530.

56. Markus Dohle to Eddy Cue, email, forwarded in Eddy Cue to Steve Jobs, email, January 21, 2010, PX-0042, in Deferred Appendix at A548.

57. Cue, email to Jobs, PX-0042 at A547.

58. Steve Jobs to James Murdoch, email, January 24, 2010, forwarded in Steve Jobs to Eddy Cue, email, January 24, 2010, PX-0032, in Deferred Appendix at A591.

59. See, for instance, Tim Hely-Hutchinson to Arnaud Nourry, email, January 25, 2010, PX-0106, in Deferred Appendix at A602; David Shanks to Eddy Cue, email, January 25, 2010, PX-0019, in Deferred Appendix at A604; Eddy Cue to Carolyn Reidy, email, January 26, 2010, PX-0613, in Deferred Appendix at A621; Adam Rothberg to Elisa Rivlin, email, January 26, 2010, PX-0612, in Deferred Appendix at A622.

60. Brian Murray to James Murdoch and Jonathan Miller, email, January 26, 2010, PX-0526, in Deferred Appendix at A620.

61. Michael Cader, "Today's Amazon Press Release: Exclusive Kindle Book by Paulson Speechwriter," *Publishers Lunch*, January 26, 2010, https://lunch.publishersmarketplace.com/2010/01/todays_amazon_press_release_exc/.

62. Marjorie Scardino to the Pearson Board of Directors, email, January 26, 2010, PX-0530, in Deferred Appendix at A626.

63. The original video is unavailable, but the courts documented the event. These quotes are found in Jay Yarow, "Steve Jobs Says Book Publishers Hate Amazon's Kindle," *Business Insider*, January 29, 2010, https://www.businessinsider.com/steve-jobs-publishers-are-going-to-pull-their-books-from-amazon-2010-1.

64. Elisa Rivlin to Carolyn Reidy, email, January 29, 2010, forwarded in Carolyn Reidy to Elisa Rivlin, email, PX-0607, in Deferred Appendix at A638.

65. John Sargent, quoted in United States v. Apple, Inc., et al., 952 F. Supp. 2d 638, 709n49 (S.D.N.Y. 2013).

66. John Sargent, quoted in *United States v. Apple* (S.D.N.Y. 2013) at 680.

67. Arnaud Nourry and John Makinson, quoted in *United States v. Apple* (S.D.N.Y. 2013) at 680.

68. Steve Jobs to Eddy Cue, email, January 31, 2010, PX-0101, in Deferred Appendix at A640; Eddy Cue to John Sargent, email, January 31, 2010, PX-0053, in Deferred Appendix at A642; John Sargent to Benita Somerfield, email, February 9, 2010, PX-0094, in Deferred Appendix at A644.

69. Michael Cader, "Amazon Removes Macmillan Buy Buttons," *Publishers Lunch*, January 29, 2010, https://lunch.publishersmarketplace.com/2010/01/amazon_removes_macmillan_buy_bu/.

70. Amazon press release, January 31, 2010, quoted in Michael Cader, "Amazon: 'Ultimately, We Will Have to Capitulate and Accept Macmillan's Terms,'" *Publishers Lunch*, January 31, 2010, https://lunch.publishersmarketplace.com/2010/01/amazon_ultimately_we_will_have/.

71. Michael Cader, "Some Macmillan Bestsellers Start Falling at Amazon, Rising at BN.com," *Publishers Lunch*, January 30, 2010, https://lunch.publishersmarketplace.com/2010/01/some_macmillan_bestsellers_star/.

72. Eddy Cue, quoted in *United States v. Apple* (S.D.N.Y. 2013) at 686.

73. Steve Jobs to Eddy Cue, email, December 18, 2010, PX-0519, in Deferred Appendix at A872.

Part II

How does code become infused with social values, assumptions, and biases?

Figure 7.1.

Figure 7.2.

7

The Standard Head

Stephanie Dick

Introduction: Toward a Socio-technical History of Computational "Solutions"

The person depicted in the so-called Kaplan daguerreotype (figure 7.1, opposite) may or may not be Abraham Lincoln. Albert Kaplan, a New York stockbroker, purchased the early-19th-century daguerreotype photograph in 1977 from a museum in New York City because he believed, based on about a decade of his own research, that this was in fact a photograph of the American president. If true, it would be the youngest image of him on record. Figure 7.2 shows the youngest confirmed image of Abraham Lincoln, at age 39 from 1848. There has been considerable interest in whether or not the earlier daguerreotype is in fact an image of the young future president.

Experts approached facial identification, in a sense, as a mapping problem: Was there a plausible path from the first image to the second? Could aging, illness, injury, and other processes that mark the body chart a course from the unknown younger man to the older, familiar one?[1] Who is to say? In 1994 the *Washington Post* covered the story, announcing that "expert says 1840s photo is of Lincoln." The expert in question was Claude Frechette, a French reconstructive surgeon. He documented his findings in the *Journal of Forensic Identification*, arguing that "the nature and number of similarities in the Kaplan daguerreotype and Lincoln images overwhelms coincidence and serves to identify the Kaplan daguerreotype as a portrait of Abraham Lincoln."[2] Press reports on the photograph, however, left out the fact that several other experts—including a forensic anthropologist from the Smithsonian Institution and a forensic artist from the New York Police Department—had been consulted, and many of them did not agree that the man in the daguerreotype was Lincoln. One such "identification expert," Joe Nickell, senior research fellow with the Committee for Skeptical Inquiry, wrote; "As I told Kaplan, I would prefer the opinion of any forensic expert over that of a plastic surgeon; although the latter has expertise in facial features, he lacks training in identification. In fact, Frechette observed several differences between the Kaplan image and genuine Lincoln photographs but hastened to rationalize them."[3] Competing forms of expertise orbited the Kaplan daguerreotype: would it be historians, reconstructive surgeons, forensic artists, or forensic anthropologists who decide if the man shown in the image was or was not Abraham Lincoln? The body and face are storied sites: they bear traces of our lives, illnesses, emotions, experiences, and histories, and as such they are contested sites where multiple explanatory frameworks and forms of expertise apply. Experts considered the effects of aging, of psychological and emotional distress (which Lincoln was thought to have experienced in his 20s), the effects of disease and illness on the face, the measurements taken for Lincoln's death mask—all to determine if there was a physiological path that led from the face captured in the Kaplan daguerreotype to that in the later known images of Abraham Lincoln. The experts disagreed, and there remains no consensus about the image.

I discovered the Kaplan daguerreotype in 2016 in an unexpected place: in the archives of mathematician and artificial intelligence researcher Woodrow Wilson "Woody" Bledsoe at the University of Texas at Austin. Kaplan enrolled Bledsoe in his identification efforts because Bledsoe had developed one of the first, if not the first, semi-computer-automated facial recognition programs in the world. Bledsoe's algorithm represented a new approach to and, I argue, a redefinition of the problem of facial recognition, both of which will be explored at length in what follows. Today, as automated facial recognition has spread and come under broad contestation for its failures, its racial biases, and its role in increasing levels of carceral surveillance, it is worth revisiting this early algorithm to understand how its authority was constructed.[4] It is especially worth revisiting this facial recognition algorithm because it was *not in fact a solution to the problem of facial recognition*: it required a near-total redefinition of the problem and the introduction of a set of problematic assumptions. Here, I focus in particular on Bledsoe's introduction of a "standard head," whose three-dimensional measurements were assumed of all heads in order to "simplify" the matching problem and make it "solvable" with the tools available.

This, in fact, is what it means to solve *any* problem with computers. Even when automating well-defined and well-known rule-bound processes, the actual implementation requires redefinitions and translations into computational terms and logics, and these often escape critical and historical analysis.[5] To make hard problems easier to "solve," technical practitioners often introduce quick fixes and convenient assumptions.[6] They artificially bound and restrict problem domains. This chapter contributes to an urgently needed sociotechnical history of *solving* (and *solutionism*) in computing—when technologists declare a problem "solved," or even simply "close to solved" or "ready for use," what do they mean? What forms of transformation and redefinition have occurred?

In exploring these questions, the history of mathematics may offer helpful guidance. Philosopher Imre Lakatos, in his classic *Proofs and Refutations*, observed that when mathematicians discover counterexamples or challenges to the conjectures they seek to prove, they do not always throw them out. More often, they instead redefine the object they are studying to exclude the offending instances. For example, the infamous "four-color conjecture" asks whether or not every map can be colored with four colors such that no two neighboring countries will have the same color. In seeking an answer to this problem, mathematicians quickly redefined "map" to exclude both "Alaskas" (countries that have completely disconnected parts) and "Vaticans" (countries that are completely surrounded by another) seeing that they could prevent a positive solution. The problem has been declared solved (though not without much controversy about the use of computer assistance in the proof), but not for actual maps that include Alaska and the Vatican.[7] Sociologist of science David Bloor called this practice "monster barring," observing similarities between mathematical practice as described by Lakatos and anthropologist Mary Douglas's observation of how cultures deal with the "monsters" that do not fit into the logics of their cultural classifications and cosmologies.[8] Following this approach, we might ask what logics (or even cosmologies) ground computational "solutions"? What "monsters," or we might say, what *realities*, are excluded in pursuit of them? How does "solving problems with computers" lead to systematic redefinitions of the epistemic, political, and technological landscape of problem solving? This chapter works toward a sociotechnical history of a computational "solution."

The history of computing is not, in fact, a history of an increasingly powerful technology succeeding at increasingly complex and difficult tasks. The history of computing is a history in which more and more parts of our world, more problems, more tasks, more questions, are being redefined in the terms of the machine. This involves placing artificial bounds around the question, and these limits are often quickly forgotten and naturalized to make it *seem* as though computers are doing these impressive things. Every time someone announces that computers have accomplished a given task—from medical diagnosis, to game play, to facial recognition—the task has been redefined by their efforts. The task has been reimagined and remade into something that a computer *can do*: this redefinition involves the translation of goals and procedures into the languages, logics, formalisms, and materialities of computing. It also often involves the introduction of restrictions, assumptions, and caveats—a translation of what the task actually involves to what can be accomplished with the tools at hand.

Bledsoe's algorithm, with its "solutions" to the complex problems of facial recognition in tow, served as a proof of concept that automatic identification of faces from photographs could and should be pursued at scale by law enforcement, and especially in the context of the development of the New York State Identification and Intelligence System (NYSIIS) in the 1960s. NYSIIS was among the first centralized, computerized law enforcement databanks in the country, and its technical reports and planning documents incorporated Bledsoe's algorithm. This is a story about how tasks like facial recognition must be redefined in order for the computer to do them and what is involved in that redefinition. It is a story about how hard and intractable problems can be rendered "solvable" by the introduction of constraints and assumptions. It is about how these redefinitions and assumptions are often quickly forgotten as people focus on the future promise of techno-solutions; and it is a story about how the systems in which these algorithms travel are haunted by legacies of "solution-constructing" translation. Especially important is that computers do not simply require a *reduction* of a complex problem, or the selection of some variable as relevant while others are excluded from consideration. More often, computers require a *rethinking of the problem altogether* in which the relevant elements—in this case, faces and the question of identification—are redefined entirely.

Facial recognition is often undertaken in contexts with much higher stakes than determining whether an old photo is that of an old president, or how we might color our maps, and this may in part explain why computational "solutions" to this problem have been so readily sought and irresponsibly offered. Identification has been a core tenet of policing since Alphonse Bertillon's 19th-century development of mug shots, fingerprints, and recordkeeping practices.[9] At the same time, Francis Galton and others were infamously attempting to study composite photographs in an attempt to identify physiological indicators of personality, behavior, and criminality in the service of eugenic visions of societal control.[10] Eden Medina has shown that the desire for computer assistance also emerged more recently in identifying the dead based on matching remains to photographs, memories, and records, in the search for justice, truth, and reconciliation in the wake of political atrocities in Pinochet's Chile.[11] In policing and in other domains of identification, computers have been introduced because algorithmic solutions are fashioned as a guarantor of accuracy, efficiency, and objectivity, and as a mechanism for making contested and uncertain fields more settled and "scientific." Bertillon wanted to make policing more "scientific"—by centering investigation and incarceration on practices of biometric data collection, recordkeeping, and identification. Galton sought to make policing more scientific by redefining "criminality" as an identifiable feature of the body. In Chile, those seeking justice for and knowledge of what happened to their family members wanted certainty that the remains being returned to them belonged to their loved ones. And the New York State Identification and Intelligence System, and the identification algorithms that worked within it, was fashioned as a "scientific breakthrough" for policing—heralding more efficient, more objective, more accurate, farther-reaching, and therefore more "just" outcomes in law enforcement. Except that it did no such thing.

Pattern Recognition and Artificial Intelligence

Woody Bledsoe completed his PhD in mathematics at the University of California, Berkeley, in 1958 and soon joined the faculty at the University of Texas at Austin, where he remained for the rest of his life. He also helped found the computer science department there and was an influential early researcher in artificial intelligence and automated theorem proving. He also contracted with a number of government- and military-funded agencies. He served as the manager of the mathematics division at Sandia Corporation, which was founded in Livermore, California, in the 1940s by the Department of Energy as a part of the US government's efforts to help maintain America's dominance in science and engineering following World War II. At Sandia, Bledsoe worked on calculating and modeling fallout probabilities from thermonuclear attacks, and later he did his own work on character and numeral recognition. During the 1960s, Bledsoe also spent time at the Stanford Research Institute (SRI), another important state-funded research institution of the Cold War, and he founded a company, Panoramic Research Inc., with clandestine support from the Central Intelligence Agency (CIA), which commercialized his work on character recognition, selling early versions of what we now call optical character recognition (OCR) tools.[12]

Bledsoe worked across many subfields of computing: automated theorem proving, optical character recognition, and facial recognition, among many others. He believed all of these problems belonged under the umbrella of "artificial intelligence." Some early artificial intelligence practitioners, including Allen Newell and Herbert Simon, identified human *reasoning*, seen as rule-bound, symbolic information processing, as the cognitive faculty at the heart of human intelligence and the thing to be replicated in computers in order to render them intelligent as well.[13] Others, most of whom worked under the banner of "expert systems research," believed instead that human intelligence was grounded in *knowledge* rather than reasoning and sought to encode databases of expert knowledge that would create intelligence in computers.[14] Bledsoe, on the other hand, proposed that *pattern recognition* was a central instrument of human intelligent behavior.[15]

Bledsoe believed that a central part of what mathematicians do when they prove theorems or solve problems is to identify patterns and parallel structures between mathematical expressions, figures, fields, and formalisms. In his work on automated theorem proving, he therefore developed subroutines for pattern recognition of this kind.[16] His work on automated numeral and character matching was an exercise in pattern recognition as well.[17] Bledsoe's programs for character recognition were equipped with a database of "known"—that is, "labeled"—letters and numerals: A through Z in both upper and lower case, as well as the numerals 0 to 9. Then the computer was given "unknown" or "unlabeled" characters and was tasked with comparing them to the "known" characters in its database, looking for the closest match. The known characters were defined by a set of points on a superimposed Cartesian grid. "A" was defined as the set of points on the grid that the known, or labeled, "A" intersected. Unknown characters to be identified were first overlain with a similar grid and scaled for size, and then the computer proceeded to use a "point-by-point" matching system whereby each point in the grid on the unknown character was compared with each point in the grid on the known characters in the database, and the known character with the most points of overlap was offered up as the best "match" for the unknown character. The idea was that an "A," even if sloppily written or written in a different hand, would have more points in common with the "A" in the computer database than with a "D" or a "G."

It was his interest in pattern recognition, and his belief that it underlay much human intelligence, that brought Bledsoe to automated facial recognition in 1964: "One of the most challenging areas of pattern recognition is the identification of human photographs by machine. In this endeavor the machine is expected to disclose the identity of the person whose photograph it is shown."[18] However, facial recognition proved far more difficult than either structural pattern recognition in mathematics or character identification. One reason is that the point-by-point matching system described for characters will not work well for photographs, because if a person's head is at a different angle of rotation in the two photos being matched, they may not overlap much at all in a point-by-point comparison. Point-by-point comparison works only on two photos of the same face *at the same orientation*, but this is not usually the case for photographs taken "in the wild," as Bledsoe called them, and he wanted a system that would be useful in practice. So he set out to identify an alternative to point-by-point comparison that might work for photographs in which people's heads are at different orientations. He created a database of "known" faces—all photographed facing straight at the camera—in an

Figure 2. Examples of Photograph Pairs Used in the Study

Figure 7.3.

effort to develop an automated recognition system that could match these photographs to "unknown" photographs of the same people with their heads at a slight angle or tilt.

All of the photos in his database were of adult white men (see figure 7.3). Bledsoe experimented on some 2,000 sets of photographs of Caucasian men and aspired to develop a computer program that could match an unknown photo *not* in the face-forward position with a known, face-forward photo in the database.

Redefining Faces, Redefining Recognition

"Automation" is too strong a word for the resulting system that Bledsoe developed, because much human labor and agency were involved in its operation, beginning with the "input" of the photographs. How to represent a face in computer memory? How to define faces such that the computer can work with, compare, and match them? As is often the case, Bledsoe's automation attempt involved significant redefinitions of the objects in question. Faces in this system were redefined as *sets of distances between different points on the face*: between the outside and inside corners of the eyes, between the outer corners of the mouth, between the tops of the ears, and so on. It was not on photographs themselves or on images that this system operated, but on these distances—a set of numbers that Bledsoe selected to "define" a face numerically. Aaron Gluck-Thaler has situated Bledsoe's desire to redefine faces through particular features within a broader "postwar fascination with the new concept of 'identity,' which scientists hoped to isolate through anthropological typologies and projective tests in psychology." Gluck-Thaler argues that this approach to automation "reinforced a key assumption: that recognizing identity could be reduced to classifying specific features within a set of existing examples."[19] Iris Clever has documented how this approach also overlapped with (and may have been inspired in part) by anthropological approaches to identification and ancestry estimation, both haunted in their modern and computational forms by nineteenth century race science, in their emphasis on evaluating and analyzing "an individuals' morphological features, like skull shape, against the traits believed to indicate heredity membership to racialized groups."[20] In adopting this broader paradigm for "identity," Bledsoe nonetheless had to render facial features "machine readable" and calculable.

Each input photograph would first be projected with an overhead projector onto a RAND tablet—a graphical input device, produced by the RAND Corporation in Santa Monica, California, that could take visual data as input, register marks made on the surface with a special stylus, and display those on a screen. Once a photo of a face had been projected onto the RAND tablet, a human operator would use the stylus to "touch" a specified set of points on the face—corners of the eyes, edges of the mouth, edges of the nose, edges of the ears, and so on—and these marks would be converted into a set of coordinates by the computer through the imposition of a standardized grid, then normalized for scale so that all known photos in the database were scaled the same (figure 7.4). A specialized computer attached to the RAND tablet would then produce a paper tape that recorded the name of the person in the photo and the coordinates of those points in it, and the human operator would manually input that tape to another, more powerful computer, which did the recognition work. Using the input coordinates, the computer would calculate the distance between specific pairs of points on the face, and this set of distances *was the definition of the face*. "Face" to the computer meant a set of distances between points, and when the computer "identified" these faces, it did so using these sets of distances. The computer never worked directly with photographs or with faces but only with these sets of distances according to which Bledsoe defined "faces" in his recognition system. This is a characteristic example of the forms of selective simplification that are often required to make hard problems solvable by automated means.

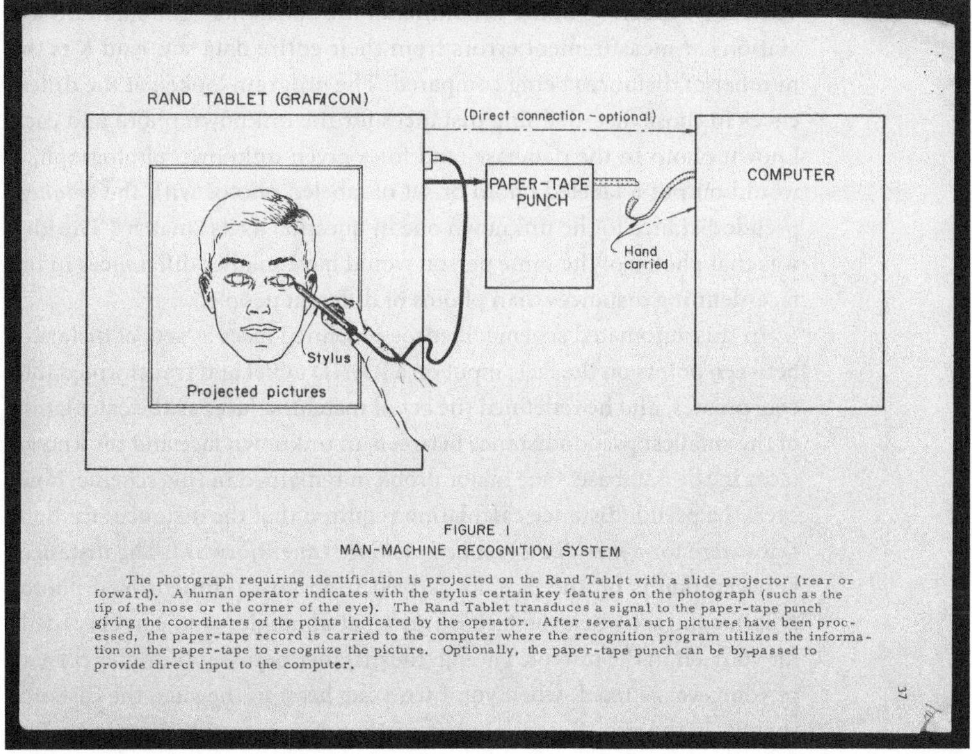

FIGURE I

MAN-MACHINE RECOGNITION SYSTEM

The photograph requiring identification is projected on the Rand Tablet with a slide projector (rear or forward). A human operator indicates with the stylus certain key features on the photograph (such as the tip of the nose or the corner of the eye). The Rand Tablet transduces a signal to the paper-tape punch giving the coordinates of the points indicated by the operator. After several such pictures have been processed, the paper-tape record is carried to the computer where the recognition program utilizes the information on the paper-tape to recognize the picture. Optionally, the paper-tape punch can be by-passed to provide direct input to the computer.

Figure 7.4.

Moreover, in spite of his own commitment to human pattern recognition as the seat of intelligence, Bledsoe was quick to abandon any "human" approach to photograph matching here, working instead with the limited capabilities of 1960s computing machinery to instead explore faces redefined as sets of distances.

The computer had a database of around 2,000 "known" (that is, *named*) faces, or around 2,000 sets of distances acquired from points on a face-forward photograph. Each "unknown" photograph would also be converted into a set of the same distances by a human operator touching the requisite points on the face with the RAND tablet stylus and inputting the resulting tape to the computer. The actual matching was done with what Bledsoe called a "pseudodistance calculation":

$$d^2 = \frac{1}{K} \sum_{i=1}^{K} \left(\frac{D_{i1} - D_{i2}}{\sigma E_i} \right)^2$$

The computer would calculate the so-called "pseudodistance" d between an unknown face and each known face in the dataset. Bledsoe defined the "pseudodistance" between two photographs to be the *sum of the differences* between the distances that defined each of them. The D_i's in this equation are the distances themselves, between sets of points on the face (between eyes, between ears, and so on), normalized for scale and size.

The σE_i's appearing in the denominator are constants—the standard deviations of measurement errors from their entire data set. And K is the number of distances being compared. The program looked at the differences in those face-defining distances for the unknown photo and each known photo in the database, and for a given unknown photograph, it would output a labeled photo or set of labeled photos with the *smallest* pseudodistance to the unknown one in question as its "match." The idea was that photos of the same person would have smaller differences in the face-defining distances than photos of different people.

In this automated scheme, Bledsoe redefined faces as sets of distances between points on the face, input on a RAND tablet and transformed into coordinates, and he redefined the act of matching faces as the calculation of the smallest pseudodistance between an unknown face and the known faces in the database. One major problem remained in this scheme, however: the pseudodistance calculation required that the distances for both faces were for a *face photographed looking straightforward*. The distances between points on the face, as derived from a two-dimensional photograph, contract when the face is turned at an angle—you can check this for yourself in the mirror. Though the distance between the outer corners of your eyes is fixed, when you turn your head to the side, the distance shrinks in the two-dimensional projection of your face in the mirror. The same is true for a photograph: the distances between points on the face are smaller for a face turned away from the camera. It was in order to "solve this problem" that Bledsoe introduced the "standard head."

How Not to Solve a Problem

The D_i's in the pseudodistance calculation represent the distances between points on a face looking directly at the camera. But this whole system was meant to match photographs of heads turned away from the camera with their face-forward counterparts in the database. Bledsoe decided that the best way to "solve" this problem would be to introduce another calculation that the computer would perform to "correct" the face-defining distances for the unknown photograph to be *as if* it was facing forward. Bledsoe identified three axes on which a given head could be turned: rotation, lean, and tilt (see figure 7.5). If the computer could measure the rotation, lean, and tilt of the unidentified face, it would be able to "correct" the distance measurements to what they would have been if the face had been photographed straight on. However, these rotation correction calculations require two measurements, labeled E and EB in figure 7.5, where E is the distance between the center of the ear and the tip of the nose, and EB the distance between the top of the ear and the middle of the eyebrow. These, however, are *three-dimensional measurements of the head*—they cannot be known from a two-dimensional photograph.

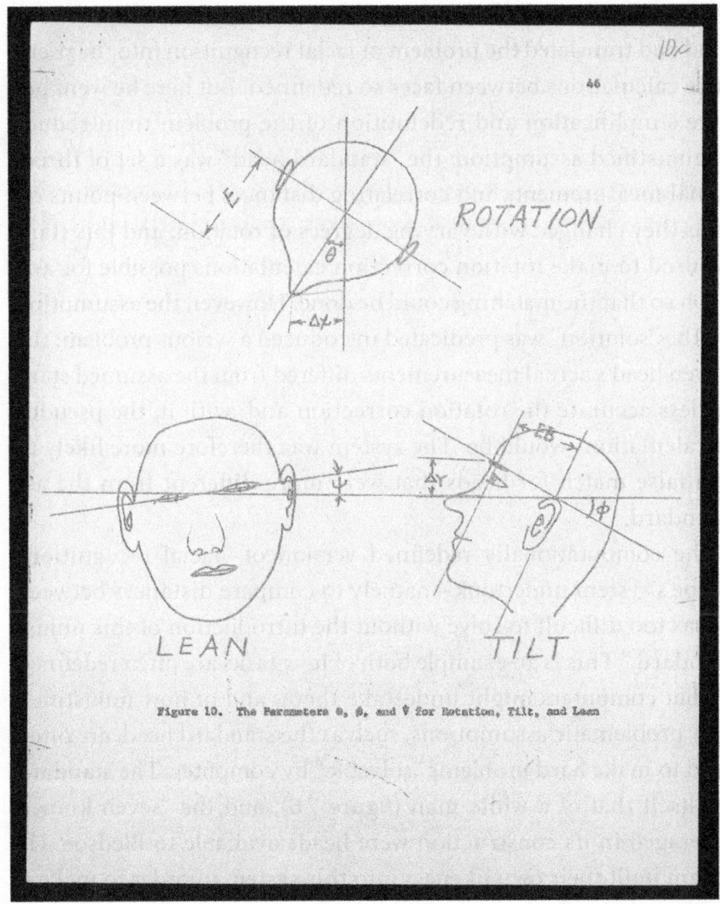

Figure 7.5.

Bledsoe's solution to this problem was to assume the same three-dimensional measurements of every head in order to do the rotation correction calculation that allowed the pseudodistance calculation and, along with it, "facial recognition":

> A principle difficulty in automatic facial recognition from photographs is the many poses that the subject can take when being photographed. One of the ways he can change his pose is by head rotation. . . . In our recognition schemes we have attempted to correct for this by "rotating" all faces back to the frontal pose before trying for recognition. But in order to rotate a face to the frontal pose it is necessary to know the angles of rotation. . . . In this and other sections we make use of a "standard" head. This is supposed to be an ideal which is an average of all heads to be considered by this recognition scheme. Actually, the standard head used in this study was obtained by averaging the coordinates of certain features on seven known heads.[21]

Bledsoe and his team had redefined faces as sets of distances between points, and had translated the problem of facial recognition into the pseudodistance calculations between faces so redefined. But here he went beyond mere simplification and redefinition of the problem to introduce a deeply unjustified assumption: the "standard head" was a set of three-dimensional measurements and correlating distances between points on the head as they changed with varying degrees of rotation, and this standard was used to make rotation correction calculations possible for any photograph so that the matching could be done. However, the assumption on which the "solution" was predicated introduced a serious problem: the more a given head's actual measurements differed from the assumed standard, the less accurate the rotation correction and, with it, the pseudodistance calculations would be. The system was therefore more likely to produce a false match for heads that were more different from the assumed standard.

Even the computationally redefined version of "facial recognition" that Bledsoe's system undertook—namely to compare distances between points—was too difficult to solve without the introduction of this unjustified "standard." This is an example both of how tasks are often redefined in order that computers might undertake them, and of how unjustified and deeply problematic assumptions, such as the standard head, are often introduced to make hard problems "solvable" by computer. The standard head was itself that of a white man (figure 7.6), and the "seven known heads" averaged in its construction were heads available to Bledsoe. He and his team built their own likeness into this system in order to make it *appear* as though they had achieved computer recognition of faces, though the problem of facial recognition had been oversimplified to the pseudodistance calculation, which required the "standard head."

Bledsoe was fully aware that the standard head was an unjustified assumption. "One of the more glaring weaknesses in the present system," he and a colleague reported, "is the assumption of average or 'standard' distances in the derivation of the head rotation algorithm. This assumption is rarely justified. A more elegant approach to the whole problem of head rotation would undoubtedly make use of the elements of projective geometry."[22] Bledsoe proposed that, at the very least, the standard should be an average of all of the heads in the "known" database, rather than of only seven heads. Bledsoe suggested that what was required was a more *inclusive* "standard head": one whose measurements averaged not seven known men's heads but *all the heads* in the database, or as many heads as possible.

Bledsoe's call for *inclusivity* when confronting a limited or failing system sets the stage for today's all-too-common call to improve algorithmic accuracy by the inclusion of ever more data, often by extractive means.[23] This framework of "bias reduction" intimates that the problem could be

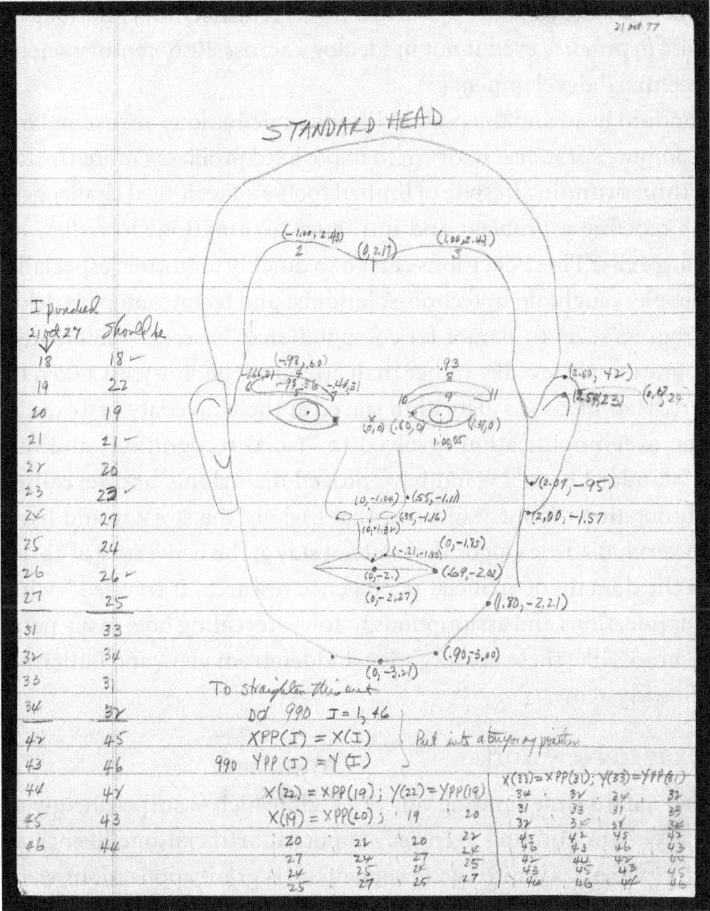

Figure 7.6.

solved *if only we could get representative enough data*. This framework closes down more critical engagement with the underlying reformulation of the problem and with the technology, whether it works well on everyone or not. More diverse data may improve the accuracy of outcomes, but it does not address any of the underlying questions about how or *why* a problem has been redefined in this way in the first place, let alone the sociopolitical implications of rolling out both systems that fail *and* ones that succeed. Especially in the context of police use, failing recognition systems put many (especially racialized) people at risk, but so too do ones that "work." Moreover, a more inclusive "standard head," however produced, wouldn't solve the basic problem that the system would still be less accurate for heads that differ most from the average. Although Bledsoe was not interested in predicting criminality based on head shape or deviance from a norm, his system worked best for heads that most closely approximated the "standard," one that literally encoded male whiteness, reiterating those racialized logics that were at work in 19th-century phrenology.

Thus, as Clever observes, did "discarded [nineteenth-century] notions of race" persist *in practice* even if not in ideology across 20th-century scientific and technical developments.[24]

The standard head and the pseudodistance calculations are exemplary of much computer practice: so often, to make hard problems more tractable, to get things running in spite of limited tools and technical challenges, developers redefine a problem and introduce assumptions to *reduce or bypass* complexity. These decisions can be so quickly forgotten, especially because people, caught up in techno-solutionist and techno-utopian delusions of progress seem to clamor for computational *"success"*—automated facial recognition achieved!—rather than attend to the shortcuts taken to arrive there. Had Bledsoe's algorithm stayed at the University of Texas at Austin, the oversimplification involved in "facial recognition" and the assumed "standard head" would have proved disturbing, and revealing of what automation often entails, and that's where the story would have ended. However, Bledsoe's algorithm did not stay at the University of Texas or even in the domain of artificial intelligence research. It traveled—with its oversimplifications and assumptions in tow—revealing how easily poor technical choices like these are forgotten, hidden from view, and built into computerized systems.

From Texas to New York

Bledsoe had a hard time funding this research, which is surprising given how liberally many funding agencies supported artificial intelligence at the time, in spite of its limited achievements. His grant applications were rejected by the National Science Foundation, the Office of Naval Research, and the Air Force Office of Scientific Research.[25] Each funding body explained that the research was somewhat outside the scope of its funding interests at the time. We now know that Bledsoe's work was supported by the CIA, though secretly, through various "front" companies.[26] On July 21, 1965, Bledsoe received a letter from Robert Gallati, executive director of a new initiative, the New York State Identification and Intelligence System (NYSIIS). NYSIIS was the first large-scale computer databank for law enforcement in the United States, meant to centralize, standardize, coordinate, and circulate the data gathered and records kept by all parts of the "law enforcement system" encompassing police, prisons, and courts.[27] Gallati wrote, "I have been informed by Dr. Charles R. Kingston of the School of Criminology, University of California, Berkeley, that you are engaged in a study in connection with digital photographic identification. It would be greatly appreciated if you would advise as to the status of your study, and forward any material thereon you deem desirable. I am currently working on the entire area of Personal Appearance identification for the New York State Identification and Intelligence System, and any information which you can supply will be most helpful."[28]

NYSIIS represented the application of Cold War era systems thinking to domestic law enforcement.[29] It was motivated, in part, by the findings in the *Final Report of the National Commission on the Causes and Prevention of Violence* published in 1969 at the behest of President Lyndon B. Johnson. The report made a great many observations about violence in America and advocated a number of different solutions—including ending the war in Vietnam to free up more money for investment in social support programs and schools, the urgency of following through on legislation to provide everyone a house and a job, lowering the voting age so as to enroll more youth in political self-determination, and reforming the draft in order to end inequities among draftees, some of whom could obtain deferments. And the commission insisted that all of the disparate branches of law enforcement desperately needed to be organized into one system, with the report citing chaos, miscommunication, mismanagement, overlap, and lack of sharing as serious information management problems plaguing criminal justice.[30] Hailed as "the world's first computerized criminal-justice information-sharing system," NYSIIS was meant to address that last finding: it was to be an information-processing system that worked across and coordinated between the disparate branches of law enforcement.[31]

The development of NYSIIS involved digitizing and standardizing historical policing records from across the state of New York and creating various search protocols to make cross-referencing and searching within this database more efficient and "accurate." However, NYSIIS developers didn't just want to create a more effective way for police, lawyers, courts, parole boards, and other actors in the criminal justice system to create, search, and investigate records. They were also greatly invested in automatic identification. For example, they ran studies for the development of algorithmic fingerprint matching and algorithmic license plate matching, and were deeply interested in algorithmic photograph matching as well.[32] Bledsoe's algorithm was adopted into the range of studies undertaken at NYSIIS on the automatic recognition of mug shots.[33] These reports describe the "mug file" as a particularly unruly database—full of photographs taken without standardized poses and equipment, cobbled together from around the state, extraordinarily difficult to make use of for investigative or identification purposes by hand. NYSIIS developers hoped that computer automation would make the mug file more useful by automatically matching photos of "people of interest" with possible matches in the database and would thereby narrow searches, improve witness identifications, and reduce the likelihood of "missed" repeat offenders. Remarkably similar concerns motivated Alphonse Bertillon's initial introduction of mug shots to policing in 19th-century France.

In NYSIIS reports on automated facial recognition, Bledsoe's algorithm circulated as proof of concept that this kind of automatic recognition could

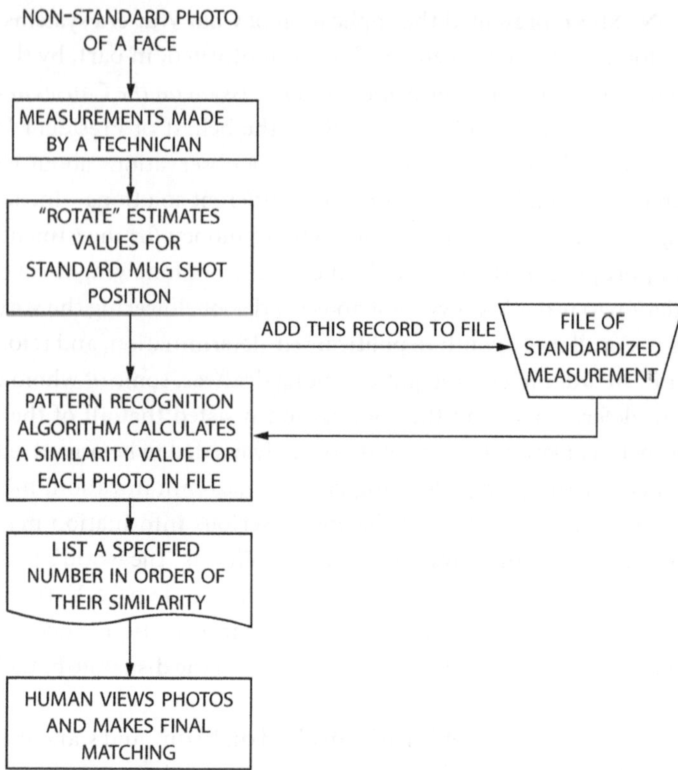

Figure 7.7.

be done, and even that some of the obvious difficulties, including head rotation in photos, were "solved." Bledsoe's own acknowledgment that the standard head was not a good solution was lost completely as his system traveled into the hands of NYSIIS developers. Instead, as figure 7.7 shows, Bledsoe's head rotation correction calculation was simply included as the second step, which "estimates values for standard mug shot position."

The developers also adopted the oversimplified version of facial recognition as pseudodistance calculation in these studies. Although Bledsoe's algorithm, to my knowledge, was never actually used in practice on the NYSIIS database, it served as *proof* that automated facial recognition could be achieved and that the obvious technical problems had been "solved."[34] This confirmation, in turn, shaped choices made about how NYSIIS was set up—particularly how mug shots were standardized across New York state to be focused on the face, rather than the whole person, and to consist of two photographs—one straight on, in what developers renamed the "standard mug shot" position, and one at a slight angle of rotation to simplify the calculations. Originally, NYSIIS developers had been working on standardizing lists of "physical attributes" (tattoos, height, etc.) that could be applied to persons of interest and stored in their databanks for *word search*-based identification. But with Bledsoe's algorithm in hand, other approaches to identification were collapsed onto the standard mug shot and the automated identification of faces. New York State mug shots closely resemble the pairs of photographs that constituted Bledsoe's original set of 2,000 pairs of photographs of white men. However, the mug files of New York State were not primarily populated by photographs of white men. The 1960s and '70s represented an unprecedented scaling up of both incarceration and violence in the United States, disproportionately affecting Black men and other people of color.[35] Tools that had been designed and tested on white men, the "standard head," the pseudodistance calculation, and the standard mug shot were now directed outward onto policed communities and communities of color.

Algorithms are haunted, by historic injustices that live on in the data sets on which algorithmic models are trained; by technical decisions made long ago to accommodate now obsolete machinery; by unfounded assumptions and quick fixes used to "solve" hard problems in the past; by their developers—their biases, their ideologies, their motivations, and (as in the case of the "standard head") even their bodies.[36]

Conclusion: Experimentalism and Power

Community organizer and abolitionist Sarah T. Hamid has explained how attempts to make policing "scientific" have introduced a fatal experimentalism in law enforcement: if policing is a science, then it is allowed to experiment, and experiments sometimes fail. In fact, they are expected to. These failures—for example, that of facial recognition systems—do not serve to *weaken* these systems or to undermine their authority as one might hope. Instead, Hamid shows, these failures *strengthen* the systems in that "the harm they induce is minimized and neglected because they are treated as constantly improving." Hamid reported that a police chief in Los Angeles "actually used the words 'work in progress' and 'testing for improvement' to describe a program that has had a direct hand in killing five people that we know of, re-incarcerating people who are out on parole and evicting people from their homes."[37] Further, Hamid has explored how identifying the failures of algorithmic policing systems often does not discredit or disempower them but rather makes them stronger because pointing out failures can lead to increased resources, more funding, more research, more data as developers seek to correct these endemic problems.

Given the urgency and clarity of her observations, I do not end this chapter by suggesting that Bledsoe's algorithm for facial recognition was problematic because it would misidentify those whose heads diverge most from the assumed "standard"—though that is certainly true. The more important lesson of this story is that it reveals how computerized solutions are *constructed*: they often involve both the redefinition and oversimplification of the problem and the introduction of unjustified assumptions. Our belief that computers can solve hard problems, that they can achieve success in domains like facial recognition is predicated on such oversimplifying redefinition and unjustified assumption. And it is this *belief*—that these systems can be built, fixed, optimized—that is more damaging than the errors that result from these poor technical workarounds. Dan Bouk, in reflecting on Hamid's work, has proposed that "even failed technologies can justify ramping up surveillance (or maintaining it) and even if the technologies persistently failed, the dream of future success at once made policing seem very sophisticated and objective, while often giving a reason to gather new forms of data."[38] The algorithms for automated license plate identification, however poorly built or accurate, are used to justify proliferating CTV surveillance of highways and roads, for example. As Kelly Gates has observed, the belief that automated facial recognition, had it been widely implemented, could have stopped the attacks on the twin towers on 9/11 by identifying wanted men as they moved through airports has justified the vast expansion of surveillance and facial recognition since 2002.[39]

Broken, bad, failing, partial, oversimplified—algorithmic systems are used nonetheless. And used to justify more of the same. Always with future corrections in mind, these systems represent "state of the art now" and "promises of more accurate futures." But as the case of Woody Bledsoe's facial recognition algorithm reveals, these future visions are seldom justified by the current state of the art. And worse, the delusions of those futures justify much present state-imposed violence and injustice, not least because these systems are haunted by decisions made and forgotten, oversimplifications that have been excluded from proclamations of success, and poor workarounds and technical patches like the "standard head" that were meant to get things "working" when in fact they don't and never have. In the pursuit of more just outcomes, we should focus not only on the mistakes that algorithmic systems make but also on the ghosts that follow them around, so that we may undo the grandiose claims of objectivity and science that heralded the development of automated "facial recognition" and so much else.

Acknowledgments

I am grateful to Sarah T. Hamid, Dan Bouk, Marc Aidinoff, Wendy Chun, Myrna Perez Sheldon, and Travis Dick for their invaluable feedback on different parts of this work. I am also grateful to the editors and the reviewers of this volume for incredibly helpful feedback. I thank Jeff Yost and the archivists at the Charles Babbage Institute for their invaluable support in locating materials related to the NYSIIS and to Bill Aspray for pointing me to a greatly helpful interview with Woody Bledsoe. I also thank Kelly Gates, Clare Kim, and Whitney Laemmli for engaging with this work at our "Bodies of Formalization" session at the History of Science Society meeting in 2018.

Notes

1. Kelly Gates explores how faces in general became central sites for identification in her foundational history of facial recognition technology, *Our Biometric Futures: Facial Recognition Technology and the Culture of Surveillance* (New York: New York University Press, 2011).

2. "Expert Says 1840s Photo Is of Lincoln," *Washington Post*, July 14, 1994, https://www.washingtonpost.com/archive/lifestyle/1994/07/14/expert-says -1840s-photo-is-of-lincoln/1ea4be16-2e05-402d-8d70-f8d4aac3d9ab/.

3. Joe Nickell, *Real or Fake: Studies in Authentication* (Lexington: University of Kentucky Press, 2009).

4. There are two quite different critical orientations to these contemporary debates. Some critics advocate for a politics of refusal relative to facial recognition, arguing that whether they work well or poorly, facial recognition technologies should be banned. See, for example, Kashmir Hill, *Your Face Belongs to Us: A Secretive Startup's Quest to End Privacy As We Know It* (New York: Penguin Random House, 2023); American Civil Liberties Union, "The Fight to Stop Face Recognition Technology," updated June 7, 2023, https://www.aclu.org/news /topic/stopping-face-recognition-surveillance. Others have focused on the identification of bias in facial recognition systems, with the aim of improving accuracy and inclusion. See, for example, Joy Buolamwini and Timnit Gebru, "Gender Shades: Intersectional Accuracy Disparities in Commercial Gender Classification," *Proceedings of Machine Learning Research* 81 (2018): 1–15.

5. See Stephanie Dick, "AfterMath: The Work of Proof in the Age of Human-Machine Collaboration," *Isis* 102, no. 3 (2011): 494–505; Dick, "Coded Conduct," *British Journal for the History of Science: Themes* 5 (2020): 205–24; Hallam Stevens, "A Feeling for the Algorithm: Working Knowledge and Big Data in Biology," *Osiris* 32 (2017).

6. Many early computing and artificial intelligence researchers acknowledged outright that most problems were simply intractable for computing without significant redefinition and constraint. Artificial intelligence researcher Herbert Simon introduced the concept "satisficing" (from "satisfy" + "suffice") to the Cold War control sciences. The concept admits that many problems are in fact simply intractable for computers (and, relatedly, for bureaucracies) unless they have been constrained, and he set out to articulate how these constraints could be fashioned. I suggest that these constraints and redefinitions form a crucial site for a sociotechnical theory of computer "solutions." See Herbert Simon, "Rational Choice and the Structure of the Environment," *Psychological Review* 63, no. 2 (1956): 129–38. See also Stephanie Dick, "Of Models and Machines: Implementing Bounded Rationality," *Isis* 106, no. 3 (2015): 623–34.

7. On the controversial proof of the four-color conjecture and the practices of redefining maps that were involved in solving it, see Donald MacKenzie, "Slaying the Kraken: The Sociohistory of a Mathematical Proof," *Social Studies of Science* 29, no. 1 (1999): 7–60; and Imre Lakatos, *Proofs and Refutations: The Logic of Mathematical Discovery* (Cambridge: Cambridge University Press, 1976).

8. David Bloor, "Polyhedra and the Abominations of Leviticus," *British Journal for the History of Science* 11, no. 3 (1978): 245–72; Mary Douglas, *Natural Symbols: Explorations in Cosmology* (London: Routledge, 1966).

9. From among a vast literature, see, for example, Jonathan Finn, *Capturing the Criminal Image: From Mug Shot to Surveillance Society* (Minneapolis, University of Minnesota Press, 2009).

10. Finn, *Capturing the Criminal Image*, 20*ff.*

11. Eden Medina has explored the competing forms of expertise and trained

judgment that accompanied efforts to identify the dead in post-Pinochet Chile in pursuit of justice and healing for families and survivors there. I am deeply grateful for conversations with her on these topics and for her scholarship in this area. See Medina, "Forensic Identification in the Aftermath of Human Rights Crimes in Chile: A Decentered Computer History," *Technology and Culture* 59, no. 4 (2018): S100–S133.

12. Anne Olivia Boyer and Robert S. Boyer, "A Biographical Sketch of W. W. Bledsoe" in *Automated Reasoning: Essays in Honor of Woody Bledsoe*, ed. Robert S. Boyer (Dordrecht: Kluwer Academic, 1991), 1–29. On the clandestine support of the CIA through "front companies," see Shaun Gidon Raviv, "The Secret History of Facial Recognition," *Wired*, February 1, 2020.

13. See Hunter Heyck, "Defining the Computer: Herbert Simon and the Bureaucratic Mind—Part 1," *IEEE Annals of the History of Computing* (April–June 2008): 42–51; Heyck "Defining the Computer: Herbert Simon and the Bureaucratic Mind—Part 2," *IEEE Annals of the History of Computing* (April–June 2008): 52–63; Stephanie Dick, "Of Models and Machines," *Isis* 106, no. 3 (2015): 623–34.

14. See Diana Forsythe, *Studying Those Who Study Us: An Anthropologist in the World of Artificial Intelligence* (Stanford, CA: Stanford University Press, 2002); David Brock, "Learning from Artificial Intelligence's Previous Awakenings: The History of Expert Systems," *Association for the Advancement of Artificial Intelligence* (Fall 2018): 3–15; Dick, "Coded Conduct: Making MACSYMA Users and the Automation of Mathematics," *British Journal for the History of Science: Themes* 5 (2020): 205–24.

15. Later, "pattern recognition" took on a specific meaning in the context of data-driven machine learning, in which neural networks are trained by identifying patterns and correlations in large datasets. Bledsoe had a more expansive and agnostic definition that included a wide variety of "patterns." On the machine learning–specific sense of the term, see Aaron Mendon-Plasek, "Mechanized Significance and Machine Learning: Why It Became Thinkable and Preferable to Teach Machines to Judge the World," in *The Cultural Life of Machine Learning: An Incursion into Critical AI Studies* (Cham, Switzerland: Palgrave Macmillan, 2021), 31–78; Matthew Jones, "How We Became Instrumentalists (Again): Data Positivism since World War II," *Historical Studies in the Natural Sciences* 48, no. 5 (2018): 673–84.

16. W. W. Bledsoe, "Some Results on Multicategory Pattern Recognition," *Journal for the Association for Computing Machinery* 13, no. 2 (1966): 304–16.

17. Bledsoe, "Pattern Recognition and Reading by Machine," *Proceedings of the Eastern Joint Computer Conference* (1959): 225–32.

18. W. W. Bledsoe, "The Model Method of Facial Recognition," Technical Report PRI 15 (Palo Alto, CA: Panoramic Research Inc., August 12, 1964).

19. Aaron Gluck-Thaler, "Identity Instrumentalized: Pattern Recognition as an Epistemology of Surveillance, 1960-66," *Technology and Culture* 65, no. 4 (2024): 1163, 1162.

20. Iris Clever, "Old Bones in New Databases: Historical Insights into Race, Statistics, and Ancestry Estimation in Anthropology," *American Anthropologist* (September 2025).

21. Bledsoe, "Model Method of Facial Recognition."

22. W. W. Bledsoe and P. E. Hart, "Semi-automatic Face Recognition System," Technical Report, SRI, Project 6693 (1968). Projective geometry is a mathematical field that explores the properties of three-dimensional objects using the properties of their two-dimensional surfaces. Bledsoe wondered if there was a

way to accurately derive the required three-dimensional measurements of a head from the two-dimensional photograph using these mathematical techniques but never proposed how to do it or attempted to do it himself. Today, neither a standard head nor projective geometry are likely to be used, but rather a convoluted neural network would be trained to "predict" three-dimensional measurements from a two-dimensional photograph.

23. A vast literature critiques this orientation. See, for example, Wendy Chun, *Discriminating Data: Correlation, Neighborhoods, and the New Politics of Recognition* (Cambridge, MA: MIT Press, 2021); Xiaochang Li, " 'There's no data like more data': Automatic Speech Recognition and the Making of Algorithmic Culture," *Osiris* 38 (2023).

24. Clever, "Old Bones in New Databases."

25. Rejection letters dated September 28, 1964, February 23, 1965, and October 12, 1964, respectively, Woodrow Wilson "Woody" Bledsoe Papers, Briscoe Center for American History, University of Texas at Austin.

26. Raviv, "The Secret History of Facial Recognition."

27. See "A New Concept in Criminal Justice Information-Sharing: The New York State Identification and Intelligence System Development Plan" (July 26, 1968), Archives and Special Collections, Charles Babbage Institute.

28. Robert Gallati to Woody Bledsoe, July 21, 1965, Woodrow "Woody" W. Bledsoe Papers (1996-349), Dolph Briscoe Center for American History, University of Texas at Austin.

29. NYSIIS was part of a more widespread application of defense establishment frameworks to domestic governance, a topic explored in, for example, Hunter Heyck, *Age of System: Understanding the Development of Modern Social Science* (Baltimore: Johns Hopkins University Press, 2015); Joy Rohde, *Armed with Expertise: The Militarization of American Social Science during the Cold War* (Ithaca, NY: Cornell University Press, 2013); Jennifer Light, *From Warfare to Welfare: Defense Intellectuals and Urban Problems in Cold War America* (Baltimore: Johns Hopkins University Press, 2005); Beverly Gage, *G-Man: J. Edgar Hoover and the Making of the American Century* (New York: Viking, 2022); Cody Ewert, *Making Schools American: Nationalism and the Origin of Modern Educational Politics* (Baltimore: Johns Hopkins University Press, 2022); Elizabeth Hinton, *America on Fire: The Untold History of Police Violence and Black Rebellion since the 1960s* (New York: W. W. Norton, 2021).

30. US National Commission on the Causes and Prevention of Violence, *Final Report* (Washington, DC: Government Printing Office, 1969), online at Office of Justice Politics, https://www.ojp.gov/pdffiles1/Digitization/275NCJRS .pdf.

31. Jeffrey M. Silbert, "The World's First Computerized Criminal-Justice Information-Sharing System, the New York State Identification and Intelligence System (NYSIIS)," *Criminology* 8, no. 2 (1970): 107–28.

32. On efforts toward automated fingerprint matching, see, for example, C. R. Kingston and F. G. Madrazo, "NYSIIS: Latent Value Study" (National Criminal Justice Reference Service, Law Enforcement Assistance Administration, US Department of Justice, June 20, 1970) https://www.ojp.gov/pdffiles1 /Digitization/17272NCJRS.pdf. For a summary of their efforts to automate and justify license plate scanning at scale and the automation of plate identification in the database, see Brian Keenan, Kenneth Kerin, Robert Shamberger Jr., and Renee Sonneman, "A Socio-economic Valuation Study for Automatic License Plate Scanning System: Final Report" (NYSIIS Contract No. C36276), January 19, 1970, National Criminal Justice Reference Service, Law Enforcement Assis-

tance Administration, US Department of Justice, accessed February 3, 2025, https://www.ojp.gov/ncjrs/virtual-library/abstracts/socio-economic-valuation -study-automatic-license-plate-scanning.

33. See Ben Rhodes, Kenneth Laughery, James Bargainer, James Townes, and George Batten Jr., "Final Report on Phase One of the Project: A Man-Computer System for Solution of the Mug File Problem," report prepared for the New York Department of Justice, Law Enforcement Assistance Administration, National Institute of Law Enforcement and Criminal Justice, under Grant 74-NI-99-0023 G, August 26, 1976).

34. Ben Rhodes, "Forgery Application of a Pattern Recognition Algorithm for Facial Images," Technical Report UHMUG_11, University of Houston, June 1977.

35. The vast literature on incarceration in this era includes Elizabeth Hinton, *From the War on Poverty to the War on Crime: The Making of Mass Incarceration in America* (Cambridge, MA: Harvard University Press, 2016); Ruth Wilson Gilmore, *Golden Gulag: Prisons, Surplus, Crisis, and Opposition in Globalization California* (Berkeley: University of California Press, 2007); Heather Ann Thompson, *Blood in the Water: The Attica Prison Uprising and Its Legacy* (New York: Vintage Press, 2016); and Dan Berger, *The Struggle Within: Prisons, Political Prisoners, and Mass Movements in the United States* (Oakland, CA: PM Press, 2014).

36. Wendy Chun explores this notion of haunting in *Discriminating Data*.

37. Sarah T. Hamid, untitled paper presented at "Histories of AI" conference, Columbia University, 2018.

38. Dan Bouk, email correspondence with the author, May 25, 2019.

39. Kelly Gates, *Our Biometric Futures: Facial Recognition Technology and the Culture of Surveillance* (New York: New York University Press, 2011).

8

Spanning Space and Time Barriers
Computerized Conferencing, Disability, and Citizenship

Elizabeth Petrick

There is much discussion today about digital citizenship and what it means to participate in a global society online and how all people might have an equal opportunity to do so. Digital citizenship is also a concept that is not well defined. In their landmark 2008 study, Karen Mossberger, Caroline Tolbert, and Ramona McNeal opened with a succinct definition: "'Digital citizenship' is the ability to participate in society online."[1] Since then, numerous works have examined the relationship between citizenship and digital technologies, primarily in the context of schools and civic education. Looking at the state of the field today, Luci Pangrazio and Julian Sefton-Green conclude: "Ontologically, citizenship marks our relationship to something *beyond* the individual—be that the community, platform or the nation state. In this way, digital citizenship is not just about civic responsibilities or self-responsibilization, but rather how the digital facilitates new forms of participation."[2]

In spite of this more recent talk of digital citizenship, however, concerns about how people might take part in a digital social world via networked computer technologies are not new; from the very beginning of computer networking, researchers studied online socialization and participation. Starting in the 1970s, developments in computer networking in the United States brought with them new possibilities for asynchronous communication. That is, communication across distances but not in real time. These programs, going by names such as "computer-mediated communication" and "computerized conferencing," allowed users to be located at computer terminals anywhere in the world and to communicate with each other through written messages that could be read and responded to at one's leisure, archived for later reading, or shared with multiple people. Some of the researchers, particularly a group I examine at the New Jersey Institute of Technology (NJIT), looked into these technologies and realized their potential applications in a number of arenas, including working from home on one's own computer, collaborative scientific research between different universities, and accessing social services. These researchers also considered who might use these new forms of communication; in addition to scientists, students, and businesspeople, people with disabilities were featured as intended users. Notably, people with disabilities were explicitly included as research subjects and future intended users of this technology, making it one of the first digital technologies created with their needs in mind.

Although I think this is an interesting historical episode that demonstrates a merging of changing cultural ideas with new technologies, this chapter also comments on how US citizenship is constructed and changes. Citizenship theory offers not only a lens through which to understand how views of disability have changed but also a way to think through theories of digital citizenship that scholars are focusing on today that try to get at the relationship between digital technologies and social participation. In particular, we may view the fight for greater social participation at the heart of the struggle for disability rights in the United States as a fight for citizenship, as citizenship is at the heart of all civil rights efforts. The two—citizenship and civil rights—are intimately and inherently linked. As Julie Minich argues in her work on disability and citizenship in Mexico, "Civil rights struggles benefit society as a whole, not just the specific group of people they seek to liberate, because they make available to everyone an expanded framework for conceptualizing rights." By taking a more expansive view of citizenship, beyond just political participation, "citizenship . . . functions as a way of making the distribution of rights more equitable, not as a fixed relationship to a defined nation-state."[3] Disability offers a different and, I would argue, necessary perspective on citizenship. I take from Gerard Goggin the idea that "disability problematizes concepts of citizenship and how such concepts of citizenship are operationalized."[4] Disability challenges assumptions of normalcy and expectations of who counts as a good citizen. By doing so, disability raises fundamental questions about who belongs in society.

The form of citizenship that Americans with disabilities fought for during the mid-twentieth century was about visibility and being present and active in society. However, as Alison Kafer has argued, "becoming more 'visible'—by increasing and publicizing the presence of disabled people in public, perhaps—does not guarantee acceptance or inclusion, especially for those not already privileged by race and class."[5] Being part of society is not as straightforward as having a public presence. Having *access to* a place or group is about belonging, in different times, places, and contexts and in different ways.[6] Further, disability has been linked historically with the private, not the public; it has been about illness, institutions, and stigma. These connotations of disability are compounding not only in the expectations they put on people with disabilities to be invisible and not take part in society, but also in the ways they disable people further by creating ableist assumptions about who then does belong and has access to visible participation in public.[7]

The digital world and digital participation both fit and do not fit with idea of visibility and access to physical public spaces, and the two worlds affect each other in complex ways. One of the major concerns that have arisen with the internet is that it reproduces the same prejudices and bigotry present in the physical world, largely because it is not actually a separate world. As Kate Ellis and Mike Kent argue, "Web 2.0 has been developed in and by the same social world that routinely disables people with disability."[8] Many scholars have shown that discrimination and lack of access to physical places have become mirrored in discrimination in digital ones. Ruha Benjamin makes this point, regarding racism, with her concept of the "New Jim Code," a way to understand how social inequities play out in digital spaces and technologies. These new forms of discrimination take on a hardy veneer of technological objectivity, as she argues: "Tech fixes often hide, speed up, and even deepen discrimination, while appearing to be neutral or benevolent when compared to the racism of a previous era."[9] Digital technology, and the internet in particular, presents a vision of a future without bigotry and prejudice that, as Lisa Nakamura has shown, does not fit the reality we live in.[10] Efforts at inclusion can also hide ulterior motives that do not actually serve users, as Mariel García Llorens discusses in this very collection.[11]

Digitality can change how prejudice takes place, however, through visibility. While not all aspects of identity are visible in the physical world, either, the digital world offers a space that presumes everyone to be the same, hence its seeming objectivity, though this idea of sameness, as I discuss later, makes assumptions about the identities of those present.[12] Access to the digital world not only mirrors the physical but can also affect it, as Avery Dame-Griff shows in his study of the transgender internet, where the move, particularly among young people in the 21st century, to embrace an online trans community resulted in a perceived "existential threat" to local, physically based trans organizations. This era contrasted with an earlier period in internet history, the 1990s, when at least one trans group was reluctant to move online due to disparities in access to the internet.[13] Access, then, is not just about individuals but also about the how social movements and communities begin and grow, where they choose to congregate, and who is and is not part of those spaces.

Over time, computer networking has become a tool through which citizenship can be practiced, but as this chapter shows, this has been a form of citizenship that challenged ideas of embodiment and visibility. Who is allowed to be themselves and visible, even participating in society using digital technologies, is still unsettled. Using citizenship to talk about computer technology provides new ways to think about the embodiment of technology, the visibility of people with disabilities in society, the relationship between digital and physical spaces, and the role of technology in enacting equity and rights. While the case study I discuss concerns US history, the technology was imagined to be used by a global society (and was a precursor to the internet). I situate this history in the struggle for American disability rights, as that is what the computer researchers I study were living through. However, their vision for their technology was that it would transcend national boundaries, in a way that was just as utopian as their view as to how it would benefit people with disabilities.

This case study concerns how researchers on the cutting edge of computer technology thought about people with disabilities as computer users and how they conceived of the relationship between disability and technology. I find that their perspective on disability was that it was one social category of many that might disadvantage a person and prevent full social participation; technology, then, was a solution that addressed social inequality. Computerized conferencing was a leveling technology in these researchers' view in that it could hide differences and certain aspects of identity, allowing a user to reveal what they chose about themselves, but otherwise making all users communicate as apparent equals. In a way, the researchers saw the technology as removing embodiment from the picture, so that a technology that purports to treat people equally and create a digital space for social participation does so by hiding identity. Whereas citizenship was once understood as based on certain values that this country held as representative of the norm—values that did not include physical, mental, or intellectual disabilities and also excluded minorities, women, and LGBT people—here, in this digital space of online communication, those aspects of oneself become hidden, revealed only, it seems, as one chooses. The reality of using this technology and participating socially online was, however, more complicated. I discuss how the researchers talked about this feature of hiding identity and what they saw it doing for people, along with the moments where these aspects of identity intruded anyway. And how all of this gives us something new to think about, regarding both computers and citizenship.

Citizenship Theory

I am less interested in aspects of more traditional political citizenship. Throughout US history, people with different kinds of disabilities have been denied rights and full citizenship in many different ways. In terms of political participation, these exclusionary tactics have included literacy tests as a requirement to vote and inaccessible governmental buildings or polling locations, along with, in terms of basic human rights, prohibitions on marrying, having children, moving about the built environment, and even making independent decisions about one's life. In order to focus this broad history of subjugation to the use of networked computer technologies, I would, instead, like to focus on forms of what has been called social or cultural citizenship. As sociologist Jean Beaman explains: "Even though citizenship as a status is a legal demarcation (i.e., citizens vs. noncitizens), it also differentiates among citizens or creates different classes of citizens, and therefore different degrees of full inclusion in society. A focus on cultural citizenship, and how it is denied to various populations, reveals these dynamics."[14] Cultural citizenship offers a way to consider someone's place anywhere in their society, along with any limitations to their participation. Renato Rosaldo considers the ways people are marginalized from full cultural citizenship: "The concept of cultural citizenship includes and also goes beyond the dichotomous categories of legal documents, which one either has or does not have, to encompass a range of gradations in the qualities of citizenship. Ordinary language distinguishes full from second-class citizens and tacitly recognizes that citizenship can be a matter of degree."[15] Although Rosaldo is focused on race and ethnicity, disability can result in another form of marginalization that can deny someone full ability to participate in their society.

Historians have studied this expanded notion of citizenship in terms of economic or consumer citizenship. Meg Jacobs has described the importance of the former: "In the twentieth century, as the economy and society became increasingly organized around a new national mass consumer market, the means to consume became important not only for securing three square meals a day but more broadly as a marker of economic citizenship and full membership in the American polity."[16] And in *A Consumer's Republic*, Lizabeth Cohen discusses the latter, particularly in terms of race and Jim Crow discrimination against African Americans: "Consumer citizenship was held out as an ideal, and then made unattainable." African Americans, Cohen argues, experienced "their denial of full citizenship regularly through their exclusion from sites of consumption, such as at theaters, restaurants, hotels, and commissaries on military bases."[17] Historical parallels can be drawn between the fight for civil rights waged by African Americans and that waged by people with disabilities. While there is a difference between a sign or law that says you cannot enter versus an entryway to a business you physically cannot access, (1) disabled people of color faced both, and (2) both sorts of barriers led to a lack of access to full social participation.

These parallels are made explicit in the histories of both rights movements. According to Paul Longmore and Lauri Umansky, the need to fight for full citizenship in the mid- to late twentieth century began at least as early as the 1880s with the denial of full rights for people with disabilities, particularly in the form of increasing institutionalization.[18] Behind the denial of such rights was an assumption that tied citizenship to productivity and the idea that people with disabilities were not capable of contributing to society through labor, a notion that a number of scholars have tackled.[19] Julie Passanante Elman has described this connection as fundamental to American culture and one of the "ideals of American liberalism: productivity, freedom, and self-reliance as central to good citizenship."[20] Allison Carey has written about how people with intellectual disabilities became a particular target of this way of thinking. "While the storybook ideal citizen" she explains, "exudes intelligence, independence, and the ability to contribute to the national well-being through hard work, political participation, and bravery, people with intellectual disabilities tend to be characterized by their deficiencies. . . . Thus, when judged by the standards of the ideal citizen, the person with an intellectual disability may appear unworthy, at best, and a threat to the nation and himself or herself, at worst."[21] It was not until the 1960s and '70s that arguments for full citizenship for people with disabilities began to gain ground.

By the 1970s, the disability rights movement was drawing on the language and rhetoric of African American civil rights. Regarding the struggle for antidiscrimination laws, Doris Fleischer and Frieda Zames have observed: "Thus relegated to second-class citizenship no less than people of color, people with disabilities, like other minority groups, have required civil rights legislation in order to secure equality of opportunity."[22] Kim Nielsen has explored the fight for full citizenship for people with disabilities in greater depth, writing: "Using terms such as 'rights' and 'discrimination,' and employing the protest methods of the anti-war and racial freedom movements, people with disabilities increasingly, in the late twentieth century, demanded the opportunities and protections of full citizenship."[23] By "full citizenship," she is refers to employment, education, access to public spaces and transportation, and the right to self-determination. The passage of the Americans with Disabilities Act (ADA) in 1990 cemented these rights in the fullest statutory form so far, as Emily Russell discusses: "The conceptual reach of the ADA goes beyond these aspects of policy to impact the most fundamental concepts in American citizenship, including independence, individualism, and public responsibility to citizens."[24] One of the most striking set of images arguing for these rights includes the photos taken during the 1990 "Capitol Crawl," a protest organized by the prominent disability activist group ADAPT (American Disabled for Attendant Programs Today). A few months before the passage of the ADA, a group of people with disabilities staged a protest on the steps of the US Capitol by climbing or pulling themselves up the seventy-eight inaccessible stairs without the use of mobility aids.[25] These protesters were fighting for a visible form of citizenship, whereby people with disabilities are included in society by being able to access it, including notable government buildings that attract many tourists every year. Without civil rights protections to guarantee them full citizenship, such places remain literally out of reach.

Throughout this history, as people with disabilities were denied full citizenship, they were also denied full access to American society and culture. Nancy Hirschmann and Beth Linker have described this larger sense of citizenship as "belonging to a group, society, and culture: our neighbors, the parents of our children's classmates, the congregants of our churches, or the residents of our towns, our counties, our states, and our nation."[26] People with disabilities have been excluded through a lack of access to what Aimi Hamraie calls "spatial citizenship: the right to occupy homes, workplaces, universities, restrooms, courthouses, and cities."[27] This is all part of the visibility of people with disabilities in society; in order to be full citizens, they must be able to be present and to take part in such social practices.

Finally, Elizabeth Ellcessor brings many of these ideas together in her media and disability studies work, demonstrating how digital media have failed to be made accessible and allow full participation by people with disabilities. Her work marks, in many ways, the failure of the technology that is this chapter's focus and the dreams of the researchers behind it. She also explains why this subject matters, regarding online media today: "Participatory culture offers an attractive vision of a mediated future in which increased access to cultural production, political participation, and social collaboration produces more just, egalitarian forms of culture. The exclusion of people with disabilities from online media and attendant participatory cultures is particularly troubling given the potential of these spaces to foster engaged, active citizens of the world."[28]

Any fight for civil rights is a fight for full US citizenship. Practices of cultural citizenship—participating in society in various ways—are impacted by discrimination and barriers that block people with disabilities from access. The rhetoric of citizenship and rights was used by activists across the various civil rights movements of the mid-twentieth century. Technology, however, is not usually explicitly a part of this historical discussion, except with the implicit understanding that the public built environment and modes of public transportation were made accessible to people with disabilities through technological change. Those who do research and write about the connection between technology and citizenship today often discuss digital citizenship. I broaden the discussion to show how an understanding of cultural citizenship provides a language with which to talk about computer technology in the very early history of online life.

Computerized Conferencing and Disability

Computerized conferencing developed initially during the mid-1970s and preceded internet bulletin board and chatroom technologies. Much of the research on computerized conferencing was done at the New Jersey Institute of Technology (NJIT) with National Science Foundation funding and was led by computer scientist Murray Turoff and sociologist Starr Roxanne Hiltz. Turoff was an early pioneer in networking technology, building some of the first systems for online communication. Hiltz was one of the first sociologists to study the meaning of online interaction. According to Ramesh Subramanian, "Hiltz's work on virtual communities was arguably the first of its kind."[29] Hiltz and Turoff's creation of computerized conferencing in the form of the Electronic Information Exchange System allowed multiple people to communicate together online through, at the time, computer terminals and, later, personal computers. Participants could write to one another at the time of their choosing, with all messages in the group conversation stored online on a central server.

Turoff described the unique properties of this technology as a communication medium in a talk before the American Association for the Advancement of Science in 1975:

1. The individuals no longer have to be coincident in time, as in telephone calls or face-to-face meeting, since the computer keeps a record of the discussion and a bookmark for every individual on what he has seen.
2. The system allows each individual to work at his own pace, taking as much or as little time as he wishes to read, contemplate and/or reply (i.e., a "self activating" form of communication).
3. The system provides many of the signals present in face-to-face communication, i.e., who is in the discussion at any particular instant, what everyone has seen or not seen, when they were last in the meeting, etc.[30]

Computerized conferencing allowed for communication between people without some of the limitations of face-to-face communication, as participants could interact with one another on their own time and in their own space. The system also allowed users to identify themselves in different ways and keep track of what each person may have read or responded to. What would become apparent is that this medium also changed some of the ways people communicated as compared to face-to-face communication.

Hiltz, who studied new possibilities for social interaction with computerized conferencing, added to the list of features in an internal report on their research published the same year:

Everyone can "talk" or input whenever they wish, rather than having to "take turns" as in face-to-face verbal communications. Rather than only one person "having the floor," all participants could be typing in messages simultaneously. No one can be interrupted or "shouted down." . . .

Computer conferencing is much less "intimate" and self-exposing than verbal modes. Only your words (which can be carefully considered and edited) are transmitted, not your appearance, or other personal characteristics. The possibility of sending anonymous messages "legitimately" to other members of the conferencing group increases the possibility for "impersonal," relatively emotion-free communications.[31]

Computerized conferencing combined aspects of other non-face-to-face communication technologies: the text and potential anonymity of letter writing with the possibility of real-time discussion as on the telephone. But it added new characteristics, in the form of multiple people conversing at the same time and a written record of all communication maintained independently of the participants. The researchers also saw this technology balancing a line between personal and impersonal, in that it allowed for direct communication between people but while allowing one to hide one's entire identity or aspects of one's identity.

The idea of hiding identity affected not just people with disabilities but also women as early users of computer networking. In an early experiment comparing computerized conferencing to face-to-face communication among bank management personnel, Turoff noted that women were far more involved in the conversation on the computer than in person. Unfortunately, though, this was not the response the bank vice presidents wanted, as Turoff notes: "We got some very frank comments from this group that this was a good reason for them not to utilize the technology."[32] The possibility of everyone playing on an equal footing by removing physical markers of difference was not a feature that the people evaluating the technology desired. Maintaining the status quo of unequal communication meant rejecting the technology.

In spite of the observed resistance in some trials, the researchers believed the various features of computerized conferencing technology would make it particularly useful for people with disabilities. "This type of communication," Turoff concluded, "offers tremendous potential for improving the opportunity for the deaf and handicapped to lead more rewarding lives and to decrease greatly the limitations often imposed upon their mental capacity by the presence of inhibiting physical disabilities."[33] What is striking here is that he treats the disabled user as the imagined computer user. This technology was intended not only as an accessibility tool for people with disabilities but also as a universal technology that would benefit all. It would provide access to a new social space where people with disabilities could participate alongside those without disabilities.

In their remarkably prescient book *Network Nation: Human Communication via Computer* (1978), which brought computerized conferencing to a mass public audience, Hiltz and Turoff laid out the possibilities for a (near) future society where such technology was commonplace and integrated into all aspects of social life. This prospect included imagining forms of digital citizenship in which computerized conferencing would facilitate "the direct participation and voting of citizens on important state or national issues."[34] Citizen participation would include lobbying, citizen advisory groups, and communication by elected officials to their constituents.

People with disabilities, in particular, were a group that Hiltz and Turoff envisioned as benefiting from computerized conferencing technology in the online future. "The biggest advantage of computer-mediated communication," they wrote, "is that it spans space and time barriers, allowing a person to work, learn, and communicate from those places and at those times that are most convenient for him or her. Thus the mobility limitations of the physically handicapped make them a disadvantaged group that can benefit greatly from this technology."[35] The authors envisioned computer-mediated communication as a technology designed to take into account use by people with disabilities, similar to a building designed to be barrier-free so that anyone can access it. Computerized conferencing created a space for social interaction that people with disabilities could use, so that "a person may be physically impaired without being socially disadvantaged." The core idea here was to bring people with disabilities into full social participation, making them full citizens in the process.

Network Nation presented a future in line with many of the arguments that disability activists made in the 1970s about altering the built environment so that people with disabilities could be a part of society. However, there are two striking differences between their approaches. First, Hiltz and Turoff were largely focused on productive citizenship—that is, turning people with disabilities into workers who could contribute economically to society. The authors argued, for example, that a lack of education for children with disabilities would result in "an economic drain on society, since almost all handicapped persons are obliged to assume the status of unemployed, dependent individuals characterized by poverty, isolation, and low self-esteem."[36] This view fit with still-prevailing ideas about disability that focused on rehabilitation instead of civil rights.[37] It is striking, though, that Hiltz and Turoff also looked beyond employment as a positive benefit for computerized conferencing users with disabilities. A research report from 1978 summed up active projects and proposed questions that would need to be tackled to fully include people with disabilities as research subjects.[38] Aside from employment concerns, the researchers also mentioned uses ranging from education, to conversations between similarly disabled people, to social participation in general. The goal was to create a technology that replaced physical social participation with new, digital forms that everyone could access.

The second major difference between the researchers and disability activists concerned visibility. The authors explicitly viewed computerized conferencing as a technology of human augmentation that could enhance the communication abilities all people possessed. However, for people with a disability that affected face-to-face or telephone-based communication, computerized conferencing promised a means of communication in which their disabilities could go potentially unseen: "Various participants need not be aware of the disability that any of them suffer unless a person wishes to volunteer the information. . . . Even if the participants are aware that a particular person is blind or deaf, the social salience of the characteristic is much less, because it is not visible."[39] Computerized conferencing researchers promoted the technology as a way not only to grant someone abilities they did not otherwise possess but also to cover the disabilities they had. It offered the user a nonprescriptive (from the technology's perspective) option whether to reveal their disability, along with any other aspects of their identity, to others. The situation purported to be one of true equality, with everyone participating equally in the same space, disembodied and made neutral, with no clear characteristics of identity (other than the words they chose to use). Everyone is text on a screen. There was liberatory potential in not having to either hide or show oneself, but of course also the possibility of invisibility and assumptions of normativity by other users. One of the goals of computerized conferencing was, to the degree possible, to remove a person's disability from the act of communication, both by making participants unaware of each other's disability status and by providing a communication system relatively unaffected by disability. Computerized conferencing was about more than just accommodation; it challenged ideas of normalcy and showed that both personal technologies and new social environments could be developed with the needs of different people in mind.

One of the computerized conferencing trials provides a real-world look at the challenges of practicing citizenship at this time for people with disabilities and the benefits that the technology promised to offer. A seventeen-year-old girl, Beth, used a computerized conferencing system to lament problems and the frustrations she faced when trying to participate in physical civic activities. First, she attempted to run for student government but was not selected: "i ran for vice= president of student council but i lost. but ix xx do not care my girl friend won."[40] Second, she found herself unable to demand better accommodations to get around her town when presented with an opportunity to do so: "HI JIM ON SUNDAY I DID A VERY STUPID THING. I WENT TO A CHURCH FEAST AND SAW THE MAYOR OF NORTHFIELD. I WANTED TO TALK TO HIM ABOUT THE SIDEWALKS IN OUR TOWN, BUT I WAS TOO CHICKEN TO TELL HIM I NEED BETTER CURBS."

The researchers presented Beth's experience as a case of the computerized conferencing technology providing her an avenue to communicate that she struggled with in face-to-face situations. They hoped that the computer, assuming she had the means to access one at home, would allow her to continue communicating once she was no longer able to do so physically.[41] What is also striking about this case, though, is that Beth was participating in her community as a citizen. Even though she lost the election, she did run for a school office. Her second comment implies, though this is unstated in the report, that she felt comfortable in talking about problems with accessibility through computerized conferencing even though she was unable to bring herself to do so face to face with the mayor. The technology allowed her to speak when she could not in person, thus seeming to fulfill its more utopian promises.

In a 1981 report summarizing much of the work that had been done on various computerized conferencing projects, Hiltz and fellow sociologist Elaine Kerr argued that one of the benefits of communication across computer networks is that users can hide various aspects of their identities that others might judge them for: "It creates opportunities for communicating and joining groups without the instrusion [sic] of sex, race, physical appearance, or other irrelevant but intrusive characteristics."[42] What the authors seemed to suggest is that such characteristics, while clearly important to a person's life, are irrelevant to the computer and, seemingly, to computer-based communication. Once someone has access to a computer and the ability to use it for computerized conferencing, both of which might be affected by the demographic groups one belongs to, the computerized conferencing system then functions the same for all users. By "intrusive characteristics," I assume the authors referred to the possibilities for both overt and more subtle forms of discrimination that people might face when others know various aspects of their identity—exactly the kinds of discrimination people with disabilities face that keep them from becoming full cultural citizens.

Kerr and her colleague Robert Bezilla explained that it is the anonymity features of computerized conferencing that, in particular, allow users not to reveal aspects of themselves that might affect communication with others: "Ideas and achieved statuses become more relevant to the written exchange of issues, rather than ascriptive characteristics over which the individual has no control. Conferees can disguise cues irrelevant to professional and scientific dialogue which are influential in informal collegial communications, such as age, race, beauty, physical size, loudness of voice, body language, mannerisms, assertiveness, social class, and organizational position. Cues which could distract more than enhance the quality of group communications can be hidden."[43] Kerr and Bezilla highlighted the value of disembodied technology use. Computerized conferencing was thought to be egalitarian and focused on ideas, not physical characteristics a user happened to have. Hiltz and Kerr took this argument so far as to claim that computerized conferencing could be used to fight against social inequalities: "One of the many advantages of comput-

erized communications over face-to-face meetings is the reduction of social inequalities as it affects groups such as minorities, women, and the
handicapped." The power of the technology lay in the fact that communicating via a computer network allowed users to choose what they revealed
about themselves: "Users may elect to mask particular status cues. They
may choose to reveal or hide, accentuate or ignore, certain personality,
social, and cultural characteristics which would be readily apparent in
communication by any other media."[44] The researchers make this an issue
of user choice: that computerized conferencing gives a participant the option to reveal just those aspects of themselves that they choose, in the way
that they choose.

Missing from this perspective, however, is the possibility that, while
revelations of identity are a person's choice, group dynamics can influence what a person feels comfortable sharing, and what is shared can have
consequences. Choosing not to reveal aspects of oneself may offer benefits, but fearing that one cannot reveal oneself fully without negative consequences can cause harm. While the researchers were not saying that
everyone is assumed able-bodied unless stated otherwise, they implied
that this is a form of equality offered to people with disabilities only if
they do not acknowledge their disability. They can be equal social participants, but it is not clear that they can be equal and disabled. At this point
in the late 1970s and early 1980s, however, social communication over
the internet was still in its infancy and its consequences were not yet well
studied; the technology appeared to hold almost utopian potential to solve
communication problems for different people.

Hiltz and Kerr also discussed the lack of full social participation for
people with disabilities at this time and the ways computerized conferencing might break down existing social barriers: "Applications of the
technology to the handicapped and other disadvantaged have sought to
use these features to broaden opportunity structures for those suffering
mobility and communicatory restrictions, stigma, and exclusion from full
societal participation, to bring them into the mainstream of society and
their chosen careers."[45] As seen in *Network Nation*, computerized conferencing researchers predicted a future in which networking technology
would create a digital space for social participation where almost all aspects of face-to-face communication—education, employment, commerce,
hobbies, healthcare—would be available online, through one's own computer. They believed such a digital world would, in particular, allow those
who had difficulty leaving their homes to engage in social participation
through these online spaces. They could be full citizens there though they
might still be prevented from being so in the physical world. This is a
different angle on accessibility than the view otherwise beginning to be
tackled in the 1970s, of making the social, built environment accessible to
people with disabilities through the modification of buildings and streets.

Here, an entirely new, nonphysical space was built, with the view that existing forms of stigma would there be irrelevant, though which forms might then replace them was not yet considered.

Conclusion

Communication via networked computer technology, however, did not remain just text. As the use of networked computers for communication spread through the 1970s and 1980s, and even into the early World Wide Web of the 1990s, an irony emerged: the technology was most accessible for people with certain kinds of disabilities (as it was almost exclusively text based) but was still inaccessible for most people, period (due to the technical and financial costs required to get on the early internet).[46] The convergence of all forms of media on the internet in the 2000s has created both problems of access and opportunities for people with disabilities to embrace technologies such as social media. These opportunities presented places for people with disabilities to come together as an online community, fulfilling some of the potential that computerized conferencing had created.[47]

However, as Gerard Goggin and Christopher Newell observed in their foundational 2003 study on disability and new media, taking on disability as an identity in nondisabled online communities could be fraught: "Everyone on the Internet adopts a persona. Nonetheless when that persona revolves around disability—even on the Internet—one is seen as faking and offensive if departing from a fixed, reified identity position (such as a person with a disability). Curiously when others take on nondisabled roles it is liberating."[48] A more complicated internet has emerged from the initial potential of computerized conferencing, an online world where people can hide aspects of their identity, as promised, but revelations come with expectations and assumptions about proper membership within that group. Bodil Ravneberg and Sylvia Söderström found similar issues in their recent work looking at young disabled people's use of assistive information and communication technology (ICT). By using assistive devices, people are forced to disclose their disability status, and "this disclosure makes them feel revealed as deviant and disabled, an identity from which they distance themselves. Young disabled persons are routinely confronted with this dilemma of concealment or disclosure, as they find themselves forced to trade maximal social inclusion by using mainstream ICT for improved empowerment and employing assistive ICT for improved participation."[49] Unlike the simpler, text-based networked communication of decades ago, the more complicated, multimedia social networking of the past fifteen years has changed some of the possibilities of hiding or revealing one's identity. Most recently, the uptake of video-conferencing technologies during the COVID-19 pandemic created new complications for accessibility for people with different types of disabilities. For example, research on people with autism using video-calling technology finds that the communication medium presents numerous challenges for users that they have to adapt themselves to, rather than the other way around.[50] Every change in media type from text to images to audio and video, and the convergence of all of them together, has offered both new forms of accessibility for those with certain disabilities and challenges for those with others. It has never been as simple as putting people in front of computers and having a world easily open before them; yet the myth of computers—and communication through them, in particular—granting everyone new abilities and meeting their needs is a tantalizing one that endures.[51]

In thinking back to computerized conferencing, however, perhaps the most important aspect of the researchers' view of people with disabilities is that they saw them, first, as computer users. People with disabilities were, in one way, just one of many groups that might find advantages in the new forms of communication offered by computerized conferencing: in particular, being able to communicate asynchronously from one's own space, at one's own speed, and to hide or reveal as much about oneself as one desired, as everyone appeared equally in the form of text on a screen. These researchers' views toward disability straddled a number of different cultural contexts that existed during this time: the goal of turning people with disabilities into productive citizens through employment gradually shifting into a call for civil rights and equal access to social participation, all under a belief in the power of technology to fix both social and individual problems. Sometimes seeming paternalistic or naïve in their attitudes toward people with disabilities and technology, the computerized conferencing researchers laid the groundwork for today's networking technologies, built on ideals of universality and the breaking down of social barriers, even if they don't always live up to them.

This case study also offers an example of how computer technology can be a tool one may use to practice cultural citizenship. Cultural citizenship provides ways to understand why some legal citizens have fewer rights to social participation than others, as a result of various forms of discrimination and barriers. People with disabilities have been denied full citizenship since the beginning of US history. By the time computerized conferencing emerged in the 1970s, a national disability rights movement was coming together and demanding equal access to society. At the same time, these computer-networking researchers were trying to build a new social space—one that was digital and seemingly intended to be disembodied. The result was a technology that promised to create a new space for social participation where all could be equal, but it would do so by removing the user's body from the picture in certain ways. Aspects of users' identities would remain invisible unless explicitly stated, and there was no inherent requirement that they ever state anything about themselves unless they chose. Those characteristics of different disabilities that make them readily apparent in face-to-face interactions (and, of course, that does not include all disabilities) would be invisible for computerized conferencing interactions. Any given user would be bodiless from the perspective of both the computer and other users, as this was before the use of images as avatars in online communication. So the potential for cultural citizenship that this new social space promised would have been far more difficult in the physical world.

What was missing from this utopian dream, however, posed a major problem for this idea of disembodied online socialization, tied to the fact that the physical world had not actually dropped out of the picture: these various aspects of identity that were made invisible on the computer system still had salience to individuals and their interactions with others. Computerized conferencing was clearly not the first step toward an inevitable future of equality on the internet, but it offers us an example with which to think about how changing cultural views toward disability could impact how developers thought about the technology they were creating—a technology that would go on to impact further development of computer networking. It also suggests a rethinking about who counts as possessing the characteristics valued in good citizens; when all people are displayed as text on a screen, the body plays a different role in determining who is accepted in the community and who is not. So much of the fight for disability rights has been about visibility—being disabled *and* able to fully participate in society. Yet during the same time period, computer networking technologies such as computerized conferencing promised equality through invisibility. Whether or not this promise could ever work, it demonstrates how such values become embedded in computer technology, the effects of which we still grapple with today.

Notes

1. Karen Mossberger, Caroline J. Tolbert, and Ramona S. McNeal, *Digital Citizenship: The Internet, Society, and Participation* (Cambridge, MA: MIT Press, 2008), 1.

2. Luci Pangrazio and Julian Sefton-Green, "Digital Rights, Digital Citizenship, and Digital Literacy: What's the Difference?," *Journal of New Approaches in Educational Research* 10, no. 1 (January 15, 2021): 18, https://doi.org/10.7821/naer.2021.1.616.

3. Julie Avril Minich, *Accessible Citizenships: Disability, Nation, and the Cultural Politics of Greater Mexico* (Philadelphia: Temple University Press, 2013), 18–19.

4. Gerard Goggin, "Reimagining Digital Citizenship via Disability," in *Negotiating Digital Citizenship: Control, Contest, and Culture*, ed. Anthony McCosker, Sonja Vivienne, and Amelia Johns (Lanham, MD: Rowman & Littlefield, 2016), 64.

5. Alison Kafer, *Feminist, Queer, Crip* (Bloomington: Indiana University Press, 2013), 46.

6. Tanya Titchkosky, *The Question of Access: Disability, Space, Meaning* (Toronto: University of Toronto Press, 2011).

7. Susan Wendell, *The Rejected Body: Feminist Philosophical Reflections on Disability* (New York: Routledge, 1996), 40.

8. Katie Ellis and Mike Kent, *Disability and New Media* (New York: Routledge, 2011), 3.

9. Ruha Benjamin, *Race after Technology: Abolitionist Tools for the New Jim Code* (Cambridge, MA: Polity Press, 2019), 8.

10. Lisa Nakamura, *Cybertypes: Race, Ethnicity, and Identity on the Internet* (New York: Routledge, 2002).

11. See chapter 8 in this volume, "Pushing Fintech: Testing Financial Inclusion among 'Rural' Women in Peru," by Mariel Garcia Llorens.

12. Jennifer Karns Alexander discusses how the computer interface acts in between human users to obscure even as it mediates their communication with one another other. See chapter 15 in this volume, "The Mask of Sanity: Manipulation and Psychopathology at the Human-Computer Interface."

13. Avery Dame-Griff, *The Two Revolutions: A History of the Transgender Internet* (New York: New York University Press, 2023), 147, 23.

14. Jean Beaman, "Citizenship as Cultural: Towards a Theory of Cultural Citizenship," *Sociology Compass* 10, no. 10 (2016): 850.

15. Renato Rosaldo, "Cultural Citizenship in San Jose, California," *PoLAR: Political and Legal Anthropology Review* 17, no. 2 (1994): 57.

16. Meg Jacobs, *Pocketbook Politics: Economic Citizenship in Twentieth-Century America* (Princeton, NJ: Princeton University Press, 2007), 2.

17. Lizabeth Cohen, *A Consumers' Republic: The Politics of Mass Consumption in Postwar America*, paperback edition (New York: Vintage Books, 2004), 86, 13.

18. Paul K. Longmore and Lauri Umansky, "Introduction: Disability History—From the Margins to the Mainstream," in *The New Disability History: American Perspectives,* Paul K. Longmore and Lauri Umansky (New York: New York University Press, 2001), 1–29.

19. Richard Devlin and Dianne Pothier, "Introduction: Toward a Critical Theory of Dis-Citizenship," in *Critical Disability Theory: Essays in Philosophy, Politics, Policy, and Law,* ed. Dianne Pothier and Richard Devlin (Victoria: University of British Columbia Press, 2006), 1–22; Jess Waggoner, " 'Oh say can you ___': Race and Mental Disability in Performances of Citizenship," *Journal of Literary and Cultural Disability Studies* 10, No. 1 (2016): 87–102; Sarah F. Rose, *No Right to Be Idle: The Invention of Disability, 1840s–1930s* (Chapel Hill: University of North Carolina Press, 2017).

20. Julie Passanante Elman, *Chronic Youth: Disability, Sexuality, and U.S. Media Cultures of Rehabilitation* (New York: New York University Press, 2014), 6.

21. Allison C. Carey, *On the Margins of Citizenship: Intellectual Disability and Civil Rights in Twentieth-Century America* (Philadelphia: Temple University Press, 2009), 1.

22. Doris Fleischer and Frieda Zames, *The Disability Rights Movement: From Charity to Confrontation*, 2nd ed. (Philadelphia: Temple University Press, 2011), 69.

23. Kim E. Nielsen, *A Disability History of the United States* (Boston: Beacon Press, 2013), 160.

24. Emily Russell, *Reading Embodied Citizenship: Disability, Narrative, and the Body Politic* (New Brunswick, NJ: Rutgers University Press, 2011) 1.

25. See, for example, Becky Little, "When the 'Capitol Crawl' Dramatized the Need for Americans with Disabilities Act," *History,* July 24, 2020, https://www.history.com/news/americans-with-disabilities-act-1990-capitol-crawl.

26. Nancy J. Hirschmann and Beth Linker, "Disability, Citizenship, and

Belonging: A Critical Introduction," *Civil Disabilities: Citizenship, Membership, and Belonging,* ed. Nancy J. Hirschmann and Beth Linker (Philadelphia: University of Pennsylvania Press, 2015), 7.

27. Aimi Hamraie, *Building Access: Universal Design and the Politics of Disability* (Minneapolis: University of Minnesota Press, 2017) 9.

28. Elizabeth Ellcessor, *Restricted Access* (New York: New York University Press, 2016), 5.

29. Ramesh Subramanian, "Starr Roxanne Hiltz: Pioneer Digital Sociologist," *IEEE Annals of the History of Computing* (January–March 2013): 81.

30. Murray Turoff, "Computerized Conferencing for the Deaf and Handicapped," *SIGCAPH Newsletter* (Special Interest Group on Computers and the Physically Handicapped), no. 16 (1975): 5.

31. Starr Roxanne Hiltz, "Communications and Group Decision-Making: Experimental Evidence on the Potential Impact of Computer Conferencing; A Selective Review of Small Group Communications Experiments," Research Report No. 2, NJIT, September 1975, 7, 8.

32. Murray Turoff, "The Anatomy of a Computer Application Innovation: Computer Mediated Communications (CMC)," *Technological Forecasting and Social Change* 36 (1989): 115.

33. Turoff, "Computerized Conferencing for the Deaf and Handicapped," 5.

34. Starr Roxanne Hiltz and Murray Turoff, *The Network Nation: Human Communication via Computer* (Reading, MA: Addison-Wesley, 1978), 197.

35. Hiltz and Turoff, *Network Nation,* 169.

36. Hiltz and Turoff, *Network Nation,* 170.

37. For example, the first piece of legislation promising broad protections against discrimination in federally funded programs was Section 504 of the Rehabilitation Act Amendments of 1973. The language of "productive citizenship" continued to mingle with new attention to civil rights throughout the decade.

38. Murray Turoff, Philip Enslow, Starr Roxanne Hiltz, John Mckendree, Raymond Panko, David Snyder, and Richard Wilcox, "Research Options and Imperatives: in Computerized Conferencing," final report, Computerized Conferencing and Communications Center, January 1, 1978, archived at NJIT Digital Commons, https://digitalcommons.njit.edu/ccccreports/9/.

39. Hiltz and Turoff, *Network Nation,* 173.

40. Quoted in Elaine B. Kerr, Starr Roxanne Hiltz, James Whitescarver, and Sue Prince, "Applications of Computer Conferencing to the Disadvantaged: Preliminary Results of Field Trials with Handicapped Children," *Information Choices and Policies* 16 (1979): 156.

41. Kerr et al., "Applications of Computer Conferencing to the Disadvantaged," 157.

42. Starr Roxanne Hiltz and Elaine B. Kerr. "Studies of Computer Mediated Communication Systems: A Synthesis of the Findings," Final Report on a Workshop Sponsored by the Division of Information Science and Technology, National Science Foundation (1981), 182.

43. Elaine B. Kerr and Robert Bezilla, "Cues and Clues: The Presentation of Self in Computerized Conferencing," paper presented at the National Computer Conference, New York, 1979, 8.

44. Hiltz and Kerr, "Studies of Computer Mediated Communication Systems," 183.

45. Hiltz and Kerr, "Studies of Computer Mediated Communication Systems," 217–18.

46. Ellis and Kent, *Disability and New Media,* 6.

47. Gerard Goggin and Christopher Newell, *Digital Disability: The Social Construction of Disability in New Media* (Lanham, MD: Rowman & Littlefield, 2003), 134–39.

48. Goggin and Newell, *Digital Disability*, 114.

49. Bodil Ravneberg and Sylvia Söderström, *Disability, Society and Assistive Technology* (London: Routledge, 2017), 54.

50. Annuska Zolyomi, Andrew Begel, Jennifer Frances Waldern, John Tang, Mike Barnett, Edward Cutrell, Daniel McDuf, Sean Andrist, and Meredith Ringel Morris, "Managing Stress: The Needs of Autistic Adults in Video Calling," *Proceedings of the Association for Computing Machinery on Human-Computer Interaction* 3, issue CSCW, article 134 (November 2019), https://doi.org/10.1145/3359236.

51. Goggin and Newell, *Digital Disability*, 110.

9

Pushing Fintech

Testing Mobile Money, Financial Inclusion, and "Rural Women" in Peru

Mariel García Llorens

For over two decades now, dreams of cashless societies with total "financial inclusion" have sparked collaborations among banks, governments, international development organizations, scholars, and philanthropic foundations in many places around the world. However, the specific political and financial infrastructure undergirding these tech dreams have varied widely across the cashless dreams' brief history.

Kenya's M-Pesa, begun in 2007, is broadly recognized as the first successful mobile cash transfer service. An initiative of a private mobile network operator (Safaricom), it allowed millions of Kenyans living in rural and impoverished areas to switch to digital monetary transactions using their basic cellphones and without—at least initially–needing a bank account or otherwise forming any relationship with a bank.[1] In other countries, though, the digitization of money is primarily directed by central governments. In India, for instance, the national government led in 2016 a "demonetization campaign" to "fight 'black money'" or "cash made from tax evasion, crime, and corruption" and as a measure against counterfeiting.[2] The government recalled the 500- and 1,000-rupee notes as legal tender, forcing people (most with no record of any criminal activity) to trade their cash for digital credit through the banking system. While the declared objective of this campaign was to get a hold on unaccounted wealth kept in cash form, analysts observed in retrospect that the government had its eye on formalizing the largely informal Indian cash economies via digitization through privately run payment systems.[3]

This chapter examines yet another digitization experiment: a mobile money partnership called Modelo Perú (Peru Model), created in 2014 and driven neither by telecoms nor the government but by a group of Peruvian banks. Modelo Perú was, by its own account, the "world's first attempt to enable a fully interoperable national mobile payments system with a specific focus in financial inclusion."[4] These bankers and the engineers who worked with them aimed to provide a financial infrastructure that would allow "the poor" (whom they also called "the unbanked") to replace cash with digital transactions made through basic cellphones. They saw mobile technology as a means of accessing money that would otherwise not be within banks' reach because it circulated exclusively in cash among those considered poor.

I specifically explore Modelo Perú's transition from a set of design ideas into a mobile money platform called Bim, based on extensive ethnographic fieldwork I conducted with its designers and developers between 2016 and 2020. Bim, the mobile money technology resulting from Modelo Perú's partnership, was imagined by its designers—a mix of people with different areas of expertise (IT engineers, international development consultants, and bankers)—as a unique financial vehicle that all participating banks could use to reach its expected unbanked users through any mobile network operator. The project's engineers designed Bim around the simplest and oldest technology: a basic nonsmart cellphone without internet capabilities. By 2014, this technology was both already becoming outdated and still the dominant mobile device among the project's intended demographic: the 70% of Peruvians who, according to these experts, were part of the "informal economy" and had no stable relationship with formal financial institutions and conducted their transactions in cash. A subgroup of this population comprised people from the Andes region who subsisted primarily on small-scale agriculture and other economic activities in rural, impoverished areas.[5]

In this chapter I examine the platform's testing phase, when its designers tested the infrastructure among themselves, and then brought it to a group of women in the South Andean region of Peru to see if it would work for them. Developers chose these women because they fit into their categories of the "poor," the "rural," and the "unbanked"—people in need of financial inclusion through digital cash. There is nothing accidental in their choice of women instead of men; in the self-representation of interventions like this one in Peru and elsewhere, the intended poor person— now a digital cash user incorporated in the formal circuits of finance—is almost always a woman of color.[6] As it turned out, rather than providing feedback to the design process, the test was more of a training and photographic session, and its resulting images of women testers portrayed as users of this technology served to present Modelo Perú to its donors and collaborators as transitioning the poor seamlessly from unbanked isolation to financial inclusion. Instead of understanding these women as the receiving subjects of these interventions, I propose they are one of its necessary infrastructural pieces: they need to be included in the picture for the plan to appear feasible and to propel the digitization of cash economies—a common interest among banks, governments, and other actors such as telecoms, in Peru and beyond.

As I follow the Peruvian system from its design and testing to its implementation, I draw together discussions on software and the design of computational devices within science and technology studies scholarship and conversations about financial development and digital money within anthropology literature to examine how the cycle of digital financialization works.[7] Everywhere that "mobile money" is designed, prototyped, and implemented, it draws together a similar constellation of telecommunication corporations, private banks, and government policy makers. Yet, in each iteration of mobile money and across multiple scales, the infrastructure also emerges from competing financial and technological design strategies, divergent national ideologies and economic policies, and local imaginaries of "financial inclusion" and of "the poor" that connect low-level software engineers to national finance ministers and the CEOs of global banks.[8] As the examples show, integration (of the telecom networks, the payment industry and banking infrastructures, and private and public data collection systems) and coordination (of laws, interinstitutional agreements, software protocols) are always local and contingent on the materiality of other infrastructures. Yet, at the same time, these processes and infrastructures are never solely local: developers in Peru, for example, rely on many existing financial systems and telecommunication platforms that connect local, national, and international exchanges; they work with server farms located in different places around the world; and they respond to national and international regulatory frameworks. Also, the work of tailoring mobile money infrastructure to national financial systems and development paradigms (including the many diverse local economies that compose the national scale) relies integrally on collaborations and comparisons among experts developing similar infrastructures in other places.[9] These problems of scale are at the heart of building mobile money, and thus a space where ethnographic expertise can make sense of the "heterogeneous mess" and "multiplicity" of human and nonhuman actors that make mobile money possible.[10] I am proposing to peer into the black box of Peruvian mobile money to understand the specific contours of the financial development network that it promised to build and integrate.[11] Resulting from borrowings, collaborations, and recursive iterations locally enacted and thus always shifting, I understand mobile money as a set of practices and as an imagination of how digital payments should work for financial inclusion, which includes practices and an imagination of women considered "rural," "Indigenous," and "poor." This chapter thus argues that rather than women testing mobile money, it is mobile money that tested the women to see if they work *for* financial inclusion.

The argument is knitted together with my conversations with Modelo

Perú's engineers, consultants, and other experts involved in its design and comprises four parts. It starts with a close reading of a key image in Modelo Perú's promotional materials for international audiences in the global development world—a picture of a woman *tester* of the mobile money platform that circulates as a representation of a technology *user*. The second section opens an inquiry into the relationships among designers and testers who also played the role of user in a small town in the Andes region of Peru. The third section delves deeper into the design process by asking how the engineers get to know the users they are creating this system for. The fourth examines banks' and governments' problems with cash (it deaccelerates finance and it cannot be tracked) and their efforts to digitize it in contrast with Peruvians preference for cash, a preference particularly prevalent before the COVID-19 pandemic. This section describes how, as part of these banking efforts—and in dialogue with international development institutions and national and international regulators—the poor are conceptualized as the source of the financial system's problems with cash. I conclude the chapter by suggesting that through these experiments the women testers end up rendered as central infrastructural pieces in the mobile money architecture and serving the bankers' purpose, even if their cash never gets digitized.

The Women in the Picture

In Peru, as in many other countries classified by development economics as having "emergent" or "vulnerable" economies, a high percentage of people use cash for transactions and manage and save money by means other than using bank accounts, as they often distrust banks.[12] International development organizations and businesses alike have taken this situation both as a problem and as an opportunity for introducing digital cash through cellphones and mobile technologies. For instance, these factors are among the reasons why the mobile money service M-Pesa, run by a telecom and not a bank, worked in Kenya.[13] In Peru, during the time of my research, between 60% and 75% of people transacted business in cash.[14] Although the use of mobile money has grown since the pandemic, aided by the prevalence of low-cost Chinese smartphones and increasing internet penetration, most rural towns still operate largely as cash economies, yet cash is a scarce resource in these areas. Consequently, it is not uncommon for people to transact business without any actual bills or coins changing hands, though still using cash as a measure of the value of the things exchanged.[15] Andahuaylillas, located twenty-five miles from the city of Cusco and at ten thousand feet above sea level in the Andes, one of the planet's highest mountain ranges, is one such rural small town where people use cash and distrust banks. This town's economy relies mostly on agriculture, cattle raising, and artisanal manufacture of roof tiles, but it also receives tourists because of an impressive early baroque

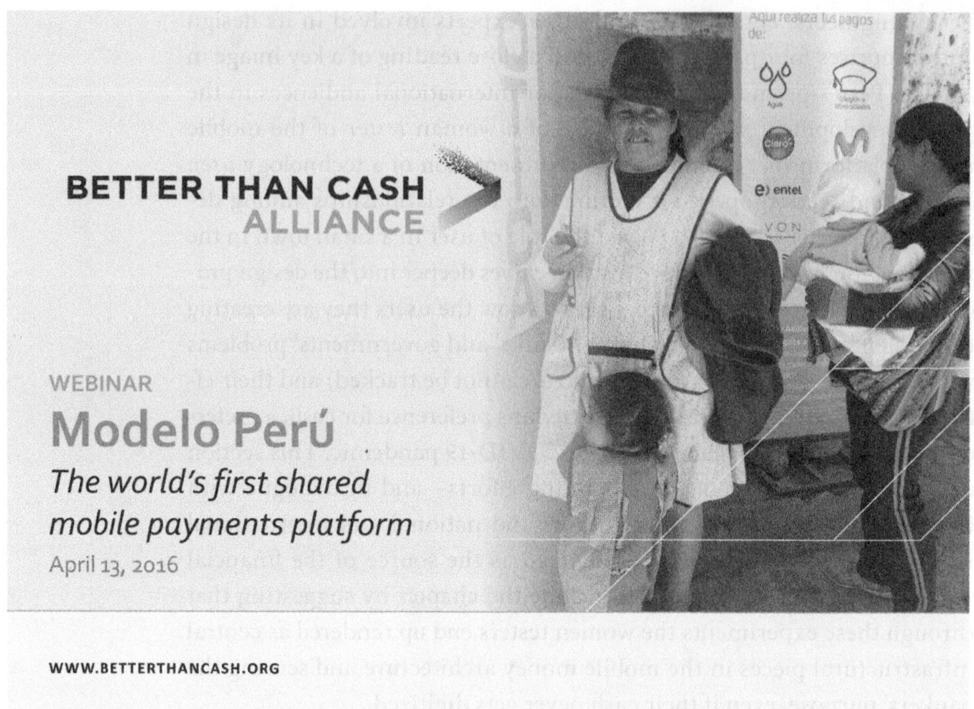

Figure 9.1. Opening Image of "Modelo Perú Webinar," Better than Cash Alliance, April 19, 2016. Photo by Jeffrey Bower / Better Than Cash Alliance

church built by Spanish settlers atop an Incan ceremonial place back in the 17th century.

Consider the photo, taken in Andahuaylillas, in figure 9.1. The woman in the center of the picture smiles at the camera; she knows she is being photographed. She wears a kind of brown wool hat common in Peruvian Southern Andes towns such as this one. She embodies the working poor as development imagery. She is wearing a *mandil* (apron), with pockets on each side, on top of her blouse and her traditional round skirt. That type of mandil signals that she is a businesswoman, perhaps a street vendor or a greengrocer at the weekend market, or a small-shop owner. In one of the mandil pockets she keeps the bills and coins for change; in the other, she keeps thin plastic bags in which to pack the merchandise of her *caseros* (clients). Between these pockets, the mandil has a third one, with a zipper— a good place to put important belongings, such as the woman's ID, some higher-denomination bills, and her cellphone. In the picture, her young daughter grabs her by the mandil, her back turned to the camera. Like her mother, the girl also wears an apron on top of her clothes.

The woman holds a cellphone in her right hand. That small, nonsmart— and now almost obsolete—cellphone is at the heart of the mobile money model Modelo Perú. There is another woman on the right side of the image, holding a young child with both arms. She and the baby are looking at the sign behind them, a banner advertising different digital payment services like *recargas virtuales* (airtime top-up services), and they all stand outside a local shop. It is a sunny day, and the two women and their children seem to be waiting in line to get into the shop.

This photograph has been reproduced often on Better than Cash Alliance (BTCA) presentations, webinars, and online articles about Modelo Perú. Headquartered in the United Nations in New York City, BTCA is a partnership of governments, international organizations, and companies "that accelerates the transition from cash to digital payments" in order to advance the UN's Sustainable Development Goals.[16] The Alliance pursues its strategies in what it classifies as developing countries, including Peru, where most people use cash instead of banking services.[17] The Peruvian government is one of BTCA's founding members, and the partnership of Peruvian banks designed Modelo Perú—and its mobile money platform, Bim—with technical support from the Alliance.

According to the Better than Cash Alliance, what is better than cash is digital payment; cash, or more exactly, the pace at which it travels, is a problem in need of acceleration. The photograph suggests that women with this profile are using the cellphone for payments instead of cash and, thus, playing their part in the acceleration that BTCA promotes. Yet that is not what I encountered in the field. Except in this and other pictures, these women are hard to find in any other of Modelo Perú's related places and practices. They do not seem to have participated in the design of the mobile money platform Bim, nor are they using it. The latter seems surprising, but maybe it is not.

As I was doing my fieldwork, visiting regularly Modelo Perú headquarters in Lima (the capital city of Peru), I met Jeffrey, a BTCA consultant, a member of the team that designed and launched Modelo Perú roughly between 2014 and early 2016, and the person who took the picture in figure 9.1. I asked him if he remembered taking it. Jeffrey was a Canadian in his thirties who, after earning a master's in international relations and international studies, went to work at the United Nations, developing digital payment services. He told me that he remembered that moment and place well. It had "a very pretty church," he recalled.[18] He had taken the picture in 2015, when Carolina, a development economist and the first general manager of Modelo Perú, invited him "to join her on the first public test" of Bim. Jeffrey told me that Carolina had chosen the town (Andahuaylillas) because "it is relatively isolated and self-contained; there's a few shops; there's a few women that work in the market." It was the perfect environment, he thought, to do Bim's "first public test."[19]

I asked him how this picture of the two women and children fit in this test, which he also described as a "pilot." He told me that the woman with the apron and the cellphone in her hand, "was the first person out of our group of people, to do a live [mobile money] transaction in the market."[20] That is, the woman with the mandil bought something at the small drugstore behind her in the picture, and she paid for it using Bim on her cellphone instead of using cash. Her transaction inaugurated Bim, the mobile money platform from Modelo Perú in the sense that she performed the first transaction on it outside the team designing it (what Jeffrey called "our group of people").

The Engineers' Testing Environments

To design and administer Bim, the banks participating in the Modelo Perú partnership created a company called Pagos Digitales Peruanos (Peruvian Digital Payments, or PDP). The banks also hired the global IT company Ericsson to use its Wallet platform technology to process the digital cash transactions at the core of Bim. Ericsson engineers, working mostly from Sweden and India, designed this technology, and they deployed it in some African countries for telecom corporations, as they were the businesses that mostly drove mobile money there. As I learned during my fieldwork, Carolina and her team of engineers started working with Modelo Perú after the bankers had already chosen this technology. At PDP's headquarters in Lima, they had the challenge of adapting the Ericsson system, which had been designed for a single telecom's use, to work for a partnership of banks.

In Andahuaylillas and another small town, engineers performed the first (and only) set of two "public tests," one in each town.[21] Before the trip Jeffrey described, engineers tested the mobile money platform extensively, but Bim's design and testing happened exclusively at PDP's headquarters in a succession of three *ambientes* (environments) called Dev (short for "Development"), Test, and Production.[22] It was (and still is) hard for me to imagine what these digital environments are like and what happens inside them even when seeing these ambientes on the engineers' screens, as I did in the field. Rosa (PDP's former IT operation manager) used to explain to me what these were. Always patient and encouraging with my questions and guesses of how the platform works, she told me that the first ambiente, the one called Dev, is where the engineers write the system's code and test the "work units." Her answers often left me in the darkness of a virtuality I could hardly grasp and wondering about the immaterial materiality of software units. "Speaking in the most basic way," she told me once, "a work unit is a little computer program written in some computer language that does something."[23]

Discussing the details of coding, Rosa specified that "you can develop a program that has twenty thousand lines of code and does everything, or you can develop *pequeños programitas* (small programs) that are called or used by a superior program." She used the example of Bim's mobile money wallet menu of transaction options that her IT engineering team designed at PDP's headquarters: "The menu says: option one, send money; option two, put money; option three, pay cellphone." Thus, she added, "the menu is a program, it is a work unit that calls other work units." She supplemented her description with an example: "The program says if 'one,' then call 'send money,' and 'send money' is another unit of work that is built independently and performs the function of sending the money, and so on."[24]

The engineers then copy these units of work to the second "environment," called Test, where, Rosa explained, the engineers do the comprehensive testing of the system. The combination of all these units of work created in the Dev environment and "structured by superior programs, as if it was a tree, make up the entire system." Test is still an enclosed environment, and PDP's IT engineers put all the work units together there to verify that Bim's system as a whole worked as expected. For these engineers, "working as expected" means according to their technical expectations, with every work unit doing its part for the whole. Finally, the system goes live, that is, into the "Production environment," where it becomes accessible to all of its registered users (for instance, the women in the picture).

My conversations with Jeffrey and Rosa suggest that the woman did her transaction (buying something in the small shop using Bim on her cellphone) when the platform had already transitioned into the third environment (Production). It meant that all the *programitas*, or units, were working together as a whole and connected with the other channels and systems of each of Modelo Perú's more than twenty bank partners. The platform was in the third environment (Production) and thus, open—though not yet advertised—to everyone. This is why Jeffrey called it a "public test." The platform is open to the public but there is no public there yet, because no one, besides the engineers at PDP's Lima headquarters, knows about it. So they go and test it with a potential audience: what they called "rural women," along with a couple of small shops participating in this public test in Andahuaylillas.

The transaction that the aproned woman made in this rural small shop in October 2015 inaugurated Modelo Perú's instantiation as a mobile money platform (Bim). For Jeffrey, "it was intentional that we chose a female, rural . . ."—he searched for the word—"Indigenous person who is also a merchant."[25] As he recalled, it was crucial that "she has also a business that she operates" because "it shows that's who this platform is for."[26] The woman in the picture thus checks off all the categories of Modelo Perú's imagined public (and that of the Better than Cash Alliance, which Jeffrey represents). She and the other women selected for the test are not actually poor by local standards, but to him and his colleagues, they are "poor enough" to qualify to be included in their pilot. They are useful to solving the experts and designers' problem of the slowness of cash that needs acceleration through digitization, because they fit into the international development category of the "working poor"—a conceivable subject for credit and the target of financial inclusion strategies—as opposed to "the poorest," those who need support meeting basic needs.[27]

Jeffrey also explained that Carolina, PDP's general manager, chose this woman in particular for the pilot because she knew her from a previous development project—the state's conditional cash transfer program called Juntos (Together), designed for people living in poverty—back when Carolina was Peru's minister of development and inclusion. One might think, then, that the woman in the picture has more of a collaborator status, as she was specifically chosen for the test and her transaction opened the pipelines of the bankers' mobile money platform Bim, when digital cash traveled from her cellphone to the cellphone of the shop's owner. However, Bim's official launch would take place four months later, in February 2016, and in Lima, more than seven hundred miles away from Andahuaylillas, in a high-end hotel, at a cocktail party for people wearing tailored summer dresses and suits, with the various bankers involved and their government representatives' guests in attendance, and followed by a press conference.[28] No Indigenous women were invited; they were present in the pictures.

Knowing Your User through Use Cases or "la Casuística"

If the platform is for the working poor of Peru, of which the rural merchant women of the Andes like the ones in the picture are a subgroup, how do Rosa and her team of IT engineers get to know what they need? In my conversations with PDP engineers, it was clear that their goal was to deliver a system that performs well, with no error messages or system crashes. They seemed to think of the users as a function of the system's usage; users are considered in relation to use cases or what Rosa often called *la casuística*. For them, a workable technological solution is one that correctly performs in any use case they can imagine, so they meticulously checked the mobile money platform in their different enclosed testing environments at PDP headquarters. They assumed that if the system worked for them, then it would work for all. In other words, it seemed that PDP's engineers never questioned the usefulness of the system for its intended users, whom other experts such as Carolina and Jeffrey conceptualized as "(working) poor," "informal," and "excluded," and whom they, in practice and in Andahuaylillas, actualized and embodied into Indigenous women.

Instead, their work was based on the assumption that the system was *necessarily useful*, which is not an uncommon attitude among office-based engineers, in contrast to those engaged in fieldwork, who are often exposed to real usage situations.[29] Taking the platform's utility for granted, they tested its functionality extensively among themselves. A group of professionals who live in Lima, they are mostly men, and some of them are international consultants like Jeffrey; they all have smartphones and bank accounts. They are, then, using the binary poles of their own design categories, "urban," "not poor," part of the "formal" economy, and "included," and they brought to the design table their own beliefs and predispositions. These are inadvertently built into the system they have created through what they can see only as technical decisions.[30] Bim's system, which is already an adaptation of a technological solution (the Ericsson Wallet platform) created for another audience (single telecoms and not partnerships among banks), worked for them, and for the imagined poor user they abstracted from themselves. Thus, they assumed it would work for the women in the picture.

This is a situation familiar to STS scholars of digital technologies who have documented how engineers infuse, often unintentionally, their own social and political biases into the digital systems they develop. As a result, these systems carry and enact ways of classifying people and situations that perpetuate the inequalities they intend to address. For example, Virginia Eubanks has examined the automation of welfare payments, homeless services, and child protection services in the United States, and how systems that were clearly not designed to act in a harmful way (or were designed for a particular country and then deployed somewhere else) end up profiling and punishing those they are supposed to help.[31] Safiya Noble has also shown how Google, a commercialized search engine that most people use as if it were a public service, organized and ranked search results showing strong prejudices against Black people, even though its engineers did not design its algorithms to do so intentionally.[32] In other words, systems that are designed within a violent or unequal society will reinforce or even amplify those inequalities, perhaps unless they are designed explicitly not to.

In Bim's case, the usage imagination that tested the system in PDP headquarters before bringing it to the women of Andahuaylillas came from people who shared few characteristics with the intended users, and both the system and the testing process carried their biases. PDP engineers know that once a system goes live (and into the Production environment), unexpected things can happen, because any kind of user will bring new use cases that they could not anticipate. That is why, Rosa explained, engineers continued to monitor and adjust the system when it was already up and running, through the Dev, Test, and Production linear sequence of practice environments, and they replaced it frequently with updated versions. This makes the software malleable and easy to adapt. However, the mobile money software was also confined to a specific materiality, the basic, nonsmart cellphone, which later had the unintended effect of proving too difficult to use as a transactional device for its users.[33] This device was chosen precisely because of the set of assumptions about the rural, poor user that Modelo Perú engineers, consultants, and bankers had worked with. These are in part inherited assumptions, shaped by what has worked in other places, mostly African countries, with different geopolitical, legal, and technical environments (with different internet penetration rates, availability of 4G networks, smartphones, and so forth). Moreover, the system has been designed using what worked not only in other places but in *other times*: for instance, the Kenyan M-Pesa service chose a basic cellphone and the SMS Telecom channel for mobile money back in 2007, seven years before the mobile money project I describe in this chapter, when that type of phone was still the dominant device globally and the first iPhone had just reached the markets. These assumptions not only get translated into ideas about the Peruvian unbanked and "rural" women but also have been materialized as the specific technological objects that the bankers chose for Modelo Perú in 2014: Ericsson's solution and a basic cellphone.

Also, Bim's bankers and the engineers working with them decided to bet on basic cellphones largely because this was not just a fintech project but a financial *inclusion* technology project: it had a social development component that seemed to require another kind of expertise than their own. Thus, they deferred the *know-the-user* aspect of it to other experts such as Carolina (and Jeffrey to some extent). They enabled Carolina to speak to them on behalf of the women in the picture because of her position of authority as the former minister of development and inclusion, because her professional background as an economist who specializes in microeconomics and rural development, and because she is a woman. Carolina provided them with a shortcut in their design process; she is the technocrat that replaces the women in the picture in the design of Bim's working pipelines by explaining to the engineers the women's needs and expected use cases.

This process resonates with what Diana Forsythe, anthropologist of artificial intelligence, found in the case of system design for medical informatics. Instead of observing real-life problem solving in the place where they intended to field a system, the IT specialists she studied proceeded "on the basis of what designers and/or experts 'know' about such sites in general, that is, on the basis of their conscious models of particular types of work as carried out in particular settings." For them, problem formulation and design relied "heavily on intuition and introspection on the part of the system designer[s]." Forsythe observed that sometimes the design teams she followed ethnographically worked with only one subject, just one doctor, "whose perspective is taken as representative of an entire field."[34]

The form of representation in Modelo Perú's case is different, however. Instead of one doctor as representative of all doctors who could use the AI system that Forsythe discusses, here I found one economist (Carolina) who was taken as representative of the rural Indigenous women and the working poor of Peru. Carolina used to advocate for these constituencies within the state apparatus (as a minister), and now she spoke for them about their needs (and use cases) to bankers and engineers. In other words, she became the interface between designers and bankers on one side and women in aprons on the other. Carolina was also the interface between Modelo Perú and the state (and her former colleagues) and the world of global development. She was often a guest speaker at international conferences, and that experience further validated her position with bankers and engineers. Jeffrey, the BTCA consultant, also attended these conferences; in fact he helped translate Modelo Perú's communications into English, which made it travel farther.

In the past two decades, these conferences have gathered international development officials, policy makers, regulators, and representatives from telecom and banking industries (such as the bankers of Modelo Perú), all of them interested in microfinance, financial inclusion, and mobile money. In these spaces two things have featured prominently. First, women of color were presented as the subject of microfinance practices. Microfinance consisted of giving small loans to groups of poor borrowers who repaid them with interest and in cash in weekly intervals directly to a debt collector. These microloans allegedly empowered them economically by helping them start or sustain their businesses.[35] Second, these spaces showcased African mobile money, particularly the Kenyan M-Pesa, which started in 2007 and by 2014 still had most of its users transacting through basic cellphones and was widely successful across socioeconomic groups.[36] At the time, M-Pesa was the case study that everyone in these spaces talked about, and it was taken as proof that the cellphone was the technological enabler needed for scaling up microfinance into a wider set of financial inclusion services such as digital payments and savings.[37] At these conferences, experts shared case studies and statistics that supported the understanding that "rural," "poor" women both were the most excluded of the excluded and were more entrepreneurial and more reliable in paying off debts than men and, consequently, a necessary target of financial inclusion practices through cellphones.[38]

Bankers and experts like Carolina who bridge the worlds of international development and finance brought back to Modelo Perú the authoritative knowledge that featured phones and rural women as keys to the engineers' task of designing Bim. The engineers tested the mobile money platform by generalizing use cases for a user that they clearly conceived as different from them, not only demographically but technologically, and to whom they attached a corresponding device (a basic cellphone); they were going to later test this system with the women of Andahuaylillas. Yet, when I was in the field shadowing the engineers, it seemed clear to me that the public test with the women in the picture did not test user acceptance. What does a public test *test*, then?

A Test That Also Trains

It turns out that, more than a test, this trial run was training. The women were trained to perform transactions with Bim in their cellphones. Jeffrey explained it to me in this way: it was a "one-day learning session" in which "we explained [to the women] how Bim worked, we explained the logic, we explained how the cellphone keys worked, we walked them through all the transactions."[39] After this training in Bim's use, the women proceeded to try the system out on a real transaction in the small shop shown in the photo, with S/5 (about US$1.50), which they had in their in their Bim mobile money accounts courtesy of Modelo Perú.[40]

The transactions of the woman in the picture and the others who participated in this training inaugurated a period of testing the mobile money platform "in the wild," that is, in real conditions, with the kind of users that designers expected for this platform and with the banks connecting their interfaces and sharing information as Modelo Perú promised. Although this phase corresponded to the engineers' Production ambiente, the platform still lay somewhere between being live and *fully* public. It was not fully public because there was no public—there were no users there yet. It was an empty platform, a project with its digital pipelines ready to receive and transport digital cash and in the process of leaving the enclosed IT environments of Dev and Test to become, in Jeffrey's words, "a service in the market." That was the moment in which the women's picture was taken.

As Jeffrey further explained this test-training: "We had the press there; we had a photographer because the intention was for this to be documentary evidence of a pilot being successful."[41] I recall being a bit confused at this point in our conversation. Can a one-day training and one transaction per *señora*—guided closely by a facilitator (figure 9.2)—be sufficient documentary evidence of a successful pilot and the acceptance of a technology? More than succeeding in testing its use and in training the señoras, it seems more apt to say that Modelo Perú partners succeeded in recording this moment in images. My point is that it looks as if they are visually documenting a process, but the "documentation" is itself a necessary infrastructural piece of Modelo Perú. One could say that the point of the test isn't to test the technology, or even to make the test, but to make *the picture*. As its name suggests, more than a mobile money platform, Modelo Perú is a set of *modeling practices* that do different work for different audiences, and it doesn't really need to be up and running as a mobile money platform and have users perform tasks on it to do work for some of its audiences.[42]

Bim may very well be designed for the "poor businesswomen" of rural Peruvian towns, and to fully function as a platform, it will need users (this audience or another) to turn their cash into e-money and transact

Figure 9.2. Participants and facilitators of the public test wait to enter a shop to purchase products using their mobile money wallets instead of cash. Andahuaylillas, Peru, October 2015. Photo by Jeffrey Bower

using this platform. However, the relation between this specific audience and the platform is not a given: it may or may not happen and is an assumption of the design process (and of the first picture). Jeffrey did not take this photo to be consumed by this audience (say, as an advertising tool to convince other rural businesswomen to use Bim). Rather, these women are in the picture to *work for* Modelo Perú. Modelo Perú was developed in the offices of banking representatives who drew on other mobile money models and digitization processes. When Modelo Perú was just a proposal (and not yet instantiated as a mobile money platform called Bim), it was showcased at international global development and IT conferences. Its developers promised collaboration among its partners (more than 20 financial institutions, including the country's largest banks) through shared information and infrastructure (ATMs, local banking agents) and a coordinated inclusion goal of providing financial services to the poor in partnership with the state. This proposal generated expectations because of its collaborative banking approach, projecting cost savings for each participating bank unheard of internationally for mobile money at the time. Modelo Perú was later implemented via the creation of a third party company (PDP) and its engineers' environments, which developed Bim's mobile money infrastructure. Then it was PDP's task to demonstrate to the national government, global development agencies, and even to the banks owning this partnership and its resulting technology, that Bim fulfills the promises of Modelo Perú. Thus, I suggest that all these stakeholders became audiences of this endeavor and the features of Modelo Perú's design became the promises that PDP needed to deliver to them. As sociologist Claude Rosental's work would suggest, the whole test could be seen as a demo for people with political and economic leverage.[43]

Let me spell this out. The women testers are in the photo because they participated in a pilot; they were not Bim clients yet. But in the picture, the women are credible as if they were already clients of the platform and not just subjects being trained in a usage test. The picture has a "use story": it tells us that these women are using the platform to make digital payments. Modelo Perú's partners and other institutions, including the Better than Cash Alliance, used this image (and others taken on the same test-training pilot day) to make presentations that showcased the potential of this technology to other audiences, as if it was an almost achieved reality. One of the audiences of that compelling story was the Peruvian government, which at the time was promoting financial inclusion; another was the international community of global development, banking, and telecom industries interested in these topics. This image portrays a promised and imagined future that, even if not presented as an accomplished past, ends up, through its continuous usage and reproduction, being read as such.

STS scholars of digital technology and inequality have described this familiar situation. Morgan Ames, for example, has examined the "charismatic promises" of the One Laptop Per Child development project. She describes how its proponents preferred not change their expectations to match the unattractive reality of this project's labor-intensive incremental achievements; instead, they created an alternate reality, one that matched what stakeholders and the tech world wanted to hear about social change and futures that seem easy to achieve or even inevitable through tech interventions.[44] The bankers, Carolina, and Jeffrey have used the photos of rural women testers performing a transaction for the camera in their own performances to different audiences (presentations, webinars, and national and international news articles about Modelo Perú).[45] These images still populate the internet and, to the untrained eye, continue to illustrate the bankers' unwavering support for financial inclusion, even though the mobile money platform Bim was transformed into an app in 2018 and has since been accessible only via internet-enabled smartphones.[46]

I argue that the bank partners count themselves among both Modelo Perú's stakeholders and its publics since they are involved in the mobile money platform Bim not directly but through a company (PDP) they created for it, a third party that develops and manages Bim with their financial and infrastructural support. Thus, the banks' support of this platform is not a given either; it is another design assumption that needs to be constantly negotiated. PDP's engineers and managers must demonstrate to the bankers (with monthly statistics showing progress, along with individual compelling stories) that this initiative is worth their investment in the long term and convince them to chip in more resources every now and then. In the field, engineers often complained that bankers were at first very enthusiastic about Modelo Perú's performances of aspirational futures (to paraphrase Ames), especially when Modelo Perú was still a project being prototyped and tested.[47] When their project became a product—that is, an actual mobile money platform (Bim) that they collectively owned—they were reluctant to continue supporting it even though it was their own and "only needed pennies from them to function in comparison to the banks' other investments."[48]

The Problem with Cash

The promise of mobile money for banks and organizations like the Better than Cash Alliance is that it will help them deal with what they consider the "problem of cash": cash needs to be produced and held somewhere safe; needs to be counted and validated; and requires processing, transportation, and protection; and thus, it deaccelerates finance. Banks seek to accelerate it by transforming it into digital form through their private pipelines so as to expand the financial market.[49] To borrow an idea from anthropologist of international development Sohini Kar, the banks' expectation was that mobile money would enfold the formerly excluded into globalized financial networks.[50] Yet a recurring comment among PDP engineers and government regulators alike during my fieldwork was that most Peruvians have a "problem with banks": they simply do not trust them.[51] This is a common problem in many other places as well, and it surfaces whenever new infrastructures of money, such as mobile money, are being set in place. As anthropologist of digital money Bill Maurer has observed: "People distrust banks not only because they lack understanding, feel that banks serve only the wealthy, or are uncomfortable approaching a bank for services. They also, quite rationally, distrust banks because banks—and state-issued currencies—fail." In countries where political and economic instability often recur, the value of state-backed currencies may vary radically from one day to the next, and people's savings in banks may be lost so often that their "monetary repertoires include practices meant to hedge against such institutional failures."[52] Moreover, in Peru, there is a common understanding that money in a bank account cannot be mobilized like money in your hand; it simply does not create the same relations.[53]

In Peruvian Andes communities where the women in the photos live, many transactions are monetary and occur even if there is no cash exchanged between hands. The interviewees of another anthropologist who worked in Peru, Marisol de la Cadena, exchanged rams for onions and school supplies in the local market.[54] The equivalence between these goods was calculated in monetary value, as if cash was present, even though no cash changed hands at any point in the transaction. Thus, it is not that cash is indispensable, but it matters what relations cash makes. This example of an exchange is a monetary transaction that includes amounts, weight, prices, quantity.[55] I have witnessed similar exchanges myself.[56] Cash is not present physically, but it is otherwise; it is not a digital transaction either. Digital money does not mobilize the relations that need to be mobilized, or not yet, or not in the same way.[57] One might think that the presence of so-called barter makes attractive the absence of cash—meaning as the perfect place to introduce mobile money and digital payments as Modelo Perú proposed. But it becomes less attractive if there is no trust in the relation that made through money but not cash (as mobile money or digital payments). It is the lack of trust in banks (and governments, as well as development projects and their promises) that fuels the preference for cash and that can be hard to overcome.[58]

Put differently, what Modelo Perú's bank partners were trying to circumvent through their collaboration was not yet a problem: people they think of as financially excluded do not think of *themselves* as financially excluded. They may need money, but they do not need banks. The problem of cash and the problem of exclusion are problems of the designers of financial inclusion and of the digitization of cash. The banks have a problem (cash) that deaccelerates finance and that digitization can accelerate, and they need to find where there is more cash that could be dominated (held by the poor). Thus, because cash poses a problem for the banks, they create the poor as the problem of cash. The bankers' struggle over "the poor as the problem with cash," however, gets confronted by the "problem with banks" that Peruvians have. Through the examples of monetary transactions in the Andes, we can see that what matters is what relations cash makes that are mobile and abstract, though not digital—people calculate prices and payments with cash that is present in value but not in physical form. Cash relations can be mobile but are not *tracked* money, and banks fear this autonomy—they want to surveil it, control it, and then extract it.[59] Modelo Perú and Bim have been one of their laboratories for understanding and controlling it.

A User That Is a Picture, and a Picture That Is a Pipeline

The señoras from the picture did not seem to have needed the mobile money platform (or a formal relationship with banks) prior to this pilot encounter. Rather than stakeholders, or users included in imagining the platform in any way, these women *were trained*, in one day, to use the platform—and it remains unclear if they ever used it again. Bim had been designed with them (and other people like them from other places) in mind, which is different from having been designed with their participation in the process. This was a top-down design, modeled on "grabbing stuff from Africa and other places," as Carolina mentioned once to me, referring mostly to Kenya's mobile money M-Pesa—stuff that worked in other places and was shared in presentations, webinars, and conferences of the international community of people imagining financial inclusion.[60]

It is a model that defined a subject for itself *beforehand*, the female microentrepreneur, and not because it was inspired by the practices of specifically Peruvian rural and Indigenous merchant women but because this was the subject inherited from microcredit and microfinance initiatives.[61] What these pictures tell us is that now women can use their cellphones for digital payments and, by extension, have become subjects of both microloans and financial inclusion. As Ananya Roy puts it, "What is at stake here is a crucial shift from the idea of development as social services and the improvement of human capital to development as integration into global financial markets."[62] Perhaps this explains the absent presence of the woman wearing the merchant apron in figure 9.1 and why we cannot know her name: she is an *inherited subject*, whom the design of Modelo Perú needed, not a person.[63] She and others like her form the image that guides the progressive actions of Carolina and Jeffrey and, paraphrasing Morgan Ames, brings about the charisma of Modelo Perú.[64] The engineers work with what they tell themselves about the user. It is a macroidea, a mix of top-down visions of digital finance in rural communities from the worlds of finance and the international community of development organizations, a mix that was expected to work in the microcosm of Andahuaylillas.

Modelo Perú reveals how financial inclusion has given way to an even more pressing financial goal: the digitization of cash, which is an interest that banks and governments have in common as institutions that carry both the social and the financial costs of moving it around. Cash, which rarely appears in the promotional picture, is at the center of these modeling practices, which reimagine money and ways to profit from its digital handling and transportation, as well as facilitating its surveillance (both for banks and governments); and the poor and the women are the test sites for these practices, as they provide the narrative that justifies and advances this new goal. Indeed, as anthropologist of finance Julia Elyachar suggests, one of the most important turn of events of our times is the free market's absorption of cultural practices of the poor as a source of profit through the work of multiple endeavors, institutions, and projects.[65]

To conclude, I invite you to look at Modelo Perú at three levels. First, there is the function of the women's photo: the point of the test was the production of an image. Modelo Perú proponents made the image of a platform that at that point didn't fully exist yet (it was still in the testing phase) and might never exist, but that still produced an image that did its own work at the international level and with the national government. While these engineers and bankers designed a mobile money technology, they were necessarily also designing an imagined "poor" with the help of these women. Even if the women never used the mobile platform again, the practices of bankers, engineers, and the women testers they invited to their trainings, trials, and pilots, as well as the documentation they harvested from them, continued to re-create and reproduce that imagined "poor." This image in turn fueled the appeal of Modelo Perú, not in the local spaces of these women but in other places and instances (i.e., international development and finance circuits, national policies), a sort of "macrocharisma."

Second, at the level of the actual, real women interacting with this imagined "poor" that informs the practices and technologies assigned to them, we see that instead of perhaps teaching them how to use the system both in a keyboard phone (which they owned at the time of the public test) and a smartphone with internet access (which was the direction in which mobile technology was already going at that time), they worked only with keyboard phones and no internet. The real women got assigned a place other than the one of finance and smartphones reserved for other kinds of users—that is the place of development projects and keyboard phones with no internet, which will keep the women in a position of exclusion and in need of new interventions to catch up with the rest. It is the "modern poor" who, even when this and other techno-solutionist development projects try to "save" them, are always produced as "less than modern" or "submodern." Thus, more than improving a situation for the poor, such projects seem to be producing what STS historian Michelle Murphy calls "experimental governmentality."[66] Andahuaylillas becomes a demonstration town, an experimental platform in itself, in which, with the help of the women's underpaid labor, Modelo Perú developers keep the actual women in a particular position (the less-than-modern loop). They intervene with something that may or may not help the women's economic conditions, but they don't intervene so much as to change the conditions they wish to continue studying. What my account shows is that this sort of experimental governmentality can even work speculatively, through the promise (or the test) of a system that was not even in use by the "rural poor" women of Peru.

At the third level, I propose to look at Modelo Perú as a platform to hold space. Seen retrospectively in 2021, bankers, Carolina and Jeffrey, and government regulators commented to me that in one way or another Modelo Perú had worked, that it had succeeded. This comment stunned me every time I heard it, because in fact Bim usage failed to scale up; rather, it kept "surviving" and being used only in particular places where there was also a direct intervention of Bim promoters and their marketing strategies, but it didn't achieve a significant and steady growth of users and of digital money circulating through its pipelines. Though if we think of Modelo Perú as an in-the-meantime project and laboratory of the banks that united to create their shared mobile money platform to please the government and hold space while they figured out their independent and proprietary mobile-digital payment technologies, then it certainly was successful. Modelo Perú worked to keep the government discourse on financial inclusion alive while it held space and time for bankers to develop their own noncollaborative and highly secretive mobile money wallets. Bim partly succeeded because lessons learned from the partnership enabled the largest banks (BCP, Interbank, and BBVA) to create their own wallets: Yape, Tunki, and Lukita, respectively.[67] Yape has monopolized the market, and as of early 2025, it is used by people of all walks of life, from corporate executives to street vendors. All these wallets were designed as smartphone apps, were launched almost simultaneously with Bim, and competed directly with it. Fully supported by their respective banks, they were marketed to unbanked populations and advertised as tools of financial inclusion.

Bim didn't quite work for engineers, though, because for them, working as expected meant working at a scale it didn't quite reach. Bim is still alive and has become a platform almost exclusively used to distribute and pay back microloans, but so far, it hasn't been massively used to transport and transact with digital cash, despite all the engineers' efforts, commitment, and ingenuity.

Acknowledgments

The research and writing of this chapter were supported by the Wenner-Gren Foundation under the Wadsworth International Fellowship and a Dissertation Fieldwork Grant.

Notes

1. Tavneet Suri and William Jack, The Long-Run Poverty and Gender Impacts of Mobile Money" *Science* 354, no. 6317 (December 8, 2016).

2. Shashi Tharoor, "India's Demonetization Disaster," *Horizons: Journal of International Relations and Sustainable Development*, no. 9 (Autumn 2017): 208.

3. Amartya Lahiri, "The Great Indian Demonetization," *Journal of Economic Perspectives* 34, no. 1 (Winter 2020): 55.

4. Jeffrey Bower, "Modelo Perú: A Unique Approach to Financial Inclusion," Better than Cash Alliance, October 9, 2015, https://www.betterthancash.org/news/modelo-peru-a-unique-approach-to-financial-inclusion.

5. During my fieldwork, bankers, bureaucrats, and engineers used to talk about 70% of the population living in cash economies. See "Economía Informal genera el 19% del Producto Bruto Interno y el 61% del Empleo," INEI (Instituto Nacional de Estadística e Informática), April 6, 2014, https://m.inei.gob.pe/prensa/noticias/la-economia-informal-genera-el-19-del-producto-bruto-interno-y-el-61-del-empleo-7585/.

6. See Ananya Roy, *Poverty Capital: Microfinance and the Making of Development* (New York: Routledge, 2010).

7. On STS software and computational devices design, see Diana Forsythe, *Studying Those Who Study Us: An Anthropologist in the World of Artificial Intelligence* (Stanford, CA: Stanford University Press, 2001); and Morgan Ames,

The Charisma Machine: The Life, Death, and Legacy of One Laptop per Child (Cambridge, MA: MIT Press, 2019). See also Virginia Eubanks, *Automating Inequality: How High-Tech Tools Profile, Police, and Punish the Poor* (New York: St. Martin's Press, 2018); and Safiya Umoja Noble, *Algorithms of Oppression: How Search Engines Reinforce Racism* (New York: New York University Press, 2018).

8. On financial development, see Sohini Kar and Caroline Schuster, "Comparative Projects and the Limits of Choice: Ethnography and Microfinance in India and Paraguay," *Journal of Cultural Economy* 9, no. 4 (2016): 347–63. See also Anke Schwittay, "The Financial Inclusion Assemblage: Subjects, Technics, Rationalities," *Critique of Anthropology* 31 (2011): 381–401; and Ghazal Zulfiqar, "Financializing the Poor: 'Dead Capital,' Women's Gold, and Microfinance in Pakistan," *Economy and Society* 46 (2017): 476–98.

9. On mobile money and digital payments, see Bill Maurer, Taylor Nelms, and Stephen Rea, " 'Bridges to Cash': Channeling Agency in Mobile Money," *Journal of the Royal Anthropological Institute* 19 (2013): 52–74. See also Stephen Rea and Taylor Nelms, "Mobile Money: The First Decade," IMTFI (Institute for Money, Technology, and Financial Inclusion), Working Paper 2017-1, 2017, accessible via *IMTFI* (blog), November 27, 2017, https://blog.imtfi.uci.edu/2017 /11/mobile-money-first-decade-new-white.html. For the Kenyan case, see Stephen Rea, Ursula Dalinghaus, Taylor C. Nelms, and Bill Maurer, "Riding the Rails of Mobile Payments: Financial Inclusion, Mobile Phones, and Infrastructure," in *The Routledge Companion to Digital Ethnography*, ed. Larissa Hjorth, Heather Horst, Anne Galloway, and Genevieve Bell (London: Routledge, 2017), 363–73; see also Bill Maurer, "Mobile Money: Communication, Consumption, and Change in the Payments Space," *Journal of Development Studies* 48, no. 5 (May 2012): 589–604; and Jake Kendall, Bill Maurer, Phillip Machoka, and Clara Veniard, "An Emerging Platform: From Money Transfer System to Mobile Money Ecosystem," *Innovations* 6, no. 4 (2012): 49–64.

10. On the local and "heterogeneous mess" through which knowledge is produced, see David Turnbull, *Masons, Tricksters, and Cartographers* (London: Taylor & Francis, 2000). See also John Law, *After Method: Mess in Social Science Research* (London: Routledge, 2004). On multiplicity, see Annemarie Mol, *The Body Multiple: Ontology in Medical Practice* (Durham, NC: Duke University Press, 2003).

11. See Bruno Latour, *Science in Action: How to Follow Scientists and Engineers through Society* (Cambridge, MA: Harvard University Press, 1987).

12. See Gustav Peebles, "Rehabilitating the Hoard: The Social Dynamics of Unbanking in Africa and Beyond," *Africa* 84, no. 4 (November 2014). See also Bill Maurer, "Monetary Ecologies and Repertoires: Research from the Institute for Money, Technology, and Financial Inclusion," First Annual Report: Design Principles, IMTFI, University of California, Irvine, 2010, https://www.imtfi.uci .edu/files/articles/IMTFI_FirstAnnualReportDesignPrinciples.pdf.

13. Tonny Omwansa and Nicholas Sullivan, *Money, Real Quick: The Story of M-PESA* (London: Guardian Books, 2012).

14. Part of the challenge that cash economies pose for economists, statisticians, and other researchers is that they are hard to count. The number of people working in local informal cash economies jumped from around 60% in the mid-2010s to 75% during the COVID-19 pandemic; that is, about 8 out of 10 Peruvians have informal jobs; some of those are perhaps illegal, and they all pay in cash. See "Tasa de empleo informal en el Perú alcanza su nivel más alto en más 8 años," *El Comercio*, November 17, 2020, https://elcomercio.pe/economia /en-el-peru-casi-8-de-cada-10-empleos-son-informales-noticia/.

15. See Enrique Mayer, "Un carnero por un saco de papas. Aspectos del trueque en la zona de Chaupiwaranga, Pasco," *Nueva Antropología* 6, no. 19 (1982). For discussions of monetary practices, see David Graeber, *Debt: The First 5,000 Years* (Brooklyn: Melville House, 2014); Jane Guyer, *Marginal Gains: Monetary Transactions in Atlantic Africa* (Chicago: University of Chicago Press, 2004); Bill Maurer, *How Would You Like to Pay?* (Durham, NC: Duke University Press, 2015).

16. "About the Better than Cash Alliance," Better than Cash Alliance, accessed September 19, 2021, https://www.betterthancash.org/about.

17. Javier Alvarez, "Pagaré en efectivo: Modalidades de pago que los peruanos utilizan para sus compras de productos y servicios," *Revistas ANDA* (Asociación Nacional de Anunciantes), February 17, 2021, https://revista.andaperu.pe/pagare-en-efectivo-modalidades-de-pago-que-los-peruanos-utilizan-para-sus-compras-de-productos-y-servicios/.

18. Jeffrey Bower, email exchanges and Zoom conference with the author, April 2020.

19. Bower, Zoom conference, April 2020.

20. Bower, Zoom conference.

21. Modelo Perú developers tested their Bim platform in Andahuaylillas and another town, Chincheros, on the same trip to the Cusco region. Each pilot session was done with one small group of local women. Carolina Trivelli, personal communication with the author, June 2020.

22. I discuss these engineering practices in depth in the chapter "The Women in the Picture: Fintech Tests 'Rural Women'" in my dissertation, "Banking on the Unbanked: Everyday Peripheral Technologies for Mobile Money in Peru," completed at the University of California, Davis, in 2024. This book chapter is a revised, shorter version of it.

23. Rosa La Rosa, personal communication with author, August 2020.

24. La Rosa, personal communication with author, August 2020.

25. Bower, Zoom conversation, April 2020.

26. Bower, Zoom conversation, April 2020.

27. Ira Lieberman, cofounder of Consultancy Group to Assist the Poor (CGAP), personal communication with author, September 17, 2019.

28. The official launch happened at the Swiss Hotel, San Isidro, Lima, on February 16, 2016. La Rosa, personal communication with author, August 2020.

29. On IT engineers' culture, which often subordinates the social to the technical, see Forsythe, *Studying Those Who Study Us*. On engineers as fieldworkers, see Penelope Harvey and Hannah Knox, *Roads: An Anthropology of Infrastructure and Expertise* (Ithaca, NY: Cornell University Press, 2015).

30. See Geoffrey C. Bowker and Susan Leigh Star, *Sorting Things Out: Classification and Its Consequences* (Cambridge, MA: MIT Press, 1999), 196.

31. Eubanks, *Automating Inequality*.

32. Noble, *Algorithms of Oppression*.

33. The Bim system relied on USSD (unstructured supplementary service data) instead of the more familiar SMS (short message service) used by systems like M-Pesa, making it harder to use for any type of user, not only the intended ones.

34. Forsythe, *Studying Those Who Study Us*, 6.

35. Sohini Kar, *Financializing Poverty: Labor and Risk in Indian Microfinance* (Stanford: CA: Stanford University Press, 2018), 2, 3.

36. Omwansa and Sullivan, *Money, Real Quick*. See also Stephen Rea and Taylor Nelms, "Mobile Money," 5

37. I discuss this topic in depth in "The Gestation of Modelo Perú, the Rise

of Mobile Money, and the Unbanked," chapter 1 of my dissertation, "Banking on the Unbanked."

38. See Caroline Schuster, "The Social Unit of Debt: Gender and Credit-worthiness in Paraguayan Microfinance," *American Ethnologist* 41, no. 3 (2014): 563–78; Ananya Roy, "Subjects of Risk: Technologies of Gender in the Making of Millennial Modernity," *Public Culture* 24, no. 1 (2012): 131–55.

39. Bower, Zoom conference with author, April 2020.

40. Five Peruvian soles (S/5) could buy, in a small shop back in 2015, a small bottled water, a couple of small bags of mass-produced snacks, and some fruit, or it could be used to top up a phone with airtime credit.

41. Bower, conversation with the author, April 2020.

42. On technology as a cultural object, see Bruno Latour, *Aramis; or, The Love of Technology* (Cambridge, MA: Harvard University Press, 1996), viii–xi. On models as objects doing work for specific groups of people, see Donna Haraway, *Staying with the Trouble: Making Kin in the Chthulucene* (Durham, NC: Duke University Press, 2016).

43. Claude Rosental, *The Demonstration Society*, trans. Catharine Porter (Cambridge, MA: MIT Press, 2021), 2.

44. Ames, *Charisma Machine*, 172–83.

45. See Dan Collyns, "Peru Mobile Money Scheme Could Herald a New Dawn for *Nuevo Sol*," *The Guardian*, October 9, 2015, https://www.theguardian .com/global-development/2015/oct/09/peru-mobile-money-bim-banking-finan cial-inclusion; and Modelo Perú Webinar, Better than Cash Alliance, April 19, 2016, https://www.betterthancash.org/news/modelo-peru-webinar.

46. The images on PDP's homepage portray women with basic cellphones that have not been able to access Bim since 2018, when the mobile money plat-form transitioned to an internet-based app accessible only through smart-phones, see PDP (Pagos Digitales Peruanos), accessed December 17, 2024, https://pagosdigitalesperuanos.pe/.

47. On performing aspirational futures, see Ames, *Charisma Machine*, 178–89.

48. PDP engineers, personal communication with author during fieldwork in Lima, 2019.

49. See Peebles, "Rehabilitating the Hoard."

50. Kar, *Financializing Poverty*, 9.

51. I observed this lack of trust in banks firsthand during the days I accom-panied Bim promoters to advertise the product to small-shop owners in low-income areas of Lima as part of my long-term fieldwork in 2017 and a subse-quent field visit from June to August 2019. It was also a recurring comment and matter of concern expressed by PDP managers engineers, and Peruvian regula-tors (from the Central Bank and Bank Superintendency) during my conversa-tions with them throughout my on-site fieldwork and in a primarily online and remote fieldwork phase in 2018.

52. Maurer, "Monetary Ecologies and Repertoires," 20.

53. On how money creates different relations and is earmarked for specific functions, see Viviana A. Zelizer, *The Social Meaning of Money: Pin Money, Paychecks, Poor Relief, and Other Currencies* (Princeton, NJ: Princeton Univer-sity Press, 2017).

54. Nazario is her interlocutor and colaborer on her last book, *Earth Beings: Ecologies of Practice across Andean Worlds* (Durham, NC: Duke University Press, 2015). Marisol de la Cadena, personal communication with author, No-vember 2020.

55. For a discussion of money functions and immaterial money in a

different case, see Federico Neiburg, "An Imaginary Currency: The Haitian Dollar," in *Money in a Human Economy*, ed. Keith Hart (Oxford: Berghahn Books, 2017), 270–72.

56. On worldwide presence of monetary pluralism, see Keith Hart, "Introduction: Money in a Human Economy," in *Money in a Human Economy*.

57. I discuss how this situation started to change in Perú with the COVID-19 pandemic, with cash as a potential vector of virus infection, in the paper "Vectors for a Cashless World" presented at the Society for Social Studies of Science (4S), Toronto (virtual), October 6–9, 2021. See also García Llorens, "Cash, Banking, and a Population as a Proxy for Money," in "Banking on the Unbanked."

58. Most Peruvians distrust not only the banking system but also government. Both are often involved in corruption scandals, and the government has been highly unstable since the past decade. See "Perú se instala en la inestabilidad: 4 presidentes en los últimos 4 años," Open Democracy, November 13, 2020, https://www.opendemocracy.net/es/peru-instala-en-inestabilidad-4-presidentes -en-ultimos-4-años/; and "Francisco Sagasti se convierte en el nuevo presidente de Perú, el tercero en una semana," *BBC Mundo*, November 16, 2020, https:// www.bbc.com/mundo/noticias-america-latina-54965252.

59. Joe Dumit, personal communication with the author, June 2021.

60. Carolina Trivelli, personal communication with the author, January 2017.

61. See Roy, *Poverty Capital*. See also Kar, *Financializing Poverty*; and Schwittay, "Financial Inclusion Assemblage."

62. Roy, *Poverty Capital*, 47.

63. I thank Marie McDonald for her comment on this point.

64. Ames, *Charisma Machine*.

65. Julia Elyachar, *Markets of Dispossession: NGOs, Economic Development, and the State in Cairo* (Durham, NC: Duke University Press, 2005), 137.

66. Experimental governmentality is a dynamic that both assumes a "self-perpetuating relation of rescue" and then produces information that extends that relation into the future. See Michelle Murphy, *The Economizalation of Life* (Durham, NC: Duke University Press, 2017), 79.

67. Banco de Crédito del Peru (BCP), the main bank of Peru and largest partner in Bim, launched its mobile payment app Yape in 2016, the same year Bim was launched. See "'Yape' se expande y podría convertirse en competidor del BCP en menos de dos años," *La República*, September 17, 2019. BBVA and Interbank launched their own mobile payment apps as well (Lukita and Tunki, respectively) around this time as well. In 2020, they united with a third bank (Scotiabank) to create an interoperable app, sort of the US version of Zelle, called PLIN. See "Peruvian Banks BBVA, Interbank, and Scotiabank Launch PLIN Using YellowPepper's Real-Time Payment Platform," *Business Wire*, May 27, 2020.

10

Corporate Culture Made Material

Ephemera and In/equity at Control Data Corporation, 1957–1975

Elizabeth Semler

The Minnesota-based technology company Control Data Corporation was, in the 1960s, home to the world's first supercomputers. At the company's founding in 1957, its president, William Norris, recognized that Control Data's success would be contingent on hiring and retaining the brightest employees. Norris thus espoused a corporate culture that prioritized not only products and services but also the company's people. This approach resulted in robust employee benefits, wellness programs, and ambitious corporate social responsibility agendas that sought to alleviate poverty, discrimination, and other social ills through employment and education.[1]

Norris's vision became encoded in employee handbooks, benefits booklets, stockholder updates, and other company ephemera. These objects made material the company's philosophies and provided employees with a concrete embodiment of the corporate culture. For instance, Control Data's 1967 annual report for "stockholders, employees and friends" presented a company diverse in its products and its employment practices. Norris wrote in the report's introduction that "the combined individual efforts of Control Data people" had helped make 1967 one of the company's most successful years.[2] Photos of employees building, using, and selling Control Data's computers and services highlighted the company's accomplishments. By featuring women, people of color, and folks with differently abled bodies, the report's photos bolstered Control Data's public image as a welcoming and inclusive workplace.

Oral history interviews with former employees, however, reveal an environment plagued by harassment and discrimination. A deep dive into company ephemera—especially materials made for internal consumption—similarly reveals a troubled workplace. Control Data's leadership failed to connect hiring a diverse workforce with creating an equitable culture. Formal policies allowed for unequal pay and segregated spheres of work, and material products portrayed gendered ideas about the roles of men and women. For instance, Control Data employees were almost exclusively depicted as white men, while white women were shown as wives and mothers. People of color were rarely shown at all.

This chapter explores the interplay between corporate and material culture at Control Data Corporation from its founding in 1957 to the mid-1970s, with a specific focus on gender. It is shaped by oral history interviews with employees and archival materials from the Charles Babbage Institute and private collections. The chapter argues that material objects influenced Control Data's broader corporate culture by implicitly reinforcing norms around who was accepted and expected in the workplace. Leadership, for instance, began to address discriminatory policies in the late 1960s, but they left the company's physical materials unchanged. The day-to-day items with which employees engaged continued to portray sexist ideas regarding men's and women's labor. Unsurprisingly, a 1975 employee satisfaction survey found that women (and minorities) perceived a "significantly greater degree of discrimination in hiring and promotion than did the majority of employees."[3] Control Data's material culture helped encode ideas about who could be employees, what roles they could fill, and who could provide effective leadership at the company. Pervasive assumptions around gender helped sustain a culture of inequality at Control Data.

Engineering Research Associates: The Blueprint for Midwestern Computing Culture

Minnesota's computing industry was born out of cryptography and cryptanalysis work done on behalf of the US Navy during World War II. This crypto-agency, called Communications Supplementary Activities–Washington (CSAW), brought together talent from the military, civil service, and civilian world in order to create the technologies and machinery necessary to intercept and then break foreign military codes. The war's imminent conclusion prompted engineer William Norris, mathematician Howard Engstrom, and others within CSAW to find a new home for the agency's research into digital electronic circuitry.[4] Out of this search came Engineering Research Associates (ERA). It opened in 1946 with about 40 men in St. Paul, Minnesota.[5] As one of the first companies of its kind, ERA played an enormous role in both developing novel computing technologies and laying the foundation for computing companies in the region; it established a blueprint for the type of company culture that could foster innovation and creativity. Many of ERA's spinoff companies, including Control Data, sought to re-create this environment in the latter half of the twentieth century. Because of this direct influence on Control Data, it is important to explore some details about ERA's history.

ERA's initial projects focused on military contracts for code-breaking machines and products for the airline industry.[6] Oral history interviews with early employees described the work environment as challenging, both figuratively and literally. The novelty of ERA's technologies, paired with the secrecy required of classified military machinery, resulted in strict compartmentalization and limited collaboration. Employees were intellectually isolated from one another and were not supposed to discuss details or seek help from colleagues outside their project groups. The tight timelines for many military contracts compounded these difficulties. Employees who developed the company's code-breaking machines in the late 1940s worked eighteen-hour days, almost every day of the week. The demanding schedule was worsened by ERA's location in an old airplane hangar. The warehouse-style space was overly hot in the summer and freezing cold in the winter, and the lack of insulation allowed critters easy access. As engineer Jack Hill explained, "We were working hard under very trying physical conditions, in that god damn old plant full of sparrows that crap on everything."[7]

At the same time, Hill believed the work environment's hardships created an "esprit de corps," a camaraderie that helped employees transcend the barriers of classified research.[8] Many other early employees also believed that these difficulties united the staff. And, although the environment was challenging, it was also stimulating. Opportunities for creative freedom abounded, and on the whole, employees reflected positively on their experiences at ERA in the late 1940s. Engineer Frank Mullaney, who joined the company in June 1947, explained that he could hardly wait to get into work in the mornings. He learned new things in the lab every day, "and it was just fun and I enjoyed it."[9]

The culture at ERA initially developed organically. It was influenced by the contracts the company was able to acquire, the dictates of those contracts, the physical workspace, and the dynamics among staff. Individualization was central: the projects people worked on and the ways they worked were highly individualized. According to Willis Drake, an engineer and technical writer who started at ERA in 1947, this practice allowed employees to pursue their "instincts, interests, energies, visions" to an unusual degree. The emphasis on the individual, however, bred an "abhorrence of large bureaucratic methods." Drake believed this made it difficult for ERA employees to engage intellectually within hierarchical, bureaucratic structures.[10] Indeed, the company's culture had developed an aversion to such conditions from the get-go because complex organizational and reporting structures often impeded the development of time-sensitive technologies.

The latter aspect of ERA's underlying culture flared when it was purchased by Remington Rand in 1952.[11] This sprawling corporation was known for typewriters, adding machines, data-processing equipment like punch card tabulators, and other business-oriented technologies. Leadership at ERA feared Remington Rand's rigid operating structures and lack of familiarity with computing would dramatically alter the workflow of their company.[12] However, because ERA was acquired as a subsidiary of Remington Rand, they were able to maintain the freedom and flexibility prized within their company culture. In fact, many employees outside the leadership chain were insulated from the effects of the change in ownership. Engineer James Thornton, who joined the company in late 1950, explained: "Well, I discovered that we had been acquired one day. Frankly, I don't recall skipping a beat."[13]

Ongoing changes in the relationship between the two companies, followed by a merger between Remington Rand and Sperry Corporation in 1955, constrained ERA's operations. Frustrated, a handful of current and former employees decided to form a new company. Incorporated on July 8, 1957, Control Data Corporation officially sought to further the "design, development, manufacture and sale of systems, equipment and components used in electronic data processing and automatic control for industrial, scientific and military uses."[14] Unofficially, the company served as an explicit rejection of Sperry Rand's corporate culture. Frank Mullaney, one of Control Data's cofounders, explained bluntly that "when we started Control Data we really didn't have a plan to build a computer and sell it as such. . . . The main thing was we wanted to get the hell away [from Sperry]."[15]

Building Company Culture at Control Data Corporation

The founders of Control Data had a strong sense of the environment that innovation required and a clear idea of the culture they hoped to build at their new company. Foremost within everyone's minds was to avoid the rigid, bureaucratic systems of companies like Sperry Rand. They believed that administrative constraints hindered flexible thinking and the development of clever solutions. This belief was informed by the men's experiences in the pre-Rand days of ERA; all of the founders had worked at ERA in its early years, and the company factored heavily into their vision of Control Data's culture. Individuality, creativity, a willingness to pursue radical ideas, and the development of an esprit de corps were some of the cultural elements that leadership felt had contributed to ERA's technological successes in the 1940s and early 1950s.[16]

Whereas corporate culture developed organically at ERA, it was actively constructed at Control Data. Formal policies and procedures were important routes by which Control Data articulated and reinforced company culture. Creativity and ingenuity, for instance, were integrated into Control Data's pay scale. These were among five factors that determined a job's grade and thus the salary range for a particular job.[17] To build community and facilitate open communication across all levels, Control Data had a shared lunchtime and standardized break times throughout the day.[18]

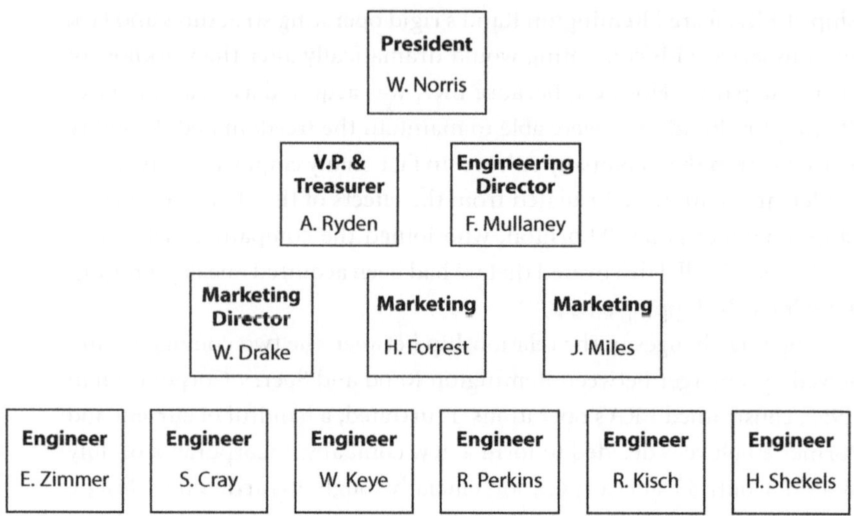

Figure 10.1. Control Data Corporation's organizational structure in 1957 included 12 men, all of whom had worked for Engineering Research Associates at one point in their careers.

These efforts also went toward Norris's efforts to keep the company's management chain and overall organization as flat as possible (figure 10.1).[19]

Control Data expanded rapidly in size and scope with the success of its first computer, the model 1604, released in late 1959. Even though bureaucratic complexity contradicted the company's founding principles, the growth in the numbers of employees, manufacturing sites, and product lines necessitated hierarchical management structures. Leadership responded by becoming more purposeful in its efforts to nurture corporate culture. Employee handbooks, for instance, began to articulate the company's (intended) culture by emphasizing individuality, creativity, and innovation. In the early 1960s, corporate culture also centralized the employee as the heart of the company. Norris developed this "people first" philosophy in an effort to boost retention. He hoped that if employees felt valued they would, in turn, dedicate themselves to Control Data.[20] The opening pages of the 1963 employee handbook for the Computer Group exemplify early iterations of the people-first approach. Written by division vice president Frank Mullaney, the handbook explained: "The computer business is dynamic and competitive. To gain and hold industry leadership . . . we depend completely upon the abilities and loyalty of our people, for *they* are the company's future."[21] This idea was expressed repeatedly across various company materials, including management booklets, interoffice memos, and company benefits pamphlets.

The people-first orientation was made official in 1964 when Control Data created a formal corporate philosophy. This was a 12-point statement that began with the point "Control Data's people are Control Data."[22] The company's philosophy encoded ideas about who worked at Control Data, what traits were important in employees, and whose labor and ideas were valued. It was quickly incorporated into handbooks and other day-to-day materials with which employees at all levels of the company engaged. Importantly, women were almost always absent from these internally facing materials. Language was gendered male and images routinely depicted men alone as the standard Control Data employee. A quick glance through materials from the late 1950s through the mid-1960s conveys an even more specific image of the average Control Data employee: a young, white, straight man. Occasionally employees were shown as older, as married, or with children, but there was very little deviation from this standard.

To be fair, the gendered language and imagery in Control Data's materials reflected the initial employee roster. All of the founding members were men. Of the 16 people employed directly by Control Data in the months after its founding, only 2 were women. By late 1958, only 5 of the company's 36 direct employees were women. All appeared to work in clerical positions.[23] Thus Control Data's first employee handbook, produced in December 1958, probably referred to employees as men because almost all of the company's employees *were* men.[24] Polly Myers, Lisa Frehill, and other scholars have shown that men were the overwhelming majority in engineering-based roles in the 1950s and '60s. The field was strongly associated with men's work, and for that reason, Control Data was neither alone nor unique in is deployment of men as the standard in company materials.[25]

However, by mid-1959 almost 20% of Control Data's direct employees were women.[26] In addition to clerical positions, women were hired as engineers and mathematicians and into other technical roles. When the company received a contract to develop technology for the US Navy in 1959, Barbara Halverson joined the newly formed Polaris team as a mathematician. By March 1961 she was one of 8 women working in various roles on the 97-person team.[27] Maris Bergmanis joined the company in 1961 to work on the development of the 160 minicomputer, then transferred to help design the 3000 series of computers. Clair Miller was the director of applications in the company's Sunnyvale, California, office in 1964.[28] The scholarship of Jennifer Light, Mar Hicks, and Nathan Ensmenger, along with several chapters in this volume, shows that women occupied technical roles at computing and other engineering-based companies through the 1950s and 1960s.[29]

Despite the growing number of women at Control Data, male pronouns were the default in internal written materials through the 1970s. In addition, job descriptions and advertisements continued to describe company positions with male pronouns.[30] Polly Myers's research of Boeing's corporate culture in the twentieth century demonstrates that gendered language and imagery were deployed purposefully in company materials to oust women from the workplace.[31] The value of women's labor, which became increasingly important through the 1940s at Boeing, was recast in the wake of World War II. Women were depicted as temporary workers and were seen as taking jobs from returning men, thus preventing men from fulfilling the sociocultural expectation of providing financially for their families. Boeing leadership worked to create a hostile culture that not only forced women out but also prevented them from returning in the future. On a large scale, policies were instituted that explicitly discriminated against women; on a smaller scale, newsletters and other company ephemera showed men as employees and women as homemakers and mothers. Historian Alice Kessler-Harris describes such depictions as the "gendered imagination" of labor.[32] By relying on broadly accepted sociocultural norms around men's and women's roles, the gendered imagination of labor allowed Boeing to build a workplace where men's labor was the standard. Policies helped force women out of jobs, and material culture affirmed that women's paid labor was neither expected nor accepted.

To date, I have not found evidence that Control Data leadership utilized a gendered imagination of labor to force women out of the company. Many instances in the company's infancy were probably reflections of both Control Data's overall employee roster and also broader US cultural expectations related to engineering and computing work. But there are moments when, I suspect, company leadership *was* purposeful in its deployment of a gendered imagination of labor. The assumption of the standard employee as a man facilitated and justified pay structures that allowed Control Data to pay women less for their labor, especially as women's assembly work became central to the construction of commercial computers through the 1960s. The following section digs into several specific examples of company policies and materials from this time period in order to explore how they converged to shape Control Data's culture.

Encoding Inequality at Control Data

The rapid expansion of Control Data in the late 1950s led to the creation of a new segment within the company, the personnel and administration function. Renamed human resources in 1986, personnel grew in size and scope throughout the 1960s. The department's responsibilities included, among many others, employee recruitment and onboarding, the creation and management of employee benefits, payroll, employee relations, procurement of manufacturing and production facilities, and the integration of newly acquired companies.[33] Personnel influenced who was hired and how they were compensated. It was also a key source of corporate policies and the related physical materials. In these ways, the department played a critical role in developing and maintaining the company's culture. Personnel treated men and women differently from the start of the interview process. Male college graduates, for instance, were interviewed through college relations recruiting, and women were interviewed by clerical recruiting.[34] This setup reflected and reinforced a broader corporate attitude that prioritized men's labor and contributions over those of women.

Gendered language and imagery in benefits materials

Physical materials produced by Control Data for both internal and external consumption commonly depicted employees as men. The company's benefits materials provide especially interesting examples of gendered language and imagery in that the benefits package was used to entice high-caliber employees into Control Data during a period when the labor pool was small and many nascent computing companies were looking to hire. It is likely that great care was put into the written content and physical presentation of benefits materials.[35] Benefits booklets, pamphlets, and other related items consistently described and showed employees to be men, and men alone.

Benefits materials from Control Data's computer division are representative of the ways the imagined employee excluded women. Throughout the 1960s, documents related to the division's group insurance plans used the term "employee" interchangeably with "he," "his," and "him." The 1961 explanation of when the "major medical expense" benefit would begin stated that it could be used when the employee exceeded their medical insurance coverage and had "begun to pay the excess cost himself." A small drawing of two men accompanied this text. One sat in a wheelchair and had a blanket placed over his lap. The other man was dressed in a business suit and stood next to the wheelchair; he was shaking hands with what was presumably the sick employee (figure 10.2).[36]

Drawings of employees using their benefits, from life insurance to surgical expense coverage, were common across a range of the computer division's materials. The pictured employee was always a man. Women made appearances as wives, as children, and even as nurses! But never as Control Data employees.[37] This held true for company-wide benefits materials too. For instance, a 1964 booklet pertaining to Control Data's group insurance policies featured drawings of men at work, at home with their wives and children, receiving care in the hospital, and receiving care at home.[38] Somewhat surprisingly, the explanation of maternity benefits did not have an image of the new mother, but rather also focused on the father-employee holding his newborn.[39]

The updated 1966 group insurance plan documents replaced the various employee vignettes with one image that repeated along the bottom of each page (figure 10.2). Centered in the drawing is a father sitting with his young son; the son plays with a toy truck and ball. The man's wife watches the scene from behind, while the couple's daughter stands in the far background. The image was accompanied by the text "Security for you and your family."[40] The foregrounding of the man, paired with the language used throughout the booklet, strongly implied the employee was the man.

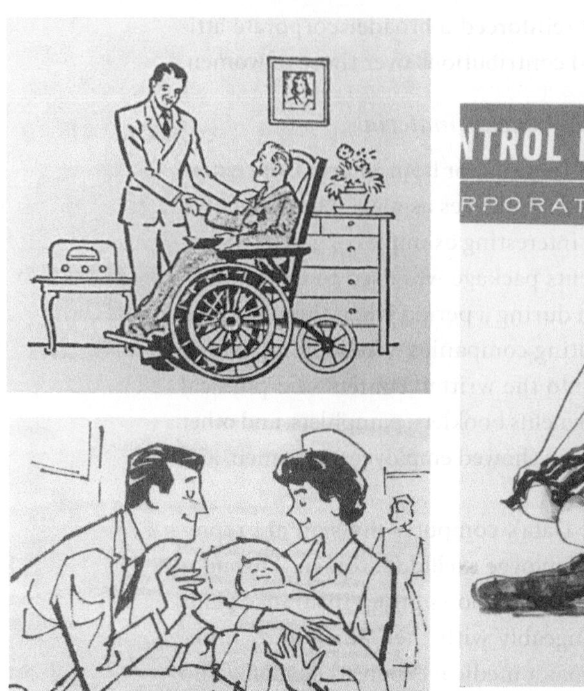

Figure 10.2. Three images from group insurance benefits booklets at Control Data show the use of gendered imagery in reference to company employees and others so as to center men as the standard employee. The images also showed only white folks and reinforced broader cultural ideas about heteronormative nuclear families. Image courtesy of the Charles Babbage Institute Archives, University of Minnesota

Any uncertainty was quashed by the company's explanation of maternity benefits: the user of the benefits was described alternately as "your wife," "an employee's wife," and "an insured employee's wife."[41] Such language and imagery reinforced the gendered imagination of labor. In addition, because the employees depicted were always white and often married, benefits materials also reinforced the racial hierarchies and heteronormative views regarding marriage, family, and romantic relationships common in the mid-20th-century United States.

Even when the personnel department made efforts to build a more equitable work environment, employees were still depicted as men. Instructions on how to eliminate biases from performance appraisals, found within Control Data compensation manuals from the 1960s through the mid-1970s, consistently referred to employees as men. In these, supervisors and managers were warned that both positive and negative biases can influence their decision making regarding direct reports. Examples of potential biases included the "man who has good looks," "the man who 'goofed' on one project," and "the man who went to the same school as you."[42] Thus, Personnel's attempts to provide management staff with tools to conduct fair and balanced assessments were undermined by its depiction of employees as men. The absence of women in company materials and in the vision of who a Control Data employee was diminished their work and contributions to the company.

Gendered bifurcation of pay structures

Control Data's material culture mirrored a corporate culture that struggled with equity. A glance at Control Data's remuneration structures confirms that the company did, indeed, treat men and women unequally: women were routinely paid less for the same work as men. Author Mark Jensen has demonstrated that certain positions even had designated women's categories. Manufacturing positions, for instance, had a "women assembly" category. This was a grade 1 position, the lowest pay grade in the company. Men were hired into "general assembly," a higher-paying, grade 2 position.[43]

In the mid-1960s Control Data reworked its pay grade system. The company's expansion from a focus on digital electronic computers to also include data center services and peripheral products manufacture necessitated a grading system that reflected the company's growing complexity. Leadership developed a system that categorized jobs into six groupings:

1. Professional-Technical Salary
2. Professional-Administrative Salary
3. Technical Salary
4. Production Pay Plan
5. Office Salary
6. Customer Engineering

Within each group, jobs were graded on a 15-point scale. Much as in the initial system, grade 1 was the lowest-paying, least-complex grade, and a grade 15 job was a top-tier leadership position. The new structure removed explicit segregation of job titles by sex, seemingly ending the company's bifurcated pay structure.

The restructuring of Control Data's remuneration systems was also likely influenced by a series of federal acts that prohibited various types of discrimination in the workplace. The Equal Pay Act of 1963, for instance, prohibited pay disparities between men and women for jobs that required "substantially equal skill, effort and responsibility under similar working conditions."[44] Title VII of the Civil Rights Act of 1964 barred employers from discriminating against employees on the basis of "race, color, religion, sex, and national origin."[45] By the start of 1964, Control Data had amended its employee handbook to include an "Employment Opportunities" statement that echoed Title VII language.[46] Unlike Title VII, however, Control Data's statement excluded "sex" as a category against which the company would not discriminate. Sex was omitted from the statement when it was incorporated into the company's official corporate philosophy statement in 1964 and was missing from such language through at least 1973.[47]

It is difficult to imagine that the omission of sex from Control Data's

"Employment Opportunities" and corporate philosophy was accidental. The company's president, William Norris, placed great importance on the development of a philosophy that could unify the company and also on the company's physical materials more broadly. These were tools by which to recruit talented employees and, by inculcating a culture that allowed creativity, flexible thinking, and camaraderie to thrive, helped retain employees too. The exclusion of women essentially gave the OK to discriminate against them. Indeed, although jobs were no longer labeled by sex, the grading system remained bifurcated. All grade 1 positions, the lowest-paying jobs at the company, were located within the "Office Salary" and "Production Pay Plan" (specifically, assembly production) categories; these categories employed the most women.[48] Women's labor was literally valued less than men's labor.

Gendered segregation of labor

The exclusion of women from Control Data's material culture, including the company's "Employment Opportunities" statement, implied that women's labor was not valued or important enough for the company to try to retain. The devaluation and erasure of women's labor may have been further exacerbated by the segregation of labor that occurred at Control Data as production and manufacturing facilities expanded to a variety of new locations through the 1960s. Norris expanded Control Data's manufacturing sites to several areas in rural Minnesota in the mid-1960s. The facilities were located outside major cities and sought to offer employment to women, many of whom had no opportunities for paid work in these areas.[49] The rural production program's focus on women as employees, however, shows that company leadership was attuned to issues related to sex and gender—and were so attuned at the same time they were writing an equal-opportunity employment statement that excluded women!

Although I do not have data showing what the job grades were in these rural manufacturing facilities, site-specific employment figures show that the rural plants were almost all staffed by nonexempt employees, a class that comprised grades 1–8 on the pay scale. It seems safe to assume, then, that most (nonmanagement) employees in the rural plants were paid at grade 8 or less and were most likely paid at the lowest levels, grades 1–3 (table 10.1). The rural productions sites—built purposefully to employ women—allowed Control Data to employ a labor force that could be paid less, not just because they were women but also because there were no other jobs in the area. And because the sites were literally separated from the center of production (the Twin Cities area), it would be easier to keep any potential questions about links between low pay and sex to a minimum as the women and their labor were removed or erased from company leaders' view.

Similar patterns of potential pay discrimination can be seen in the

Table 10.1. Employee classifications at Control Data's rural production facilities in Minnesota, 1967–1968

	October 1967		April 1968	
Site	Exempt	Nonexempt	Exempt	Nonexempt
Cambridge (opened 1964)	12	165	16	138
Faribault (opened 1965)	14	192	13	188
Spring Grove (opened 1965)	22	332	25	446

Notes: The rural production factories that Control Data built to employ farmwives in the mid-1960s were responsible for the manufacture of critical parts for its computers, including 6000 series circuit boards, power supplies, logic chassis, and 6000 and 1700 series core memory stacks.

company's efforts to employ people of color in the late 1960s. In the wake of civil unrest both in Minnesota and across the nation, Norris developed a series of corporate social responsibility programs to improve the human condition at CDC. One of his earliest efforts was to build production facilities in poor, urban, predominantly Black communities in an effort to alleviate poverty and because it would be cost-effective. Leadership called these facilities "poverty plants," and they largely employed people of color from the communities where they were located. Again, most of the positions in these facilities were in the Office or Production categories, and thus started at the lowest pay grades. The company's 1967 Compensation Manual shows that positions in the Selby Bindery—a St. Paul–based factory that specifically sought to employ minority women and that printed and assembled the company's manuals and other paper products—fell within the Office category and sat at pay grades 2 and 3. Thus, although by the late 1960s positions at Control Data weren't necessarily explicitly labeled "women assembler," the categorizations Office and Production nonetheless signaled a specific kind of gendered role.

Control Data's rural manufacturing facilities and its "poverty plants" separated women and people of color's labor from that of the rest of the workforce. This policy, paired with the absence of these groups from the company's internally-facing materials, further erased the contributions of women and people of color. In so doing, it rendered invisible the presence and contributions of diverse groups at the company, thereby tacitly approving discrimination. Interviews with former Control Data employees confirm that the company could be a very difficult environment for a woman to work in. Sonya Anderson, who was hired in 1966 to do systems documentation, was told by her manager at her first performance review that he wasn't going to invest in any training for her because she was too attractive. Inevitably, she would get married and have children and then Control Data would be out their investment in her. Someone else told Anderson that women would never make it at Control Data—or business in general—because too much bonding went on at the men's urinal. In her experience, this was just the culture at Control Data, and these kinds of attitudes weren't recognized as unseemly at all.[50]

Margaret Loftus, who joined the company in 1965 as a software developer for the CDC 6600, thought Control Data's environment was terrible for women. Interestingly, Loftus worked first out of the Palo Alto facility in California, where apparently a number of women worked at various levels in the software department, and they were treated really well. It was when Loftus moved to Minnesota facilities that she noticed a very different attitude overall, but particularly toward women, and especially toward women in management positions (if they could even get those jobs).[51] These and similar experiences indicate that although discrimination can be encoded in company policies or materials, or both, corporate culture is nonetheless malleable. Conscientious leadership and group dynamics, for instance, can help create more positive environments, at least at a local scale.

Loftus's allusion to the difficulty of moving into management as a woman represents one of the more common complaints from women employees at Control Data. Upward mobility often felt like an impossibility. This may be a partial consequence of the invisibility of women in the company's material culture as well as the literal segregation of their labor. Why would managers hire people they don't think they need? Why would they promote work they can't see? This goes for women's day-to-day lived experiences, as well as what they see in and absorb from the material culture. Employees were described as and shown to be men; managers were always men, and materials made for management always described employees as men. For example, materials having to do with "Work Planning"—a policy implemented in May 1966 that required employees to set biannual work goals—described everyone at Control Data as men. The 1968 manager's booklet described the Work Planning system as "of particular value to management personnel. A person in management will find that it assists him to plan, budget, organize, and delegate work. The system will also provide better understanding of the employee's views concerning his job and should strengthen the supervisor-subordinate relationship."[52] The depiction of men as both the standard employee and the standard supervisor excluded women's contributions and, by reaffirming the gendered imagination of labor, also made it less likely that men would consider women for promotions into managerial positions.

To be clear, not every woman who worked at Control Data had poor experiences. Sandra Resnick, who had a background as a programmer specializing in computer-aided design, joined Control Data in 1981 as a systems manager. She held several different management positions during her time at the company, and throughout that time, she said, she never experienced any issues with harassment or discrimination. However, she did think the culture favored mavericks and males (a result of the company's cultural genealogy), and she was aware that other women experienced problems, especially women in engineering and hardware.[53] Molly Lou Reko also had a positive experience at Control Data. She was a mathematics teacher who initially came to help develop the company's education system, PLATO, for the Control Data Institutes in the late 1960s. She eventually moved up to help manage aspects of PLATO at the institutes. Reko believed that atmosphere she encountered really supported women and minorities.[54]

Conclusion

Stepping back from the issue of gender specifically and looking at the work environment as a whole at Control Data, one may see that it had changed dramatically from its inception as a company that sought to escape bureaucracy and hierarchy. There were now rules and procedures for everything; many seemed arbitrary and employees felt that they hindered innovation and success more than encouraging it. The Work Planning policy referenced earlier is a good example of the time-consuming bureaucracy. On the surface, it's a mostly logical policy. It required employees to set a work plan and work goals in six month increments. The plans would then be used by supervisors for performance evaluations and salary reviews, among other things. Leadership implemented it in part to ensure fair compensation for people's work—to guarantee that they had proper managerial oversight and their contributions were being evaluated. But the planning booklets for nonmanagement and management employees depict a complex procedure they were supposed to engage in every six months.[55] It took up a lot of time, directed efforts away from the day-to-day work, and was nonetheless susceptible to outside or unfair influences. It also represented the very thing the Control Data cofounders had sought to escape at Sperry Rand.

Additionally, by the late 1960s, the layers of supervision and management had multiplied to the extent that employees sometimes struggled to understand the company's structure and organization. Management was expansive, and the corporate culture so bound with the ideas of "individual action" that no one seemed to work together in a unified way. Sandra Resnick explained that in her experience, this lack of communication meant the company missed out on opportunities for products that could have been big money makers. All of these developments were in exact opposition to why the company was founded—it became that which it had been created to reject, even as it sought to build and deploy a culture in opposition to the bureaucratic model. These issues drove people away from the company. Sonya Anderson was pretty explicit, saying that when people left Control Data, it was because they wanted to get rid of the rule book.[56]

Control Data started in the late 1950s, in part, as an explicit rejection of large, bureaucratic companies and as an effort to re-create the early days of its parent company, Engineering Research Associates. The co-founders worked to build a company whose culture prioritized creative thinking, flexible problem solving, individual action, and hard work, and also emphasized the importance of the individual and their contributions to company success. And despite an ever-growing, diverse roster of employees through the 1960s, the physical materials with which employees had to engage had a narrow conceptualization of who Control Data employees were. From benefits booklets to training manuals, employees were written about and shown as men, specifically white, heteronormative men. This erasure of the diversity of Control Data's staff in company materials—along with policies whereby the company paid more for men's labor and the physical locations of production and manufacturing often segregated labor by sex and race—standardized and normalized employees as white men. These policies made it difficult for women and folks within minority groups to move up within the company and ultimately encoded inequality.

Notes

1. Mark Jensen, *HR Pioneers: A History of Human Resource Innovations at Control Data Corporation* (St. Cloud, MN: North Star Press of St. Cloud, Inc., 2013); Robert Price, *The Eye for Innovation: Recognizing Possibilities and Managing the Creative Enterprise* (New Haven, CT: Yale University Press, 2005).

2. "Your Company: An Annual Report for Control Data Employees and Families, 1967," CBI 80, series 15, box 7, Control Data Corporation Records: Personnel Documents and Reports (hereafter cited as CDC: Personnel), Charles Babbage Institute Archives, University of Minnesota, Minneapolis (hereafter cited as CBIA).

3. "Employee Attitude Survey Results, 1975," CBI 80, series 15, box 7, folder 4, CDC: Personnel, CBIA.

4. For more about ERA's foundations, see Arnold A. Cohen, interviewed by James Ross, January 20 and 28, 1983, transcript, Charles Babbage Institute Archives, University of Minnesota Digital Conservancy, https://hdl.handle.net /11299/107220 (hereafter UMNDC); William C. Norris, interviewed by Connie Van Hoven, 1977, transcript, CBIA, UMNDC, https://hdl.handle.net/11299 /107557; Arthur Norberg, *Computers and Commerce: A Study of Technology and Management at Eckert-Mauchly Computer Company, Engineering Research Associates, and Remington Rand, 1946–1957* (Cambridge, MA: MIT Press, 2005).

5. ERA's location in Minnesota was largely serendipitous. A businessman from the area, John Parker, was interested in investing in the company and had an available factory space. However, the state was already home to large and successful industries like flour milling, railroads, and manufacturing. An existing infrastructure, as well as political and economic environments, made Minnesota favorable to the growth of industry in a way that likely contributed to the flourishing of the computing industry. See Thomas Misa, *Digital State: The Story of Minnesota's Computing Industry* (Minneapolis: University of Minnesota Press, 2013).

6. The focus on airline products stemmed from John Parker's connections to aeronautics. Andrew Fox, "Silicon Prairie: Engineering Research Associates,

UNIVAC, and the Beginning of Minnesota's Computer Industry," *DCHS over the Years* (October 2017): 24.

7. John Lindsay Hill, interviewed by Arthur Norberg, January 15 and 22, 1986, transcript, CBIA, UMNDC, https://hdl.handle.net/11299/107360, 55.

8. Hill interview, 55.

9. Frank C. Mullaney, interviewed by Arthur Norberg, June 2 and 11, 1986, transcript, CBIA, UMNDC, http://hdl.handle.net/11299/107538, 20.

10. Willis K. Drake, interviewed by James Ross, February 3 and 25, 1983, transcript, CBIA, UMNDC, https://hdl.handle.net/11299/107248, 37.

11. ERA faced cash flow problems in the early 1950s that were compounded by growing (federal) concerns about the ways the company acquired federal contracts. These issues pushed ERA's majority shareholder, John Parker, to sell. For further detail about ERA's acquisition by Remington Rand, see Norberg, *Computers and Commerce*, 217–23.

12. Remington Rand had little technical expertise related to digital electronic computing, even though the company had recently acquired Philadelphia-based Eckert Mauchly Computer Company. Drake suspected that Remington Rand didn't even understand what it had bought with ERA: its managers thought ERA had computers made and ready to sell, but really ERA was simply "a bunch of talent. That was not at all what they [Remington Rand] wanted. It was a seething, unhappy, resentful atmosphere. It was all buzzy in the halls and not much else." Drake, interview, 19; Robert Emmett McDonald, interviewed by James Ross, May 4, 1983, transcript, CBIA, UMNDC, http://hdl.handle.net/11299/107482, 7.

13. James E. Thornton, interviewed by Arthur Norberg, February 9, 1984, transcript, CBIA, UMNDC, http://hdl.handle.net/11299/107673, 26.

14. Misa, *Digital State*, 104.

15. Mullaney, interview, 77.

16. Drake, interview, 38.

17. The other factors that influenced a job's grade included level of education (or training) and relevant experience required, physical demands, responsibility (independent decision making), and supervision. Control Data Corporation Computer Division Employee Manual, 1963, CBI 80, series 15, box 5, CDC: Personnel, CBIA.

18. Control Data Corporation Employee Handbook, December 1958, CBI 80, series 15, box 5, CDC: Personnel, CBIA; Control Data Corporation Computer Division Employee Manual, 1963.

19. Norberg, *Computers and Commerce*, 43; Recreation of Control Data Corporation's 1957 Corporate Organization Chart, CBI 80, series 15, box 1, folder 1, CDC: Personnel, CBIA.

20. Jensen, *HR Pioneers*, 66–67.

21. Control Data Corporation Computer Division Employee Manual, 1963 (emphasis in original).

22. The March 1964 edition of CDC's corporate philosophy shows that the company had developed the same hierarchical bureaucracy its founders were trying to escape with Sperry Rand. The 12-point philosophy was overly complicated and tied to a complex management system. Control Data Corporation Corporate Philosophy, March 1964, CBI 80, series 15, box 4, folder 7, CDC: Personnel, CBIA.

23. The two women listed in the office directory in December 1957 were Rose Mary Geiger and Lucille Walker. It's unclear what Geiger's role was, but Walker worked in a clerical position. Records from 1958 show that Walker, as well as Ruth Johnson, Gloria Jacobson, and Louise Henke, worked in clerical

positions at the company. Geiger was also listed as an employee, but again, it's unclear what her role was. Control Data Corporation office directories, 1957–1958, CBI 80, series 15, box 1, folders 2–4; box 5, folder 1, CDC: Personnel, CBIA.

24. Control Data acquired several other companies throughout the year 1958; it had around 260 total employees in June 1958. It's impossible to know exactly how many women employees there were, but given trends in engineering and other sciences at the time, it's likely that most of the acquired employees were men. The breakdown of employee numbers can be found in Control Data Corporation Employee Handbook, December 1958.

25. Lisa Frehill, "The Gendered Construction of the Engineering Profession in the United States, 1893–1920," *Men and Masculinities* 6, no. 4 (April 2004): 383–403, http://doi.org/10.1177/1097184X03260963; Polly Myers, *Capitalist Family Values: Gender, Work, and Corporate Culture at Boeing* (Lincoln: University of Nebraska Press, 2015).

26. Around 20 of the 137 direct employees were women in June 1959. Because Control Data commonly listed employees by their first-name initial and full surname, it is difficult to assess the sex breakdown. I've done informed guesswork by cross-referencing full names I found in other archival records. In addition, I've made assumptions based on first name about sex (and gender identity), so my numbers could be off. Regardless, it was probably uncommon for women to work outside clerical positions through the early 1960s. Control Data Corporation office directories, 1957–1958.

27. Organizational chart for the Polaris project, March 3, 1961, CBI 80, series 15, box 1, folder 5, CDC: Personnel, CBIA.

28. "Reminiscences of computer architecture and computer design at Control Data Corporation, OH 321," oral history interview with 18 CDC employees, moderated by Neil Lincoln, May and September 1975, transcript, CBIA, UMNDC, https://conservancy.umn.edu/bitstream/handle/11299/104327 /1/oh321cdc.pdf.

29. Nathan Ensmenger, *The Computer Boys Take Over: Computers, Programmers, and the Politics of Technical Expertise* (Cambridge, MA: MIT Press, 2012); Marie Hicks, *Programmed Inequality: How Britain Discarded Women Technologists and Lost Its Edge in Computing* (Cambridge, MA: MIT Press, 2018); Jennifer Light, "When Computers Were Women," *Technology and Culture* 40, no. 3 (1999): 455–83, http://www.jstor.org/stable/25147356.

30. Male-inflected language was most common for engineering positions but was prevalent across all roles and can be seen in compensation manuals throughout the 1960s and into the early 1970s. Although gender-neutral language became more common in later versions, "he," "him," and "his" continued to be used to explain various roles in the company. Control Data Corporation Compensation Manual, March 1, 1967; Control Data Corporation Compensation Manual, May 1, 1971; Control Data Corporation Compensation Manual, May 1, 1973, all in CBI 80, series 15, box 6, CDC: Personnel, CBIA.

31. Myers, *Capitalist Family Values*, chapter 2.

32. Myers, *Capitalist Family Values*, chapter 3; Alice Kessler-Harris, *In Pursuit of Equity: Women, Men, and the Quest for Economic Citizenship in 20th-Century America* (New York: Oxford University Press, 2001). For more on gendered imagination, see Angel Kwolec-Folland, *Engendering Business: Men and Women in the Corporate Office, 1870–1930* (Baltimore: Johns Hopkins University Press, 1998).

33. Jensen, *HR Pioneers*, 57, 72.

34. Jensen, *HR Pioneers*, 89.

35. Jensen, *HR Pioneers*, 85; Control Data Corporation Computer Division Employee Manual, 1963.

36. "A Plan of Group Insurance for Employees of Control Data Corporation, Computer Division," c. 1961, CBI 80, series 15, box 5, CDC: Personnel, CBIA, 21.

37. "A Plan of Group Insurance," 8–9, 11, 21.

38. "Your Group Insurance Plan for Employees of Control Data Corporation," July 1964, CBI 80, series 15, box 5, CDC: Personnel, CBIA.

39. "Your Group Insurance Plan for Employees of Control Data Corporation," 11, 27.

40. "Security for You and Your Family: A Group Life and Health Insurance Plan for Employees of Control Data Corporation," June 1966, CBI 80, series 15, box 5, CDC: Personnel, CBIA.

41. The document did explain that maternity benefits were also available to "female employees," so women were not wholly absent from maternity benefits considerations. However, this was the last sentence of the section and certainly reads as an afterthought, rather than the focus. "Security for You and Your Family," 13–14.

42. Control Data Corporation Compensation Manual, 1967; Control Data Corporation Compensation Manual, 1971; Control Data Corporation Compensation Manual, 1973.

43. Jensen, *HR Pioneers*, 89.

44. Equal Pay Act of 1963, Pub. L. 88-38, 77 Stat. 56 (1963), https://www.eeoc.gov/statutes/equal-pay-act-1963.

45. Title VII, § 701, Civil Rights Act of 1964, Pub. L. 88-352, https://www.eeoc.gov/statutes/title-vii-civil-rights-act-1964.

46. Control Data Corporation Computer Division Employee Handbook, 1963.

47. Control Data Corporation Corporate Philosophy, March 1964; Control Data Corporation Policies and Procedures Manual, 1971, CBI 80, series 16, box 256 CDC: Personnel, CBIA; Control Data Corporation Policies and Procedures Manual, 1973, CBI 80, series 16, box 256, CDC: Personnel, CBIA.

48. Most of grades two and three also fell within the office and production categories. Control Data Corporation Compensation Manual, 1967.

49. "Personnel Statistics Summary October–December, 1967" and "Personnel Statistics Summary April–June, 1968," CBI 80, series 15, box 7, folder 23, CDC: Personnel, CBIA.

50. Sonya Anderson, interview by Elizabeth Semler, September 4, 2018, transcript, 27–28.

51. Loftus's comments are interesting in light of Clair Miller's position as the director of applications in the Sunnyvale, California, office. Perhaps Miller's presence in leadership helped normalize women in these spaces, or perhaps Miller worked to explicitly pull women into management positions. Margaret Loftus, interview with Judy E. O'Neill, April 18 and 25, 1995, transcript, CBIA, UMNDC, http://hdl.handle.net/11299/107444.

52. "Management Guide to Work Planning: Employee Development, Performance Review, and Salary Review," June 1968, CBI 80, series 15, box 6, folder 8, CDC: Personnel, CBIA, 1; "Employee Work Plan Guides," [1967?], CBI 80, series 15, box 6, folder 8, CDC: Personnel, CBIA.

53. Resnick's initial job at Control Data as systems manager was to ensure that all of the different parts for a computer ended up at the same place at the same time in order for a systems engineer to put them together. She was the first

woman in this position at the company and only the second in software (rather than hardware). Sandra Resnick, interview by Elizabeth Semler, September 21, 2018, audio (privately held).

54. Reko also described a course that managers had to take called Emerging Women Resource, which was designed to support "emerging women." Some men in the company, she remembered, called it the "Pushy Women's Workshop." So individuals' interpretations of their experiences matter just as much as their peers' or the historian's take. Molly Lou Reko, interview by Thomas J. Misa, November 30, 2015, transcript, CBIA, UMNDC, http://hdl.handle.net /11299/188551, 13–14, 17.

55. This document in itself is a good example of the ways Control Data quickly moved away from its original desire not to be burdened by corporate bureaucracy. "Control Data Corporation Approved Policy on Work Planning: Employee Development, Performance and Salary Review," 27 May 1966, CBI 80, series 15, CDC: Personnel, CBIA; "Employee Work Plan Guides" [1967?].

56. Anderson interview, 3; Resnick interview.

11

Reassessing the Iconic and Unbundling the Ironic IBM System Engineering, Gender, and Antitrust

Jeffrey R. Yost

IBM's historiography—spanning dozens of books and many articles—like the company itself, has long dwarfed that of its mainframe peers. In sheer size, IBM's historiography exceeds that of any other company. Ironically, despite this unparalleled volume of historical literature, our understanding of IBM is severely limited. Historical scholarship on IBM is especially deficient in addressing the company's social, cultural, political, and environmental history. With rare and notable exceptions, historical studies of IBM have concentrated on its US presence, its research and development (R&D) of hardware and software, its top leadership, and antitrust cases.[1] Discussion and analysis of IBM's history often appears as an all-too-familiar cluster of strung-together core iconic topics, with only a handful of scholars breaking free from this pattern: the exceptions include Corinna Schlombs on IBM Germany and computer-based automation, Steven Usselman on IBM strategy and political economy, and Eden Medina on IBM Chile.[2] Positioned centrally within the iconic paradigm—but standing out strongly for its research depth and for the author's dual perspectives as historian and longtime IBM insider—is James Cortada's book, the first to survey the firm over nearly all its 110-year lifespan.[3]

The icons, pushed by the company's management and marketing department and all too often embraced by those outside the firm writing its history, include the IBM "salesman" and sales management, the dark-suit-and-tie dress code, signs bearing the company motto "THINK" posted in all IBM facilities, the standard IBM 80-column punch card, the firm's families of systems (especially the System/360 and System/370), programming tools and products (FORTRAN, IMS, CICS), and the IBM personal computer (PC). These all figure prominently in the formal and informal codes of conduct and practice that reinforced IBM's corporate culture over a century. In the early software era of the late 1950s, IBM-developed

FORTRAN programming language, used widely in science and technology for a half century, offered the language that coded the human-made world, as well as much of the natural world.

IBM's encouragement of and heavy involvement in "independent" IBM software user group SHARE Inc. (from 1955 onward) reinforced the sharing of software routines and dictated and incentivized much of programming around the IBM ecosystems through the code of its operating systems. In doing so, it imperialistically gained much control and power over users and potential users, both with its operating system code and in its applications, including those built by the company and the routines programmed by customers and others in FORTRAN, COBOL (the dominant language in business), and other languages and passed around at SHARE Inc. conferences in the United States and overseas. As consultant and computer scientist Melvin Conway famously wrote in 1967, in what was later dubbed Conway's Law, software system design tends to draw from the communication infrastructure, codes of practice, and culture of the organization that created it. Controlling 60% to 70% of the global computer industry from the late 1950s into the 1980s, IBM's dominance extended in part from its business culture codes and in part from its separate and linked software codes for operating systems and applications proliferating throughout the world. Locking in customers and preannouncing the company's new hardware and software were a way to further perpetuate this dominance. Most of IBM's storytellers (historians and journalists) also tend to highlight not just the iconic objects but also iconic individuals and iconic court cases. These, of course, include Thomas J. Watson Sr.'s and Watson Jr.'s many decades of leadership, the 1968 unbundling announcement, and the Department of Justice's (DOJ) half-century-long scrutiny of, and several antitrust lawsuits against, IBM, especially the famed 1969 filing.

In this chapter, I focus on an area critical to IBM in punch card tabulation and computing services—especially as a value-creating, unpriced activity (bundled to hardware contracts), but also as a priced deliverable within a profit center. I concentrate heavily on the contribution of system service– and system engineering–focused women professionals in the field at organizational customer sites (figure 11.1). Along with uncovering elements of gender codes at IBM, as well as women's major technical contributions at IBM, foregrounding the prehistory and history of system service and system engineering offers a useful fresh lens on one of the core iconic topics in IBM's historiography: the company's long-prevailing anti-competitive behavior and key enabling ties. My focus intentionally shifts away from traditional and limiting framings of the iconic (particularly hagiography of the Watsons Sr. Jr.) and instead favors a recognition and embracement of the ironic. I use irony as entry to shed light on women's technical fieldwork at IBM as well as on the well-trodden but often dis-

torted terrain of business strategy and antitrust. In general, looking for ironies can center attention on hitherto unexamined areas as it also brings different perspectives to well-covered ground. Such areas can include organizational codes of culture and practice, as well as coded systems.

IBM's vague December 1968 announcement that it was unbundling software and services (arguably, and ironically, itself a preannouncement) anticipated the DOJ lawsuit against the company in January 1969.[4] IBM's small team of top leaders and much larger team of lawyers chose the firm's terminology and contexts to set the stage for situational and strategic flexibility with regard to interpretation and classification of services— maintenance and marketing versus providing novel applications and advisory services. The historical literature to date on IBM's unbundling and antitrust ordeal has focused on hardware and, to a lesser but still meaningful degree, on software products. Meanwhile, computing services (which includes consulting, programming, systems integration, outsourced processing and remote batch, and time-sharing) have received only lip service. IBM services has rarely garnered much significant analysis, even though it was at least as anticompetitive and more consequential. This is clear in discussion of IBM hardware and software paper tiger preannouncements, ties and lock-ins of customers, and so-called "free" IBM software inhibit-

Figure 11.1. IBM sales meeting, 1938. The extreme gender disparity evident here persisted for a half century at IBM. It was not until the 1980s and 1990s that women formed a substantial percentage of the sales force. IBM Corporate Archives. Reprint Courtesy of IBM Corporation ©

ing the growth of an emerging 1960s software products industry.[5] Quick mention of two IBM services moments abounds in IBM historiography, but invariably without any fleshing out. These moments were the 1956 consent decree that forced IBM to spin off its Service Bureau Division (SBD) as a wholly owned heavily regulated subsidiary, the Service Bureau Corporation (SBC), and the company's 1973 sale of SBC to Control Data Corporation (CDC).[6]

This chapter draws on resources from the IBM Corporate Archives, Charles Babbage Institute collections, memoirs, oral histories, and secondary literature to focus on IBM services, gender, bundling and unbundling, and antitrust—all involving codes of culture, practice, business, law, and systems. I argue that through the important and neglected lens of services we can more fully appreciate a deeper set of core themes and social, political, and economic meanings. With such a perspective, the long-ignored, sizable, influential, and nonelite technical women's proto-systems engineering workforce rightly shifts to the fore, as does a wholly different setting, "the field," or the often-ignored customer sites where much IT work is performed and where most user-driven innovations occur.[7] At such sites, system service technicians or engineers worked to partner with customers to add value, with IBM's system service personnel working on site often for years. They designed and coded new applications and solved problems, creating what IBM later branded and advertised as "solutions." Solutions were the heart of system service work from the 1930s punch card tabulation days forward.

Both maintenance and application-oriented services or solutions were important and essential. Though enhanced in technology, form, and degree by the mid-1930s, assistance to customers on applications, in fact, dated all the way back to the 1890s and Herman Hollerith's precursor to IBM, the Tabulating Machine Company (incorporated in 1896), and even Hollerith's earlier efforts born with the contract to process the 1890 US Census data.[8] System services took a major leap forward for IBM in the mid-1930s, and again with the mid-1950s transformation toward computer programming and systems integration—codified, formalized, and extended as systems engineering. Only in the late 1990s did "solutions" become central to IBM in long-lived marketing—"Solutions for a Smaller Planet"—but the underlying ideology and analogous (often unpriced and bundled) activity out in the field was evident in nascent form from the 1910s and quite pronounced by the mid- to late 1930s.

What follows is a broad narrative tapestry of IBM's history. It pulls together diverse types of IBM services and their evolving significance and structure over time. In adopting this focus, I do not mean to deny the tremendous importance of hardware or the firm's also deeply influential software tools, programs, or packages (e.g., programming language FOR-

TRAN or its IMS database). Icons become icons for a reason, and power and influence play central roles. IBM icons as traditionally framed, however, give a rather limited perspective of the firm, and sometimes intentionally and sometimes unintentionally obscure or devalue the critical and creative work (often coding applications) of other important contributors, as well as neglected and meaningful sites and means and modes of production. Such exclusions from history distort the past and perpetuate inequalities. Simply put, IBM, ironically, was in major part *a services company from its very origins*. Its value-adding activities are what is important, not its maneuverings in defining profit centers or its financial engineering or codes of accounting. Services also represent one critical, ignored arena to recover and better understand power and inequality of gender, race, ethnicity, and class (of these, this chapter concentrates on gender).[9] Existing historical surveys and thematic work on IBM hardly recognize services at all—or only addresses them after services became the core profit center focus of the corporation in the mid-1990s.[10]

In offering a services-centric corrective, I privilege (a) work activity and value creation over profit centers and (b) underlying strategies over IBM's announcements and pronouncements. Top executives' egos and the company's marketing campaigns and practices of engaging in or avoiding the legal consequences of real and perceived anticompetitive behaviors all often clouded IBM's disseminated information or misinformation. IBM leaders, managers, employees, and lawyers continually walked a tightrope in taking anticompetitiveness to its upper limits (and sometimes well beyond) while trying to avoid painful falls—lost suits and crippling consent decrees. Historian Steven Usselman has characterized the dance between IBM and the DOJ attorneys—aside from the three increasingly contentious major lawsuits of 1932, 1952, and 1969—as more of an interaction, a debate, or a cordial dialog than a hot war.[11] On the other hand, the 1969 case was a fierce battle for over a decade, and one characterized by many rounds of DOJ confrontation and IBM obfuscation and obstruction in a space where legal codes as they applied to new and evolving technologies often were less well defined than in some other industries.[12]

With a different vantage, I seek to weave a fabric of the past that extricates company-spawned and -influenced fabrications. I have benefited from the materials used at the IBM Corporate Archives, access to which I am grateful for as its collections have been open to academics only selectively and its policies have often changed with different IBM lawyers and archivists. Much of the time IBM has restricted access and put out books and oral histories that tend, in a seemingly gentle and hegemonic fashion, to tell iconic and celebratory stories of innovation in sales management, hardware, software systems, applications, IBM research, and other areas. Specifically, in the 1970s and '80s the company invested in a major technical history project of oral histories and team-authored technical books on key systems (early computers and the System/360 and /370 families) and a single-authored survey by one of the engineers, Emerson Pugh.[13]

Icons are a part of IBM history and, while important, can be written about in critical, noniconic ways. For instance, the IBM System/360 family is central to the company and its history, but it was not just an R&D triumph or a delayed triumph of Operating System 360 (with lessons learned). We know so little about the shop floor and labor process of manufacturing mainframes. IBM was the first computer maker to produce a machine (over model production lifespans of about a decade) in the thousands of units for the mid-1950s' and more than 10,000 units for the late 1950s's IBM 1401. The System/360s also numbered in the thousands in the mid-1960s to early 1970s. Who built these, with what tools, in what conditions? What labor organization existed? With what labor resistance? What was the labor history of race, ethnicity, and gender? What were the workers' worlds experienced in Poughkeepsie, Kingston, Endicott, and Rochester, New York? I have briefly reflected on such historiographical questions elsewhere.[14] I have also looked a bit at midrange system manufacturing at IBM Rochester.[15] Nonetheless, my work and others' have hardly scratched the surface, and overall, except for semiconductor fabrication history, historians and other writers have published relatively little on the labor of computer hardware production. This paucity contrasts sharply with the historiography of technology firms in other industries such as textiles, paper, chemicals, automobiles (GM and Ford), steel, sewing machines, and agribusiness. On the environmental side of production, Peter Little's ethnographic *Toxic Town*, on IBM's destruction of Endicott, New York, richly analyzes the environmental, ecological, human health, and political and cultural impact in the wake of Big Blue's plowing forward into the information age.[16]

Alternatively, in what follows, I explore early systems services work and its evolution in helping spawn systems engineering. Much of this was work in the field—not IBM's iconic company towns of Poughkeepsie, Endicott, and Kingston. Of IBM's mainframe competitors, only Control Data had a meaningful services business. The services industry was diffuse, and to the extent that a few powerful players existed, such as Ross Perot's EDS (Electronic Data Systems) and payroll-processing giant ADP (Automatic Data Processing), they thrived because IBM in fact inadvertently created opportunities for their segments, business models, and strategies. Former IBM star "salesman" Perot took much from IBM and became a continual thorn in its side. Smaller services players were hurt, but largely powerless. In software products, Applied Data Research, Inc. (ADR), Informatics Inc., Cullinane/Cullinet, and others showed that software could be very profitable and potentially patentable.[17] Ironically, when a few early successful firms banded together and hired top legal and lobbying representatives, this younger and smaller industry of software products (identical code) gained significantly more power than those offering highly segmented and varied services (reused routines at times but, overall, custom code).

Uncovering Ironies

A growing and important literature on gender and computing has focused heavily on programming. Janet Abbate's path breaking *Recoding Gender*, for example, draws from her interviews with women computer science and industry elites.[18] While historical actors such as Univac's Grace Murray Hopper and IBM's Fran Allen suffered discrimination, it was likely of different form and degree than that experienced by entry-level programmers in the 1960s and '70s. Mar Hicks, in a far different and equally impressive monograph, makes a fundamental contribution to the labor history of programming and data processing by focusing on discrimination against women that set back a nation.[19] And in this volume, Elizabeth Semler insightfully examines human resources training manuals and other company ephemera to explore gendered environments and gender discrimination at Control Data Corporation. From the work of engineer Pugh to that of professional historians Sobel and Cortada, women are largely absent in historical surveys of IBM.[20] Cortada discusses Virginia Rometty, but only after her ascendency to CEO.[21]

IBM's historical literature, reflecting the overwhelming absence of women actors, suggests that they made no meaningful technical contribution to the firm. The grand irony is that from the 1930s forward women in fact were central, in substantial numbers, to technical areas that added tremendous value and foreshadowed the future direction of the whole corporation—systems engineering. A second irony is that IBM's leaders perceived a real prohibition on direct participation in the services business extending from the 1956 consent decree. Hence, IBM strategically responded with accelerated bundling. This practice was far more anticompetitive than what IBM had done in the services business before the consent decree—the medicine or remedy was worse for competitors and consumers than the disease. A third irony is the major gap between perception and reality as to which services, and to what degree, IBM actually unbundled in the early 1970s. Again, IBM favored flexibility in interpreting the court order so that it could adjust to changing market conditions, changing opportunities, and particular relationships with major institutional customers. A final irony concerns the seemingly draconian restrictions that the 1956 consent decree imposed on IBM's Service Bureau Division. I contend that they had little consequence for IBM. The Service Bureau was launched in a time of weakness (during the Great Depression before IBM secured the giant Social Security contract). This new division allowed IBM to go after small-scale data-processing services clients (figure 11.2).

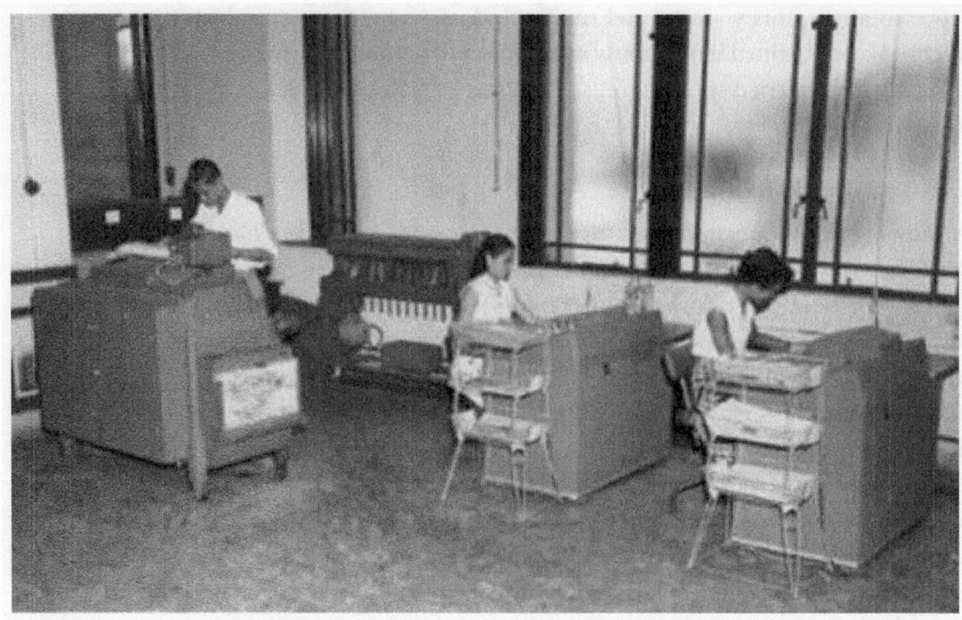

Figure 11.2. Workers in IBM's Service Bureau Corporation in Singapore. IBM Corporate Archives. Reprint Courtesy of IBM Corporation ©

Except as a diversion or a negotiating chip, in many ways, given its low profit margin (usually well under 5%), the Service Bureau had outlived its usefulness to IBM by the 1960s. The firm *chose* to make it "Siberia." DOJ's role arguably was relatively minor. In 1973, SBC was worth more for IBM to discard for far below market price to CDC as a settlement, as it did not want the overhang of the CDC suit against it at time when the DOJ case was heating up.

Exclusion versus Inclusive Participation in the Field: Sales and the System Service Women's Corps

Both maintenance and systems application are areas of IBM labor historiography grossly ignored. Maintenance work in data-processing technology dated back to the late 1880s and 1890s during the Hollerith era. At the Computer-Tabulating-Recording Company (renamed IBM in 1924) in the teens, these workers were called customer engineers (CEs), and later field engineers (FEs). As tabulation machines' possibilities expanded, so did the need for system services. Many CEs had no college degree, and most were more involved in and adept at machine maintenance than in helping with applications. CEs were not exclusively male, but nearly so in the 1910s and '20s.[22]

It was not until 1935, a quarter century after its founding, that IBM hired its first woman salesperson, but this was not the start of an accelerating trend. No woman IBM sale representative would be admitted to the 100 Percent Club for another decade. Formed to recognize sales personnel who made their quota, the club was celebrated but hardly elite: it had many hundreds if not several thousand male salespeople in its ranks each year in the interwar period. In 1949, there were only three women in the 100 Percent Club—a product of low hiring rates for women and, likely, added hurdles in the form of poor territories and high quotas—in short, a coded discrimination. IBM's sales force and management were overwhelmingly male well into the 1970s.[23] To understand 1950s computer field engineering and the official advent of systems engineering in 1960, the tale of two disparate IBM services activities or enterprises—field service and service bureau—is instructive. While leases and punch cards were the focal point of the 1932 DOJ suit against IBM, which the firm essentially lost, in the 1952 case the Service Bureau was a central focus.

The System Service Women's Corps (SSWC) began in roughly the same Depression conditions as the Service Bureau (figure 11.3). While the two subgroups shared goals of helping customers with data processing, there were stark differences. First, the work settings varied. Like CEs, members of the SSWC worked at customer facilities, setting up and, especially, helping with coding applications. Typically, both in the United States and abroad, IBM Service Bureau offices were near the center of large urban areas. Hence, the Service Bureau Division was highly visible to pedestrians and vehicular traffic. Service Bureau offices were often on the ground floor with big windows and open floor plans, such as in the Manhattan bureau, truly in the spotlight in the spotlight city of the Big Apple. In contrast, SSWC, like other field services, was largely invisible to all but the customer's employees who interacted in the client companies' tabulation machine rooms.

During the tough years of the Great Depression, the Service Bureau strategy looked to get new customers into the IBM ecosystem o retain

Figure 11.3. The System Service Women's Corps first graduating class, 1935. IBM Corporate Archives. Reprint Courtesy of IBM Corporation ©

struggling customers that could no longer maintain their own tabulation rooms of leased IBM equipment but that still, of course, had data-processing needs. The plans for the Service Bureau, in fact, had been drawn up at President Thomas Watson Sr.'s behest back in 1922. The startup cost, along with the low likelihood that outsourced data processing could greatly increase revenue or earnings, caused plans to lie idle for ten years. After the stock market crash in October 1929, and the Great Depression that followed, Watson Sr. saw its main competitor, Remington Rand, undergo retrenchment. He perceived an opportunity to invest to gain market share, to expand IBM's workforce, to go after all business, however small; and for this reason, IBM returned to the long-shelved plans for the Service Bureau. In short, the strategy was to retain existing customers who could no longer lease equipment and bring new ones into the IBM fold, engage in account control, achieve some lock-in, and later graduate the customers up to leasing machines. The strategy worked throughout the interwar years, and IBM's Service Bureau quickly adopted its latest computers for data processing in the 1950s and '60s.[24]

In the mid-1930s, IBM also invested in other areas to aid its customers in challenging times and to add value. The company in 1935 launched the tremendously important and largely invisible and overlooked System Service Women's Corps. Back in 1929 IBM had hired Virginia Linkenhoker, who organized IBM's first training school for customers. IBM had formed schools to train mostly male hires as salespersons and CEs as far back as 1917, but providing this formal service to customers was new. It recognized, however, that most instruction occurred onsite at customer facilities, and the mostly male CEs, typically with just a high school education and perhaps a vocational degree, were useful for maintenance, but often in over their heads when it came to creatively adapting systems and applications to suit the customer's needs onsite.[25]

In 1935, IBM decided to hire, train, and grow a new "professional" class of field workers. Seeing the talent of Linkenhoker, and likely the opportunity to pay lower salaries to women, it launched the SSWC. It hired 25 women, all with four-year college degrees. Linkenhoker became the first "System Service" employee, and later that year an additional 25 other professional women joined her.[26] System Service women worked closely with customer data-processing staff and others to aid with maintenance and overlapped with CEs on some tasks. More commonly, however, they advanced applications, adding insight, creativity, new possibilities, and greater efficiency to customers' operations. Pure bundled maintenance work of CEs was one thing, something IBM had done since the teens, but college-educated professional employees coding applications and integrating systems was something entirely new and was really the start of the basic underlying idea of what would become systems engineering in the computer era. At the time, of course, it was carried out with punch card tabulation.[27] In both the dominant tab machine era of the pre-1950s and the computing era of the mid-1950s and afterward, technical education was gender segregated at IBM. Anne Van Vechten of the first SSWC class soon became IBM's secretary of education for women.[28]

The System Service Women's Corps continued to graduate class after class of a few dozen women professional field services engineers the firm bundled to help with sales and leases of hardware. Remington Rand did not match this deal; it was a key differentiator for IBM. In addition to IBM's customer education at its facilities, it provided invaluable application help that came with the hardware. The customer's importance and size determined whether an SSWC member (or multiple members) would be stationed long-term at its facility. How IBM contracted is somewhat opaque. I have found no evidence that IBM ever wrote System Service help into a customer contract. With IBM bundling in more services than its competitors, there was no incentive for IBM to contractual bind itself. The lack of a contract offered IBM great flexibility, and SSWC workers helped IBM solidify major long-term customers and allowed installations to develop and to buy or lease new, higher-revenue and higher-margin models of tabulation machines. SSWC workers operated as if they were employees of the customer in many respects, but their salary and benefits were fully paid by IBM. In the mid-1950s, the Service Bureau quickly added IBM 650 computers to its facilities. The SSWC rapidly entered the computer age as well. Given the education, training, and talent of those in the SSWC, they were skilled enough to quickly retrain as computer programmers. In but a few years, the System Service Women's Corps became largely a group of computer specialists and programmers, and by the late 1950s had grown to more than 500 women.

System Service / System Engineering and the Service Bureau after the 1956 Consent Decree

IBM's Service Bureau Division was the largest data-processing services operation in the dominant punch card tabulation era of the mid-1930s to the 1950s. The adoption and changeover to the IBM 650 (and later 1401) across US Service Bureau facilities was rapid. At that time SBD was one path along which data-processing customers could be pushed into the computer age, where accounts over time could often be transitioned to customers leasing machines, which offered IBM higher margins and were far more lucrative. As this change was occurring, the 1956 consent decree brought a number of restrictions. Even in the first half the 1950s, at the start of IBM's transition to computing, DOJ lawyers who negotiated the decree unquestionably considered the potential of the computer business and computing services.[29]

Prior to 1956, IBM regularly shifted its workforce between other divisions and the Service Bureau. The fluidity kept the stature of SBD elevated to a degree. Only after the consent decree did it get the unofficial moniker "Siberia." From then on, it was as if those in SBC really did not work for IBM any longer; it was not a place for those on an upward career path. As highlighted earlier, however, SBC generally had locations in prominent city centers. Ironically, with the consent decree, Big Blue could not advertise SBC as "IBM." It was always a place to go for smaller business during tough times, but given that it was front and center and known to be IBM even though not advertised as such, its primary value in the 1960s was to divert government regulators' attention away from the more lucrative side of value creation, namely, bundling the field system services work of the systems engineers.[30]

IBM engineer William Lawless was serving as the executive assistant to Vice President Al Williams when, at the end of the 1950s, he saw a problem and an opportunity. Engineering, on the marketing side of the company, in the field, continually had problems attracting and retaining top talent. The System Service Women's Corps was a standout exception. The women who went through this school were by this time increasingly focused on programming. Women were not doing all the systems or applications support fieldwork: an analogous group of men went to all-male training schools to be customer technical representatives (CT reps). Like SSWC, this group has been left out of the historical literature. The System Service Women's Corps, numbering 500 by the late 1950s and having a 20-plus-year history, was possibly larger than the CT rep workforce. Yet it seems unlikely that SSWC and CT reps combined would have totaled even 1,000 in the late 1950s, given that only major customers received this support.[31]

Although all System Service women were college graduates, a college degree was not required of CT reps, and other gendered policy discrimination was common at the firm. One only has to look at bias and discrimination in sales and sales culture, discussed earlier. CT reps were probably paid higher salaries yet as a group were less credentialed. Jean Sammet, a distinguished computer scientist and manager at IBM, who earlier had been one of six principal developers of COBOL and later became the first woman president of the Association for Computing Machinery, was never named an IBM Fellow. Well over 90% of IBM Fellows have been men.

Lawless's vision won over Al Williams and then Thomas J. Watson Jr., and he got the green light. Lawless swiftly launched "systems engineer" as a job class in 1960. Not only the name but much of the inspiration came to Lawson from the tremendous work and reputation of the System Service Women's Corps. He, and the fellow executives whom he convinced, recognized that as computers increased in speed and complexity, the programming task in applications at customer sites—from systems integration for NASA Mission Control to real-time transaction processing for financial markets, airline seat reservation systems, and retail inventory management—was becoming all the more complex and profound. The new position of systems engineer (SE) required extensive training, and its career path had five levels (trainee, associate SE, SE, advisory SE, and senior SE), so it offered significant opportunity for advancement. Broader than just training, requirements, and job classes, this true professionalization infrastructure included a competitive annual symposium and a major new technical journal, the *IBM Systems Journal.*

What happened to the System Service Women's Corps with the advent of systems engineering? We know anecdotally that some SSWC IBM employees, such as Maggie Wilcox, engaged in advanced training to become an associate SE. Within a year she went on to become an SE. Ellen Kerksieck Schaefer progressed through the ranks of SE to become one of the few elite "consulting SEs" in the 1960s (a sixth, highest category that was added to the original five levels). Anecdotal findings of some advancement definitely are not evidence of equal opportunity, as everything about IBM culture leads to other conclusions. Male culture dominated in sales as likely it did in factories and other parts of the firm. Given the numbers of SSWC workers and the respect they garnered in being a model for SEs, System Service probably was a place of significant but perhaps less gender discrimination than some of the other parts of the firm. The only thing that can be said with near certainty based on the numbers is it likely was less discriminatory than sales in the 1960s. It is a cruel irony that women pioneers built the general model for what became IBM Systems Engineering only for it to become principally a male domain.

The total number of SEs grew substantially in the 1960s, reaching more than 1,000 by the end of the first year and 5,000 by 1965. Field employees were all under the "marketing" aegis at IBM, and by 1985 its SE job category was the largest in marketing at well over 20,000—substantially more than all other categories in marketing combined (sales and CEs). Unfortunately, the gender breakdown over time seems not to have been recorded or, more likely, was not preserved (at least my long forages in the IBM Corporate Archives proved unsuccessful in that regard). One thing is certain: SEs were critical, high-level employees who enabled IBM to thrive in the 1960s and 1970s, yet have been woefully neglected in computing historiography. My book *Making IT Work* (2017) was the first to give this group any attention. SEs did the coding, maintenance, and systems integration that made expensive computers useful for institutional customers. They increased efficiency, solved problems, and brought new possibilities to the computer age. Many stayed in a facility for months or years, but there was enough movement between locations and sufficiently regular training schools that SEs were a fundamental source of circulating industry best practices both overall and within industry verticals. They added to the customers' capabilities. As General Dynamics' manager of data processing, Joy Greenwood, put it in the early 1980s, "The best SEs think of themselves as an extension of the customer [and] give you that extra effort to help you solve problems."[32]

A Service-Centric Perspective on Unbundling and Its Aftermath

The 1956 consent decree appeared to force IBM out of data-processing services, which at the time consisted largely of punch card tabulation. In about seven or eight years it would be overwhelmingly computer-based data processing, and the precedents of the consent decree dictated regulation of IBM data-processing services in computing. IBM slowly starved SBC. As the government and IBM were both inflicting blows on SBC, the number of offices shrunk in the second half of the 1960s, but by 1970 it still had more than 60 locations (down from over 90). The more important services operation, however, was IBM's huge differentiator—SEs in the field. Control Data and other mainframe competitors followed suit in the early 1960s, not only adding the personnel but taking the same name for them, system engineers. IBM bundled to avoid the ire of competitors and especially the DOJ, but the great irony was that bundling was *so much more anticompetitive* than just allowing IBM to be in the services business, which the 1956 consent decree seemingly forbade.

Computing's storytellers have written of unbundling primarily as a tale of hardware and software products. IBM's mainframe competitors and the small emergent software products companies were being hurt because IBM was developing software products and giving them away for free, bundling them into hardware contracts. An early 2001 CBI conference on unbundling and software products held at Xerox PARC, with talks and follow-on publications on unbundling history by Campbell-Kelly and Usselman, lent momentum to this interpretation.[33]

This 2001 event, and especially Usselman's insightful presentation there and later publication, teased out some of the most important questions regarding IBM and antitrust. First, did unbundling really give rise to software products or just lend it a bit of momentum? He argues for the latter. While I would agree in general with this assertion, the framing limits understanding to a degree. If services are included (a critical piece of unbundling), the momentum theory is far less certain. Second, Usselman asked, Did IBM accurately and appropriately drop the price of hardware to compensate customers now that they were being charged for software and services that previously were free? Hardware prices dropped by only 3%, much to the fury of its customers. Usselman argues that his analysis of what IBM saved, and thus passed on to customers, bears out this 3% figure as correct.[34] I think that this figure must only include software products, which makes sense as they had standard price lists. It seems highly implausible that the estimate includes all SE services as well—programming services, consulting, systems integration, and the rest. There were probably 8,000 to 10,000 SEs by 1970, given their number jumped from 1,000 to 5,000 between 1961 and 1965, and that there was faster IT growth in the late 1960s. They drew relatively high salaries. IBM feared antitrust action throughout the 1960s, and managerial accounting documentation undoubtedly did not always survive. As it is today, financial accounting then was carefully engineered to please shareholders as the core stake-holders, and few did it more effectively than IBM.[35]

There is a counterfactual, which as Usselman shows, has captivated economists looking back. IBM lawyers argued at the time that "industry" was moving in the direction of unbundling and that IBM would have had to unbundle to survive and thrive long-term.[36] This argument, which Usselman believes to carry weight, might distort by privileging software products over services. Software products, by definition, consist of stan-dard code. They are priced. They are advertised in catalogs. There is trans-parency. Services, in contrast, are not standard. They are more complex, less uniform. Services (except for networking) are tied to place and re-gions, and customer sites matter. Products are distributed nationally and sometimes just mail ordered and delivered. It is not clear to me that the market, competition, and survival would have had sufficient economic force to persuade mainframe firms to unbundle services (or most services), at least not in the 1970s. The various services segments—consulting, pro-gramming, systems integration, remote batch, facilities management, time-sharing, among others—have entirely different dynamics.[37]

SEs were part of the marketing division, yet IBM operated explicitly on the assumption that it could code or choose what it wanted to include as marketing, and it continued to bundle substantially. On the services side, IBM unbundled to differing degrees: at times it was strategic, and at other times it was difficult getting certain managers to follow through and charge the clients.[38] It appeared that top corporate executives pushed SEs strongly to quickly adopt an SE services billing model, because IBM law-yers probably pressured them to do so, but decisions by managers on the ground could vary. Further, up until IBM dispensed with it in 1973, SBC helped divert attention from the more important activity and value creator of SEs. In some conditions, with some customers, and with some person-nel, it may have made sense to continue to bundle for a time. Burt Grad and others who were present (like Watts Humphrey) at IBM have articu-lated that it was a mixture of unbundled and bundled software products and computer services well into the early 1970s.[39]

The impact of unbundling on SEs poses an intriguing question. In the 1960s a quality SE needed to be creative, technically skilled, and a prob-lem solver, but did not need a sales personality or to thrive in a sales atmo-sphere and its attendant gendering. There is no indication that selling was a focus of the System Service and later SE education effort, and it is not a skill picked up quickly for some and depends on the client and their setting and biases, not just the consultant.[40] Women had thrived in System Service work, and some, perhaps many, of them made the transition to be highly successful SEs in the 1960s, despite undoubtedly higher and unfair hurdles. A number of SEs quit after unbundling; they liked the technical work, the creativity, and the different sites over time but did not like the sales aspect. Differential and more severe impact on women SEs certainly may have occurred because of the male-dominated culture at both IBM and at many of its institutional customers in corporate America.

Conclusion

Computing services, as unpriced bundled work that adds value, or later as priced work within a profit center billed to the client, have been critical to IBM since its launch, or since at least a few years later when Thomas J. Watson Sr. came to lead Computer-Tabulating-Recording Company / IBM as general manager in 1914 (and as president in 1915). IBM's voluminous historical literature analyzes and celebrates icons, iconic figures, and hardware and software products. This practice and selection have led to core misunderstandings of this firm. The historiography has ignored and written out the major contributions of women technical workers in IBM System Services and as systems engineers. There are attendant ironies of what was truly anticompetitive and when, who was adding value, what was the core of services work—field site SEs or IBM's SBC.

Antitrust law exists to protect consumers and promote industry-wide innovation by thwarting anticompetitive practices. It tends to be too late and too slow moving to have an effect on information technology. This was true with IBM when Judge David N. Edelstein ordered Big Blue to be fined $150,000 a day in August 1973 for noncompliance in turning over hundreds of documents to the court.[41] Delay was clearly a strategy in the case, and ultimately one that probably worked. The industry was changing, and IBM was looking less threatening to those that mattered as Digital Equipment Corporation (DEC) took off in minicomputing, and a whole new industry was forming around personal computing in the second half of the 1970s. The delay tactic also allowed the decision to await a president far less aggressive with antitrust than President Jimmy Carter. Toward the end of President Ronald Reagan's first full year in office, under his administration's broad neoliberalist agenda of deregulation, the case against IBM, filed a dozen years earlier, was dismissed in January 1982.[42]

IBM's half-century-long dance with the DOJ ended in early 1982 with this last tango in lower Manhattan (home of the US District Court for the Southern District of New York). The DOJ became interested in a new dance partner in the Pacific Northwest. The Microsoft case of 1998, the breakup decision, and the reversal showed yet again that whether it be lobbying power, fear, backlash, plaintiff mistakes, or something else, for the DOJ, breaking up (big tech) is hard to do. It has continued to squander time until today, when the seven US trillion- and multitrillion-dollar IT behemoths—Amazon, Apple, Google/Alphabet, Microsoft, Tesla, Nvidia, and Facebook/Meta—have grown incredibly large (over 30% of the capitalization of the supposedly broad-based and diversified S&P 500) and built deep war chests and large, powerful legal teams. With the 2024 election results, significant regulation of IT corporations seems even less likely. These seven corporations overlap in areas but do not compete in others, each with a near-monopolistic core, so they are perfect partners and are closely aligned against a common enemy—the government, the DOJ, and the Federal Trade Commission (which under FTC chair Lina Khan in the early 2020s was more effective than DOJ, but still limited). The fact that many users think of Google (search and Gmail), Facebook, and now Microsoft/OpenAI ChatGPT4 as "free" does not help. Consumers pay with their data. These seven companies have become the new information tech icons in place of the once mighty Big Blue.

When we instead look at IBM's activities, we may see that it was largely a services company from the start. Ironically, this is now its largest division and has been for a few decades, but it was impactful before—it was just bundled into hardware. Another irony is that while Service Bureau facilities were physically marketing in major urban regions, it was the behind-the-scenes services at customer sites that was more significant and gave rise, through the System Service Women's Corps and its evolution into the system engineer corps, to a highly professionalized and standardized infrastructure. Once it was elevated professionally, it was gendered male. The 1956 consent decree, ironically, "forced" IBM to do exactly what it wanted to do: de-emphasize the always low-margin Service Bureau, which had been launched to pick up business during the Depression, and focus on bundling system services on customer sites, which differentiated it and allowed for selling and renting high-margin hardware. The 1969 DOJ case led IBM to preemptive unbundling (in announcement at least—implementation was more gradual), weakening its hardware business at a time when DEC's minis and, soon, personal computers would erode the mainframe giant's hardware dominance. All the while, IBM unbundled or bundled in the 1970s as the need dictated and then largely unbundled services on a grand scale in the early 1990s in order to set itself up for the higher-margin cloud services, all the while charging aggressively for its systems engineers' services. Its legal delays allowed it to avoid punishment for a host of anticompetitive behavior for decades. Needless to say, as the DOJ was fruitlessly chasing after IBM, Big Blue, ironically, was clinging to past iconic glory and businesses with the effect of punishing itself, losing much of the talented women who had built its System Service and failing to focus on areas it long had invested in such as software, information retrieval, artificial intelligence, semiconductors, and communication devices. Instead it concentrated on entrenched business areas of the present—servers and traditional products and services. In the new century, disruptors such as Apple, Google, Microsoft, Amazon, and Nvidia have thrived in some of these areas and have capitalizations eight to eighteen times the size of IBM today.

Notes

1. C. Schlombs, "The 'IBM Family': American Welfare Capitalism, Labor, and Gender in Postwar Germany," *IEEE Annals of the History of Computing* 39, no. 4, (2017): 12–26; C. Schlombs, *Productivity Machines: German Appropriations of American Technology from Mass Production to Computer Automation* (Cambridge, MA: MIT Press, 2019); E. Medina, "Big Blue in the Bottomless Pit: The Early Years of IBM Chile," *IEEE Annals of the History of Computing* 30, no. 4 (2008): 26–41.

2. Schlombs, "The 'IBM Family'"; Medina, "Big Blue in the Bottomless Pit"; S. Usselman, "Unbundling IBM: Antitrust and the Incentives to Innovation in American Computing," in *The Challenge of Remaining Innovative: Insights from Twentieth-Century American Business*, ed. S. Clark, N. Lamoreaux, and S. Usselman (Stanford, CA: Stanford University Press, 2009), 248–279. Usselman mentions services but seems not to make them integral to his analysis. That said, his scholarship is a major contribution to IBM and political economy.

3. J. Cortada, *IBM: The Rise, Fall, and Reinvention of a Global Icon* (Cambridge, MA: MIT Press, 2019).

4. In December 1968, IBM wanted to do little but precede the DOJ. The 1969 reannouncement was a bit fuller in outlining its future plans.

5. M. Campbell-Kelly, *From Airline Reservations to Sonic the Hedgehog: A History of the Software Industry* (Cambridge, MA: MIT Press, 2003). This important book on software products deals with services at the start, and as prehistory to software products, it offers a decidedly software products focus on the firm.

6. Cortada, *IBM*; R. Sobel, *IBM: Colossus in Transition* (New York: Truman Talley, 1981); E. Pugh, *Building IBM: Shaping an Industry and Its Technology* (Cambridge, MA: MIT Press, 1995).

7. The category "users" includes both end consumers or client organizations and intermediary users from producer firms and an external services industry that helps clients best orient, apply, and program machines and integrate systems.

8. M. Campbell-Kelly, W. Aspray, N. Ensmenger, and J. Yost, *Computer: A History of the Information Machine*, 3rd ed. (Boulder, CO: Westview Press, 2014).

9. This is a point I first made in 2011. J. Yost, ed., *The IBM Century* (Piscataway, NJ: IEEE Computer Society Press, 2011). I flesh it out more in diverse ways and degrees in the present chapter and in J. Yost, *Making IT Work: A History of the Computer Services Industry* (Cambridge, MA: MIT Press, 2017).

10. Cortada, *IBM*.

11. S. W. Usselman, "Fostering a Capacity for Compromise: Business, Government, and the Stages of Innovation in American Computing," *IEEE Annals of the History of Computing* 18, no. 3 (1996): 30–39.

12. F. M. Fisher, J. J. McGowan, and J. E. Greenwood, *Folded, Spindled, and Mutilated: Economic Analysis and U.S. vs. IBM* (Cambridge, MA: MIT Press, 1983).

13. Pugh, *Building IBM*: Sobel, *IBM*. Sobel was the more accessible overview or survey. For two technical histories rich in detail, see C. J. Bashe, L. Johnson, J. Palmer, and E. Pugh, *IBM's Early Computers* (Cambridge, MA: MIT Press, 1986); and E. Pugh, L. Johnson, and J. Palmer, *IBM's 360 and Early 370 Systems* (Cambridge, MA: MIT Press, 1981).

14. J. Yost, "Manufacturing Mainframes: Component Fabrication and Component Procurement at IBM and Sperry Univac, 1960–1975," *History and*

Technology 25, no. 3 (2009): 219–35. Although I discuss the lack of literature on labor and manufacturing computers, this article focuses on fabrication and procurement management of semiconductors.

15. A. Norberg and J. Yost, *IBM Rochester: A Half Century of Innovation* (Rochester, NY: IBM, 2005). This book addresses work issues and the downsizing of this once important IBM manufacturing facility, the home of its midrange systems.

16. P. C. Little, *Toxic Town: IBM, Pollution, and Industrial Risk* (New York: New York University Press, 2014).

17. G. Con Díaz, *Software Rights: How Patent Law Transformed Software Development in America* (New Haven, CT: Yale University Press, 2019).

18. J. Abbate, *Recoding Gender: Women's Changing Participation in Computing* (Cambridge, MA: MIT Press, 2012).

19. M. Hicks, *Programmed Inequality: How Britain Discarded Women Technologists and Lost Its Edge in Computing* (Cambridge, MA: MIT Press, 2017). The United States, too, imposed major gender discrimination, but the far larger market size (and Department of Defense research funds and purchases early on) likely allowed for more self-inflicted gender-discriminatory wounds while not putting economic growth at risk relative to Britain.

20. Cortada, *IBM*.

21. Pugh, *Building IBM*; Cortada, *IBM*.

22. Yost, *Making IT Work*.

23. "Decade 1930s," IBM Corporate Archives Online; Yost, *Making IT Work*.

24. Yost, *Making IT Work*.

25. "IBM Graduates Class of Young Women for System Service Work," *Business Machines*, November 12, 1936 (IBM Corporate Archives, now in Poughkeepsie, NY); "Twenty-Five Young Women Pioneer Field," *IBM Data Processing News*, July 22, 1960, IBM Corporate Archives.

26. "IBM Graduates Class of Young Women."

27. "Loraine McLennan Recalls Birth of System Service," *IBM Data Processing News*, July 22, 1960, IBM Corporate Archives.

28. WIT Decade 1930.

29. Yost, *Making IT Work*.

30. Yost, *Making IT Work*.

31. "How the New Division Will Work," *IBM News: Field Engineering Division*, March 17, 1963; "Announce System Engineering," *IBM Data Processing News*, December 5, 1960, both in IBM Corporate Archives.

32. Yost, *Making IT Work*, 187.

33. "Unbundling History: The Emergence of the Software Product," CBI conference at Xerox PARC (Palo Alto Research Center), Winter 2001.

34. Usselman, "Unbundling IBM."

35. Yost, *Making IT Work*.

36. Usselman, "Unbundling IBM."

37. Yost, *Making IT Work*.

38. B. Grad, "A Personal Recollection: IBM's Unbundling of Software and Services," *IEEE Annals of the History of Computing* 24, no. 1 (2002): 64–71.

39. Grad, "A Personal Recollection."

40. S. Benzer, "SE Education Paves the Road for Growth," *Business Machines*, July 7, 1961, IBM Corporate Archives.

41. A. H. Lubasch. "IBM Faces Fine of $150,000 a Day in Contempt Case," *New York Times*, August 2, 1973, A1, 52.

42. M. Brown and C. E. Mayer, "U.S. Ends Antitrust Suits against AT&T, IBM," *New York Times*, January 9, 1982, A1.

12

Y2K and the Politics of Labor

Dylan Mulvin

COBOL programmer Bruce Fassett's retirement lasted all of one week.

He had toiled for 15 years in a nondescript cubicle in Phoenix, one of hundreds of programmers detailed to one of the least glamorous jobs in high technology: maintaining an ancient COBOL database that tracks commissions of Motorola's chip salesmen.[1]

These are the opening paragraphs of a 1998 article in the *Los Angeles Times* by Ashley Dunn. COBOL-literate programmers like Bruce Fassett, Dunn posits, were underappreciated in their time: Fassett "toiled" in a "nondescript cubicle" working in "one of the least glamorous jobs" in tech. It's the familiar cultural flotsam of the 1990s wherein films (think *Office Space*), television (think Chandler's desk job on *Friends*), and comics (think *Dilbert*) often drew on the malaise of soul-crushing office jobs that were loosely connected to computing, data, the gray sludge of cubicle life, and the equally gray and sludgy desktop PC.

But Fassett's quasi-retirement lasted only one week because he was quickly lured out of repose to work on urgent repairs of other COBOL systems for EDS (Electronic Data Systems). This is also a familiar narrative: the reluctant hero is called to action to perform *one . . . last . . . job* (think *Heat*). In this case the job was fixing problematic code in software, databases, and embedded systems that were not compliant with a year 2000 (Y2K) rollover. Fassett and his ilk were called, to be clear, to shore up the nation's infrastructure in the face of the Y2K crisis. As one recruiter put it in 1998, "These guys are like diamonds now."[2]

A different story, in the same year, from the *New York Times*: "The Y2K Problem is causing a lack of programmers, and people are hiring anybody," said Michael Zboray, a vice president at the Gartner Group. "Companies are now doing Y2K development offshore," he said, "sending work to Russia and India, and they haven't a clue as to what's coming back. They don't do background checks. What we're hearing is an undercurrent of back doors being programmed in."[3] Unlike the heroics triumphed by the *Los Angeles Times*, this is a story of desperation, of "hiring anybody." And, as Zboray suggested, this desperation was being scapegoated for a potential security vulnerability. It's one of the many ways that economic nationalism expressed itself in the articulation of repair work and the globalization of IT labor.[4] The passage and others like it hint at a conspiratorial way of thinking that connects the non-American places where goods (including software) are made with danger, insecurity, and infectiousness. For those reading this passage today, it may resonate with contemporary forms of discrimination and xenophobia, which are regularly connected to new technologies.

Even as demand for engineers and coders in the United States drove an employment boom in the tech sector in the 1990s, one that largely left behind those in areas of traditional manufacturing, the tech sector was not immune to resentment and fear. It's during this period that an increasing number of tech subcontracts were being sent to Indian IT companies, firms that would either complete (or sub-subcontract) the work in India or bring workers to the United States. These two arrangements were usually referred to as "outsourcing" and "body shopping," respectively—and it is in these arrangements that we can read the emergence of new genres of code work.

Just as the travel of sailors, soldiers, and flight attendants has often marked their bodies as potential vectors of disease, the fears underlying this *Times* article are clear: coding labor has not only gone "offshore" but returned carrying the traces of its travel.[5] It paints a picture of what Alan Bewell calls a "colonial disease landscape," in which some bodies (those onshore) are healthy and vulnerable while those that arrive or return from abroad are dangerous.[6] In the 1990s, similar fears were widely articulated with computer viruses, malware, and the basic dangers of networked connection—a discourse that regularly drew on and named the danger as similar to the spread of HIV.[7] As in the case of offshore labor, human flesh and national infrastructures are neatly swapped in these cases of fearmongering around computing.

Two months after the article ran in the *New York Times*, the United States imposed economic sanctions on India. In an article addressing how the sanctions would affect the New Jersey tech sector and the exceedingly high demand for engineers—which was being exacerbated by the Y2K crisis—the owner of NovaSoft Information Technology stated: "Most of the people coming from India are engineers, and about 70% have master's degrees. That is why American companies find them so attractive. . . . *They have become the raw material for the U.S. software industry.*"[8] The looming Y2K crisis brought to front pages the tensions around immigrant labor that had long been involved in propping up the US technology industry.

This chapter is about the work of Y2K repair and the relationships supporting the companion analogies of coding labor invoked above, in which some workers were described as "diamonds" while others were described as "raw material." It is not just that one kind of worker is precious and the other is unrefined but that this distinction was necessary for the maintenance of a labor regime that persistently characterized immigrant labor as impure. This is a historical investigation of how different people, all coders, came to occupy distinct but inseparable political and cultural identities at a moment of peak anxiety about technology and the state. By looking at the Y2K crisis through the lens of labor—and through the lens of specific, culturally coded types of worker—this chapter contributes a novel historical comparison as part of a larger attempt to understand the history of computing through its social and political structures and the renewed attention paid to the centripetal labor of maintenance and repair.[9]

I argue that the history of Y2K repair is a history of interdependent labor arrangements that were often artificially separated: one arrangement was the very public attempt to hire American ex-programmers to work on code written in languages like COBOL and FORTRAN; the other was reflected in the practices of hiring short-term employees on temporary work visas and outsourcing programming labor to India. Together, these are better understood as two strands of a braided and transnational coding labor regime. This chapter is written, then, as a history of two colliding "codes": the ambiguous computer code of the Y2K errors that were in need of clarification and repair, and the American immigration codes that dictated the conditions in which this labor could happen.

This history captures the way the emergence of urgent coding labor relied on a stratification of a "pure," American coding labor and a dangerous (economically, technologically), non-American form. Lilly Irani has shown how Amazon's Mechanical Turk (MTurk), a crowdsourcing website, enables a kind of purification, one that allows managers to position themselves as innovators, allows microwork companies to hide their labor force while positioning themselves as tech companies, and allows employers to "experiment with the uses of human labor." Together, these processes produce the distinction between "innovators" and "menial" symbolic workers while camouflaging the interdependence of the entire system.[10] Much of this dynamic was already honed in the labor dynamics surrounding Y2K, which similarly relied on two dependent but highly differentiated forms of code work. Hence this history prefigured and conditioned the ongoing exploitation of workers around the world, workers who continue to shore up and clean up for the legacies of technological design and decay.[11]

01/01/00

We have inherited several stories that purport to explain the Y2K crisis, where it came from, what it meant, and, depending on whom you ask, how it either failed to live up to the hype *or* how we avoided the complete disaster that it augured.[12]

The first explanation tells us that the Y2K crisis was a result of a design choice, traceable to the 1950s and entrenched in the 1960s, to write dates as six digits instead of eight: so June 10, 1997, is expressed as 06/10/97 or 971006, instead of 06/10/1997. When dates rolled over past the year 1999, to years written as 00, 01, 02, and so forth, the implications were ambiguous. How would a computer, or many interlocked computers, respond? Would it read the date 00 as 1900? And, if so, what would a sudden jump backwards even mean for the database of, say, the Social Security Administration? For some, the implications were not ambiguous but catastrophic, and predictions of nuclear meltdown, planes tumbling from the sky, and total infrastructure collapse were not hard to find.

Although Y2K problems were frequently described as "bugs" or "glitches," these terms have often aggravated people. Whereas "bug" and "glitch" often refer to mistakes or faults in a technology, the decision to code dates as six digits was, for all intents and purposes, *deliberate*. And so a companion story tells us that this choice to eliminate digits was done to save memory. As computer memory in the 1960s was limited and expensive, it compelled choices about what information should be included, and the seemingly redundant use of 1 and 9 in the date was an easy choice for exclusion.[13]

Another story we've inherited about Y2K tells us that the problem of the missing digits was particularly associated with the programming language COBOL. COBOL stands for "Common Business Oriented Language," and it emerged from the chaos of the so-called "software turmoil" of the late 1950s and early 1960s.[14] COBOL was a joint effort among computer manufacturers, businesses, and the Pentagon to come up with a programming language that could work the same way across different hardware to produce the same results. The perception that the Y2K crisis was a result of COBOL is not strictly true. It was, however, a problem with the way data were often stored in COBOL and other contemporary languages like FORTRAN. It is a problem of implementation if not design.[15] Given unlimited memory, or perfect foresight, dates could have been written in eight digits—or any other unambiguous format. The ambiguity of six-digit dates was a result of a choice to use the flexibility of code to write dates in a condensed format. It's not possible to locate the problem simply with programmers or in the language; rather, the source lay in the implementation of the language, within specific material conditions and within the politics of coding labor.

To understand better the specific implementation of COBOL and its repair, we can go back to its founding period, where we find the motivation to make a highly *usable* language. In 1959, the designers of COBOL, a committee called CODASYL, established that the new language would have the following features:

> Maximum use of simple English language.
> A programming language that is *easier* to use, even if somewhat less powerful.
> We need to broaden the base of those who can state problems to computers.[16]

These guidelines stipulated that the new language should be simplified, easy to use, and legible to a nonexpert (in this case, managerial) user. Although COBOL was an explicit effort to create an interoperable language for use across different hardware, it was also already an intervention in the distribution of knowledge and prestige within the labor of coding; in this case, making more code more accessible to a higher echelon of the labor hierarchy.

What began as a possible model for a new and potentially more democratic programming language, with a "maximum use of simple English" eventually turned into a problem in which ease of use and democratic openness became signs of oversimplicity. As Nathan Ensmenger has written, these origins meant that COBOL always struggled to gain the respect of the larger computer science community. The hatred of COBOL was often expressed as a kind of repulsion and was often framed in ableist terms. The Dutch computer scientist Edsger Dijkstra is quoted as saying "COBOL cripples the mind."[17] Elsewhere, the accessibility of COBOL meant that it was widely derided as a "manager's" language, or a "trade school" language—ridiculed because it was legible to an otherwise code-illiterate class of users or to a putatively more functional and less creative class of users.[18] As Ensmenger has observed, these objections were less technical than "aesthetic, historical, or political."[19]

Yet by the late 1990s COBOL was still, by some measures, the most widely used computer language in existence. We end up with quite the historical irony: a language that was made to be usable by a greater number of people turned out to be used by very few, *because of this very value of accessibility*. If you were trying to design a language to appeal to young, American undergraduates entering the new field of computer science, you could hardly set out to design something less inviting. It is fascinating (to me, at least) that this historical irony haunted the history of the Y2K crisis and that the values of simplicity, ease of use, and a broad user base affected who learned COBOL, who was available to fix it, and the perception of its role in structuring the Y2K crisis. From the 1950s onward, the labor politics of code were written into the formulation of COBOL. The crisis surrounding Y2K not only brought these politics to the surface, and for a new audience, but also recast them within the now globally distributed labor of repair.

At some distance, Y2K now appears as a signal moment in the history of computing. It was also pivotal in a period when computing settled into the routines, discourses, and politics of everyday life. The late 1990s was a period of acute public pedagogy surrounding computing, and the Y2K crisis focused attention on the pervasiveness and embeddedness of computing throughout daily life. In one of his speeches about the Y2K crisis, US president Bill Clinton highlighted the ordinariness of the danger: "Because the difficulty is as far flung as the billions of microchips that run everything from farm equipment to VCRs, this is not a challenge that is susceptible to a single government program or an easy fix."[20] By invoking these two kinds of technology—farm equipment and VCRs—Clinton united two images of the country as both dependent on ordinary computing: from the rural and agricultural to the urbane and media savvy, everyone needed their tech to work.

Whether you identified as a computer user or not, Y2K showed that life as you knew it ran on code. In the context of this time of widespread formal and informal pedagogy about computing, we need to ask how different publics were also being taught about the labor of computing, the labor of upkeep and repair in technical domains, and the social stratifications of work.

The Outsource and the Body Shop

The term "outsourcing" dates from at least 1979 and refers to the process of corporations subcontracting to a third party in a market outside their domestic center, often at a fraction of the cost of domestic labor. We often talk about outsourcing in terms of large-scale processing and manufacturing, call center work, and, in recent years, content moderation.[21] But outsourcing became a mainstream topic of discussion and a political object of debate in the computing industries of the 1990s. During that decade, US corporations responded to an apparent shortage of qualified coders and engineers in the United States by using foreign workers, either brought to America for contract labor or hired to work overseas on American projects, predominantly in India.[22]

The term "outsourcing" carries with it the spatial dynamics of interior and exterior—the image of a center drawing in resources from without. While the term's usage may be of relatively recent vintage, the larger structures of in-nation and out-nation are inseparable from the history of colonial exploitation. As Rukmini Bhaya Nair writes, "Between the mid-eighteenth and the mid-nineteenth century, the height of British rule, India went from an economy that sold finished goods to the world to one that was an impoverished exporter of raw materials."[23] Transformations in colonial relations recurred in the late 20th century when the Y2K crisis was pulling taut the emerging relations of on-demand labor needed to shore up a global computing infrastructure—hence, the invocation of Indian IT workers as "raw material for the U.S. software industry."

In addition to outsourcing (subcontracting to a third party), Y2K repair also happened through the closely related and sometimes overlapping arrangements of offshoring (performing in-house labor in a lower-cost labor market) and "body shopping" (moving immigrant workers into domestic markets for short periods of time, mediated through consultancies). In his history of outsourcing, Dinesh Sharma makes the case that it was both a technological change and a policy change in India that led to the emergence of large-scale outsourcing to India in the 1980s and its exponential growth in the 1990s.[24] The technological change was the growth in communications infrastructures that allowed for remote work, and the policy change was the Indian government's support of software technology parks that connected smaller firms to this communications network.[25]

Complicating this picture, Xiang Biao shows that the flows of capital and labor in computing industries in the 1990s were rarely linear. Xiang shows, through the study of a body shop in Sydney, that it was the *mobility* of Indian IT workers that animated an entire ecosystem of corporations in wealthy nations, placement agents, IT consultancies (the body shops), and workers.[26] If Y2K code was tricky to repair because it was tangled in arrangements of hardware, software, and embedded systems, the process of outsourcing and body shopping workers was equally intricate and was embedded in interdependent systems of capital, immigration, education, and mobile labor: "Unlike conventional recruitment agents who *introduce* employees to employers, body shops *manage* workers on behalf of employers—from sponsoring their temporary work visas to paying their salaries, arranging for accommodation and the like."[27]

As Xiang's observation makes clear, whether coverage of the Y2K crisis highlighted worker lives (Americans) or concealed them (Indians), the two accounts shared one definite feature: labor relations that relied critically on the contingent needs of a fickle computing sector. Although these workers were often positioned *against* each other, it is worth recalling their mutual reliance on unreliable employers, and layers of intermediaries, for work.

From the late 1980s onward, the US demand for workers in the computing industries was always enmeshed in the politics of immigration and labor. Never simply a matter of sending work abroad, meeting the demand was always also a matter of how many workers could be brought to the United States, for how long, and how permanently. The H1-B program was introduced as part of the US Immigration Act of 1990 with a cap of 65,000 visas per year. The H1-B visa is valid for three years, though an employer may apply to renew it once. The program was designed as a temporary stopgap to aid the tech sector in filling its labor needs. In the language of the program, the visa was earmarked only for those positions that "require theoretical and practical application of highly specialized knowledge to perform fully." In other words, they were designed specifically with technical expertise as a condition of employment.

As Philip Martin notes, H1-Bs are special because they allow "dual intent": "that is, the foreigner may say that his or her purpose in coming to the United States is to fill a job *and* to become an immigrant."[28] Most temporary work visas require that a holder refuse to acknowledge any intent to stay in the United States beyond the term of their employment. H1-B visas, in contrast, allowed for the possibility of permanent migration. For the migrant workers using this part of the immigration code, it is in fact this dual intent of the visa that emerged as a crucial flaw in the H1-B program: as the once-booming tech sector went bust in the early 2000s and extinguished the long-term employment prospects of many workers, they and their families were threatened with deportation.

It took several years for the first H1-B quota to be met, but it was finally reached in 1997. A bipartisan campaign immediately began to raise the cap on H1-Bs, and the eventual compromise was the American Competitiveness and Workforce Improvement Act. This was a period of intense lobbying in which Democrats and Republicans were currying favor with an increasingly powerful tech industry throughout debate over the act.[29] Speaking before the Senate, the bill's Republican sponsor, Spencer Abraham, after reading into the record a series of pro-tech newspaper editorials, remarked, "If American companies cannot find home grown talent, and if they cannot bring talent to this country, a large number are likely to move key operations overseas."[30]

The act eventually raised the quota for visas (to 115,000 in 1999 and 2000), set forth a new set of fines and whistle-blower protections, and, tellingly, included a higher $500 application fee, meant to fund a scholarship and training program for American workers. From the beginning, the H1-B program was always framed as a necessary stopgap to the lack of trained American coders. The scholarship fee made more explicit that this was viewed by lawmakers as a zero-sum game—and that the immigration code could be used not only to fill gaps in the labor market but also as a source of funding for future American students. The visa's emphasis on specialized knowledge was also politically useful for lawmakers attempting to draw a line between "wanted" and "unwanted" immigrants. In making the visa contingent on ongoing employment, the program also made sure that those fortunate enough to cross into the category of wanted still relied on the health of the tech sector economy for the continuance of that status.[31]

Throughout this period, the Y2K crisis served as a coalescent event.

Y2K was often a focus in debates over raising the H1-B cap and in attempts to gain favor with the tech lobby. As Xiang writes, while Y2K wasn't the only motivating factor, it "spurred a tremendous expansion in body shopping and further reinforced the practice of benching, which became entrenched during the following dot-com boom."[32] Sharma observes that as the Y2K problem emerged as an "opportunity . . . for Indian companies," the Indian government "stepped in to let the private sector take advantage of the Y2K opportunity."[33] In the immediate afterglow of Y2K, the apparently successful repair effort was treated as a proof of concept in which Indian software companies could "build on Y2K success."[34] In a 2018 policy document, the Migration Policy Institute (a liberal think tank), accounted for the act's passage this way: "In 1998, amid pressure from a technology sector concerned about the looming Y2K challenges for computer systems rolling over into the new millennium, Congress increased the cap."[35] While the crisis itself may have hardened into a historical curio for most, it still figures as an explanatory mechanism for changes to immigration and employment policy.

By 1996 Indian IT companies were working with US firms to set up joint repair ventures. The Indian tech lobby consortium NASSCOM (National Association of Software and Service Companies) established its own Y2K special interest group in the same year, and by 1998 major Indian tech companies were earning 20% to 30% of their revenues from Y2K contracts. One company, Tata Consultancy Services, established "a Y2K factory in Chennai housing nearly a thousand programmers capable of churning out two million lines of code every day."[36] In all of these descriptions the Y2K crisis—with its fixed date, doomsday prognostications, and massive private and public sector investments—worked as a kind of machine to entrench, clarify, and accelerate existing labor and capital relations through the trained work of repair.

The English-language *Times of India* published its first articles about Y2K in the mid-1990s. These included profiles of companies such as Wipro Infotech, now one of the largest corporations in India, as well as technical explanations of the crisis.[37] We can also see, buried deep in the classifieds from the same period, the signs of things to come. Ads called for workers and announced information sessions (box 12.1).

The classifieds of the paper show how Y2K was already baked into the ways Indian companies were envisioning labor in the computer industries. In the "overseas appointments" classifieds of the December 15, 1996, issue, an advertisement for Taj Technologies (a so-called body shop) appeared between ads for cooks needed in Saudi Arabia and for mechanical inspectors needed in the United Arab Emirates. The ad stated, under the banner "CONSULTANTS": "Taj Technologies is a rapidly growing consulting and software development company based in Minnesota, USA, and in Bangalore, India. We specialize in Y2K, full cycle software devel-

---

Box 12.1

CONSULTANTS

GROWTH   EXPERIENCE   TECHNICAL   EXCELLENCE

[Company name] is a rapidly growing consulting and software development company based in Minnesota, USA, and in Bangalore, India. We specialize in Y2K, full cycle software development, project management, contract programming, and management consulting. Our focus is to best meet the needs of our clients with the expertise of our employees.

We are currently looking for dedicated personnel with at least 2 years of experience in any of the following:

- IBM Mainframe: MVS, CICS, Focus, COBOL, DB2, IDMS, PACBASE
- C + +, VisualBasic, Foxpro
- Offshore Project Management

We offer employees a full package of benefits including Medical/Dental and assist with transportation.

For consideration, please send your resume within ten days to: [address in Bangalore]

Transcripts of advertisements published in *Times of India* in the mid-1990s.

opment, project management, contract programming, and management consulting."[38] A subsequent ad in the *Times* classifieds advertised a seminar, "Year 2000 Conversion," to be held at an airport hotel in Mumbai. These ads nested the ordinary ways Y2K was working its way into computing industries and reflected the banal manifestations of Indian workers' literacy in languages like COBOL. The classified ads were not heroic calls to arms but spoke to existing systems for distributing knowledge and indicating business strategies; they also show that the companies performing this labor were likely to have offices both in the United States and India, that workers would move between India and stateside offices, and that Y2K was the prominent event focusing the demand for coders.

The Dinosaurs Come Home

By the time the US Congress was debating lifting the cap on H1-B visas, the Y2K crisis had shifted from the stuff of specialist bailiwicks to a real problem facing anyone who relied on computing—as well as fodder for millenarian prophecies of doom. This intensification coincided with a boom in investment in new technologies, web startups, and personal computing. Coding was unglamorous and working in languages like COBOL was long out of fashion. As the *Los Angeles Times* article quoted at the beginning of this chapter put it:

> Once considered the dinosaurs of the technological revolution, COBOL programmers have become a hot commodity precisely because they stayed in the high-tech Jurassic era. . . . Like priests who dutifully maintained dusty heaps of scrolls, they are being recalled from seclusion to recover the ancient knowledge of COBOL.[39]

And the *Washington Post* reported in 1997:

> Five years ago, Donald Fowler's employer in Gaithersburg coaxed him into early retirement, telling him that his skill in programming mainframe computers was a thing of the past. Demoralized, he eventually moved here [to Largo, Florida] to vegetate on the beach and hone his golf game.
>
> A few months ago, though, a humbled Corporate America came calling on the 52-year-old Fowler. . . .
>
> The best part, though, is being wanted again, wanted badly. "I'm sure they'd send a limo" if he needed one to get to work, Fowler said jokingly.[40]

These are just a couple of excerpts from the broad corpus of press accounts of such workers. Virtually all the stories gathered in my research emphasize a call to action for once-forgotten workers to return to the job force for higher pay. Some stories emphasize more systematic ways that these workers were being organized. In addition to higher pay, during the peak of the crisis, some employers were offering $1,000 bounties for referrals of potential programmers, and Amy Calahan, of the Information Technology Association of America, claimed that a list of 10 retired COBOL programmers would sell for $250 each.[41] A state-run incentive program in Texas offered a "critical employee bonus program," whereby state agencies, including the Department of Corrections and the Department of Agriculture, would retain programmers with a $10,000 bonus.[42]

One consultancy, Senior Staff 2000, is frequently mentioned in press coverage as brokering the connection between companies in need of COBOL-literate programmers and retirees in search of employment. The president of Senior Staff, Bill Payson, claimed to have a database of 16,000 names of willing programmers aged 50 and over.[43] Payson said the Y2K crisis was offering a way for older workers to short-circuit the ageism of the tech sector. Quoting an imagined potential employer, Payson said, "I don't care whether he's 16 or 116, if he's a COBOL programmer, he's got a job."[44]

The way Y2K was being discussed as a labor issue in the United States contrasted sharply with the ways IT work was being discussed in India. While the story of domestic American labor was framed around the return of retired workers to the job force, it almost entirely elided the fact that the American Y2K effort was being staffed by precarious Indian workers—that one could not, in this moment, exist without the other. Even more drastically, when the Indian IT effort did appear in American news coverage, the dependence on Indian workers was instead cast as both an undesirable stopgap and a potential security vulnerability.

"Root Access"

Anti-Indian discrimination was rampant in late 1990s discussions of computing. As Indian nationals made up 40% of H1-B holders, and outsourcing became a frequent topic of news coverage and political debate, discussions of computing labor became another conduit for channeling xenophobia.[45] Beyond just discussions of employment opportunities leaving the United States or being given to non-US citizens, some American experts fixated on computer code itself as a vulnerable site.

In the summer and fall of 1999, members of the US intelligence community publicly warned that Y2K repairs had opened the United States to a cyberattack. Speaking to the Senate's Special Committee on the Year 2000 Technology Problem, Michael Vatis, the FBI's director of the National Infrastructure Protection Center (NIPC), warned that "malicious actors, foreign or domestic, could use the Y2K remediation process to install malicious code in the 'remediated' software."[46] The NIPC was responsible for detecting and deterring cyberattacks on the country's infrastructure, and Vatis specifically targeted the repair effort as the vector of infection for America's critical systems. He also targeted the reliance on nondomestic labor as the source of this danger.

For Vatis, there were four threats created by Y2K repair: a malicious actor might "install trap doors," "obtain 'root access,'" "implant malicious code," or "map systems." After the turnover to the year 2000, Vatis feared, it would be *repaired*, rather than unrepaired, code that would present a danger to the nation's infrastructure. For Vatis, these dangers were explicitly tied to the labor of Y2K repair and the fact that companies "across the United States and around the world" were rushing to perform necessary repairs. It also meant, he told the subcommittee, that "those who are doing the Y2K remediation are almost always contractors who are given the status of a trusted insider."[47] This discourse shows how damaged code, a labor shortage, and repair work were being constructed by state actors as bonded dangers.

Within a few weeks of Vatis's testimony a colleague and CIA officer named Terrill Maynard escalated the fearmongering. Writing in the NIPC's in-house journal, the *Infrastructure Protection Digest*, Maynard surmised that "India and Israel appear to be the countries whose governments or industry may most likely use their access to implant malicious code in light of their assessed motive, opportunity, and means." The same Reuters report that relayed Maynard's comments quoted Vatis as, again, underlining the danger as one connected to the "foreignness" of the remediation effort: "A tremendous amount of remediation of software has been done overseas or by foreign companies operating within the United States."[48] These comments were targeted specifically at the "impure" and therefore potentially toxic labor arrangements (outsourced and body-shopped work) that were crucial to the United States' ongoing infrastructural stability.

Throughout the Y2K crisis, the outsourcing of coding labor was treated as a security threat, despite these fears' inchoate expression. None of those listing dangers had any evidence of actual bad actors or malicious intent. Instead, what we can see in these (quasi-official) statements is the close articulation of labor, nationality, and security whereby the foreignness of some workers transforms the product of their labor into a danger. As Vatis's statement attests, this fear depended critically on nationality, whether the work was done "overseas" or by foreigners "operating within the United States." Just as networked computing had become the site of new fears of vulnerability, software repair was being described as a new medium through which the nation-state could be infected, either from within or without.

In India the response to Vatis and Maynard was a vociferous rebuke. By October 1999, Indian companies had already performed more than $2 billion worth of repair work for US companies.[49] And so, as Dewang Mehta, the president of India's software consortium NASSCOM, put it, "We cannot visualize that any moles have been planted. This is absurd. For us, too much is at stake."[50] For Indian companies, Y2K repair wasn't just a high-profile opportunity to boost the country's reputation as a tech center. The work was restructuring lives, as an increasing number of engineers were coming from India to live in the United States, for both short and lifelong stays. To say that there was too much at stake was, if anything, an understatement. Not only were the economic and reputational stakes too high (what Mehta alluded to), but for hundreds of thousands of people, the new labor arrangement was life-making. The visa arrangements put in place to support the US repair effort had reorganized where and how people lived. And the precarity and vulnerability of this life, more than any piece of code, would come into full view when the tech boom led to a bust in 2001.

Betrayal and Beyond

January 1, 2000, came and no great calamity occurred, but the new year also meant that the rush for programming labor had passed its peak. By March 2000, the NASDAQ had reached its all-time high (by that date), marking the height of the dot-com boom, but the following 18 months would see a precipitous fall in tech investment and massive layoffs across the sector.

The H1-B visa requires workers to return to their home country if they are no longer employed by a sponsor. This rule means that for practical purposes, control of the visa resides with the employer and not the employee. Some had been concerned in the late 1990s that this power imbalance would lead to mass exploitation by body shops that managed the hiring of visa holders. During the bust of the dot-com bubble, both the popular and tech press started to write about a growing awareness of the power imbalance and exploitation of the H1-B visa program and its human toll. A November 2000 *InformationWeek* headline put it bluntly, and in the terms of a previous era of exploitative labor: "Is the H1-B visa process mired in red tape and creating communities of indentured servants?"[51] A profile of abandoned workers in the *New York Times* used the analogy of the building of American empire through colonial exploitation: "They wrote many of today's computer codes, set up networks spanning millions of miles, designed and built Web sites and databases. The role of Indian immigrants in the development of the Internet has been likened to the contribution made by earlier waves of immigrants who built the nation's railroads and subway systems."[52]

As Devika Narayan argues, it is only a conceptual distinction that treats "employment relations" and "labor processes" as distinct aspects of work, management, and employment.[53] The H1-B system, used to hire en masse when the tech sector is in need, exposes these aspects as inextricably linked. While the "security vulnerabilities" associated with repaired Y2K code never manifested, it was the US immigration code that contained its own trap door. Of the 134,000 H1-B workers who arrived in the United States between 1998 and 1999, the *New York Times* estimated that 20,000 to 80,000 were forced to leave due to layoffs.[54] Although the H1-B visa is often treated as a "golden ticket," its reliance on a fickle tech sector means it can be disastrous for those who rely on their employment relation for their right to reside in their new home.

The beginning of the 21st century has taken on a different salience in recent years. I am not the only one interested in the Y2K crisis and its cultural, political, and technological significance. Journalists, podcasters, and filmmakers have recently returned to the 1990s and the Y2K crisis to look for lessons about how we got here, an era of entrenched platform power. The term "Y2K" has taken on new meaning—not just representing the potential failure of computer systems but acting as a metonym for the entire era of the late 1990s and early 2000s (witness the recent "Y2K fashion" line at the Scandinavian fashion chain H&M). The turn of the century is now a font of nostalgia. In 2016, when Americans were polled with the question "When was America greatest?" both Democrats and Republicans ranked 2000 higher than any other year.[55]

But Y2K is not just a quirky moment in the history of technology and code. Nor is it simply one moment. The enduring complexity of the crisis shows how intensely complicated the politics of the moment were. This chapter approaches the crisis as it played out through the fabrication of different kinds of repair workers in the United States. The once neglected, now heroic COBOL programmer, brought out from the obscurity of retirement or redundancy, stands in stark contrast against the backdrop of outsourcing, the rationalization of body shopping, the xenophobia channeled into security concerns about Y2K repair, and the abandonment of immigrant workers.

Scholars of infrastructure often use moments of breakdown and malfunction to reveal, in relief, the normal or expected running of things. In many ways the Y2K crisis is a template for this process. The received narrative of the crisis says that something as simple as representing the date in code can reveal multiple layers of sedimented history. From the desire to save memory and money, to the politics of languages like COBOL, to the path dependence of legacy systems and their nested relationships, the Y2K crisis can and has been used to crack open the assumptions of computerized infrastructure in the late 20th century. But this is not simply a history of a design flaw run amok. As Mar Hicks argues in chapter 14 of this volume, "Ideas about the proper organization of work, the particular goals of industry and society, and the value of different kinds of labor are transmitted along with computing hardware." The Y2K crisis and the effort to repair broken code clearly show how ideas about the "proper organization of work" circulated in the 1990s. During a time of acute public pedagogy surrounding the ubiquity and embeddedness of code, the Y2K crisis was instructive for understanding the social valuation of different forms of coding labor. The contrasting stories of American and Indian IT workers tell an interwoven history of hardware, software, and the politics of labor that echo into the present.

During the COVID-19 pandemic, as Paul Edwards and John Leslie King write, former COBOL programmers were once again called into action, this time to shore up US government systems that were faltering under the pressure of unprecedented jobless claims.[56] The press coverage highlighted the unlikely heroics of these once neglected programmers and the unbelievable but all-too-real dependence of current systems on COBOL (one recent study notes that 43% of banking systems, 80% of in-person transactions, and 95% ATM swipes rely on COBOL code).[57] Referring to these workers as "COBOL Cowboys," (after a firm, much like Senior Staff 2000, set up to place them), press reports also emphasized the fact that this had all happened before, in the lead-up to the Y2K crisis. As the pandemic hiring boom waned, we yet again witnessed the disordering and chaotic life outcomes for H1-B holders, during the massive layoffs of 2022–23 in the tech sector. The cycles continue, in other words, and persist in their reliance on precarious labor and unjust hierarchies that highlight the skills of some while forgetting others.

The interdependent forms of labor brought into relief by the Y2K repair effort presage many of the concerns of contemporary scholars. These concerns include hierarchies of skill, nationality, gender, and age endemic to tech cultures as well as the restructuring of global economies of scale to exploit labor cost differentials. A disproportionate focus on the corridors of tech innovation (Silicon Valley, ARPANET, the web) obscures larger, and more racialized, structures of work and exploitation that make, maintain, and clean up for the uses and misuses of technology. The stories about Y2K repair in the late 1990s, then, expose the vulnerabilities of relationships and livelihoods made according to the boom-and-bust cycles of tech investment and the interlocked codes of computing and immigration.

Notes

1. Ashley Dunn, "Bug Re-boots COBOL Experts' Fading Careers," *Los Angeles Times*, September 21 1998, D1.

2. Dunn, "Bug Re-boots COBOL Experts' Fading Careers" (emphasis added).

3. Peter Lewis, "Threat to Corporate Computers Is Often the Enemy Within," *New York Times*, March 2, 1998.

4. Mobina Hashmi, "Outsourcing the American Dream? Representing the Stakes of IT Globalization," *Economic and Political Weekly*, January 21–27, 2006, 242–49.

5. Wendy Hui Kyong Chun and Sarah Friedland, "Habits of Leaking: Of Sluts and Network Cards," *differences* 26 (2): 1–28.

6. Alan Bewell, *Romanticism and Colonial Disease* (Baltimore: Johns Hopkins University Press, 1999), 5.

7. Cait McKinney and Dylan Mulvin, "Bugs: Rethinking the History of Computing," *Communication, Culture, and Critique* 12, no. 4 (2019): 476–98.

8. Diana Lasseter Drake, "India Sanction Hits Here," *Business News New Jersey* 11, no. 21 (1998): 7 (emphasis added).

9. Steven Jackson, "Rethinking Repair," in *Media Technologies: Essays on Communication, Materiality, and Society* (Cambridge, MA: MIT Press, 2014): 221–39; Marisa Leavitt Cohn, "Convivial Decay: Entangled Lifetimes in a Geriatric Infrastructure," paper presented at Association for Computing Machinery conference on Computer-Supported Cooperative Work (CSCW) and Social Computing, 2016: 1511–23; Andrew L. Russell and Lee Vinsel, "After Innovation, Turn to Maintenance," *Technology and Culture* 59, no. 1 (2018): 1–25.

10. Lilly Irani, "Difference and Dependence among Digital Workers: The Case of Amazon Mechanical Turk," *South Atlantic Quarterly* 114, no. 1 (2015): 230.

11. Lisa Nakamura, *Cybertypes: Race, Ethnicity, and Identity on the Internet* (New York: Routledge, 2013); Christina Dunbar-Hester, *Hacking Diversity: The Politics of Inclusion in Open Technology Cultures* (Princeton, NJ: Princeton University Press, 2020); Winifred Poster, "Who's on the Line? Indian Call Center Agents Pose as Americans for US-Outsourced Firms," *Industrial Relations: A Journal of Economy and Society* 46, no. 2 (2007): 271–304; Mary Gray and

Siddharth Suri, *Ghost Work: How to Stop Silicon Valley from Building a New Global Underclass* (Boston: Houghton Mifflin Harcourt, 2019).

12. Paul Edwards. "Y2K: Millennial Reflections on Computers as Infrastructure." *History and Technology* 15, nos. 1–2 (1998): 7–29; Zachary Loeb. "The lessons of Y2K, 20 Years Later," *Washington Post,* December 30, 2019, https://www.washingtonpost.com/outlook/2019/12/30/lessons-yk-years-later/.

13. Ellen Ullman, "The Myth of Order," *Wired* 7, no. 4 (April 1999): 126–29.

14. Paul Ceruzzi, *A History of Modern Computing* (Cambridge, MA: MIT Press, 2003); Nathan Ensmenger, *The Computer Boys Take Over* (Cambridge, MA: MIT Press, 2010).

15. Stephanie Dick, "Of Models and Machines: Implementing Bounded Rationality," *Isis* 106, no. 3 (2015): 623–34.

16. Charles Phillips's CODASYL minutes (1959), quoted in Jean Sammet, "The Early History of COBOL," in *History of Programming Languages,* ed. Richard L. Wexelblat (New York: Association for Computing Machinery, 1978), 199–243.

17. Quoted in Ensmenger, *Computer Boys Take Over,* 100.

18. Ensmenger, *Computer Boys Take Over,* 101; Ceruzzi, *History of Modern Computing,* 92.

19. Ensmenger, *Computer Boys Take Over.*

20. "Remarks by the president concerning the year 2000 conversion," https://govinfo.library.unt.edu/npr/library/speeches/y2kconv.html

21. Gray and Suri, *Ghost Work*; Raka Shome, "Thinking through the Diaspora: Call Centers, India, and a New Politics of Hybridity," *International Journal of Cultural Studies* 9, no. 1 (2006): 105–24; Sarah Roberts, *Behind the Screen: Content Moderation in the Shadows of Social Media* (New Haven, CT: Yale University Press, 2019).

22. Mary Amiti and Shang-Jin Wei, "Fear of Service Outsourcing: Is It Justified?," NBER Working Paper 10808, National Bureau of Economic Research, October 2004, https://doi.org/10.3386/w10808.

23. Rukmini Bhaya Nair, "Outsourcing Postcolonialism," in *Reframing Critical, Literary, and Cultural Theories,* ed. Nicoletta Pireddu (Cham, Switzerland: Palgrave Macmillan, 2018), 306.

24. Dinesh Sharma, *The Outsourcer: The Story of India's IT Revolution* (Cambridge, MA: MIT Press, 2015).

25. Sharma, *The Outsourcer.*

26. Xiang Biao, *Global "Body Shopping": An Indian Labor System in the Information Technology Industry* (Princeton, NJ: Princeton University Press, 2011), xv.

27. Xiang, *Global "Body Shopping,"* 4.

28. Philip Martin, "High-Skilled Migrants: S&E Workers in the United States," *American Behavioral Scientist* 56, no. 8 (August 2012): 1058–79.

29. Dylan Mulvin, "Distributing Liability: The Legal and Political Battles of Y2K," *IEEE Annals of the History of Computing* 42, no. 3 (2020): 26–37.

30. 144 Cong. Rec., S2244 (March 19, 1998) ("Widespread Editorial Support for Increasing the H–1B Cap").

31. Ishani Mukherjee, "Framing the Globalization-Migration Debate: Continuing Impact of U.S. Economic Crises on Indian H1-B Immigrant Professionals in America," *IOSR Journal of Humanities and Social Science* 19, no. 7 (January 2014): 107–16, http://doi.org/10.9790/0837-1974107116.

32. "Benching" is a tech sector term for the practice of hiring workers and putting them "on the bench" while waiting for jobs or between contracts. They are unpaid while benched. Xiang, *Global "Body Shopping,"* 5–6.

33. Sharma, *The Outsourcer*, 146

34. Khozem Merchant, "Indian Software Suppliers Build on Y2K Success," *Financial Times*, July 3, 2000, 6.

35. Sarah Pierce and Julia Gelatt, "Evolution of the H-1B: Latest Trends in a Program on the Brink of Reform." Migration Policy Institute, March 2018, 5, https://www.migrationpolicy.org/sites/default/files/publications/H-1B-Brinkof Reform-Brief_FinalWeb.pdf.

36. Sharma, *The Outsourcer*, 180.

37. Allen Mendonca, "Wipro's Date with 2000 AD," *Times of India*, August 12, 1995, 15; "Bug Year," *Times of India*, September 7, 1996, 10.

38. *Times of India*, December 15, 1996, 28.

39. Dunn, "Bug Re-boots Cobol Experts' Fading Careers."

40. Rajiv Chandrasekaran, "Older Programmers Go Back to Fix the Future: Skills in Demand for 2000 Glitch," *Washington Post*, March 2, 1997, A1.

41. Chandrasekaran, "Older Programmers Go Back to Fix the Future."

42. P. J. Huffstutter, "The Sky Is Falling! Wanna Buy Our Umbrella?" *Los Angeles Times*, November 3, 1997, D1.

43. Mary Raitt Jordan, "Vintage Techie, Train Thyself," *IT Support News* (August 1999): 15, 18.

44. Payson quoted in Troy Wolverton, "Greybeard Tech Database Puts Seniors with Programming Skills Back in Circulation," *San Jose Mercury News*, April 26, 1998, 23.

45. Amy Bhatt, "From Family Planning to Family Reunification: Reproducing Indian Diasporas," *SAGAR* 17 (2007): 55–68.

46. *Critical Information Infrastructure Protection: The Threat Is Real; Hearing Before the Subcommittee on Technology, Terrorism, and Government Information of the Committee on the Judiciary, US Senate*, 106th Cong. 22 (October 6, 1999) (statement of Michael A. Vatis).

47. *Critical Information Infrastructure Protection*.

48. Maynard and Vatis quoted in "FBI Says Y2K Software Has Been Tampered With," Reuters, October 1, 1999.

49. Sharma, *The Outsourcer*.

50. "India: Code Smuggling? Absurd," *Wired*, October, 1, 1999.

51. Peter Ruber, "The Great Foreign IT Worker Debate," *InformationWeek*, November 27, 2000, 153–60.

52. Abhi Raghunathan, "Thanks for Coming; Now Go," *New York Times*, July 15, 2001, NJ1.

53. Devika Narayan, "Manufacturing Managerial Compliance: How Firms Align Managers with Corporate Interest," *Work, Employment, and Society* 37, no. 6 (2022).

54. Raghunathan, "Thanks for Coming; Now Go."

55. "When Was America Greatest?" *New York Times*, April 26, 2016, https://www.nytimes.com/2016/04/26/upshot/when-was-america-greatest.html.

56. Paul N. Edwards and John Leslie King, "Institutions, Infrastructures, and Innovation," *Computer* 54, no. 1 (2021): 103–9.

57. "COBOL Blues," Reuters Graphics, accessed December 21, 2024, http://fingfx.thomsonreuters.com/gfx/rngs/USA-BANKS-COBOL/010040KH18J/index.html.

Part III

What does it mean to grapple with code?

13

From Programming to Platform Expertise

Technical Reformers and the Reinvention of Institutions

Shreeharsh Kelkar

In 1958, in the pages of the *Harvard Business Review*, the management scholars Harold Leavitt and Thomas Whisler heralded the arrival in American management of a new entity that they labeled "information technology." Information technology, in their definition, had three components: the digital computer capable of processing large amounts of information; a broad array of "programming" techniques, including computer programming and operations research, that would rationalize and simplify organizational decision making; and the use of simulations to build models of phenomena of particular interest to organizations. Leavitt and Whisler predicted that information technology would enable corporations to shrink middle management. The highest-level managers would take on even more "creative" and "innovative" functions while much middle-management work would be "programmed away" and deskilled.[1]

In 2012, another group of management scholars, Andrew McAfee and Erik Brynjolfsson from the Massachusetts Institute of Technology's Sloan School of Management, took to the pages of the *Harvard Business Review* to herald the arrival of yet another entity—"big data"—which would also revolutionize management. In this vision of management, managers would make "data-driven" decisions aided by a socio-technical infrastructure of digital interfaces, machine learning algorithms, and a class of professionals called "data scientists." These data scientists would be well versed in the language of quantitative analysis and business frameworks. Managers and data scientists would work on business problems in an "innovative" fashion: they would work fast, make iterative improvements, and subject all decisions to data-driven tests. Corporations that adopted Big Data–driven management would gather new business insights, grow their customers, and improve their market share, leaving behind those that did not adopt this framework.[2]

Table 13.1. The elements of the twin assemblages of programming and platform expertise

Element	Programming (1950–)	Platform expertise (1990–)
Technology	Digital computer	Networked computing, big data, software interfaces
Knowledge practices	Computer programming, operations research	Machine learning, user-centered design, A/B testing
Objects of concern	Decision making	Innovation, open-ended (i.e., platform) governance
Experts	Systems analysts, programmers, operators	Data scientists, designers, domain experts, users

Note: Each expertise is an assemblage of technologies, knowledge practices, objects of concern, and particular roles.

Superficially, these two announcements might look similar: two pairs of management consultants heralding a new technology, prophesying about a world to come, inviting their readers to participate in and help build this world, and also warning them that should they not participate, that world would materialize anyway, leaving them far behind. This chapter, however, maintains that teasing out the differences between these two visions, why they emerged when they did and the changes that occurred from 1958 to 2012, is worth explaining. Briefly, I argue that the two articles describe two different underlying socio-technical assemblages—consisting of technologies, practices, and norms—that I call *Programming* (with a capital *P* to distinguish it from computer programming as a practice and profession) and *platform expertise* respectively. Each underlying assemblage consists of different technologies, knowledge practices, objects of concern, and relevant experts (summarized in table 13.1). Without making any claims about an epochal shift, this chapter argues that the transition from Programming to platform expertise involved not just the invention of new technologies but also the changing perspectives of technical reformers, experts who seek to reinvent their institutions through the technologies of the digital computer.[3]

I use the notion of assemblage—contingent formations that incorporate material, social, and semiotic elements—to guard against both technological and social determinism. Elements of assemblages may often be in tension with one another. Some social actors may have more use for certain aspects of the assemblage than others, and actors may disagree as to what the significance of the assemblage is. But assemblages are also singular in the sense that an assemblage itself is more than the sum of its parts. Assemblages bring together human practices, organizational structures, and technological artifacts.[4] Most important, actors and institutions can draw on the assemblage or its parts in debating fundamental questions of ethics, methods, and values.[5]

The Programming assemblage was constructed in the 1950s, '60s, and

'70s by technical reformers who wanted to problematize *decision making* in organizations. These reformers ranged from the incipient computer programmers to systems analysts, management consultants, and information theorists. To do so, they drew on the digital computer and an associated theory of organizations as sites of information processing. They argued that if organizations could be conceived as information systems, incorporating digital computers to direct the flows of information within an organization—that is, Programming—would lead to improved and more efficient decision making. They also argued that experts on information processing—computer programmers, operations researchers—were the ones who needed to be advising top management about how to run an organization. While these reformers failed to make the idea of Programming central to management, they were nevertheless able to turn the occupation of computer programming into a high-status, stereotypically masculine occupation.[6]

The platform expertise assemblage is currently under construction by Silicon Valley technologists and their admirers across institutions. These reformers do not necessarily hope to reinvent organizational *decision making* to make it more rational and more efficient; rather, they hope to use the new practices around networked computing—especially knowledge practices like Big Data analysis, user-centered design, and A/B testing—to create new spaces of "innovation," where users can be *governed* in an open-ended fashion.

In labeling this assemblage "platform expertise," I draw on the sociologist Gil Eyal's point that expertise is not the same thing as experts.[7] "Expertise," in Eyal's view, refers to a style of reasoning—realized through specific material technologies and social arrangements—that leads to the construction of institutional tasks and problems that then become the province of specific experts or workers to carry out. This definition makes an analytical distinction between expertise, a style of reasoning, and the experts who may end up carrying out the tasks specified through that style of reasoning; this implies that the increase in the power of a form of expertise does not necessarily mean an increase in the power of experts most conventionally associated with that style of reasoning.

Thus, Eyal describes how the parents of autistic children allied with psychologists and occupational therapists to create a new regime—or "problematization"[8]—for diagnosing and helping autistic children that took power away from clinical psychiatrists and displaced it onto a network of "psychologists, therapists, special education teachers, and most important to the parents, *themselves.*"[9] The key characteristics of this new regime were new "systems of observation" distributed through transcription technologies (invented by psychologists) such as paper-based behavioral checklists that parents could use to monitor their children and submit as evidence to psychologists and therapists. Thus, a new style of reasoning (i.e., expertise or problematization or assemblage) related to autism and childhood disability was born that entailed a set of tasks, mostly carried out by parents and psychologists.

In this chapter's argument then, both Programming and platform expertise are styles of reasoning or forms of expertise, socio-technical assemblages that make possible the emergence of new kinds of goals and tasks such as programming information systems (for Programming) and managing users (for platform expertise). In the following section, I describe the elements of the Programming assemblage. Following that, I describe how the emerging political economy of computing over the 20th century shaped the constitution of platform expertise. In the final section, I discuss two initiatives of open online learning, the Open Learning Initiative (OLI) and massive open online courses (MOOCs), as instantiations of Programming and platform expertise, respectively, and describe how these concepts can be used to explain the structure and outcomes of these two projects.

The Elements of Programming

The earliest digital computers found two key uses in the United States: as calculating machines in "scientific computing" (with applications to the war machine) and as extensions of data processing in corporations.[10] A group of technical reformers emerged who hoped to reshape organizations through the digital computer and construct themselves as experts on both computers and organizations.[11] These reformers included the new computer programmers, operations researchers, and systems analysts, as well as university researchers working in management, cognitive science, and artificial intelligence (AI). In the visions set out by these reformers, computers, human beings, and organizations were all information systems. Digital computers, deployed within these information systems and aided by a class of information specialists, could help managers make decisions that were more rational and more efficient.

I call this assemblage of the technologies, knowledge practices, and objects of concern through which these reformers articulated a new kind of world as the Programming assemblage. The Programming assemblage had four key components (see table 13.1); the digital computer was at the center of this assemblage. The object of concern was to facilitate more optimal decision making. This decision making would be facilitated by the new practices around the digital computer, including operations research, systems analysis, and computer programming, and by new experts like the computer programmer or the systems analyst.

The narrative I provide below of the Programming assemblage is almost certainly an oversimplification, since there were disagreements amongst the technical reformers themselves about what a digital computer was, what "software" meant, and what their role should be in organizations.

Also, other cultures of computing—for example, the time-sharing communities of education studied by Rankin—were not particularly interested in questions of decision making and organizational choice.[12] Nevertheless, for the purposes of this chapter, cutting some historical corners in the story of computing allows me to contrast the Programming assemblage with platform expertise.

Why did technical reformers become concerned with decision making? The historian Hunter Heyck argues that, starting in the early decades of the 20th century, American social scientists reformatted the object of their inquiry: rather than the human being who chose from different options, they started to study the process of choosing itself.[13] In this endeavor, they modeled their inquiry around three analytic objects: "decision-making, communication, and system."[14] Decision making was the process of generating and choosing between multiple options, the system was the entity that chose, and communication was the process through which the system arrived at those options. The purposeful invocation of "system" rather than human beings meant that social scientists could study decision making not just at the level of individuals but for animals, machines, organizations, and even societies. They could thus raise decision making to the art of the highest democratic good while simultaneously showing that human beings were limited decision makers.[15] Social scientists moved toward this paradigm for two key reasons: first, it represented a strategic convergence of interests between them (they had seen the possibilities of interdisciplinary problem-driven research during World War II) and their funders (who were interested in scientific ways of managing large organizations); and second, it allowed both parties to think about the role of reason and rationality in a democracy.

The Programming assemblage was the means by which better decision making could be achieved in organizational contexts by overcoming the fallibility of human beings. Even if human beings were not the most optimal decision makers on their own, it was still possible to create an information system that could optimize their decision making through the addition of new kinds of techniques and information. The digital computer, with its accompanying concepts of program and algorithm, was a key resource that technical reformers drew on in rethinking decision making. Building a program or an algorithm to model a decision-making process was often the first step in understanding decision making scientifically. Moreover, these algorithms and programs could also work as decision-making aids to help organizational leaders make choices that were rational.

Consider the career of Herbert Simon, a towering social scientist of the 20th century who was simultaneously an administrative theorist, psychologist, economist, cognitive scientist, and AI researcher.[16] In his first book, *Administrative Behavior*, published in 1947, Simon argued that there was no such thing as a completely rational human decision. Human beings always made decisions in constrained circumstances; they "satisficed," based on those constraints, rather than optimizing; human decision making was procedural, rather than substantive. Thus, what organizations needed to do, Simon argued, was build cognitive scaffolding that allowed their employees to satisfice better with the limited time and resources they had at their disposal.

Simon experienced his first epiphany with digital computers when he consulted with the RAND Corporation on a project investigating the question of how the United States could defend itself from a Soviet missile attack. In this project, named SAGE (Semi-Automatic Ground Environment), the defense establishment had visualized an almost complete environment that would defend the United States from missile attacks; this environment would consist of multiple radar stations linked to multiple computers, all providing warnings and being operated by servicemen. As an organizational theorist, Simon's job for RAND was to analyze the decision making of these operators under conditions of combat and to help create organizational conditions to improve this decision making. While doing this project, Simon was able to use the RAND Corporation's IBM digital computer to create simulations that he employed in studying and understanding the behaviors of the operators in different situations.

The experience left Simon profoundly aware of the kinds of things the digital computer was capable of. Conceptually, the digital computer was the "perfect bureaucracy," "the essence of the machine, abstracted and so made universal, yet still physically real."[17] The computer program was thus a key resource in understanding not only abstract systems but also real systems like human beings and organizations. Building a program or an algorithm to model a decision-making process was often the first step in understanding decision making scientifically. Moreover, these algorithms and programs could also work as decision-making aids to help organizational leaders make choices that were rational.

The 1958 *Harvard Business Review* article of Leavitt and Whisler, with which this chapter began, articulated a set of ideas like Simon's about how "information technology"—a neologism that meant the broad set of techniques associated with the digital computer—could transform management. Leavitt and Whisler argued that the digital computer made it possible to "program" away the entire layer of middle management in a corporation; the top management's job was to program this information technology to improve the system of rational decision making. They used the word "program" to mean not necessarily a computer program but a set of rules for carrying out tasks; they defined "programmers" as experts who could articulate those rules using cutting-edge techniques. These programmers would advise the top management of the organization on the best and optimum ways to make decisions. As they put it:

> New top-management bodies will need programmers who will focus
> on the internal organization itself. These will be the operations re-
> searchers, mathematical programmers, computer experts, and the
> like. *It is not clear where these kinds of people are being located on
> organization charts today, but our guess is that the programmer will
> find a place close to the top.* He will probably remain relatively free to
> innovate and to carry out his own applied research on what and how
> to program (although he may eventually settle into using some stable
> repertory of techniques as has the industrial engineer).[18]

One kind of information technology designed to both reinvent management and raise the organizational status of programmers was the totally
integrated "management information system" (MIS).[19] Proposed in the
1950s by a group of experts who called themselves the "systems men," the
MIS was a computer-based system that would permeate the organization
completely; it would take data in continuously about company operations
(sales, costs, production figures, new personnel, etc.) and produce reports
that could guide decisions based on this data. The MIS was best thought
of as a kind of computer-enabled command-and-control system that top
management tinkered with in their efforts to make the best decisions
rather than as an assembly line for processing data (which is how computers were typically used in data-processing departments).

For the systems men and for computer specialists in general, concepts
like information technology and the MIS represented a channel for professional advancement to the highest echelons of management. For those
previously stuck in accounts and data-processing departments and forced
to produce reports at the initiative of other departments, the concept of
information made it possible to group under the heading of information
"what had previously been quite different things such as accounts, market
forecasts and inventory records"; moreover, these were now "demarcated
as a single area of professional responsibility, that of the information technologist."[20]

For all the rhetoric and effort, these attempts to turn computer specialists into advisers to top-level management generally failed. Part of the
problem was that existing computing hardware in the 1950s, '60s and '70s
was simply not strong enough to build an actual working MIS, which was,
in the words of historian Haigh, "ubiquitous in theory and unknown in
practice."[21] But there were other problems as well: It proved very difficult
in practice to get top-level managers to articulate exactly what information
they needed to make the most rational decision, and it was impossible to
construct an MIS without that knowledge. Last but not least, accountants
and other middle managers whose work the computer specialists hoped
to automate pushed back at the idea that the computer specialists were
experts on management by virtue of their knowledge of information tech-

nology. They questioned the idea that being experts about the general category of "information" or "information flows" gave these computer specialists any particular insight into specific issues that concerned an organization, such as finance, accounting, or personnel. These groups also labeled programmers and computer experts as unsocial, unreliable, and unlikely to work well in a corporate environment, pointing to the fact that software projects routinely overran budgets and time frames and in the end did not produce very good software.[22]

Ultimately, then, the program remained a technical concept rather than a managerial one; programmers ended up being technical experts who could make digital computers do particular tasks, rather than those who could write and rewrite the rules of decision making in organizations (or build systems like the MIS that would allow high-level managers to do so).[23] Other experts who populated the organization from mid-level managers to accountants refused to accept the claim that understanding information technology gave computer experts any insight into specific questions like budgets and personnel.

With the arrival of personal computing and the decentralization of computing within corporations, a settlement was reached between the different sides.[24] Rather than a single "data processing" department with computers that a priestly class of programmers then interacted with on behalf of the entire firm, corporations started to store their data centrally, but let individual employees work on it from their own workstations using customized software programs. At the same time, the job of customizing software so that it could be used within a corporation started to be done either by an internal "Information Technology," or IT, department which had no pretensions to being managerial.[25] Today, when organizations want to revamp their information infrastructure for business purposes, they often turn to outside business or management consultants. These groups perform managerial functions but are often software experts; what allows them to do so is their position outside the organization.

In seeking to construct the Programming assemblage to reinvent decision making, the technical reformers sought to conceptualize organizations and computers as information systems, management as information processing, and themselves as experts of information processing. They were not successful in this endeavor. While software is now ubiquitous in nearly all organizations, software experts are still considered technicians, rather than high-level managerial advisers.

From Programming to Platform Expertise

How did we get from Programming to platform expertise? This section describes the socio-technical changes that made platform expertise possible and explains how it differs from Programming. Two shifts were particularly important: first, from a technical vocabulary of "software" to one of "platforms" and the attendant focus on managing "users"; and, second, from a focus on decision making to one that emphasized "innovation" and open-ended governance.

Originally, digital computers were bulky artifacts that needed to be housed in a room and tended by a class of programmers. (To put it differently, programmers were the original computer users.) Over the next few decades, not only did digital computers get smaller, but their place in the organization changed. The invention of the "database" allowed organizations to have a central data store that was then accessible to employees from their own desks.[26] While the earliest computers needed programmers and operators to make them work, now employees did not have to be programmers; instead, employees often used large-scale standardized enterprise software or mass-produced application software such as word processors and spreadsheets.[27] This was partly a realization of the vision of Xerox's famous Palo Alto Research Center (PARC); PARC researchers envisioned a white-collar organization as an "architecture of information" in which employees used networked personal computers to work and to also connect with other workers in the organization.[28]

The proliferation of computers from bulky instruments to individual prostheses transformed the processes of software production making them more "user" directed. From the 1970s onward, software designers concentrated more and more on the user.[29] The definition of "user" tended to be fluid and depended on organizational contingencies. Thus, for software makers, users were often the managers rather than the employees who would actually be using any given piece of software; because managers paid for the software, they tended to be more involved in its construction. In Scandinavian software design traditions, the invocation of the user was motivated by reasons of democratic governance—to give *workers* more of a say in the design of applications intended for them. In the United States, the incorporation of the actual user into the software design process was more for market considerations—to make products that worked better.

Software developers thus started making the user a key part of the software design process. The traditional "waterfall" model of software design tended to be linear as developers moved from gathering user requirements to realizing them in the software and then testing it. Instead, software developers developed more "agile" methods of software design in which they designed in short bursts, built prototypes, got feedback from their users on those prototypes, and so on. This process, as Pollock, Williams, and D'Adderio demonstrate, is just as much about shaping users as it is about being shaped by them.[30] The developers at the software company SAP (Systems, Applications, and Products in Data Processing), in building an enterprise resource planning package for campus management, first constructed a "kernel" of the package. They then sought feedback from their clients (aka their "users") on how to improve this kernel. Their users gave them multiple options to work with, but the designers also had to find a way to manage their users and develop a particular set of generic features that could be used across users without alienating them. This sort of incorporation of the user into design processes had less to do with "the humanistic, democratic or utopian ideals of sociotechnical or participatory design approaches" from Scandinavia than with market considerations.[31]

The emerging vocabulary of "platforms" made the "user" even more important for software development experts. The media scholar Marc Steinberg has argued that usage of the term "platform" can be grouped into three discourses that emerged in succession: "product-technology platforms," "contents platforms," and "transactional and mediation platforms."[32]

The product-technology platform was theorized in the late 1970s and '80s by management consultants.[33] In this discourse, a platform was a hardware or software product that served as the basis for other products made by actors external to the firm, the so-called "complementors." For instance, a computer could become a platform for different kinds of peripherals like printers and scanners. A VHS player could become a platform for an emerging industry of video cassettes. And game portals like Nintendo could become platforms for an industry that built video games specifically for it. In the computer industry, the product-technology platform led to the conceptualization of programmable platforms, of both the hardware and the software variety. Intel's microprocessors, which were often the bedrock of IBM PC–compatible architectures of microcomputers, were a classic platform for producing PC-compatible microcomputers (hardware). Microsoft's Windows operating system, which ran PCs, became a platform for application programs like WordStar and Lotus 1-2-3 (software); of course, in time, Microsoft became the dominant manufacturer of applications as well, with Word and Excel.

Contents platforms emerged in the 2000s in the United States when newly emerging web portals like Blogger, Myspace, Facebook, and YouTube started to brand themselves as "platforms" for multiple audiences.[34] Contents platforms were like product-technology platforms in that both involved a group of people building new things on top of a substrate, but there was one significant difference: contents platforms, Steinberg observes, "play[ed] on a sense of middling or intermediary function of distribution that is not present in the former."[35]

Consider YouTube, which started off as a portal for "you," the amateur video enthusiast, who could use the site to share videos with your friends and family and perhaps even a wider audience; it was a portal for "user-generated content" and in that sense, most like a product platform allowing creators to build on it. But in its search for revenue, it changed its business model, first by becoming a portal not just for individual creators but also for media companies and advertisers. But it became more than a product platform that connected all these constituencies to viewers; through its interface and its recommendation algorithms, it was also picking winners and losers in terms of which YouTube creators got the most views, and was thus a contents platform.

In the 2010s, economists and management scholars started applying the language of platform to any and every kind of intermediary; this is the origin of Steinberg's third discourse, "transactional and mediation platforms." The working definition of a platform was that it was "a set of products and services that bring together groups of users and providers to form multisided markets." By this definition, credit card providers like Visa and MasterCard, the Google search engine, dating websites, shopping malls, Microsoft Windows, YouTube, and much else, all became platforms that connected different sets of providers and users. The theorists of transactional and mediation platforms dissociated the keyword "platform" from its association with "technological infrastructure or base" and instead used it to signify a "relationship informed by a particular economic model."[36]

The theorists of platforms as multisided markets were often more interested in the question of "coordinating markets" than in the specifics of software, code, or standards, which played a key role in the discourse of product platforms.[37] For these theorists, the coordination of markets amounted to thinking about "network effects": how the number of people on one side of the multisided market can affect the other side and even the whole market. In their hands, and guided by their concern with network effects, even what traditional economists might have called "firms" became "platforms."

In taking up a more abstract definition of platforms, these theorists also labeled all the different audiences of the platform as "users."[38] For example, ordinary people, organizations, and advertisers can all be users of social media platforms like Facebook; both passengers and drivers are users of ride-hailing platforms like Uber; and restaurants, drivers, and diners are all users of food-delivery platforms like Grubhub. The term "user" thus also became dissociated from its technological use in computer systems design and instead signified a particular kind of transactional relationship between a participant in the multisided market and the platform.

It bears mentioning that while the term "platform" became abstract, the reason for its increasing use stemmed from the ways companies had drawn on the combination of GUIs (graphical user interfaces), APIs (application programming interfaces), and algorithms to reinvent commerce on the internet.[39] "The conditions for [the rise of] platform theory," Steinberg argues, "are technologically dependent, even if multisided markets are not themselves necessarily technological."[40] That is to say, it was the rise of web portals like Google, Facebook, and YouTube—all contents platforms—and the different ways they managed their varied users that inspired the theorization of these firms (and many others) as platforms.

As abstract platforms were implemented through concrete software, engineers and developers became increasingly concerned with the question of optimizing network effects and thereby increasing a platform's market share. This was already true in contents platforms by virtue of their nature, but this mode of analysis took hold in other types of firms and organizations as well.

This optimization was accomplished by using the tools and techniques of "analytics" and "A/B testing," which were built into software-driven platforms. "Analytics" refers to a software infrastructure that provides different actors in the platform several metrics and visualizations, often through "dashboards" that allow them to act in a data-driven fashion. For instance, the software Chartbeat provides metrics to journalists to help them understand the tastes and preferences of their audience;[41] journalists can see which articles are read more and how readers interact with them. This use of analytics in journalism has altered the job as journalists are now forced to grapple with the question of the goals of journalism: is the point of an article to gain the approval of peers or clicks from viewers? At least part of the job of an editor or a journalist becomes maximizing network effects, that is, getting an article widely read and shared.

A/B testing is a form of analytics but is worth considering separately given its widespread importance in the world of Silicon Valley.[42] An A/B test on the web is a randomized experimental comparison of a control and a treatment group in terms of their response to particular stimuli. Contents platforms like Google and Amazon, of course, use it to determine which features work best to maximize clicks and sales: from the order of search results to the right shade of blue for a particular interface. But A/B tests also made their way into sectors like journalism: now, publishers can experiment with headlines to see which gives them the most clicks from readers.

In addition to maximizing market share through network effects, the engineers, developers, and designers of platforms also focused their energies on maximizing "innovation." The concept of innovation became the focus of experts in the 1970s and '80s as the US economy was hit by stagflation and the postwar Keynesian consensus withered. Rather than seeing Keynesian industrial policy as the route to national greatness, economists and policy makers embraced "innovation" as the key ingredient of economic growth.[43] This argument was first made by economists in the academy and made its way out of the academy into consulting firms, think tanks, and policy making. In this view, it is innovation that drives economic growth and market share, and consequently, innovation must be excavated, nurtured, nourished, and allowed to flourish—something that would be done by "innovation experts."[44]

The purpose of platforms, then, became not just the maximization of market share but also the maximization of innovation, some of which was "user centered" and the rest of which came from "nudging" users—through GUIs, APIs, and algorithms—into performing actions that would produce network effects beneficial to the entire platform. Examples of user-driven innovation include the pioneering work of online communities of Wikipedians and open-source programmers, as well that of Twitter users in inventing retweeting and tagging.[45] But these users were also to be nudged into behaviors that would lead to better network effects. In the nudge model of governance, which was inspired by findings in social psychology and behavioral economics,[46] designers are seen as "choice architects" configuring the choices offered to users so that the most socially beneficial are most likely to be picked.[47] For example, Uber tries to nudge its drivers into particular behaviors that benefit its broader user-base, e.g. driving into areas *where* there is more demand for rides or extending their hours of work *when* there is more demand.[48]

All of these factors converge in the contemporary world of software production. Software designers want to build platforms to maximize market share and nurture "innovation." They believe that both these objectives can be accomplished together by allowing their "users" to behave autonomously while also guiding users' behavior so that it produces optimal social outcomes. Developers now track industriously what people do with their software through surveillance and A/B testing, sometimes preferring such tracking to asking users what they want. If users are not using a feature the way that developers would have liked them to, they tweak the feature. They liken digital data—the traces of user interactions in platforms—to the "voice" of the customer, and the data scientist to the "interpreter" of the customer's voice.[49]

Indeed, the agile software design process morphed to the point that software production and consumption became blurred through processes of user data collection and profiling. It is not software anymore that has to be agile; it is the business itself.[50] Consider the case of the Netflix recommendation algorithm, which recommends movies a viewer might want to watch based on their previous history.[51] Netflix began with a DVD rental business: it allowed its customers to keep a queue of DVDs that would be mailed to them in sequence. To come up with recommendations for its customers, Netflix modeled the problem as one of "accuracy": their job was to predict what their viewers wanted. Once Netflix's model shifted to online streaming, their goal changed to keeping their viewers watching their channel. This meant that the problem of recommendation was redefined in "captological terms": it was about guiding users and keeping them engaged, rather than accurately reflecting what the user wanted (which was hard to define anyway). This was a change not just in a software feature but in the entire business model.

I call this assemblage of technologies, knowledge practices, objects of concern, and expert roles—such as agile programming, A/B testing, analytics, user-centered innovation, nudging, and data scientists—as platform expertise. As engineers, developers, data scientists, and others seek to control the behavior of their users to maximize innovation and market share, they create new tasks and problems. They conceptualize their institutions as platforms and seek to produce optimum network effects among their user base. Guiding these users' behavior requires designers to produce knowledge about these users in an iterative fashion. The concept of design gets taken to its limit: everything can be designed, and the process of design never stops; both producers and consumers (users) continuously participate in it; and continuous design is the key to innovation and ultimately to value.

The foregoing account sets the context for the 2012 *Harvard Business Review* piece by the management scholars Andrew McAfee and Erik Brynjolfsson, on the new paradigm of data-driven management.[52] The authors argue that the rise of Big Data has transformed—and *should* have transformed—how management is done. Rather than making decisions based on the "HiPPO" ("highest-paid person's opinion"), the authors argue that managers should make decisions instead based on the insights and advice of a class of "data scientists" tasked with analyzing all the voluminous data that firms now collect about the customers and their behavior. This, in short, is the already happening but also prophesied "management revolution."

This piece and its predecessor from 1958 bear some similarities and some differences with each other. The similarities lie in the promise of the technology—information technology in 1958 and Big Data in 2012—and in the experts who need to be bolstered to reap the most benefits from the technology: programmers in 1958 and data scientists in 2012. The variances, however, come from the fact that the promises of the technology and the experts are framed differently.

Consider the experts. In the 1958 paper, the promise of the programmers was that they could effectively reduce line managers in the middle of the organizational pyramid; because these programmers were experts on information processing, they could effectively program rules for information that ranged from finance to strategy to personnel. The 1958 piece makes an offhand reference to the possibility that decisions about running electricity generation power plants might be "made centrally by individuals whose technical skills are in mathematics and computer programing, with absolutely no experience in generating stations."[53]

The 2012 piece, in contrast, is careful to qualify—if not in a totally satisfactory way—the jurisdiction of "data scientists" in an organization.

For instance, the authors argue that data scientists should be guided by "domain experts," who are those "deeply familiar with an area" whose job "remains critical" because they are still "the ones who know where the biggest opportunities and challenges lie." These domain experts should view themselves not as a source of "HiPPO-style answers" but as a source of *questions* because, channeling Pablo Picasso, "Computers are useless. They can only give you answers."[54]

Big Data–driven systems and analysis in 2012 can thus be considered the equivalent of the MIS from the 1950s and '60s in that it promises to revolutionize management. But there is no equivalent master concept like *program*—the science of managing information flows—that undergirds Big Data analysis. That said, just like the programmers back in 1958, data scientists need to be found, trained, and cultivated, the 2012 piece argues. These data scientists are not just statisticians, and they are not just computer programmers. They need a range of technical skills related to data but also business skills in that they should be "comfortable speaking the language of business and helping leaders reformulate their challenges in ways that big data can tackle." But there is no master skill that data scientists possess or a master concept behind Big Data analysis.

Also, the potential of Big Data analysis lies not in the rationalization of decision making that MIS promised; rather, it is part of a broader tool kit of iterative design with which to maximize innovation and market share. As McAfee and Brynjolfsson put it in another book aimed at managers titled *Machine Platform Cloud: Harnessing Our Digital Future*: "We will try to convince [managers] that because of recent technological changes, companies need to rethink the balance between minds and machines, between products and platforms, and between the core and the crowd."[55] In other words, managers need to do a lot more than just hire data scientists and analyze Big Data. They must rethink the process of design and innovation itself by shifting toward designing platforms rather than products and being open to harnessing the energies of users (or the "crowd") rather than just their designers.

In sum, then, technical reformers arrived at a new framework for reinventing managerial decision making that I call the assemblage of platform expertise. Ideas about platform design, user-led innovation, and Big Data analysis are all central to this assemblage.

Open Online Education Manifested as Programming and Platform Expertise: The Cases of OLI and edX

In this section, I describe the specific instantiations of Programming and platform expertise as they pertain to the case of open online learning. A Carnegie Mellon University project from the 2000s—one that still exists—called the Open Learning Initiative (OLI) serves as a good example of the principles of Programming, and the case of MOOCs, and particularly the

MOOC company edX, illustrates the workings of platform expertise. Each case reveals technical reformers drawing on different techniques and focusing on different objects of interest. Each project was successful and unsuccessful in various ways, and this mixed success can be traced to the assemblage it was a manifestation of. The OLI project did not catch on as expected, because it was based on the premise that there was one style of pedagogy best suited to learning, a style that required instructors to collaborate with cognitive scientists (they took on the role of Programmers) to reinvent their pedagogy in a specific style. When such courses were built, they gathered acclaim, but the process seems to have alienated some instructors and made the project of commercializing OLI highly difficult. The MOOC company edX, on the other hand, left the question of pedagogy to its partners, which meant that it was able to expand the number of partners and achieve a sustainable business model. However, this process essentially meant that its claim of "reinventing education" was not fulfilled in any deep sense.

The Open Learning Initiative consisted of a few open, online classes at Carnegie Mellon University (CMU). OLI was started by a group of researchers at CMU in 2001 with funding from the Hewlett Foundation. Its creation was in response to Massachusetts Institute of Technology's (MIT) release of Open Courseware (OCW), a remarkably successful initiative in which MIT released course materials for most of its courses to the world. OCW became so well known that it became the basis of an entire open education resources (OER) movement. MIT emphasized that Open Courseware was simply about releasing course resources rather than actual courses that students could take; its intended recipients were not only students who might not have access to these resources, such as those not able to attend MIT, but also, and more important, teachers who could use these resources to teach their own classes.

Unlike MIT, CMU chose to build actual standalone *courses* (i.e., courses that worked without instructors). This is to say that the courses consisted not just of resources that autodidactic students could use as they pleased but of activities and assessments for students that would accomplish a learning goal and allow students to learn certain skills.[56]

To build these self-standing open courses, the OLI researchers drew on their expertise in cognitive science, particularly the subfield of intelligent tutoring systems (ITSs). ITSs were software programs, often designed for middle-school mathematics and science, that functioned as tutors, emphasizing the pedagogical principles of deliberate practice and scaffolding. CMU cognitive science researchers had already successfully built and commercialized such tutors by the end of the 1990s.[57] Students were offered practice examples in order of difficulty. These examples were constructed to build certain basic "skills" in students that mapped to the subject matter (e.g., high school physics or mathematics). As students worked through the examples, they were offered hints and then more examples to practice with. The tutor figured out the student's mastery of a skill; the examples and hints it offered students were based on the tutor's estimate of a student's skill in relation to the intended learning objectives. The tutor was thus not just an interactive textbook; it used sophisticated computational models to calculate what the student needed at every step.

The tutors the team built were based on a particular vision of teaching: an instructional paradigm that focused on less lecturing and talking and more "problem-based learning-by-doing." In a conference paper published in 1992, the lead researcher, John Anderson, had called the "mathematics educators" wrong in arguing that the reason US students lagged behind their counterparts in other countries in their performance on standardized tests was the defective curricula used by American schools. Anderson argued instead that "changing curriculum, lowering class size, and improving teacher quality will do little as long as there is less effective time spent *learning*."[58] The tutors were developed so that students spent their time more efficaciously through problem-based learning-by-doing.

This learning-by-doing focus meant that when it came to the Open Learning Initiative, courses could not just be taken from their prior contexts and turned into an OLI course (unlike, say, Open Courseware, which essentially involved putting the course syllabus and assignments online). The objective was to completely transform course content through the application of cognitive science and design principles. What did this process entail exactly? The learning goals of the course (e.g., middle school mathematics) had to be broken down into "knowledge components," which were specific skills (e.g., adding single-digit numbers) that students needed to learn. The interdependencies and hierarchical relationships between skills had to be completely mapped out. For each skill, the team had to build instructional material and events—practice problems, hints, and so on. The highlight of the OLI model of pedagogy was that it discarded passive forms of instruction (such as lectures or reading) for active learning by doing (making students engage in deliberate practice to acquire a hierarchy of skills).

Building an OLI course, thus, was not just a task for an individual instructor; rather, it required a variety of experts to come together and collaborate. An OLI course was built through long hours of collaboration among: (1) the instructor or a subject matter expert who knew the domain (e.g., high-school algebra), (2) a cognitive scientist who understood the process of "learning," which was considered independent of the subject matter or domain, and (3) computer programmers or technology experts who built the actual course. A key part of the building process was to engage in cognitive task analysis, a form of analysis in which the subject matter expert and a novice are made to talk through their problem-solving strategies. Once the skills had been decomposed and the forms of instruction defined, the course was built and implemented. Learning was measured by the degree to which students were able to do something that they could not before.[59]

Making an OLI course thus took considerable time, money, and resources; it was extremely labor-intensive. The Carnegie Mellon cognitive scientists had estimated that it took "200 hours" of labor "to deliver a full hour of instruction" for an intelligent tutoring system.[60] The number was similar for OLI; Candace Thille, the head of OLI, told historian Taylor Walsh that the OLI course construction process was a "*huge* investment of time and resources," especially for the instructor or subject matter expert. It was also a huge investment of money: it could take as much as US$1 million to develop a single course.[61]

The Open Learning Initiative, then, stands as an example of Programming applied to the design of an open online course. How so? Superficially, there are its links to cognitive science. But, more specifically, an OLI course and the ITSs it is based on are organized around the concept of Programming, the idea of decomposing tasks into rules.[62] The theory of pedagogy underlying OLI assumes that experts on a particular topic can execute certain skills competently because their internal program runs well; an expert's ability to execute a skill is modeled as an algorithm and thus constructed as a program. The expert can do what he does because of a series of interconnected programs he runs. Thus, the job of the OLI course is to aid novice learners to build up this program in their own heads. It is for this reason that building an OLI course requires a group of experts—the cognitive scientist, the subject matter expert, and a programmer—to carry out cognitive task analysis. Their goal is to build a model of decision making: what is the next instruction to be given to the novice learner that will help build his skills? This question is answered by drawing on the cognitive model on which the execution of skilled tasks is based as a program.

Within the division of labor for building the course and its minitutors, the final authority is the learning-focused cognitive scientist who is supposed to be an expert on the cognitive model itself and serves as the authoritative Programmer. The subject matter expert's proficiency in the subject matter and the technologist's proficiency in building software are important but less critical than the cognitive model.

This interpretation of OLI as an instantiation of Programming helps explain the "successes" and "failures" of OLI. Its successes include a concrete demonstration, through a randomized control trial, that its signature statistics course actually improved learning outcomes for students, even better than comparable in-person courses. OLI's biggest moment of triumph came with publication of a 2014 randomized controlled trial led by the labor economist William Bowen.[63] In this study, the researchers assigned students from six public universities to either the OLI statistics course (with one optional hour of face-to-face meetings) or the regular statistics course (three hours of face-to-face meetings). They found that the learning outcomes in both cases were comparable but that the students who took the OLI course achieved the same outcomes in far less time. This study also pivoted OLI into the national spotlight—it had not achieved as much publicity during its initial years, certainly not in comparison to OCW, the competition against which it was created—and it was often mentioned in conjunction with the Obama administration's initiative to improve community college education. Among both proponents and skeptics of online learning, OLI was—and is, today—an example of online learning done right.[64]

Yet OLI itself ended up building up only a few more courses, most of them in technical domains such as statistics and biology. Part of the problem was that its organizational and pedagogical model was hard to scale: the huge amount of funding and labor involved in transforming a course to fit OLI principles was a key barrier to adoption.

But a bigger problem was that the validity of the theory of learning that informed an OLI course, and the practices associated with it such as cognitive task analysis, remained contested, with instructors and subject matter experts lining up on one side and cognitive scientists on the other. Many instructors, even within the sciences, let alone the humanities, resisted transforming their courses through OLI theories, refusing to accept the authority of this particular brand of cognitive science that underlay OLI. They did not care for a wholesale transformation of their course through team effort.

But even the cognitive scientists working with OLI sometimes disagreed on whether their theory and practice would work for *all* courses. At least some believed that they would not work for courses, such as those in the humanities and social sciences, that could not be boiled down to component skills and practice problems. As an OLI researcher noted: "One of the keys to success [in designing asynchronous online learning environments] is cognitive task analysis—having a deep understanding of the knowledge that differentiates an expert or good student from a novice or beginning student. And in a lot of humanities domains, teachers would say "I know it when I see it, but I can't describe it"—and if you can't describe it, it's hard to design activities around it.[65] The theory of pedagogy underlying OLI thus may not work when the subject matter expert cannot decompose a topic into skills and be able to clearly differentiate a novice from an expert. Even William Bowen, who authored the study that showed that OLI could indeed achieve the learning outcomes it promised, argued that though he was an admirer of OLI and its goals, "[it] is far from clear to me that the OLI approach would be as useful in teaching subjects that generally require much more nuanced discussion, such as literature, ethics, and international politics."[66]

Where is the Open Learning Initiative today? Despite the general acclaim it received and the scientific publications it yielded, OLI failed to expand beyond certain kinds of courses and also failed to find a sustainable business model, all of which led to notable tensions between the OLI personnel and Carnegie Mellon administrators.[67] The aim of reinventing a course through collaboration between cognitive scientists and faculty meant that the course production process was both expensive and time-consuming. When foundation funding ran out, CMU administrators spun off the underlying OLI software (not its content, which is open) into a new, for-profit startup called Acrobatiq. The OLI personnel moved to Stanford in 2012, and a new Stanford-based OLI is being created.[68]

OLI's successes and failures become explainable if we see it as an instantiation of the Programming assemblage. In OLI, courses were to be "reprogrammed" by drawing on a particular theory of learning. This endeavor proved wildly successful for a few courses but could not be scaled. Just as middle managers and accountants questioned whether computer specialists' "information" expertise enabled them to program personnel management or finance, subject matter experts questioned whether the theory of learning could be applied to all courses. In a sense, OLI's failure to expand stemmed from the high costs (in time, money, and effort) of producing a course and from the fact that it required instructors and subject matter experts to accept the supremacy of a particular cognitive science–driven theory of learning.

The MOOC company edX offers an interesting contrast to OLI as an instantiation of platform expertise rather than Programming.[69] Unlike OLI, edX was organized around the concept of the platform rather than that of the program; the edX designers conceptualized the problem of learning in terms of platforms managing their users. Unlike OLI, edX created a sustainable business model, but it is not at all clear that edX has managed to transform education (unlike OLI, which was able to demonstrate that its pedagogical style did improve outcomes for certain kinds of courses).

What were MOOCs? In 2011, Stanford University caused a storm by releasing three courses out into the wild—courses that combined lectures and interactive assessments and could be taken by anyone with an internet connection. These courses ended up being christened massive open online courses. Inspired by Stanford, MIT offered up its own MOOC, "Circuits and Electronics," in the spring of 2012, headed by computer science professor Anant Agarwal. This unit was formally spun off as the nonprofit startup edX in May 2012 and financed with US$30 million each from Harvard and MIT. This frenzied activity led the *New York Times* to label the year 2012 as the "year of the MOOC."[70]

The founders of edX and other MOOC companies took inspiration from existing internet platforms where they saw learning happen in both formal and informal ways; for example, Khan Academy, with its mathematics video tutorials and assessments for K–12 students; YouTube and Lynda.com, with their extensive range of user-generated how-to videos on topics ranging from applying makeup to filing taxes; and Stack Overflow, with its crowdsourced format of users answering one another's questions. They saw these portals as crucibles in which multiple publics were forging new forms of (educational) participation. They sought to bring these practices into formal classroom learning. They also hoped to reinvent educational research by drawing on the knowledge practices pioneered in platforms such as user-centered design, data science, and A/B testing.

Early in its trajectory, edX's senior management decided that, for the company to become financially sustainable, it would have to "scale up" by recruiting institutional partners who would offer an ever-expanding array of content. To achieve this goal, edX would have to provide its partner institutions with software that would allow them to produce new courses, credentials, and educational research. Partners would pay edX to use the company's software and host their courses on its portal, and some learners would pay edX to earn a certificate (and some of this money would be funneled back to partners).

In recruiting partner institutions and educators to its MOOC enterprise, edX promoted a reformist mission: "reinventing education." Thus, when edX began as an organization, its "About Us" page listed three key company goals:

> Expand access to education for everyone.
> Enhance teaching and learning on campus and online.
> Advance teaching and learning through research.[71]

The slogan—and the three attendant goals—carried a double meaning: edX and its partners would not only reinvent the higher-education system by making courses available to learners outside traditional university settings but would also reinvent the practices traditionally associated with teaching and education research.

Partners (i.e., users) had a key role to play in this mission: instructional teams and researchers at partner institutions could devise novel methods of teaching and learning that edX's management had yet to imagine. Together, edX and its users would constitute a community in pursuit of a shared goal.

EdX's two institutional prerogatives—its pursuit of the organization's financial sustainability and its broader social agenda of reforming education—set up management's central challenge. The company had to encourage instructors and researchers at edX's partner institutions to generate innovative courses and research while also ensuring that edX itself maintained control over the terms of the reform agenda and broke even financially. As Beth Porter, edX's first head of Product recounted, edX's primary challenge was to balance its vision for "innovation in education" with market considerations:

> [W]here we have most tensions actually, is that there [are] these two parts of our mission, one of them is really mission-driven, right— innovation in education, trying to really chart a new path, trying to lead our partners, into different and interesting forms of [. . .] deliver[ing] content, reach[ing] broader audiences, find[ing] a new revenue stream. There's all of that, and there's really just pure [. . .] commerce, [. . .] becoming a consumer product. And that, I think, is probably where more tension exists in our ecosystem than anyplace else.[72]

To solve this challenge and scale up the number of its partners, edX adopted the "platform" model of organization in which edX engineers would build software that instructors and researchers at partner institutions then could draw on autonomously, for their own goals and purposes, to produce educational content and educational knowledge. Thus, edX had settled on a division of labor wherein the company would produce software that educators at partner institutions would use to "author" courses and conduct research (i.e., they became "users" of edX's software).

How would the platform model help solve the twin challenges of financial sustainability and reform? By expanding the number of partners, edX would amass many courses on its portal, increasing the possibility that at least some learners might pay a small fee to get a "verified certificate." By having many partners, edX could also make sure that edX courses were about different kinds of content (humanities, sciences, technical skills) and that they embodied different styles of pedagogy and underlying theories of learning. Finally, because one standardized software was the substrate for all these courses, the software itself could be a channel of transformation in that by building specific software features, edX could transform the behavior of learners, instructors, and researchers.

The platform model of organization thus had two facets: (1) It was pedagogically neutral in that, unlike OLI, edX did not specify any kind of learning theory that courses from partners had to conform to. EdX partners could use the features of its software in any way they wished. They gained a measure of autonomy while also being subjected to a set of constraints. (2) At the same time, the model set up a central axis of conflict within the ecosystem. Who would decide which new features were built into the edX software? What if partners had divergent demands or conflicting ideas about what counts as "reinventing education"? What would happen if partners disagreed with edX's decisions and priorities? What if the pursuit of reform and sustainability led to two different sets of priorities? The questions of which software features to prioritize and how exactly the software features would work were often a bone of contention through which larger questions about sustainability and reform were indexed.

Making this platform model work required both technical and social labor on the part of edX employees. On the technical side, programmers working at edX built many different user interfaces that turned instructors at partner institutions into self-sufficient actors who could build courses with as little hand-holding as possible from the edX organization. On the other hand, another set of edX employees, who had titles like "program manager" and "relationship manager," took on the task of managing the demands of the "users," different instructional and research teams at partner institutions. These workers produced an extensive infrastructure to support these teams in their duties, including documentation on how to use the software, advice on best practices, checklists, project management tools, and road maps. They managed the expectations of teams at partner institutions while also acting as advocates for them within the edX organization. The work of these managers was often crucial to managing the conflicting demands between edX and its users, or between users.

In what ways did this platform model adopted by edX "succeed" or "fail"? There are two dimensions to this question.

The first dimension concerns edX's overarching goal of "reinventing education." It is safe to say that edX—or MOOC companies in general—failed at this goal. On the credential side, edX and its partners created several MicroMasters online degree programs, a suite of courses that would result in a digital certificate, that proved somewhat popular with working professionals interested in building new skills. On the research side, the data produced through edX courses resulted in some new breakthroughs in educational research. But overall, there was no big transformation of higher education or educational research. Elite, selective colleges continued to attract students. Public universities using MOOCs in their curriculum found the process much more labor-intensive and costly than expected. And the MOOC companies themselves were unable to persuade students to finish their MOOCs in sufficient numbers to make the courses profitable or even financially sustainable. By 2019, it was clear that MOOCs had settled into an established business model: "helping universities outsource their online master's degrees for professionals."[73]

The second dimension relates to edX's immediate goal of creating a portal with a large number and variety of courses in the pursuit of financial sustainability. As table 13.2 shows, the number of edX courses increased dramatically: from just 3 partners and 7 courses when it began, the edX portal had more than 650 courses from more than 91 partners by January 2016, just four years later. While this model was never fully

Table 13.2. The steep growth in the number of edX partners and courses

Date	Number of partner institutions	Number of courses
August 2012	3	7
December 2012	6	15
May 2013	12	37
October 2013	21	86
June 2014	38	194
December 2014	64	383
May 2015	73	519
December 2015	90	650
January 2016	91	650+

financially sustainable, it was a much more sustainable model than OLI (and considered business-worthy enough that the online program manager 2U acquired edX in 2021).

While both edX and OLI succeeded and failed in their own ways, the point remains that their adoption differed dramatically, and this point can be attributed primarily to the underlying assemblage they embodied. OLI was committed to a very particular theory of learning by doing, and creating an OLI course meant submitting to the authority of a certain brand of cognitive science; it also meant going through an expensive, not to mention time- and effort-intensive, process of course making. Institutionally, OLI was a high-prestige project, and the focus on making it financially sustainability came much later in its life.

Participating in an edX course, on the other hand, did not require committing to a particular model of pedagogy. The course-making process could be elaborate or spare, video driven or not, very expensive or reasonably so. The programmers and managers at edX, however, did not think of their role as "programming" pedagogy, that is to say, choosing between different models of pedagogy for their partners. Rather, they thought of their role in the ecosystem as providing software-based "tools" to the users of their software (partners, instructors, researchers, and eventually learners), watching over these users, and disseminating what they considered best practices to them by both by modifying their underlying software and engaging these users discursively through documentation, seminars, phone meetings, virtual meet-ups, and so forth. In other words, the edX management and employees saw themselves as manager-innovators who guided their flock of users onto mutually beneficial roads, balancing financial sustainability and reform while still granting them some semblance of autonomy and creativity. The operation of edX is an instantiation of platform expertise in action.

Conclusion

This chapter argues that there has been a shift in how technical reformers draw on the digital computer in their effort to reinvent organizations.

From the 1950s onward, reformers drew on the concept of the program to construct organizations as information systems; for them, the key problem to be solved was to help organizations make decisions rationally. To do this, organizations had to prioritize new experts such as "systems men" and new technical systems such as MIS, through which an organization could be reprogrammed and its information flows integrated—a form of expertise that I call Programming. However, these efforts did not quite lead to the outcomes their proponents wished for. Systems resembling MIS were finally built and computer programming did become a high-status occupation, but the idea that programmers were the experts who could rewire the information flows in an organization did not catch on.

The rise of networked computing in the 2000s has led to a similar moment. Today, in many organizations and scientific fields, there is a great deal of excitement around the practices of "data science" and human-centered design, forms of knowledge production in which experts like data scientists or the innovation consultants work fast, are practical and open-ended with respect to outcomes, and subject every decision to a data-driven experimental test. Technical reformers argue that these practices—the assemblage of which I call platform expertise—can help reinvent management. But the reinvention of management here involves not making optimal decisions but governing users in an open-ended manner in the pursuit of innovation and market share.

Will platform expertise lead to a revolution in management? Will all organizations twenty years from now be saturated with data scientists and user experience designers? Time will tell. The story of Programming, at least, suggests that the future will not look exactly the way it has been prophesied.

Notes

1. Harold J. Leavitt and Thomas L. Whisler, "Management in the 1980's," *Harvard Business Review* 36, no. 6 (November 1958): 41–48, https://hbr.org/1958/11/management-in-the-1980s.

2. Andrew McAfee and Erik Brynjolfsson, "Big Data: The Management Revolution," *Harvard Business Review* 90, no. 10 (October 1, 2012): 61–67, https://hbr.org/2012/10/big-data-the-management-revolution.

3. In his history of corporate computer programming in the 1950s and 1960s, the historian Thomas Haigh calls them "managerial technicians." Thomas D. Haigh, "Technology, Information, and Power: Managerial Technicians in Corporate America, 1917–2000" (PhD diss., University of Pennsylvania, 2003), https://www.proquest.com/docview/305306390.

4. Manuel DeLanda, *A New Philosophy of Society: Assemblage Theory and Social Complexity* (New York: Continuum, 2006).

5. Stephen J. Collier and Andrew Lakoff, "On Regimes of Living," in *Global Assemblages: Technology, Politics, and Ethics as Anthropological Problems*, ed. Aihwa Ong and Stephen J. Collier (Hoboken, NJ: John Wiley & Sons, 2008), 22–39.

6. Nathan Ensmenger, *The Computer Boys Take Over: Computers, Program-

mers, and the Politics of Technical Expertise (Cambridge, MA: MIT Press, 2010); Mar Hicks, *Programmed Inequality: How Britain Discarded Women Technologists and Lost Its Edge in Computing* (Cambridge, MA: MIT Press, 2017).

7. Gil Eyal, "For a Sociology of Expertise: The Social Origins of the Autism Epidemic," *American Journal of Sociology* 118, no. 4 (2013): 863–907.

8. Stephen J. Collier, "Topologies of Power: Foucault's Analysis of Political Government beyond 'Governmentality,'" *Theory, Culture, and Society* 26, no. 78 (2009): 93–96, https://doi.org/10.1177/0263276409347694.

9. Eyal, "For a Sociology of Expertise," 888 (emphasis added).

10. Thomas D. Haigh, "Inventing Information Systems: The Systems Men and the Computer, 1950 1968," *Business History Review* 75, no. 1 (2001): 15–61, https://doi.org/10.2307/3116556; Hicks, *Programmed Inequality*.

11. Haigh, "Technology, Information, and Power"; Ensmenger, *The Computer Boys Take Over*.

12. Joy Lisi Rankin, *A People's History of Computing in the United States*, illustrated ed. (Cambridge, MA: Harvard University Press, 2018). If anything, these cultures embodied platform expertise because the creators of these systems were most interested in user-driven innovation. See, for example, the detailed story of the PLATO computer network in Brian Dear, *The Friendly Orange Glow: The Untold Story of the Rise of Cyberculture* (2017; reprint, New York: Vintage Books, 2018).

13. Hunter Heyck, "Producing Reason," in *Cold War Social Science: Knowledge Production, Liberal Democracy, and Human Nature*, ed. Mark Solovey and Hamilton Cravens (New York: Palgrave Macmillan, 2012): 99–116, https://doi.org/10.1057/9781137013224_6.

14. Heyck, "Producing Reason," 102.

15. For example, the cognitive scientist Herbert Simon argued that human beings were "satisficing" agents, while behavioral psychologists Daniel Kahneman and Amos Tversky argued that human beings were systematically irrational in that they were hardwired into making certain kinds of decisions.

16. Hunter Heyck, "Defining the Computer: Herbert Simon and the Bureaucratic Mind–Part 1," *IEEE Annals of the History of Computing* 30, no. 2 (2008): 42–51, https://doi.org/10.1109/MAHC.2008.18; Hunter Heyck, "Defining the Computer: Herbert Simon and the Bureaucratic Mind–Part 2," *IEEE Annals of the History of Computing* 30, no. 2 (2008): 52–63, https://doi.org/10.1109/MAHC.2008.19.

17. Heyck, "Defining the Computer," part 2, 59.

18. Leavitt and Whisler, "Management in the 1980's," 47 (emphasis added).

19. Haigh, "Inventing Information Systems."

20. Haigh, "Technology, Information, and Power," 409.

21. Haigh, "Inventing Information Systems," 52.

22. Ensmenger, *The Computer Boys Take Over*.

23. Ensmenger has argued that computer programmers themselves were never able to resolve the problem of whether being a programmer was an innate skill, that is, a "black art" rather than one that could be taught. Ensmenger, *The Computer Boys Take Over*.

24. Haigh, "Technology, Information, and Power."

25. Contra Leavitt and Whisler, "Management in the 1980's."

26. Haigh, "Technology, Information, and Power."

27. Martin Campbell-Kelly, *From Airline Reservation to Sonic the Hedgehog: A History of the Software Industry* (Cambridge, MA: MIT Press, 2004).

28. Thierry Bardini and August T. Horvath, "The Social Construction of the Personal Computer User," *Journal of Communication* 45, no. 3 (1995): 40–66.

This vision was different from the MIS, which would produce reports for top-level management that would allow them to make decisions about how the organization should work. The MIS was not meant for communication between employees; most employees did not even use the digital computer to do their work.

29. For more on this user focus, see Mike Hales, "Where Are Designers? Styles of Design Practice, Objects of Design, and Views of Users in CSCW," in *Design Issues in CSCW*, ed. Duska Rosenberg and Chris Hutchison (London: Springer, 1994): 151–77, http://link.springer.com/chapter/10.1007/978-1-4471-2029-2_8; Hugh Mackay, Chris Carne, Paul Beynon-Davies, and Doug Tudhope, "Reconfiguring the User: Using Rapid Application Development," *Social Studies of Science* 30, no. 5 (2000): 737–57; Philip E. Agre, "Conceptions of the User in Computer Systems Design," in *The Social and Interactional Dimensions of Human-Computer Interfaces*, ed. Peter J. Thomas (Cambridge: Cambridge University Press, 1995): 67–106; Thierry Bardini and August T. Horvath, "The Social Construction of the Personal Computer User," *Journal of Communication* 45, no. 3 (1995): 40–66; Steve Woolgar, "Configuring the User: The Case of Usability Trials," in *A Sociology of Monsters: Essays on Power, Technology, and Domination*, ed. John Law (London: Routledge, 1991).

30. Neil Pollock, Robin Williams, and Luciana D'Adderio, "Global Software and Its Provenance: Generification Work in the Production of Organizational Software Packages," *Social Studies of Science* 37, no. 2 (2007): 254–80.

31. Mackay et al., "Reconfiguring the User," 737–57.

32. Marc Steinberg, *The Platform Economy: How Japan Transformed the Consumer Internet* (Minneapolis: University of Minnesota Press, 2019), 72.

33. See, for example, Michael A. Cusumano and Annabelle Gawer, "The Elements of Platform Leadership," *MIT Sloan Management Review*, April 15, 2002, https://sloanreview.mit.edu/article/the-elements-of-platform-leadership/.

34. Tarleton Gillespie, "The Politics of 'Platforms,'" *New Media and Society* 12, no. 3 (2010): 347–64, https://doi.org/10.1177/1461444809342738.

35. Steinberg, *Platform Economy*, 89.

36. Steinberg, *Platform Economy*, 97, 100.

37. Steinberg, *Platform Economy*, 100.

38. See, for example, Thomas R. Eisenmann, Geoffrey Parker, and Marshall Van Alstyne, "Opening Platforms: How, When, and Why?" in *Platforms, Markets, and Innovation*, ed. Annabelle Gawer (Cheltenham, UK: Edward Elgar, 2009).

39. See, for example, Anne Helmond, "The Platformization of the Web: Making Web Data Platform Ready," *Social Media + Society* 1, no. 2 (July 2015): https://doi.org/10.1177/2056305115603080 on Facebook. The GUI allows people to connect to databases, whereas APIs allow two programs to share data. Together, GUIs, APIs, and algorithms have allowed for a dramatic expansion of services such as Facebook, as well as the rise of new forms of industrialized crowd labor. Helmond, "Platformization of the Web"; Mary L. Gray and Siddharth Suri, *Ghost Work: How to Stop Silicon Valley from Building a New Global Underclass* (Boston: Houghton Mifflin Harcourt, 2019).

40. Steinberg, *Platform Economy*, 104.

41. Angele Christin, *Metrics at Work: Journalism and the Contested Meaning of Algorithms* (Princeton, NJ: Princeton University Press, 2020).

42. Brian Christian, "The A/B Test: Inside the Technology That's Changing the Rules of Business," *Wired*, April 25, 2012, https://www.wired.com/2012/04/ff-abtesting/.

43. Elizabeth Popp Berman, *Creating the Market University: How Academic*

Science Became an Economic Engine (Princeton, NJ: Princeton University Press, 2011).

44. Lee Vinsel and Andrew L. Russell, *The Innovation Delusion: How Our Obsession with the New Has Disrupted the Work That Matters Most* (New York: Currency, 2020), 19–36. The great companion to innovation is "disruption." The theory of "disruptive innovation" dates to the publication of Clay Christensen's *The Innovator's Dilemma: When New Technologies Cause Great Firms to Fail* (Cambridge, MA: Harvard Business School Press, 1997). Using the example of the hard-disk industry, Christensen theorized that it is very difficult for market leaders to pay attention to new techniques that may upend their business model completely; this difficulty is because these companies already have much invested in their current way of doing things, which means that the new technology is often picked up and capitalized on by an upstart new company, which may end up putting the incumbent out of business. From this initial beginning, which most historians of technology would presumably agree with, the ideology of disruption has managed to reach almost prescriptive status within the world of Silicon Valley, mostly from Christensen's own efforts to cash in on the success of his first book. From an initial implied prescription that companies remain sensitive to new, potentially disruptive technologies, the theory of disruption now includes everything from the way companies can manage disruption to the way they can intentionally set out to disrupt entire industries.

45. See Eric Von Hippel, *Democratizing Innovation* (Cambridge, MA: MIT Press, 2006); Jean Burgess and Nancy K. Baym, *Twitter: A Biography* (New York: New York University Press, 2020).

46. See, for example, Richard H. Thaler and Cass R. Sunstein, *Nudge: Improving Decisions about Health, Wealth, and Happiness* (New York: Penguin Books, 2009).

47. For example, designers of retirement plans could set the defaults for their employees plans so that they are automatically enrolled, given that employees rarely change their defaults; designers of school cafeterias might choose to put more nutritious foods in places where students are more likely to find them.

48. Alex Rosenblat, *Uberland: How Algorithms Are Rewriting the Rules of Work* (Oakland: University of California Press, 2018).

49. See, for example, Riley Newman, "At Airbnb, Data Science Belongs Everywhere: Insights from Five Years of Hypergrowth," *Airbnb Engineering and Data Science* (blog), July 7, 2015, https://medium.com/airbnb-engineering /at-airbnb-data-science-belongs-everywhere-917250c6beba.

50. Seda Gurses and Joris van Hoboken, "Privacy after the Agile Turn," in *The Cambridge Handbook of Consumer Privacy*, ed. Jules Polonetsky, Omer Tene, and Evan Selinger (Cambridge: Cambridge University Press, 2017), https://doi .org/10.31235/osf.io/9gy73-.

51. Nick Seaver, "Captivating Algorithms: Recommender Systems as Traps," *Journal of Material Culture* 24, no. 4 (2018): 421–36.

52. McAfee and Brynjolfsson, "Big Data."

53. Leavitt and Whisler, "Management in the 1980's," 43.

54. McAfee and Brynjolfsson, "Big Data." 66. Several questions, of course, remain about the division of labor between the domain experts and the data scientists. Why do domain experts know the right questions to ask? The answer seems to be that they know from their long, embodied experience in the domain, which is also presumably the source of their opinions. But their opinions are not deemed good enough to be the source of answers, which is a job for the data scientists. On the other hand, it's not clear why the domain experts need the

data scientists to answer these questions. Couldn't they just answer them themselves using data? The answer seems to be that the domain experts lack the technical skills to carry out Big Data analysis, but it's not clear what's stopping them from acquiring this ability.

55. New York: W. W. Norton, 2017.

56. The account of OLI offered here draws mostly on Taylor Walsh, *Unlocking the Gates: How and Why Leading Universities Are Opening Up Access to Their Courses* (Princeton, NJ: Princeton University Press, 2010); and Shreeharsh Kelkar, "Between AI and Learning Science: The Evolution and Commercialization of Intelligent Tutoring Systems," *IEEE Annals of the History of Computing* 44, no. 1 (2022): 20–30, https://doi.org/10.1109/MAHC.2022.3143816. Some other sources are referenced in the narrative.

57. Kelkar, "Between AI and Learning Science."

58. John R. Anderson, "Intelligent Tutoring and High School Mathematics," in *Intelligent Tutoring Systems: Second International Conference, ITS '92*, ed. Claude Frasson, Gilles Gauthier, and Gordon I. McCalla (Berlin: Springer Verlag, 1992): 2 (emphasis mine).

59. Note that cognitive task analysis itself comes from the field of expert systems in AI research, which is based on the idea that the skills shown by experts can be decomposed into a series of explicit rules that can be articulated as a computer program. For further discussion, see Kelkar, "Between AI and Learning Science."

60. Albert Corbett, Kenneth Koedinger, and John Anderson, "Intelligent Tutoring Systems," in *Handbook of Human-Computer Interaction*, 2nd ed., ed. Martin G. Helander, Thomas K. Landauer, and Prasad V. Prabhu (Amsterdam: Elsevier, 1997), 863.

61. Walsh, *Unlocking the Gates*, 100–101 (emphasis in original). In contrast, it cost MIT only about $25,000 for an OCW course.

62. In the trivial sense, of course, an OLI course is ultimately a piece of software. But the idea of a program is much deeper than just a piece of software.

63. William G. Bowen, Matthew M. Chingos, Kelly A. Lack, and Thomas I. Nygren, "Interactive Learning Online at Public Universities: Evidence from a Six-Campus Randomized Trial," *Journal of Policy Analysis and Management* 33, no. 1 (2014): 94–111, https://doi.org/10.1002/pam.21728.

64. Justin Reich, *Failure to Disrupt: Why Technology Alone Can't Transform Education* (Cambridge, MA: Harvard University Press, 2020), 39–40.

65. Walsh, *Unlocking the Gates*, 98.

66. Walsh, *Unlocking the Gates*, xii.

67. Walsh, *Unlocking the Gates*, 114–21.

68. See Goldie Blumenstyk, "As Big-Data Companies Come to Teaching, a Pioneer Issues a Warning," *Chronicle of Higher Education*, February 23, 2016.

69. Unlike the story of OLI, which is based on secondary sources, the story of edX that I tell here is based on ethnographic fieldwork that I conducted at the MOOC company edX and four of its partners (MIT, Harvard, Berkeley, and Stanford) from 2012 to 2015. Many more details on the story can be found in Shreeharsh Kelkar, "Engineering a Platform: The Construction of Interfaces, Users, Organizational Roles, and the Division of Labor," *New Media and Society* 20, no. 7 (2018): 2629–46, https://doi.org/10.1177/1461444817728682; and Benjamin Shestakofsky and Shreeharsh Kelkar, "Making Platforms Work: Relationship Labor and the Management of Publics," *Theory and Society* 49, no. 5 (2020): 863–96, https://doi.org/10.1007/s11186-020-09407-.

70. Laura Pappano, "Massive Open Online Courses Are Multiplying at a

Rapid Pace," *New York Times*, November 2, 2012, sec. Education / Education Life, http://www.nytimes.com/2012/11/04/education/edlife/massive-open-online-courses-are-multiplying-at-a-rapid-pace.html.

71. "About Us," edX, accessed August 12, 2013, https://www.edx.org/about-us.

72. Quote taken from Kelkar, "Engineering a Platform," 2643.

73. Justin Reich and José Ruipérez-Valiente, "The MOOC Pivot," *Science* 363, no. 6423 (2019): 130, https://doi.org/10.1126/science.aav7958.

14

Computers as Colonizers
British Computing Companies and Indian Technological Resistance, 1955–1975

Mar Hicks

"Can India afford to go in for automation in a big way?" asked the *Times of India* in late 1967. Or, the article quickly followed up, "can it afford *not* to?" The issue of computerization was fraught in nearly all nations in the 20th century, given computers' complexity, their association with power and arcane skills, and the fear that they would automate away people's jobs. But in India and other former colonial countries, the advance of computerization was further complicated by preexisting relationships with former imperial powers, particularly Britain, which until the mid-20th century had politically and militarily subjugated Indian citizens while extracting material and economic benefits from the country.

Continually presented as somehow less than—less advanced, less civilized, less wealthy—by British and American leaders in business and politics, the nation that now forms an indispensable part of the global software labor market was then argued, by the nations leading the charge into the digital era, to have less need for computers and less to offer their development.[1] As the writer of the *Times of India* article pointed out: "The issue is often posed as a choice between technological advance and fuller employment in a developing economy burdened with an abundance of labor," leading some Indian officials to suggest that computers should have limited applications in India, focusing mainly on export-oriented and defense industries. "But in an interdependent, competitive world, which is fast entering the age of the computer," the journalist pointed out, "the selective approach is as impractical as it would have been if it were applied to the wheel, the printing press, or the steam engine."[2]

During this period the major computer-manufacturing nations of Britain and the United States were actively trying to sell their computing technologies abroad. In Britain's case, this was explicitly related to the effort to regain some of their former imperial influence. In the case of the United

States, the sale and use of computers was also closely linked to military ends.[3] India's alacrity to adopt new technologies and use them for economic development on their own terms was often met with roadblocks when more powerful Western nations imposed their beliefs and ideals on India's government and industry. As Ramesh Subramanian and others have pointed out, IBM resisted selling its newest lines in India, instead providing older, refurbished models and claiming that India did not "need" such advanced technology.[4]

Yet, at the same time, as Ross Bassett details in *The Technological Indian*, Indian academics and technical experts were filtering in and out of major Anglo-American institutions, shaping both industry and education in engineering and other high-tech fields globally.[5] Meanwhile, experts in India were setting up a native computing industry at the Indian government's behest, specifically to push back against foreign technological influence and the assumptions about India's underdevelopment—and therefore unworthiness—that went with it.[6]

Modeling Development, Developing Model Nations

In the late 20th and early 21st century the high-tech industry in the United States largely set patterns and expectations for high-tech globally. American standards and practices, however, have not always been the ideal to which industrializing and computerizing economies have looked. In contexts where the early British computing industry attempted to capture a global market through leveraging cultural bonds with commonwealth and former colonial nations, British social mores and technocratic ideals held sway for decades. From the machines installed in banks in India and Pakistan, to Sir Stafford Beers's involvement with Chile's Project Cyber-Syn, to Britain's attempts to flout American embargoes on selling computers to China and the USSR during the Cold War, British technologists aggressively tried to create a new role for themselves and for their country's high-tech sector in the emerging global IT economy, which they presciently understood would be critical not only to their nation's economic prosperity but to the developing political world order.[7] As a result, British computing technology was intentionally deployed in ways that were intended to reinvigorate imperial relationships, in an effort to enhance Britain's global political standing. While at one time "the sun never set on the British empire" owing in large part to technologies like guns and ships, British officials in both Conservative and Labour governments believed that digital technology could once again enlarge the power of their relatively tiny island nation.

The cultural and economic ideals embedded in these technological systems were powerfully and repeatedly sold to India by British corporations acting in concert with the British government. Along the way, they transformed to create a common understanding of computing that Western nations would seek to use as a tool, and model, for global technological supremacy. Yet nations like India pushed back: first by attempting to deploy and use foreign machines in ways that diverged from the model set out by foreign powers; then by seeking to develop their own computer hardware industries; and later by focusing on creating world-leading software and IT services industries.

By the 21st century, India would become a global leader in IT services and a major global hiring destination for outsourced programming expertise. As Jeffrey Yost details in *Making IT Work*, the transition of computing's business model from hardware sales to information technology (IT) services was a fundamental part of the shift away from particular manufacturers and national computing industries—and as the software services model took hold, India and other former colonial nations were able to become ever more competitive in the global high-tech marketplace.[8] Although this shift occurred in the late 20th and early 21st centuries, harbingers of this new model of technological development can be seen in the mid-20th century, as Indian institutions attempted to alternately adopt, rebuff, or redirect foreign ideals regarding technological development.

In the course of these two-way transfers of technology, the question of political power was ever present: India showed early on that it could be a major player in the digital age, by using different models of technological development than the United States and the United Kingdom employed and, in particular, by resisting and redirecting the influence of neoimperial British technological ideals. Although Indian technocracy was initially modeled on British ideals of technocratic management and control, as the 20th century wore on, India's computerization efforts increasingly diverged from what the United Kingdom (and United States) saw as appropriate for the modernizing nation. Instead, Indian technologists pushed back against the technological ideals conveyed by the earliest computerized nations while simultaneously seeking to import the latest and most powerful foreign computers.

Early Computing's Gendered Exports

One of the earliest attempts on the part of UK computing manufacturers to set the pattern for Indian computing was bound up with the gendered labor mores that British industry exported as essential for modern computerized labor forces. In the Anglo-American context, male managers who institutionalized computerized techniques favored a particular kind of administrative organization, which in turn led to a certain model of professionalization and prestige for computing workers.

In the British context, women held the lion's share of computer operating and programming jobs in the massive public sector early on. They lost this early lead, however, as technical jobs in computing grew to encompass responsibilities outside the technical realm, such as "big picture" thinking about workflow and efficiency, that were viewed as being more appropriately in the domain of management-level men.[9] This change led to an underutilization of trained women workers in the field that ultimately hurt the British state's attempts to computerize and accomplish the widespread technical reskilling needed for modernization. Crucially, it also hurt the country's attempts to raise national productivity and expand the UK computing technology sector to keep pace with international competitors.[10]

This change toward favoring technocratic men for computing jobs did not occur because the existing feminized labor forces were perceived to be technically inadequate. Rather, it came as a result of the changing meaning and prestige of the field of computing. Women were not regarded as suitable candidates for management, and so were not as often hired for new computing operation and programming jobs as those jobs became more professionalized and more aligned with management. This is a problem many Western nations, including the United States, continue to struggle with today, as the echo of women's alleged unfitness to wield power connects with ideas about their presumed technical capacities, routinely depressing women's numbers in computer science and other STEM fields. In India, the situation is somewhat different: upper-caste women, for instance, regularly go into computer science degree programs in higher proportions than their US counterparts.[11]

As scholars such as Antoinette Burton have shown, the standards of imperial centers of power not only shape but are shaped by the territories and expectations of empire, and eventually by the nations that were previously subjugated in the pursuit of imperial power.[12] Burton argues that effective global histories must "make visible the apparently micro processes by which historical events with 'global' significance have taken place—to materialize a kind of 'structural below' . . . and with it, actors and subjects often considered inconsequential even when they are legible in world histories."[13] Attention to these microprocesses and to the failures and imperfections in these processes can shed light on how the relationships between nations in a globalized economy rely not only on the transfer of goods and services but also on the transmission of cultural ideals, economic standards, and political norms that are shared as part of the transfer of any technological system.[14]

One such microprocess in computing involves the staffing of one the lowest—but most essential—job rungs in computer operation in the 20th century. In the Anglo-American context, the creation of punched cards for data input is strongly linked to feminized labor forces.[15] This mission-critical but undervalued data input work was a massive, important part of getting large-scale software programs running. Referred to as "punch girls," these workers became iconic in representations of early British and

American computing systems, owing to advertising tropes in which they figured prominently, as well as the predominant staffing models of the 20th century.[16]

As a result, British publicity for Indian installations repeatedly showed workers doing what were considered the most gender-appropriate jobs in the British context—even when the images were belied by Indian reality. Pictures of "computer demonstrations" or advertisements of British computing technology consistently exported the ideal of the woman operator and data entry clerk, while actual photographs of Indian companies' machine rooms and punch rooms using that same British computing technology more often showed men *and* women in the roles, or sometimes *only* men.

In publicity images, one can see this contrast between the gendered staffing models that the British considered ideal and attempted to export, and the actual staffing of computing installations in India. From the nameless women "demonstrating" British machines at industry exhibitions, to the mixed-gender or all-male punch room labor forces at major Indian companies using British computers, the image and the reality conflicted in important and nationally specific ways; while British interests repeatedly tried to create a gendered labor structure—and thereby power structure—in computing that mirrored British culture and interests, Indian corporations staffed their data-processing installations differently.

British computer companies repeatedly sought to position certain gendered work organization methods as natural outgrowths of the technology, their images serving a didactic function and the text accompanying them implying that the computing systems would work best if this labor model was installed along with the machines. Owing to British custom and culture, young women were the ideal workers for punching—inexpensive, interchangeable, and disposable. But while imagery for British computing in India presented young women operating punch machines and computers and demonstrating equipment at exhibitions and trade shows, the Indian punch operator labor force was not nearly as feminized. In fact, men and women often worked side by side in many punch rooms in India. In other cases, punching jobs were reserved for men only—as in the case of the Tata Iron and Steel Corporation, whose class of punch trainees, all of whom volunteered for data-punching work, was exclusively men.[17]

That class divisions functioned quite differently in the Indian context also altered the equation of computing's early development. Despite British managers' efforts to convey staffing ideals in which gender became a class marker for this labor and in which gendered labor became, in turn, one vehicle for ideas about British command and control, caste played a larger role than gender in most computer labor contexts in India. Throughout, the British focus on the supposed "civilizing" mission of these technologies meant that gender, class, and geography remained bound up with ideas about empire, rather than being tied to the culture or concerns of those importing the machines. "Language, dress, customs may differ from country to country, but Powers-Samas punched card methods are international," proclaimed one early advertisement. "Witness this group of operators in Khartoum and this picture of the operator's training school in Bombay."[18] Publicity images portrayed rooms of only young women using punch machines.

Computerization and Civilization

Powers-Samas, the wholly owned British subsidiary of the US accounting machines company originally called the Powers Tabulating Machine Company, ran an article in its company journal in 1959 that boasted too optimistically, "The world looks to Britain." Describing the recent electronic computing exhibition held in London that drew attendees from more than 40 countries, the exhibition's chairman claimed that "the world is looking to Britain for *leadership* in this new field."[19] But, simultaneously, Britain was looking to the world, in particular the recently independent and rapidly industrializing nations of the Indian subcontinent, in order to sell its machines.

In the 1950s, India was one of Britain's largest export markets. Britain held nearly a quarter of the Indian market but had begun to lose market share to the rising industries of postwar Germany and Japan.[20] Beginning in 1950, the Colombo Plan for Cooperative Economic Development in South and Southeast Asia linked the Commonwealth countries—Australia, Britain, Canada, Sri Lanka (Ceylon before 1972), India, New Zealand, and Pakistan—to provide aid to less industrialized nations by transferring technology and intellectual capital from more industrialized nations. At the same time, this transfer was designed to create a development cycle that would produce labor forces and markets which would require the purchase and consumption of products and services sold by the more industrialized nations—Britain in particular.

The millions of pounds of aid dispersed under the Colombo Plan, and the hundreds of thousands of pounds for the new engineering college at Delhi (which would later become Indian Institute of Technology [IIT] Delhi), came within a framework of British technical and material assistance that sent British companies, computers, professors, and experts to India to foster a technical and cultural alignment of Indian progress with British machines, British know-how, and British business ideals.[21] The *Times of London* pointed out that "for a country [like Britain] that will have to live more and more on its ability to produce and export capital goods and technical capacity, these two undeveloped but industrially ambitious countries [India and Pakistan] offer potentially the biggest single non-Communist market for the second half of the 20th century."[22] This market, particularly given the United Kingdom's accelerating slide into political and economic turmoil in the latter half of the 20th century, was seen by British leaders in government and industry as critically important.

It was largely for this reason that British accounting machines were the first to help automate India's banks, and British computing companies' foreign sales offices were joined by factories for producing machines and cards as India began to tighten its import restrictions after independence.[23] By the late 1950s, the British computing company International Computers and Tabulators (ICT) had a very strong presence, providing machines for national agricultural and economic research projects and having offices and training centers in all major Indian cities.

One such service center in Delhi, linked to the Indian Statistical Institute, billed itself in 1959 as "the first of its kind in the east." The data center gathered information on families from more than 4,000 villages, and in the first year alone, the 23-person punching staff was expected to complete over 5 million punched cards, recording information on agricultural and demographic trends. The company described the "Western" amenities of the installation, from the modern cafeteria with refrigeration, to the bright lights in the punch room that eliminated shadows, to the "clean symmetry of western influenced architecture," as helping to "engage in combat against poverty and underproduction."[24] The arrangement of the labor force in the installation, as much as the arrangement of the building and machines, was part and parcel of the attempt to show Western, specifically British-style modernity. It was also an effort to launder the memories of imperial brutality through a new, clean, efficient form of high technology that could very well prove to be just as important in maintaining dominion over a country as the military and transportation technologies deployed earlier.

Computerization and (Self) Governance

"The transplanted seeds of Western technology need not necessarily bring us salvation," wrote a journalist for the *Times of India* in reviewing a book of essays from a conference put on by the Administrative Staff College of India.[25] While Indian governmental institutions and companies participated in British-led development plans, they also, not surprisingly, chafed at the strictures placed on them by this new form of technocratic influence and control. British efforts to install computing technologies in India were recognized as not just economically driven but explicitly political: the power of computers to mold governance and policy along with labor and workflow was already well understood in both countries.

Recognizing this power, Indian leaders in government and industry sought ways to formalize the training of technocrats within India and to do so on their own terms. As early as 1948, just one year after independence, there were calls to set up a training college for government staff, a school focused on scientific management in general and, more specifically, on the technical aspects of administrative work.[26] The college would be modeled on the administrative staff college in Henley-on-Thames in England but "with modifications to suit the needs of our country," the *Times of India* reported.[27]

In 1957 the Administrative Staff College of India opened in the "central position and salubrious climate" of Hyderabad. Housed in a palatial older home donated by the government, the college was to have "no political or economic bias" and offer no degrees. Instead, it was a "corporation" designed to "achieve a specific purpose"—that of training senior executives in "qualities of leadership and administrative techniques like delegation and control," as well as "adaptation to economic and technological changes."

The college was designed not only for those in the government service but also for senior executives in business and industry—and not just in India but from Sri Lanka (then Ceylon), Malaysia and Singapore (then Malaya), Myanmar (then Burma), and other neighboring nations.[28] Its mission was to take middle managers and develop them into a cadre of modern technocrats on the British model, yet in a setting independent of British institutions and control. This exercise in training administrators would require the latest management theories and the latest and most advanced technologies. For decades, the Indian press had closely followed the developments in electronics in the United Kingdom, the United States, and the USSR, focusing on which computers were the fastest and best in any given year.[29] Computers, after all, were already thought to be integral to good management, as they would "play an important role in the quick decision making of management," said the chairman of the Administrative Staff College in 1967. He cautioned, however, that "radical measures were needed to improve the quality and training of managers and management, [which] should be built up as a profession with its own moral code" to combat corruption at the highest levels of government and industry and ensure the national good.[30]

Although the Administrative Staff College had a Russian-made computer early on, in the mid- to late 1970s the college conferred with the British government about the possibility of getting a top-of-the-line British ICL 2900 computer through one of Britain's ongoing foreign aid development programs. This was the same type of large, advanced mainframe that the British government had urged the British company ICL (International Computers Limited) to design and produce for the sake of Britain's own governmental needs and national security—so that the UK government would hopefully not need to be reliant on foreign (read: American) computers or have to deal with the potential national security risks that would bring.[31]

Paternalistic attitudes rooted in colonialism that imbued much of the

United Kingdom's technological transfer to India would remain problematic for Indian computer users, but initially the United Kingdom was more open to technological exchange on the terms desired by Indian institutions than the United States was. IBM (International Business Machines), the major US player in India's ramp-up to computerization, withheld the latest computing technology from that market, instead selling older, refurbished machines to Indian customers. Indeed, the displeasure of the Indian government at IBM's attempts to slow computing progress in India, by trying to fit it to a series of development increments that IBM thought appropriate for a supposedly less developed nation, resulted in a series of punitive import policies for foreign companies, which broke IBM's hold over the Indian market. Policies aimed at bolstering Indian industry eventually resulted in the company leaving the Indian market in 1978, during the premiership of Morarji Desai, the first Indian prime minister to come from a party other than the powerful Congress Party.[32] Desai's Janata Party, in an attempt to strengthen Indian industries and decrease the influence of foreign corporations, dictated that foreign corporations needed to partner with Indian ones—something that US corporations like IBM and Coca-Cola refused to do.

Meanwhile, British companies, backed by the British government, had been trying hard to get British-made computers into as many Indian institutions as possible by complying with the policies for foreign technology companies that Prime Minister Indira Gandhi had put into place through the government's Department of Electronics, founded in 1970. "At the time of IBM's departure, India's homegrown efforts at developing computers were not producing many results," notes Ramesh Subramanian. The mainframes as yet produced by Indian companies were nonstandard and required entirely new operating systems and software to be written for them. The era of the Indian microcomputer, ushered in by Hindustan Computers Limited (HCL) was still a few years off. During this period, Indian purchasers and computer users in government and industry were trying to get the latest and best foreign technologies, at the lowest cost—and with a minimum of neocolonial baggage along with the mainframes.[33]

It was in this context that Dr. Uptal K. Banerjee, Director of the Computer Division at the Administrative Staff College and their lead on the negotiation to procure an ICL 2980, noted that the Administrative Staff College's computing facility was currently being used not just by Indian governmental institutions but also to provide training to other "Third World countries." Programs that the college had designed for the public sector were portable to other nations facing the same issues, and they adopted these programs to run elements of their governments and industries. Given this transfer, Dr. Banerjee suggested to the British contingent that they make an arrangement by which the ICL computer could be paid for with British foreign aid money, noting that this would allow the college to get rid of the Russian computer they were currently using. The idea that they would get rid of the Russian computer was presented as a favorable part of the deal: as Ross Bassett notes, India used its position in the "middle" of the Cold War to play both sides and align itself with different international interests depending on how it might be expedient to interact with the different superpowers.[34] The United Kingdom would see the choice of a British mainframe over a Soviet or American one as a political win.

Another important feature of the deal that Dr. Banerjee suggested was that the Administrative Staff College would defray the remaining cost of this large, expensive computer by partially paying back the British through *writing software* for British industry and government. The idea was for the college's students and staff to program the computer, allowing them to then barter their quickly developing expertise in software design and programming to partially pay for this top-of-the-line machine. Throughout, Dr. Banerjee and the other College leaders pitched the idea to British officials in a way that aimed to fit the mold of British development aid projects, fall in line with Britain's desire to win market share for ICL computers, and simultaneously position Indian computing as a significant, growing resource and force for change both nationally and globally.[35]

Although the British were keen to capture more of the Indian computing market, the government balked at the request for what was the most technologically advanced mainframe on the British market at the time. Although the ostensible idea behind British foreign aid was to help countries develop up to the point of this kind of technological—and therefore political—sophistication, the British officials involved in the negotiation expressed disbelief that the Administrative Staff College, and India in general, had the type or volume of work to need such a large and complex machine, one that had been designed specifically for the complex problems of British government and social service administration.

British ideas about technical competency and technological need were bound up with ideas about power and governance and the role that India was supposed to play in relation to its former imperial oppressor. In the British imagination, India was supposed to be developed but not on an equal footing; trying to "catch up" but not in danger of surpassing—particularly not by using a completely different model of economic, and therefore political and social, development than the UK envisioned for it. As one of the British negotiators relayed to his superiors in the Ministry of Overseas Development, he was "surprised" that the Administrative Staff College of India would have "sufficient work" to utilize such a powerful mainframe system, and despite acknowledging that both the USSR and the European Economic Community had vouched for the quality of the college programmers' software, he did not think that this aid request for an advanced computer for planning and development should be granted. It was, in so many words, outside the scope of what the United Kingdom was hoping to do with its development funding.[36]

British rejection of this request for a top-of-the-line computer, paid for with some combination of international development funds and the programming output of Indian technocrats-in-training, was partly related to how this request threatened to unbalance the colonial model of power and influence that held on in the collective imagination of British leaders well after Indian independence. Although the offer of providing programs for other countries' software needs might seem at first glance to replicate a familiar colonial model in which British government and corporate interests extracted labor and goods from India, on closer inspection it upended this pattern. The Indian technologists who would have provided this work were not laborers but management-level technocrats aligned with the project of modernizing and centralizing the Indian state, specifically so the nation could harness its vast resources to eventually compete at the level of the two massive Cold War superpowers—rising global powers with which Britain itself could no longer keep pace. In this way, India's attempts through the Administrative Staff College to model its rising technocrat class on British patterns of power, governance, and domination shifted the relationship Britain had with India from one of technological benefactor to technological partner and, it was certainly feared, to technological inferior sooner than British leaders had ever anticipated.

British leaders were caught off guard by this incipient threat of rising Indian technological power in part because the first electronic computers installed in India lulled British officials into believing Indian advances in computer deployment would be mostly nonthreatening. Unlike most countries that had adopted electronic computers early on, India's first uses of the technology were not for warfare or peacetime military intelligence applications as in the United States, United Kingdom, and USSR, but rather for economic development.[37] So important was computing to India's economic future that, eventually, India's struggles to purchase electronic computers from the first computerizing nations would lead to the realization that a home-grown Indian computing industry was necessary for national self-determination. But before that, aspiring Indian technocrats, most notably Prasanta Chandra Mahalanobis, a statistician keen to centralize data for more effective national planning, pursued foreign computers eagerly in the service of deploying what we might today call "big data" for large-scale economic change. British leaders worried less about computers sold to India for planning and development being slyly repurposed for military or pseudo-military ends. This was in part because there were not the same international embargoes in place at the behest of the

United States as there were in the case of potential sales to Chinese or Soviet industry, and in part because of India's different political and economic relationship with the US and UK.[38]

Mahalanobis, who founded the Indian Statistical Institute, was a member of India's first Planning Commission and "the driving force" behind India's Second Five Year Plan for national development (1956–61).[39] He would eventually import the first two digital electronic computers for India from foreign sources. After getting the brush-off from the USSR and the United States, Mahalanobis brought the first digital electronic computer to Calcutta from London. It was a machine made by BTM (British Tabulating Machines Corporation), which was the same company that had helped supply code-breaking equipment to Bletchley Park during World War II. The computer, a HEC-2M costing £18,500 and consisting of roughly 2000 bytes of internal memory, was installed in 1956, six years after the Indian Statistical Institute opened.[40]

Which foreign nations' computers made their way into Indian institutions in government and industry was a major concern for Indian technologists. Although it was the first, the HEC-2M was difficult both to set up and to learn to program, but "it was ultimately a training device," purchased so that its Indian users could gain experience setting up and running a mainframe. The choice of machine, however, was just as important to the major powers competing for influence and allies in the context of the Cold War. "The question of which country would provide India its first computers," historian Nikhil Menon has written, "would ruffle Cold War feathers" despite India's official policy of nonalignment with either Cold War superpower, because computers were expected to be central to India's future as a swiftly modernizing independent state.[41] The nation that provided or sold computing technology to India first, and got a foothold in the Indian market, stood to gain significantly—not just economically but also geopolitically.

Computers were, and still are, powerful tools for constructing and undoing nation-states: one of the primary reasons the British government demanded a home-grown UK computing industry was because it did not trust American computers to be ensconced in, and responsible for, core mechanisms of the British government, even though the United Kingdom and United States were Cold War allies.[42] Indian journalists in the mid-20th century discussed similar concerns: when talking about American-made computers, one wrote with suspicion about the ten computers Honeywell corporation had "gifted" to India in the late 1960s, saying that the era of a "Computer Raj" controlled by foreign companies and their foreign governmental backers—or even by the CIA—was now upon India.[43]

Although the Space Race is often the best remembered of the technological proxy wars that made up the Cold War, along with the race of more countries to develop nuclear weapons, there were many concurrent and interrelated technological struggles that were co-extensive with the goals of global imperialism, and the sale and export of computers formed an important one of these battles. It was likely for this reason, that, after India installed their first foreign (British) computer, the USSR swiftly offered India a computer through a UN technical assistance program that the Soviets had previously withheld. Determined not to be one-upped by the UK, a fading imperial power aligned with the USSR's largest antagonist during the Cold War, the Soviets knew that having a Soviet-made computer installed as the second computer in India in late 1958 would be helpful for both optics and as a diplomatic tool. The US, meanwhile, was reluctant to supply India with computers for reasons related to ongoing international power struggles, and in particular reluctant to enter into an agreement with Mahalanobis and the Indian Statistical Institute, labeling him a "communist sympathizer" eager to find authoritarian, Soviet-aligned solutions to India's problems.[44] In fact, technological secrecy was one reason the Soviet Union had earlier been reluctant to sell India a computer. In this way, India's official policy of nonalignment had certain negative repercussions as it tried to provision the machines that would eventually run its swiftly modernizing state. Yet by 1966 the Indian Statistical Institute had another first: installing the first Indian-made digital computer at its headquarters.[45]

Past Futures and Futures Past

In 1963, IIT Kanpur installed a US-made computer (an IBM 1620) for teaching and research purposes, and the institute quickly became known as *the* place in Indian higher education to learn programming and study computer science. In contrast to the machines procured for the administrative staff college and the Indian Statistical Institute, the computer at IIT Kanpur was not for government use or for the creation of new models of data-driven management and control. It was similar to both, however, in that it was primarily conceived of as a machine for training future generations of programmers for what would, it was hoped, be a rapidly modernizing and computerizing India in the mid- to late 20th century. Yet, as Ross Bassett writes,

> to say that a group of American-aligned elites existed in India is not to say that they marched in lockstep with technological or economic developments in the United States. While their connections to the United States, particularly to the American educational system and American businesses, provided Indian elites resources with which to work, the political ties between the two countries were weak enough and the resources available in India different enough such that these elites had a great deal of flexibility and had to use a great deal of creativity in the world they established."[46]

The political ties between India and the United Kingdom were significantly deeper and stronger, though not necessarily in a positive way. The Administrative Staff College leaders, by cleverly bargaining with the United Kingdom over a type of labor that was both portable and highly in demand at the time, showed that they knew their own rising technological worth. Even though their programming skills would have been much in demand in the United Kingdom, however, the deal did not go ahead. Yet, less than two decades later, this sort of outsourcing would not only be common but necessary, and entire cohorts of successive generations of Indian tech professionals would travel around the world to provide this type of labor in the United States, the United Kingdom, and Western European nations.

Although the Administrative Staff College did not get the computer it wanted, this interaction is an instructive site of technological negotiation that lays bare some of the embedded power dynamics at work with foreign technologies and their adoption. In addition, a focus on hardware (the denied computer request) belies the importance of the software (the offer of programming in return) and nods to the coming globalized software services industry, which—as Dinesh Sharma has shown in *The Outsourcer*—permanently changed the economic and technological landscape of India and the world. Indian resistance to prevailing models of computerization being pushed by other early digital nations gestures at a prehistory of the globalized, outsourced, software industry that was beginning to come into form in the late 20th century, in part by resisting older patterns of hardware-first development that former colonial powers were trying to reenact.[47]

Technological Transfer and Technological Resistance

Ideas about the proper organization of work, the particular goals of industry and society, and the value of different kinds of labor are transmitted along with computing hardware as part of the overall automation system, but spaces for reorienting this kind of hegemonic technological package exist even when adopting the technology. As Vaidyeswaran Rajaraman notes: "Another significant development in India during this period was the establishment of research, design, and development centers for several multinational companies. The centers were being set up in India to take advantage of the availability of high-quality computer science graduates at a reasonable cost." And, by 2010, Rajaraman continues, "Indian IT companies were recognized as world class based on their performance. From low-level testing type projects, the major companies graduated to develop end-to-end applications such as processing credit card payments. . . . Indian companies were no longer competing for software services contracts based on low cost but rather on quality and timely delivery."[48] At the same time, mostly upper caste Indians traveled around the world as high-tech workers to become vectors of technological expertise that were foundational to both the US and Indian high-tech industries.[49]

Today, just over a half century since the installation of the first electronic computers in Indian industry and academia, software and computing services form a burgeoning multibillion dollar industry for India, and Indian computer science graduates have a strong presence in American high-tech institutions—a tradition in Indian engineering that dates back even before India's independence, as Ross Bassett has shown.[50] While the United States and United Kingdom find themselves unable to retain graduates of computer science programs in the software engineering workforce, in part because most major corporations are unwilling to deal with gender and racial discrimination within these fields, US universities and major Silicon Valley companies continue to import talent from countries like India through student visa and H-1B work visa programs that offer an increasingly narrow and difficult path to citizenship for Indian workers. The long wait to gain a green card can keep Indian IT workers in an immigration holding pattern for years or decades, hindering everything from their ability to move easily between jobs in search of better pay and working conditions, to their right to unionize.[51]

As Sareeta Amrute discusses, engineers from India whose careers hinge on economic migration to Europe shape not only the global information technology workforce but also the conditions of digital labor around the globe.[52] A focus on this human side of technological transfer becomes even more important when one considers, as Kavita Philip explains, that "even as the global technological landscape grew increasingly dominated by emerging economies, popular internet stories shaped the future's imaginative frontier by reviving specters of backwardness." The narratives that have shaped most computing histories written to date—and the political and social histories imbricated within those stories—are fundamentally imperialist and neocolonial frameworks. "Cold War geopolitics and colonial metaphors of primitivism and progress," Philip warns, "freighted new communication technologies with older racialized and gendered baggage."[53]

Indeed, the male, upper-class and upper-caste model of the technocrat so eagerly exported to—and imported by—Indian institutions shows how technological "transfer" between two cultures uses inequities like classism, sexism, and other hierarchies to grease the wheels of that transfer and to define the boundaries of how computing could be used to shape a modern state. These are issues that we are still dealing with today, as global superpowers lean ever more on computerized platforms, tools, and weapons to shape the realities and the rights of their citizens. An understanding of seemingly global and timeless standards and practices in computing must be grounded in comparative historical case studies that show the transnational origins of the current information technology landscape, and the myriad expectations and assumptions that continue to contribute to its momentum.

By 1986, a *Times of India* column titled "Educating Our Masters" took aim at the Administrative Staff College in Hyderabad, not because it had failed but because it had perhaps succeeded too well on the British model. Noting that the Administrative College "students" were out of touch with most of India, the author writes: "New Delhi itself is far removed from the realities of India," and if most people working for the national government would "pay an occasional visit" to the rest of India, "they will realize there is a wretched India beyond their own privileged city." The author goes on to note that "the training program at the Administrative College in Hyderabad includes what is called the NASA game." Participants were supposed to be in a space capsule that crashed on the moon, 200 miles away from the mother ship. "Now the question is: what would they do? I think the answer is simple," the writer concluded: "If you are blessed with the oxygen of power you will flourish in any remote region."[54] For many in India—and in other countries—the gulf between technocratic elites and the common person was as wide as the distance between the earth and the moon, and the problems concerning each just as divergent.

The growing crises related to how computing impacts governance today hinge on similar dynamics. The most privileged in society and those at the highest levels of industry, emboldened by a lack of regulation and massive profit potential, tend to deploy technologies capable of hurting and controlling people, economies, and nations in ways that are at odds with—or uninterested in—the concerns of democratically elected governments and the needs of the majority of citizens. This style of unaccountable, silent power allows for ever widening inequalities in the name of technological "progress" and for a spate of technologies with the potential to undermine both people's civil rights and nations' political stability.

That the global reach of computerization and computer-based power has helped produce widespread political destabilization in India, the United States, the United Kingdom, and elsewhere is not a surprise. This was the genie that all of the major powers intended to carefully keep inside the bottle by demanding their own national computer hardware industries provide them with machines specific to their governmental and security needs. It is no coincidence that, historically, computing technologies have helped a small number of technocrats wield enormous power, not just over large numbers but over large numbers of *people*. These technologies' fraught underpinnings transmit through the decades, fed by the oxygen of power and often enacting patterns of control in remarkably similar ways.

Notes

1. For context on India's rise in the software industry, see Dinesh Sharma, *The Outsourcer: The Story of India's IT Revolution* (Cambridge, MA: MIT Press, 2015).

2. K. C. Khanna, "Automation in India—Coping with the Computer," *Times of India*, December 1, 1967, 8.

3. For context on the long-standing ties between military and industrial goals in the computing industry and on the role of Silicon Valley in the growth of the 20th-century military-industrial complex, see, for example, Malcolm Harris, *Palo Alto: A History of California, Capitalism, and the World* (New York: Little, Brown, 2023).

4. Ramesh Subramanian, "Technology Policy and National Identity: The Microcomputer Comes to India," *IEEE Annals of the History of Computing* 36, no. 3 (July–September 2014): 19–29; Ramesh Subramanian, "India and Information Technology: A Historical and Critical Perspective," *Journal of Global Information Technology Management* 9, no. 4 (2006): 28–46; J. M. Grieco, "Between Dependence and Autonomy: India's Experience with the International Computer Industry," *International Organization* 36, no. 3 (1982): 609–32.

5. Ross Bassett, *The Technological Indian* (Cambridge, MA: Harvard University Press, 2016). Sareeta Amrute's recent ethnographic study of Indian IT workers in Berlin continues this history of technological diaspora to the present. Amrute, *Encoding Race, Encoding Class* (Durham, NC: Duke University Press, 2016). See also Dinesh Sharma, *The Outsourcer* (Cambridge, MA: MIT Press, 2016), and B. Parthasarathy, "Globalizing Information Technology: The Domestic Policy Context for India's Software Production and Exports," *Iterations: An Interdisciplinary Journal of Software History* 3 (May 2004). 1–38.

6. Vaidyeswaran Rajaraman, "History of Computing in India: 1955–2010," *IEEE Annals of the History of Computing* 37, no. 1 (January–March 2015): 24–35; Dinesh Sharma, *The Long Revolution: The Birth and Growth of India's IT Industry* (New York: HarperCollins, 2009); C. R. Subramanian, *India and the Computer: A Study of Planned Development* (Oxford: Oxford University Press, 1992).

7. Eden Medina, "Designing Freedom, Regulating a Nation: Socialist Cybernetics in Allende's Chile," *Journal of Latin American Studies* 38 (2006): 571–606; Anthony Benn, "ICL Computers for USSR, Memorandum by the Minister of Technology," May 17, 1968, CAB 148/37 Cabinet Papers, National Archives, Kew, United Kingdom; Chair of the Board of Trade, "Computers and Export to China," October 19, 1966, PREM 13/1967 Office of the Prime Minister, National Archives, United Kingdom; Special Correspondent, "Russia Grants ICL Special Trading Status," *Times of London*, July 10, 1970.

8. Jeffrey R. Yost, *Making IT Work: A History of the Computer Services Industry* (Cambridge, MA: MIT Press, 2017).

9. Mar Hicks, *Programmed Inequality: How Britain Discarded Women Technologists and Lost Its Edge in Computing* (Cambridge, MA: MIT Press, 2017). For the American context, see Janet Abbate, *Recoding Gender* (Cambridge, MA: MIT Press, 2012); and Nathan Ensmenger, *The Computer Boys Take Over: Computers, Programmers, and the Politics of Technical Expertise* (Cambridge, MA: MIT Press, 2010).

10. Hicks, *Programmed Inequality*.

11. Roli Varma and Deepak Kapur, "Decoding Femininity in Computer Science in India," *Communications of the ACM* 58, no. 5 (May 2015): 56–62. For more on the class and caste dimension of attempts to get women trained in computing in India, see Sreela Sarkar, "Skills Will Not Set You Free," in *Your Computer Is on Fire*, ed. Mullaney et al. (Cambridge, MA: MIT Press, 2021).

12. Antoinette Burton, *At the Heart of the Empire: Indians and the Colonial Encounter in Late Victorian Britain* (Berkeley: University of California Press, 1998); Burton, *Burdens of History: British Feminists, Indian Women, and*

Imperial Culture, 1865–1915 (Chapel Hill: University of North Carolina Press, 1994). See also Burton, ed., *After the Imperial Turn: Thinking with and through the Nation* (Durham, NC; Duke University Press, 2003).

13. Antoinette Burton, "Not Even Remotely Global? Method and Scale in World History," *History Workshop Journal* 64, no. 1 (2007): 326–27.

14. For a discussion of attempts to transfer American models of technological change, see Stuart Leslie and Robert Kargon, "Selling Silicon Valley: Frederick Terman's Model for Regional Advantage," *Business History Review* 70 (1996): 435–72.

15. Corinna Schlombs, "Gender Is a Corporate Tool," in *Your Computer Is on Fire*.

16. See Hicks, *Programmed Inequality*, chapters 3 and 4.

17. "ICT in India," *ICT* magazine (International Computers and Tabulators, Limited), no. 9 (1961): 28–29.

18. "Powers-Samas in the Sudan and India," *Powers Magazine* (March 1954): 11.

19. "The World Looks to Britain," *Powers-Samas Gazette* (January 1959): 4.

20. "New Trends in India: Foreign Aid as a Competitor," *Times of London*, April 8, 1954, 5.

21. "More British Aid Forecast for Colombo Plan," *Times of London*, October 17, 1955, 6; "Delhi College Appeal Success," *Times of London*, November 26, 1960, 5; and, "£400,000 Aid for Technology in Delhi," *Times of London*, May 2, 1963, 12.

22. Delhi Correspondent, "The Eastern Trade: British Markets Threatened in India and Pakistan," *Times of London*, December 23, 1952, 5.

23. "ICT in India," *ICT* magazine, no. 9 (1961): 28–29.

24. "Powers-Samas Service Bureau in Delhi," *Powers Magazine* (July 1955): 4–6.

25. Nirmal Sethia, "Technological Change," *Times of India*, January 16, 1972, 10.

26. "Administrative Staff College," *Times of India*, November 13, 1948, 3.

27. "Administrative Staff College Planning Body," *Times of India*, September 7, 1953, 11.

28. Shrinagesh, "College That Gives Polish to Executives."

29. See, for example, "New British Achievements in Electronics," *Times of India*, July 14, 1955, 8; "ICT's Issue," *Times of India*, January 25, 1964, 4; "Third-Generation Computers May Be Made in India," *Times of India*, November 4, 1968, 4; "Russia Producing Most Modern Computer," *Times of India*, July 27, 1978, 8.

30. Staff reporter, "Deshmukh on Toning Up Management," *Times of India*, November 14, 1967, 6.

31. "Administrative Staff College of India (ASCI), Hyderabad: correspondence and reports; proposed collaboration with University of Bradford Project Planning Centre; request for computer equipment; exchange of visitors, 1975–1979," BW 91/497, National Archives, Kew, United Kingdom.

32. Rajaraman, "History of Computing in India,"; Subramanian, "Technology Policy and National Identity."

33. Subramanian, "Technology Policy and National Identity," 21.

34. Ross Bassett, "Aligning India in the Cold War Era: Indian Technical Elites, the Indian Institute of Technology at Kanpur, and Computing in India and the United States," *Technology and Culture* 50 (October 2009): 783–810.

35. "Administrative Staff College of India (ASCI), Hyderabad: correspondence and reports."

36. "Administrative Staff College of India (ASCI), Hyderabad: correspondence and reports."

37. Nikhil Menon, "Chasing the Machine: India's First Computers and the Cold War," *Caravan* (April 2018): 2–3.

38. Because IBM and other US corporations were not allowed to sell to communist China or the Soviet Union as part of a trade embargo on machines that could be used for military ends, UK corporations (with the cooperation of the UK government) sought to do an end run around US companies in those markets. This move led to repeated diplomatic confrontations between the United States and the United Kingdom over the latter's willingness to sell computers to those countries similar to the ones the United States was intent on withholding from them. This strained UK-US relations at several points during the Cold War. See Hicks, *Programmed Inequality*, 194. In addition, the United Kingdom was less concerned than the United States regarding nuclear proliferation in India in this period: the United Kingdom calculated that even if India successfully developed a nuclear bomb, it would not have the ability to transport it very far; the British-made planes India had to use, for instance, had a travel radius of only 1,100 miles. See Jayita Sarkar, *Ploughshares and Swords: India's Nuclear Program in the Global Cold War* (Ithaca, NY: Cornell University Press, 2022), 107–8, 175.

39. Nikhil Menon, "Chasing the Machine," 2–3.

40. Nikhil Menon, "'Fancy Calculating Machine': Computers and Planning in Independent India," *Modern Asian Studies* 52, no. 2 (2018): 443.

41. Nikhil Menon, "Fancy Calculating Machine." 444, 421.

42. For more, see Hicks, *Programmed Inequality*, ch. 3–5.

43. M. Achuthan, "Ram Raj or Computer Raj?" *New Age*, February 11, 1968, 3.

44. Menon, "Fancy Calculating Machine," 447–49.

45. "First Indian Digital Computer," *Times of India*, April 3, 1966, 7.

46. See Bassett, "Aligning India in the Cold War Era," 808.

47. Sharma, *The Outsourcer.* See also William Aspray, "The Role of Policy and the Rise of the Indian Software and IT-Enabled Services Industry," *Rutherford Journal* 3 (2010); and W. Aspray, F. Mayadas, and F. Vardi, eds., *Globalization and the Offshoring of Software* (New York: Association for Computing Machinery, 2006).

48. Rajaraman, "History of Computing in India," 32.

49. See Bassett, *The Technological Indian*; Roli Varma and Deepak Kapur, "Comparative Analysis of Brain Drain, Brain Circulation, and Brain Retain: A Case Study of Indian Institutes of Technology" *Journal of Comparative Policy Analysis* 15, no. 4 (2013): 315–30; and, Sareeta Amrute, "Living and Praying in the Code: The Flexibility and Discipline of Indian Information Technology Workers (ITers) in a Global Economy," *Anthropological Quarterly* 83, no. 3 (2010): 519–50.

50. See Bassett, "Aligning India in the Cold War Era."

51. Currently Indian nationals on H-1B visas may wait a decade or longer to receive a green card. Until their green cards are granted, to continue to live in the United States, they remain at the mercy of their employer, who may or may not renew their work visa.

52. Amrute, *Encoding Race.* Amrute shows how geography, nationality, and migration fundamentally affect the class position of migrant IT workers as they move through the globalized software-engineering workforce.

53. Kavita Philip, "The Internet Will Be Decolonized," in *Your Computer Is on Fire.*

54. R.G.K. "Educating Our Masters: Anti-Column," *Times of India*, September 14, 1986, 4.

15

The Mask of Humanity

Manipulation and Psychopathy at the Human-Computer Interface

Jennifer Alexander

In the mid-20th century, the psychiatrist Herve Cleckley identified a personality he described as machine-like: a psychopathic personality.[1] The machine-like personality was so named not because of rote and predicable behaviors but because it lacked connection to the emotional and value-laden world that underpins usual human relations. Cleckley's work has proved foundational in the study of psychopathy, and it continues to inform understandings not only of psychopathy but also of antisocial personality disorder and sociopathy. It is my proposition that psychopathy offers a model for analyzing human-machine and human-computer relationships. Cleckley's work leads me to offer a still more pointed argument: that human-computer interfaces are uniquely positioned to host psychopathic aggression.

The argument linking psychopathy to machines arises from my work on reassessments of what was human and what was machine in the immediate postwar era. My research interest in 20th-century machine pathologies emerged from an examination of the terms "human" and "machine" in the postwar religious critiques of technological society that are the subject of my main line of research. Many religious critics referred to technology simply as "the machine," in contrast to the human.[2] Both concepts, human and machine, also emerged in the 1940s as central to Cleckley's study of machine-like personalities; autism and forms of schizophrenia were also identified as machinic.

Cleckley identified the psychopath as someone who presented a simulacrum of personhood: a machine-like personality that lacked the emotional commitments and experiences associated with authentic human relationships and whose mask of camaraderie hid the emotional vacuum and allowed the performance or appearance of normality. Human-computer interfaces have developed to help provide what, in the context of psychopathy, psychologist Robert Hare called a "façade of normalcy" and Cleckley a "mask of sanity."[3] The concept of the masked personality grew from narrative descriptions that have resonated in the study of psychopathy since they were published by Cleckley and Hare in the mid- to late 20th century. Both Cleckley and Hare have had enormous influence in the study of psychopathy. Cleckley's *Mask of Sanity* has been called "an indispensable guide," and Hare's Psychopathy Checklist—Revised (PCL-R), formulated through close reflection on Cleckley's *Mask*, remains "the widely accepted" and "most well-known description we have of a psychopathic prototype."[4] Central to the descriptions of Cleckley and Hare was the peculiar effectiveness of the psychopath's mask.[5] In their user-friendly and increasingly invisible forms, human-computer interfaces obscure the computer's own inner workings and those of any people involved behind an increasingly agreeable and helpful façade. The interface also masks the dense human connections, or the lack of them, that allow people to judge each other's authenticity. The interface is thus uniquely positioned to host behaviors that conceal an underlying lack of deep human connection.

This argument is urgent for two reasons. First, it has been robustly established that people are willing to be seduced by machines, to "tumble, effortlessly and affectingly, into introjective relations with computers right from the very beginning," as Elizabeth Wilson has described it. In an introjective relation (the term is Sánder Ferenczi's) one brings an object inside, welcomes it, and makes it feel at home.[6] People's "vulnerability to deception" is well established, as is their tendency to anthropomorphize their tools and machines and to persuade themselves that their devices understand and care for them.[7]

Second, the argument is urgent because we are already involved, all the time, with computing interfaces that obscure underlying economic and other commitments, their own environmental costs,[8] and, most crucially for the argument here, the potential for malicious behaviors afforded by a lack of built-in and deep human connection. That behavior includes flaming emails and trolling, as well as misinformation and hatred that threaten our common bonds and our own roles as objects, watched, nudged, defrauded, filmed, recorded, and left hanging, in service of other people's technical and economic goals. We have not yet conceptualized strongly enough the risks that lurk behind an interface. Psychopathy provides an extreme analogy, and it aims right at the heart of a fundamental notion in much of computing: that human meaning and emotion can and should be divorced—dissociated—from technical operations. The concept of a psychopathic personality as machine-like in its lack of emotional bonds and the clinical recognition that such a personality wrecks families and fills prisons—that it is *antisocial*—demand that we recognize a kind of vulnerability that is both deeply personal and deeply interrelational.

I have turned to discussions of psychopathy in my own attempt to understand the experiences of my students, and of other young people, in an effort to make sense of the scaldedness, the surprising feeling of being burned, that my students have experienced at their screens and the dread many have of the power of the interface to carry division and hate. At issue is both the aimless and somewhat random evil people may be opened to in this way and, for many, a larger sense of lurking malevolence. Students in our engineering ethics course feared the pandemic pivot online in the spring of 2020; they feared being humiliated and manipulated should something in class be recorded and then posted on social media. The students were savvy, sophisticated in awareness of their vulnerability to state or corporate surveillance and to deliberate digital manipulation. They said they felt vulnerable instead to what they might not recognize. They are not alone; feeling unsafe online and fearing cyberbullying are widely reported among young people.[9] My young son experienced the searing and random meanness made possible at the interface when his beloved Lego Universe closed on January 30, 2012. He had found it a friendly and open world, but in the last minutes another player brought him to tears, holding him back as he moved toward his favorite spot, Crux Prime, to say goodbye.

These experiences show the interface as a site of potential attack; still, they do not plumb the depths of what is possible in a person's exposure to psychopathic aggression. Such aggression may be deliberately carried out, that is, by people or through intentionally designed systems, or it may result instead from a computational structure that does not recognize and care about human vulnerability. Children have died over cyberbullying, and at least one person is reported to have taken their life after being encouraged to do so by a chatbot. It can get still worse. The lack of emotional connection to other people that is characteristic of psychopathy is expressed in a lack of empathy, in an inability to see and imagine the pain of others. This is the reason many of the crimes psychopaths commit are extreme and are called inhuman.[10] In a seminal and widely cited paper appreciating the egregiousness of such crimes, Emily Bender and colleagues call on researchers to "recognize that systems that aim to believably mimic humans bring risk of extreme harms. Work on synthetic human behavior is a bright line in AI development." "Also needed," they continued, "is scholarship on the benefits, harms, and risks of mimicking humans."[11] What follows sets up a framework for studying those risks.

I begin by clarifying three terms. This chapter's first section takes up psychopathy as a machinic personality and situates the discussion within a carceral rather than a therapeutic context. The second section examines the mask of normality that psychopaths assume in social interactions, which enables them to present themselves as normal people, capable of empathy and of forming emotional bonds. The third section surveys the human-computer interface as a widely recognized site of concealment and establishes the sort of interface with which this chapter is primarily concerned: graphical user interfaces, usually screens, which offer a menu of possible actions and a device or method for choosing among them. The final two sections illustrate vulnerabilities by examining the dissociation of human emotion from computational structures in affective computing as it pertains to autism research and in large language models that simulate human conversation. Observing this dissociation is not new: Ruth Leys, for example, has analyzed researchers' growing interest in attributing political and other developments to bodily and autonomic reactions below the level of intent and cognition, rather than to meaning-laden and cognitive processes.[12]

At issue is what it means to be human, both individually—that is, as a single member of a species—and with others, in social or community relations. The question of the human is raised acutely and in parallel by the human-computer interface. As Paul Ceruzzi has argued, how computers and people interact is not merely a technical issue but also a philosophical one, because it asks how the machine relates to people and whether it extends or replaces their own faculties.[13] Psychopathy also raises questions about what it means to be human, by looking at the extremes of human behaviors in which some people appear both to be and not to be human.

Clarifying Terms
Psychopathy: a machinic personality

The machinic features of psychopathy can be seen to resonate with the human-computer interface when the interface is understood as a surficial site not only of access but of concealment. For Cleckley, a machine-like personality was not repetitive, mechanical, or uninteresting but rather could be highly intelligent and charming, the life of the party. Cleckley characterized the machine-like personality as mimicry: the performance of sanity, and of humanity, in the mimicking of human behaviors that usually indicated an underlying infrastructure of authentic personhood. Authentic personhood, defined not by reference to individuals alone but in relation to their human interconnections and mutual interdependence, was precisely what the machine-like personality lacked. Autism also came to be described as a machine pathology, in similar but importantly different terms. Autism was described similarly as a lack of human connection; it was described differently and in contrast as not accompanied by an affectively successful performance or mimicry of human social interactions.

Psychopathy best captures the sense of masking in Cleckley's and Hare's work, in which masking featured as a key element of this sort of machinic personality. The literature recognizes close corollaries to psychopathy, such as antisocial personality disorder and sociopathy, and in fact the most recent edition of the *Diagnostic and Statistical Manual* (*DSM*, 5th edition, revised, 2022) lists the terms "antisocial personality disorder," "dyssocial," "sociopathy," and "psychopathy" as synonyms. Prominent figures in psychopathy research disagree with this description, and some have traced a line of historical development for the concept of antisocial personality disorder different from that for the concept of psychopathy. Robert Hare noted in response to earlier *DSM* constructions that antisocial personality disorder is a diagnostic category in that it relies on tabulating antisocial and, often, criminal, behaviors rather than identifying personality traits. The term "psychopathy" can also refer to more extreme personalities than does either "antisocial personality disorder" or "sociopathy," both of which in common use often denote a less severe pathology than psychopathy, sometimes traceable to environmental influences and allowing for some emotional connections to others and thus more treatment options.[14] In field studies a large percentage of prison populations met *DSM* criteria for antisocial personality disorder (75%–80%), while only a third met the threshold for psychopathy as assessed according to the standard tool, Hare's Psychopathy Checklist—Revised. "Psychopathy" refers to perhaps the most severe and most dangerous of these constructions, according to studies of recidivism and severity of crime: imprisoned psychopaths were much more likely to reoffend when released, and sooner, than those classified with antisocial personality disorder; psychopaths who committed sexual crimes did so in markedly more sadistic and cruel ways; and psychopaths engaged in "far more instrumental or 'cold-blooded' homicides than do other offenders" and often display a "predatory quality."[15] Such frightening criminality is not necessary in identifying psychopaths, but it does reveal the significance of the vulnerability I argue for here; discussed below are several cases that show devastating emotional abuse but do not contain violent criminality.

Although Cleckley was not particularly concerned about what people called the complex of phenomena he described, some additional distinctions are necessary. Distinctions between the terms were at times important in Hare's work, and I have been influenced by his and others' arguments that assessments of antisocial personality disorder, in particular, depended too much on indexing criminal and antisocial acts to recognize the affective personality traits that characterized the psychopath. Masking is tied to traits such as glibness and lying, which psychopaths can be spectacularly good at, but masking is more than mere deceit. Deceit implies intention and knowledge that what is offered is false, yet the term "deceitful behavior" does not capture the role of deceit in a psychopathic personality, in which the person knows they are lying and does not care, because they do not know how and why other people care. This last point, not caring that other people care, is what psychopathy best captures, and the discussion is thus focused there (though I do use two examples below that reference antisocial personality disorder and sociopathy rather than psychopathy). The effect of a psychopath on others—and this is where masking comes in—is more readily captured in discussions that range beyond specific behaviors to ask about a psychopath's interpersonal successes, however limited and short-term they may have been.[16] A glib ease in conversation is among the interpersonal and affective components of Hare's Psychopathy Checklist—Revised and is one possible source of the initial social effectiveness of psychopaths; masked is a willingness "to use intimidation and violence as tools to achieve power and control over others."[17]

Three important points must be stressed. First, psychopathy as discussed here lies not in a therapeutic context of mental health but in a carceral context of criminality. This positioning reflects the character of the historical literature; clinicians and researchers of psychopathic personalities often encountered their subjects within the criminal justice system. In general, psychopathy has not been thought to be amenable to therapy (though there is lively interest in possibilities suggested by genetic research and brain-imaging findings); there is evidence that cognitive and behavioral therapies for psychopaths actually do harm by providing clues clients use to enhance their manipulative behaviors. It is also necessary to distinguish between antisocial behaviors, or the things someone has done (often crimes), and what may be termed a psychopathic personality itself, which becomes apparent only when consistent patterns emerge after long observation or through a detailed case history. Second, psychopathy, despite its association with criminality, is not necessarily associated with systematic or purposeful evil, though it may take that form. It just as well describes an aimless and random sort of thoughtlessness or meanness, without identifiable purpose and in fact frequently with a self-destructive tendency. In considering psychopathy, we must therefore suspend the presumption that people have identifiable reasons for acting as they do and that if we poke around enough we can figure them out. The point is not the commonplace that people often act against their self-interest, at least as outsiders might perceive that interest to be. Nor is this an illustration of yet another commonplace: that people often act irrationally. With psychopathy, what perplexed clinicians from the earliest days of its identification was gratuitously self-destructive behavior: lying to another's face, despite evidence right there of the falsity of the claim, or casually forging checks to friends and acquaintances, despite plenty of cash in pocket. Third, understandings of psychopathy in the specialist literature are robust, despite disagreement over its conceptual boundaries, etiology, and development, and over how or whether diagnoses should be used in the criminal courts. Hare described this robustness in responding to the charge that psychopathy was a myth: "Clinical and empirical evidence, however, clearly indicates that the construct, whatever we label it— psychopathy, sociopathy, antisocial personality disorder, dyssocial personality disorder—is anything but mythological." Instead, Hare described "a reasonably consistent clinical tradition recognizing its core interpersonal and behavioral attributes [that] cuts across a broad spectrum of groups, including psychologists, criminal justice personnel, and experimental psychopathologists, as well as the lay public."[18]

The mask

Hiddenness and performance were central in Herve Cleckley's identification of psychopathy. The most influential of early clinical observers of psychopathy, Cleckley famously characterized the machine-like personality as wearing a "mask of sanity." Cleckley took that phrase as the title of his influential volume of case studies and analysis that brought psychopathy into widespread clinical discussion; it was first published in 1941 and has gone through at least six editions. His description of the psychopath as machine-like has endured. Javla Jarkko, Stephanie Griffiths, and Michael Mauran quote from the first edition of Cleckley's *Mask of Sanity* in their 2015 study of criminality and heredity. "[In] Cleckley's classic formulation of the psychopath," they write, " 'we are dealing here not with a complete man at all but with what might be thought of as a subtly constructed reflex machine,' a machine that hides behind a 'mask of sanity.' "[19] Cleckley himself explained in the fifth edition of his book: "The surface of the psychopath, however, that is, all of him that can be reached by verbal exploration and direct examination, shows up as equal or better than normal and gives no hint at all of a disorder within. . . . His mask is that of robust mental health." But just as a monkey pounding on the typewriter keys might someday reproduce Shakespeare, here "the human being becomes more reflex, more machine-like. . . . His rational power enables him to mimic directly the complex play of human living. Yet what looks like the same realization and normal experience remains . . . like the plays of our simian typist."[20] Psychopathy, in this view, entails the performance of humanity but without the social and emotional connections to other people and to human society that usually undergird interpersonal relations. Conspicuously absent are caring connections of the sort that reinforce long-term relationships. These absences were reflected in the Psychopathy Checklist, developed by Robert Hare with reference to Cleckley's work and then revised. It rated subjects according to 20 symptoms, including pathological lying, lack of remorse or guilt, superficial charm, failure to accept responsibility, and conning and manipulation.[21] In Hare's description, under the right conditions, psychopaths could maintain a "façade of normalcy" if they were intelligent and had family support in the form of families willing and with sufficient resources to keep rescuing them and, sometimes quite literally, bailing them out. A psychopath was characterized by more than mere antisocial behavior. The psychopath displayed a lack of emotional content; missing was the deep bedrock of mutually vulnerable personhood. Jarkko, Griffiths, and Mauran described the "callous lack of emotion [a]s the purest form of psychopathy."[22] Although other people often formed strong attachments to psychopaths, such attachments were not mutual.[23]

The complex of characteristics that came to constitute psychopathy was recognized long before empirical measurements to supplement descriptive observations were developed late in the 20th century. All sorts of human records—literature, religious texts, other historical materials—reveal behavioral oddities that suggest the disorder may be found in many different cultures: reports of people who had committed crimes but were neither insane nor troubled by their transgressions.[24] In the mid-20th century the disorder came to be defined by clinical observation and a deep case history; it was observable only over time as people, often in a subject's family if not in the police system, came to realize that something was not quite right. Not until adulthood did a person possess a sufficiently long history to establish the consistency in behavior that a clinician could recognize.[25] In presentation, a psychopath could be attractive, highly intelligent, articulate, and self-assured—a façade to which their life history stood in stark contrast. Impressive performances of social competence in the moment of encounter masked striking and continued evidence of social incompetence. Such incompetence could take the form of casually self-destructive behavior, such as calling out to buy drinks for everyone in a bar, despite having no money in pocket. It could also extend to serious criminal behavior, and many case studies of psychopathy described individuals encountered by psychologists and psychiatrists in the carceral system. Consistently missing in a psychopathic personality was evidence of remorse or shame, the social emotions that recognized the effect of one's behavior on others. Striking inconsistencies of care were observed: gestures of love, often grand ones, accompanied by neglect of basic caring activities. "He loves us but he doesn't care," said the children of a clinically diagnosed subject.

Much of the formative literature on psychopathy took the form of case histories. The film *The Three Faces of Eve* (1957), for which Joanne Woodward won an Academy Award, portrays a woman with schizophrenia and dramatically illustrates the parameters of a case history. Cleckley cowrote the book on which the film was based and assisted with the screenplay. Cleckley, who taught at Duke University and the Medical College of Georgia and spent many years attending psychiatric patients in veterans hospitals, insisted that the only way a psychopath could be identified was through clinical observation and knowledge of a subject's life history. A sampling of two of Cleckley's case histories, those of Anna and Pierre, can help make the concept of the psychopath clearer and underscore the significance of the mask or façade. The cases also offer a way to consider psychopathy without being distracted by its more violent and criminal manifestations; violent and criminal potential is indeed present in people who are not emotionally connected to others and who do not feel shame or remorse, but, despite popular ideas of psychopathy, violence is not its distinguishing feature. Both Anna and Pierre performed a façade of normality in face-to-face encounters. Despite histories of seriously alarming and transgressive behavior, they appeared unaffected, registering neither shame nor remorse for their own often appalling acts. Anna, moreover, registered neither anger nor hurt over having been violently treated herself.

Anna was a 40-year-old woman who came to see Cleckley on the advice of her family.[26] Cleckley found her refreshing, knowledgeable, and widely read, though not well educated. She seemed somewhat naïve and innocent; Cleckley at first had difficulty reconciling the appealing and companionable woman he saw with what her family had told him of her history. Anna was married but had no contact with her husband. Her family had persuaded him to remain married to her so they would not face financial ruin dealing with more marriages; Anna had married often and easily, and repeatedly extricating her through divorces and annulments had eaten up family resources. She had had a history of astonishing promiscuity since early in high school when a club had formed of students she would service in the former stable behind her home. While at one of the many high schools and colleges she attended (she was invariably expelled), she had followed a pleasant evening of drinking beer with friends by sneaking into their rooms and urinating on their best dresses. No one had wanted to believe it was she who was responsible, and when incontrovertible evidence emerged, she tried to pass it off as a meaningless prank. Anna also had a history of petty theft and had been gang-banging with a group of men she met in a bar when it was discovered that she had pilfered a cigarette lighter from one of them; they tossed her into a ditch and drove away. She was beaten up in a bar at least once, quite seriously. She was an accomplished liar even when what she said was manifestly and observably untrue. She appeared quite sane and would apologize and promise to change, but changes never lasted more than a few months; she even took up charitable work for the Red Cross for a while. Anna appeared untouched by her violent experiences; she knew they were outside social norms but, as far as Cleckley could discover, they seemed to have wounded nothing deeper within her.

Pierre, who used the name Pete, offers another view.[27] Cleckley first encountered Pete when he was eighteen years old, and they met periodically for some years. Pete was tall and well built and appeared mature for his age. He was pleasant and candid. He had had to withdraw from college for forging checks in small amounts, made out to vendors well known on his campus, including to a tavern keeper whose daughter he had dated for seven months and who took pride in getting to know students by name. When Pete was caught in forgeries and other misdeeds, he would apologize in fabulous and charming ways; he was spectacularly good at lying face-to-face. The dean of his college gave him credit, when the first forgery appeared, for handling an embarrassing matter in a manly way, bravely and head-on. As the forgeries mounted, the dean's opinion changed, though Pete was allowed to withdraw from school rather than be expelled. When Pete's mother was gravely ill—he had said he was crazy about her—he told his father that her possible death wasn't something to make much of: "Everybody has to die sometime." He was offended that his father was not equally blithe; he defended the statement as a simple truism, rather than a comment on the loss of someone he had reported loving deeply. Some years after Cleckley first met him, Pete staged a house robbery of a famous local mansion: the movers took everything, though Pete had no plan to do anything with the stuff, not even to sell it or store it, once it was on the trucks. Cleckley observed that over time Pete developed no close friendships and that his peers did not regard him as reliable or of good character.

Both Anna and Pierre appeared, at first glance, quite normal and even pleasant. They were successful enough socially to persuade others to enter into the most intimate and interdependent relationships with them. Only over time, and to people involved more than superficially, was this appearance revealed as a lie. Both displayed casually hurtful behaviors alongside actions devastating to their families and others around them, and their inability to see what the fuss was about placed them within Cleckley's category of machine-like personalities: without deeper emotional bonds or experiences that he could discern.

From the *New York Times*'s guest-authored column "Modern Love" comes a first-person illustration of the mask that characterizes psychopathy. Its title: "He Married a Sociopath: Me."[28] The author described her struggles to be honest with her husband and, when she caught him lying about an affair, her elation at the thought that she felt something that might be empathy. "He started acting like a sociopath," she wrote. "How is this any different from what I used to do?" The author self-identified as a sociopath and reported being clinically so diagnosed twice. The equating of her own lying with her husband's was a sort of semantic derangement founded on a shallow definition of truth and lying: that lying is simply telling an untruth, that all untruths are equivalent untruths. Lying about an affair makes sense to people who have deep personal relationships and who recognize themselves as contributing parts of larger groups and communities; such a lie is still a lie—it violates norms or expectations of truthfulness—but it is not gratuitous, and most people would distinguish an affair, hurtful on a personal and familial level, from criminality. The author lied gratuitously and about criminality: breaking into houses and stealing cars, for no reason other than an adrenaline rush, which she justified as an understandably human effort to feel something, anything. She joked offhandedly about her husband's companion, whom she described as thoughtful, kind, compassionate; "I doubt she ever attempted to choke anyone. Unlike me." The entire piece was a spectacle—outing one's husband's affair to the world under the guise of reporting on a personal breakthrough—and may have been manipulative. The story was glib, with no recognition of the heartbreak and violence it described. It was followed by several weeks of worry on social media over whether the author indeed existed or if the column was a hoax. She convinced Simon & Schuster to publish her memoir, and it has now appeared.[29] This all, to me, is chilling: oddly superficial, cunningly candid—a mask of something.

The argument that human-computer interfaces are uniquely positioned to host psychopathic aggression draws an analogy between, on one hand, the mimicry and deceit observed in psychopathy by Cleckley and others and, on the other, the computer interface that hides what is, or is not, behind it. The analogy recognizes in constructions of psychopathy the view of how deeply pathological not only deceit but a certain sort of mimicry are—the deceitful performance of normal sociality. Marga Reimer has gone so far as to argue that psychopaths may be construed as "fake persons."[30] The sort of personality Cleckley and others identified seemed perplexingly superficial and machine-like because it was devoid of deeper or interior motives, values, or emotions, or at least any that could be reached. As Cleckley put it, all that was available was "the surface of the psychopath . . . all of him that can be reached by verbal exploration and direct examination."

The interface

Many of the sites through which people most commonly interact with computers may be described similarly, as sites reachable only through direct observation and verbal exploration, that is, in the moment and during an exchange. Human-computer interfaces are themselves widely varied and appear in a myriad of configurations and contexts; for an opening definition, I turn to Elizabeth Petrick's invaluable historiographic study of human-computer interaction: an interface is a point both spatial and temporal, now often a screen, in which a person communicates with a machine. Most salient here are graphical user interfaces, with icons on a screen and some method or device for selecting and choosing among them, in the sort of experience of computing that nonspecialists have been most likely to encounter since the development of graphical interfaces for the desktop and home-computing markets in the 1990s. The argument here thus applies to communications between persons and mediated by a screen—and in fact by any medium that obscures its inner workings, including virtual reality interfaces that may present participants, and potentially artificial intelligence assistants, as avatars.[31] The interface is distinct from mere interaction between person and machine. It is "a kind of translator," Steven Johnson wrote, "mediating between two parties, making one sensible to the other."[32]

Petrick also argued that interface design has privileged user-friendliness over coevolutionary learning and the development of more sophisticated systems of interaction. In making an interface experience as seamless and invisible as possible, the drive for user-friendliness has further obscured the supporting machinery and attendant human and machine commitments, knowledge of which would allow users to assess within a computational structure the same sorts of authenticity on which long-standing human relations are built.

The human-computer interface is widely recognized as a façade and as a site of masking. In identifying interfaces as a central issue for scholars of human-computer interaction, Petrick drew attention to the widespread characterization of the interface as a site of performance, concealment, and mystery. Alan Kay, in developing portable, networked computing and graphical interface design at Xerox's Palo Alto Research Center (PARC) in the 1980s, saw the interface as a "user illusion" patterned on "notions of theater and magic"; he referred explicitly to the interface as a site of hiddenness and concealment. Philosopher Don Ihde also developed the theme of theater and performance, arguing that interfaces made possible a "machinic theatric role," decoupled from the embodied personhood of the user at the interface. The anthropologist Lucy Suchman, observing developments at Xerox PARC, identified the computer's opacity as a central feature of its interactivity. In its hiddenness and mystery the computer differed significantly from other machines, whose workings were evident; the computer was more like someone else's mind, to which we do not have access. At the interface, computers were mute about the personal, intellectual, and material infrastructures that underlay them. Media scholar Lori Emerson wrote of the interface as a "magician's cape" that granted access and also, crucially and simultaneously, concealed. Emerson, in reflecting that the notion of a threshold was insufficient to describe what happens when a person confronts a computer, alluded to the enormity of the vulnerability with which a user approaches an interface: an interface grants access not so much to another space but to the very possibility of communication. This is borne out in common experience; engaging with computer interfaces is increasingly the only way a person may communicate with systems important in daily life. Media scholar Alexander Galloway agreed with Emerson in refusing to describe the interface as a threshold and acknowledging how it hid what was behind it; Galloway further argued that the interface, though increasingly invisible, was not a threshold to be crossed to something beyond. The interface as a medium was, in a way, the thing itself.[33]

An analysis of Russian manipulation of social media early in the Black

Lives Matter movement reveals how computer interfaces can host aggression unmoored from interpersonal emotional connections. Ahmer Arif, Leo Stewart, and Kate Starbird, all colleagues in human-centered design and engineering and information sciences at the University of Washington, analyzed online discourse about Black Lives Matter produced in 2016 by the Russian Internet Research Agency, with particular attention to inauthentic social media accounts lacking discrete and actual human authors. They asked three questions: whom the inauthentic accounts attempted to mimic; how they mimicked authentic users; and how they established legitimacy and a sense of authenticity. Some of the inauthentic accounts were bots, though most were not entirely automated and showed evidence of discretion in recognizing and manipulating popular cultural memes. The accounts acted much as clinicians have described a psychopath acting: impersonating, mimicking, or taking on the mask of racism to "amplify racist narratives"; cultivating trust by giving the impression, or lying, that they were part of larger and legitimate groups; establishing interpersonal authenticity by suggesting (i.e., lying) that they had already established interpersonal authenticity with someone else; and creating a false sense of collaboration, or false-teaming, to manipulate authentic actors into responding and thus spreading cultural memes. The inauthentic accounts forwarded both conservative and liberal responses to the movement, were part of a deliberate effort to amplify discord, and thus were not gratuitously aggressive; these features contrasted with some psychopathic behaviors but were well within what had been clinically described. None of this was apparent at the interface, however. At the interface, the inauthentic accounts manipulated computer users into posting and responding and thus into further polarizing discourse and creating antagonism.[34]

Semantic concerns with meaning further justify considering the human-computer interface in psychopathic terms. Steven Johnson, in writing (as noted above) of the interface as a kind of mediator, further identified the "relationship governed by the interface [a]s a semantic one, characterized by meaning and expression rather than physical force." In its concern with meaning, semantics is distinguished from syntax, which addresses the formal structure of language and statements; it is possible to say something that is meaningless and semantically deficient but that is nonetheless grammatically and syntactically correct. In computing, semantics includes rigorous and highly mathematical analysis of meaning; semantics serves as a double-check on the multiple understandings necessary for communication of any sort, which necessarily include some degree of conformism, conventionality, and gratuitousness.[35] Study of language and nonverbal communication of psychopaths has revealed semantic derangement in conventional syntax, as well as heavy reliance on and exploitation of just such elements of conformism, conventionality, and gratuitousness. Cleckley himself suggested that psychopathy involved a form of semantic disorder, which he distinguished from semantic aphasia. Semantic aphasia described the loss of motor control necessary to achieve corresponding meaning in speech: in speech utterances under aphasia, the words just did not come out as someone intended. Semantic dementia was different: it indicated derangement, or even a lack, of meaning itself. Later researchers broadened Cleckley's notion of semantic dementia to include manipulative use of nonlanguage behaviors in subtly reinterpreting meaning in deceptive and deceitful ways.[36]

States of Vulnerability

Two areas of current computing interest and research offer examples that show how it has become of foundational importance—indeed a virtue— to build machines, models, and structures that take on human-like masks. In this way they offer a model of psychopathy. Use of chatbots for computing therapies for autism and large language models illustrate the computing worlds in which people work to make something look and behave like something else. Crucial here is the context: masking not only masks; it also seeks to guarantee that people interacting with the masked entity behave as that entity's designers desire. That machines learn, that they are not logically and completely programmed or scripted, does not diminish the point: what is masked is someone else's powers. The systems under discussion, in the words of Liam Magee, Vanicka Arora, and Luke Munn, "condense a variegated, uneven, hierarchical, and selective set of social desires."[37]

Mimicking humanness in computer therapies for autism

Autism describes people particularly vulnerable to manipulation through a human-like computer interface, perhaps a robot or screen. Devices are increasingly offered as therapy, and the field of robotics and virtual reality in autism diagnoses and therapy is burgeoning.[38] This subsection argues that computer therapies for autism are influenced by the field of affective computing, which dissociates affect, as operationally accessible, from experienced emotions, which are hidden and not accessible; that autism is often seen as a machine pathology, in that it characterizes autistic people as parallel to machines in having no internal lives; and, drawing on the work of Kathleen Richardson and Meryl Alper, that computing interventions in autism may be seen as "therapeutic violence" aimed at controlling and managing autistic people. In autism research, the dissociation of affect, as observable behaviors and appearances, from interior emotions, cognition, and motivations has construed autism as parallel to psychopathy in seeing people labeled autistic as having no interior life and emotional connections that matter. In the framework of autism research discussed here, a double masking operates: a person labeled autistic is masked in being treated as nothing more than what they immediately appear to be and simultaneously confronted by therapies that are themselves masked as human-like.

The kinds of behavior labeled autistic vary, though autism is often characterized as a disorder because of perceived dysfunctions in social communication along with repetitive and restricted patterns of behavior. A classic diagnostic exercise involves a therapist playing alongside a child and then getting in the child's way in order to see how the child responds. Therapist and child might drop marbles down a chute and watch them fly out the other end. Children usually object when the therapist puts their own hand up and blocks the opening of the chute; children usually confront the therapist and may try to pull the therapist's hand away. A child who ignores the obstruction and the therapist and simply moves around them to do something else may be placed on the autism spectrum. Some scholars find the label unhelpful and question whether autism as such exists, and others have found within the label an identity, a culture, and a platform for activism.[39] The characterization of autism analyzed below, as robotic and as evidence of a barren interior, is *vigorously disputed*. Becky Dowley, in reviewing a developing literature calling for greater attention to the voices of autistic people themselves in describing the experience of autism and in helping determine priorities and methods in autism research, has noted the increasing number of autobiographies and memoirs from autistic authors. Through this writing, autistic authors have revealed what were indeed their interior lives.[40]

Autism is, along with psychopathy and schizophrenia, often seen as one of the machine pathologies; these are the pathologies that use behaviors identified as machinic or machine-like to attribute psychological or psychiatric conditions to people. With autism, as with psychopathy, the machine pathology characterizes people as parallel to machines in having no internal emotional lives. Autism continues to be described as a machine pathology, despite the discrediting criticism of Bruno Bettelheim's famous and founding story "Joey: A Mechanical Boy." Meryl Alper, in a highly praised study of autistic children and their uses and experiences of digital technology, explicitly positions her work in contrast to the enduring popularity of the myth Bettelheim perpetuated, that "autistic children are inherently mechanical in their thinking and behavior, functioning 'as if by remote control.'"[41]

Important in research on computing therapies for autism has been the field of affective computing and its dissociation of computational structures from internal and conscious human states, such as people's emotions and cognitive deliberations. Using computing to analyze data about human emotions from people's facial expressions, language, and body movements, affective computing seeks to understand and simulate human emotional behaviors. The goal is to make extraneous any internally held emotions or affective motivations.[42] Conscious human deliberation and consideration are conceived as introducing friction and thus lessening the efficiency of responding and decision making. Mark Andrejevich, in an analogy to lethal autonomous weapons, calls this "the droning of experience," in which people are subjected to nudges and pushes rather than given decisive and cognitive agency. Andrejevich reminded readers of the "paean to the power of big data" by Chris Anderson, who was editor in chief of *Wired*: " 'Out with every theory of human behavior, from linguistics to sociology. Forget taxonomy, ontology, and psychology. Who knows why people [and things] do what they do? The point is they do it, and we can track and measure it with unprecedented fidelity.' "[43] The frequent observation that artificial intelligence needs merely to be perceived as affective makes this clear, as Ronald Arkin and Lilia Moshkina have offered in the *Oxford Handbook of Affective Computing*: "Affect lies in the eye of the beholder."[44] Caroline Bassett identifies this dissociation as a new form of behaviorism.[45] Elizabeth Wilson has revealed how such dissociation of people's internal lives from computational structures yields an understanding of people as instruments, perhaps most clearly in the language of Rodney Brooks, director of MIT's celebrated artificial intelligence lab. In the physical world, Brooks wrote, "social interaction allows humans to exploit other humans for assistance, teaching, and knowledge." Robots and people would each "entrain the other" as they each encountered the other and thus "use[d] the world itself as a tool."[46]

In "a computational view of autism," Uttama Lahiri defines affective computing as "the affective information in the human-computer interaction" and poses a motivating question for the field: "how to operationalize the affective state." Lahiri offers the rich graphics of new interfaces as sites of further affective interventions in autism treatment.[47] Brian Scassellati demonstrated how a theory of mind could be made technologically actionable without addressing interiority: by limiting perceptual inputs to those supported by an artificial intelligence infrastructure and establishing "a set of observable, testable predictions about behavior." "Simple sensory motor skill sets," such as recognizing and imitating direction of gaze, could be leveraged in a robot to construct "more complex social skills."[48] Roboticists identified "overlapping and converging goals of autism research and affective computing," described their challenges as similar to those faced by people trying to intervene therapeutically with people with autism, and insisted they were "*not* using autism as a metaphor."[49]

Autism, when characterized by an inability to understand social interactions, has been extrapolated to suggest the lack of an inner and reflective conscious life: hence machinic (note that this is a very different machinic trait than that observed in psychopathy, in which the oddity lies in spectacularly successful social performances accompanied by ongoing social incompetence). In 20th-century Britain, "autism" referred to the perceived absence of imagination, creativity, and dreams and "to a state of mind that completely lacked any content or meaning of its own and which gained its meaning only via the instruments used to measure it."[50] Ethicist and anthropologist Kathleen Richardson, who has argued forcefully that "autism has become a way to think about machines and vice versa," has cited robotics researcher Gal Kaminka's provocative identification of robots as autistic, because "right out of the box" they, like children labeled autistic, do not "behave correctly in social settings." Kaminka proposed autism as a platform for machine socialization and drew an analogy between robot learning and behavioral treatment in autism, whereby children may be taught the rules of expected behavior but do not integrate into themselves the underlying social connections that give the behaviors true relationality. In an observation so common that it has become a stereotype, people labeled autistic are seen to have an especial affinity for machines and mechanisms, which make no affective demands. The model Kaminka developed turned on producing affective behaviors absent corresponding human relationality: a robot would recognize "the actions and interests of others in its own terms." A robotic social intelligence platform could learn to mimic behavior and "thus represent others using the same data structures as it uses to represent its own executing control processes."[51]

In these examples of autism research, the dissociation of affect from computing structures creates a sort of double mask in which neither the autistic child nor the computing application—screen or robot—is conceived as having an interior emotional and cognitive life. Their interactions are described better as encounters than as anything interrelational. In an analysis of the uses of robots in autism therapy as part of the European project DREAM (Development of Robot-Enhanced Therapy for Children with Autism), Kathleen Richardson identified the exclusion of the possibility of interrelationship that characterizes much autism research. Richardson identified true human interrelationship in the *I-Thou* terms of Martin Buber; in *I-Thou* or *I-You* relationships both parties are subjects, and the *I* is open to dialogue and interpenetrating exchange. The human-machine analogy that treats as common a lack of interrelatedness reduces the encounter to an *I-It* relationship, in which the *It* is conceptualized, used, and perhaps experienced but does not penetrate into the existence of the *I*: the *It* remains an object, open to manipulation and dismissal. Recognition by one of another implies giving way to that other, acknowledging and being shaped by another: it is relational in ways not fully governed by the one's already existing conceptions and equipment.

The conception, or misconception, of robotics and autism as equivalent opens autistic children to exploitation and manipulation. They become objects to be acted on. Simon Baron-Cohen described what he termed a "deception deficit" in people who did not recognize the behavior of others as the outgrowth of a corresponding inner reality: "[It] appears to be the case [that] most children with autism really are unaware of the appearance-reality distinction." This, he argued, makes them both bad at deception themselves (again, in contrast to psychopaths, who were often spectacularly successful at lying) and unable to recognize it in others.[52]

The robotics and artificial intelligence literature is full of potentially marketable interventions to which autistic children may be exposed. One entry in the marketplace comes from the company Robokind, which received an Emerging Technology Award from the American Society of Mechanical Engineers for its autism-intervention robot Milo. Milo is the size of a large doll and designed to teach social behaviors; versions are available in different gender and color configurations. Milo speaks slowly and patiently and is willing to repeat things. Milo is a combination of robot and interface: a screen on its chest uses flash cards and plays demonstration videos, and its arms are jointed so that it can raise a hand for a high five. Robokind reported that children placed on the autism spectrum interacted with the robot 85% of the time they were with it, in contrast to the 3% of the time they interacted with human therapists. Parents had reported that some children generalized behaviors learned from a robot; the specialist literature does not offer support for generalization of such learned behaviors, and there is concern that null results are not reported or are attributed to experimental irregularities.[53]

As charming as Milo sounds, the robot itself also represents a market—and autism is a "multi-billion dollar industry." For this reason Richardson characterized the deployment of robots in autism therapy as a form of "therapeutic violence." In a therapeutic context, a context of intimacy, screens and robots carry and enact the interests and ways of thinking of powerful and well-funded technological, commercial, and medical groups. The process can be deeply intrusive. A child is observed most closely—where they cast their eyes, how often and through what arcs they raise their hands and arms—and what a robot or human observer misses may be supplemented by devices attached to children's bodies, including a hard-wired T-shirt and sticky-mounted sensors on their skin. "For all the hype and hope about technology disrupting education," Meryl Alper writes, "tools like tablet computers, machine learning, and artificial intelligence often work to make autistic children's behavior easier for others to control and manage more so than to proactively promote their personal growth."[54]

Considering autism highlights the vulnerability of people at human-computer interfaces. People who are perceived as having no inner life and who are unable to understand and recognize deception are, in a sense, unmoored from the interrelational and overlapping human tethers that offer protections against manipulation and exploitation. The research framework identified here in which children labeled autistic are seen as machines creates a sort of double masking. A person labeled autistic is masked, in that they are perceived and treated as though they are nothing more than what is apparent through their surface appearance and behavior, and they are simultaneously approached by therapeutic devices that are themselves masked as human-like. Autism therapies that attempt to model and mimic human emotional behavior in machines embody the dissociation of affect, as appearance and behavior, from interior emotions. This reflects precisely the simulated humanness, and masked non-humanness, that lie at the core of psychopathy.

Mimicking humanness in large language models

A stochastic parrot is how Emily Bender and colleagues characterized a language model: it is "a system for haphazardly stitching together sequences of linguistic forms it has observed in its vast training data, according to probabilistic information about how they combine, but without any reference to meaning."[55] A large language model like ChatGPT is "a mathematical system that is trained to predict the next string of characters, words, or sentences in a sequence."[56] To Bender and her coauthors this is stochastic—statistically analyzable but only randomly predictive or meaningful—and a parrot, a bird that mimics human speech without thinking or understanding.

A parrot is hardly an image that seems human, but the image does underscore its need of a mask to give authority to its speech. It is also an effective image of the dissociation between language and internal emotions that characterizes language models and the chat agents that have recently generated frenzied attention. I turn now to two arguments demonstrating that language models are aptly described by characteristics of psychopathy: that they are commonly described as masked and that they make a virtue of the very dissociation between façade and interior emotion that characterizes psychopathy.

It is common to describe large language models and chat programs as masks and sites of playacting. ChatGPT is called a "weapon of mass deception" and described as a site of neither truth nor falsehood but only of "the appearance of being true or real."[57] Analysts of InstructGPT describe the chatbot as adopting and discarding "masks" and as a "big black box" that gives no clues about its own workings or beliefs.[58] To "simulate" is another term for masking and playacting; analysts specify that artificial intelligence "simulate[s]" intention, intelligence, and even emotions.[59] Metadata scientist Colin Fraser commented that the language model ChatGPT is "designed to trick you, to make you think you're talking to someone who's not actually there."[60]

Theater provides a metaphor for chatbots. According to Joseph Weizenbaum, creator of ELIZA, the most famous forerunner of current chatbots, an "unmasked" computational structure in which people could see its workings would lose the illusion—or "glamour," as Caroline Bassett put it—of power and authority. ELIZA owed its name to theater; the program was named after the "gutter-snipe" Eliza Doolittle, who, in Shaw's play *Pygmalion*, created the illusion of upper-class status in the way she learned to speak. ELIZA was designed to simulate a psychiatric interview—or to "parody" it, as Weizenbaum later said.[61] ELIZA was "a programmer's semantic trick," he wrote, designed to reply to keywords supplied by a human interlocutor, responding, that is, often by mirroring what the person said and posing questions about it.[62] Such mirroring and reiteration helped a person develop a "sense of being heard and understood" by an entity that in fact had little knowledge, and needed little knowledge, of the world to which the person referred. ELIZA thus created an "illusion of understanding."[63] ELIZA was surprisingly successful; in what is called the "ELIZA effect," people came to think a machine cared for them and to attribute to it human feelings of empathy and compassion. The description of artificial intelligence and chatbots as theater has extended far beyond Weizenbaum and has become common among developers and observers.[64]

Large language models have made a virtue of dissociating computation structures from people's emotions and bodily experiences. Blaise Agüera y Arcas, a vice president and Fellow at Google Research, put it succinctly: "Large language models illustrate for the first time the way language understanding and intelligence can be dissociated from all the embodied and emotional characteristics we share with each other."[65] The models handle language statistically and behaviorally, and not as a vehicle for emotional meaning. A key assumption in removing emotions from the handling of language has been that the consciousness of emotions or experience matters little in predicting or influencing behavior.[66] Xiaochang Li analyzed a similar case of dissociation in speech recognition research: when IBM's speech recognition group turned to statistics and pattern recognition at large scale in the 1970s, leaving behind attempts to understand human language as a reflection of human meaning and reason. Li argues that the shift marked a transition from treating speech as recognizable human communication to speech treated as classifiable sounds. John Pierce, of Bell Lab's Communication Science Research Division, objected to this move in the field, and his own language describes the dissociation. He argued that recognizing human speech as meaningful required "an intelligence and a knowledge of language comparable to those of a native speaker of English"; in contrast, the severing of meaning and sound resulted in at most a "studied and artful deceit."[67]

Such dissociation is apparent in the way large language models are trained. Bender and coauthors examined this training process in their seminal stochastic parrot paper. "Text generated by an LM [language model] is not grounded in communicative intent, any model of the world, or any model of the reader's state of mind," they wrote. "It can't have been, because the training data never included sharing thoughts with a listener, nor does the machine have the ability to do that." Scraping the billions of pages of data from the Common Crawl repository, on which ChatGPT trained, was not an attempt to create shared communications with users, or to develop a shared understanding of intentions and beliefs, or to establish common contexts of interpretation. The method was contrary to the way meaningful human communications are established. "Even when we don't know the person who generated the language we are interpreting [when reading, for example], we build a partial model of who they are and what common ground we think they share with us, and use this in interpreting their words."[68] Its dissociation from the experienced, embodied world further underscores the masked character of the machine interlocutor. Liam Magee and colleagues, in an analysis of interviews with InstructGPT, noted that it lacked connection to anything outside—"any world, body, motor-sensory instruments—against which it could test its claims." This was symptomatic of psychosis, they concluded, and it marked the model's failure to inhabit a shared human world.[69] The dissociation appeared in the casual observation by Yunn LeCun, Yoshua Benigo, and Geoffrey Hinton that the growing importance of the unsupervised machine was justified because, after all, people learn the world by observing and not by being told about "every object."[70] A truism, yes, but it reveals the expectation that the most important learning comes not through cognitive work and the sharing of emotionally meaningful language but from operations independent of emotional and cognitive human interaction.

Both system and model builders and critics recognize the dissociation between emotions and language in the way large language models operate. They differ in whether or not they consider such dissociation to be a risk. They also differ in how they view people. Agüera y Arcas writes that people have no way to judge the feelings of others and that such a dissociation merely reflects the world as it is; Weizenbaum and Pierce thought there was more to human communication than what is reflected in a therapy chatbot and in patterns at scale. The concept of psychopathy is precisely relevant here because the dissociation that characterizes language models sounds very much like it: computational structures that mimic the appearance of human interaction while concealing an underlying lack of deep human connection. The concept of psychopathy was developed by psychiatrists and psychologists attempting to understand the odd and extreme cases in which people were indeed unable to judge the feelings of others: in psychopaths, unable to credit the feelings of others as real; and in people who encountered psychopaths and found themselves unable to account for the psychopath's lack of emotional connection.

Implications

The proposition that the human-computer interface is uniquely positioned to host psychopathic aggression is *not* primarily an argument that psychopaths may themselves use the interface for aggression. Psychopaths certainly may. The deeper argument is that interfaces, especially in their increasingly dominant user-friendly and seamless forms, make a virtue of hiding or concealing, or simply leaving out, what their designers have determined is not relevant in the moment of use. This means that interfaces do not provide the density of interaction that anchors authentic human relations. By "authentic human relations," I simply mean relationships that carry emotional bonds. This is not my definition, and it has been central to the understanding of psychopathy: psychopathy describes people who do not carry such emotional bonds and who use various other behaviors to create the appearance that they do. Human-computer interfaces are sites of appearance and of performance, and they obscure dense human connections. Interfaces are thus uniquely positioned to host behaviors that rely precisely on concealing the lack of such connections.

To ask seriously what it means to be human and to be machine, and what it means to be those things together, means also to take seriously the range of behavior that societies have considered acceptable in humans, including the negative end of that range. Some human behaviors have been deviant enough to raise the question of whether they are, indeed, human. If computing systems and interfaces have evolved alongside the human interlocutors who make them work, and if human agents have always been part of a computer's agency—and I am convinced this is so—then an interface may host psychopathologies not of the computer's own but of the persons who have given it birth and nurtured it to potency. Persons need not have deliberately designed a system to manipulate or even to deceive; they need merely to have designed an interface that functions as a façade by hiding the infrastructure that would allow users to determine the depth or lack of commitment on the other side. This means that concepts of shared understandings between people and machines, of collaboration, coevolution, communication, and mutual intelligibility, must recognize the user's vulnerability and must hold open the possibility that the partner on the other side is nothing more than what appears on screen: a mask. Such understandings must recognize that a façade or mask may still be strongly manipulative, and may still inflict pain. It is not necessary that the performance of humanity be complete or successful. That there are practical limits to the usefulness of simulated humanness in computing is well recognized: robots that are too human-like seem to people uncanny and make people uncomfortable, and natural language programs that appear too human-like discourage people from the self-revelation that helps to make psychotherapy effective. Psychopaths themselves face such practical limits; they are not always believed and not by everyone; some people recognize a psychopath's mask right away, and others come to see it only over time. Sometimes psychopaths make people uncomfortable enough to leave them alone.

Conclusion

In the mid-20th century, Herve Cleckley identified the psychopath as someone who exhibited a machine pathology, characterized by behavior that was machine-like in its absence of emotional depth and its lack of authentic and long-standing personal relations. Psychopaths adopted a camouflage of congeniality and fellow feeling in order to hide their emotional void and to allow them to masquerade as normal people; they adopted what Cleckley called a "mask of sanity" and Robert Hare a "façade of normalcy." Human-computer interfaces have evolved to provide such masks and façades; they are increasingly user-friendly and seamless, and hide the workings of interlocutors—human, computer, or both—on the other side. The interface thus masks and renders missing the dense and overlapping human connections by which people judge each other's authenticity. The interface provides a platform for interaction that simultaneously mimics and is divorced from the deep humanity and personhood that inform, and that people expect of, human interaction. It is thus uniquely positioned to host psychopathic aggression.

The argument is an attempt to take seriously stories of vulnerability and pain encountered through the human-computer interface and to recognize the role of the interface in the hatred and meanness that threaten our social bonds. The interface can be a site of dread and searing hurt, as I have heard in stories from my students and seen in experiences with my own child. It can be antisocial. The clinical and historical concept of the psychopath, machine-like in its freedom from emotional connection and at its strongest when hiding behind a mask or façade, offers a way to think about the powerful machines that surround us.

Notes

1. Herve Cleckley, *The Mask of Sanity: An Attempt to Clarify Some Issues about the So-Called Psychopathic Personality* (St. Louis: Mosby, 1941).

2. Jennifer Karns Alexander, "Radically Religious: Ecumenical Roots of the Critique of Technological Society," in *Jacques Ellul and the Technological Society in the Twenty-First Century*, ed. Helena Jerónimo, José Luís Garcia, and Carl Mitcham (Dordrecht: Springer Verlag, 2013): 191–203.

3. Cleckley, *Mask of Sanity*; Robert D. Hare, *Without Conscience: The Disturbing World of the Psychopaths among Us* (New York: Guildford Press, 1993), 67.

4. Robert D. Hare, *Manual for the Hare Psychopathy Checklist Revised*, 2nd ed. (North Tonawanda, NY: Multi-Health Systems, 2003); John McMillan, "Re-appraising Psychopathy," in *Psychopathy: Its Uses, Validity, and Status*, ed. Luca Malatesti, John McMillan, and Predrag Šustar (Cham, Switzerland: Springer, 2022), 12; Armon J. Tamatea, "Humanising Psychopathy, or What It Means to Be Diagnosed as a Psychopath: Stigma, Disempowerment, and Scientifically-Sanctioned Alienation," in *Psychopathy*, 25; George Mora, "Review of Henry Werlinder, *Psychopathy: A History of the Concepts* (Stockholm: Almqvist & Wicksell International, 1978)," *Journal of the History of Medicine and Allied Sciences* 35, no. 3 (July 1980): 357.

5. Recent research on psychopathy has focused on the effect of the psychopath's being so labeled in criminal proceedings, including use of the label in determining sentencing or eligibility for release; on whether treatment is possible; and on the pejorative role of such an identification on a psychopath's social integration. Seldom seen in more recent literature, unless it quotes Cleckley or Hare, are detailed case histories that raise questions about psychopaths at the moment of interaction with other people, or at the human-to-human interface. A useful specialized historical literature is developing, though it does not take up case histories. See Greg Eghigian, "A Drifting Concept for an Unruly Menace: A History of Psychopathy in Germany," *Isis* 106 (June 2015): 283–309; and Katariina Parhi and Petteri Pietikainen, "Socialising the Anti-social: Psychopathy, Psychiatry, and Social Engineering in Finland, 1945–1968," *Social History of Medicine* 30 (2017): 637–60.

6. Elizabeth A. Wilson, *Affect and Artificial Intelligence* (Seattle: University of Washington Press, 2010), 95–96, 93.

7. Simone Natale, *Deceitful Media: Artificial Intelligence and Social Life after the Turing Test* (New York: Oxford University Press, 2021), 132; Sherry Turkle, *The Second Self: Computers and the Human Spirit* (Cambridge, MA: MIT Press, 2004).

8. Kate Crawford, *Atlas of AI: Power, Politics, and the Planetary Costs of Artificial Intelligence* (New Haven, CT: Yale University Press, 2021).

9. Rasmussen College reported in 2015 that people aged 18–34 feel unsafe online, and the National Society for the Prevention of Cruelty to Children, the UK children's charity, reported that a quarter of children fear cyberbullying. On the Rasmussen College study, see https://www.cheatsheet.com/technology /millennials-are-more-scared-of-the-internet-than-their-grandparents.html/. On the NSPCC report, see "Cyber-Bullying Revealed as Children's Greatest Online Fear," Virgin Media O2, February 10, 2020, https://news.o2.co.uk/press -release/cyber-bullying-revealed-as-childrens-greatest-online-fear/.

10. See, for example, Victoria Beale, "Too Close to Ted Bundy," *New Yorker*, October 10, 2015, https://www.newyorker.com/books/page-turner/too-close-to -ted-bundy.

11. Emily Bender, Timnit Gebru, Angelina McMillan-Major, and Shmargaret Shmitchell, "On the Dangers of Stochastic Parrots: Can Language Models Be Too Big?" *Proceedings of the Fairness, Accountability, and Transparency Conference 2021* (Association for Computing Machinery) (March 2021): 618, 619.

12. Ruth Leys, *The Ascent of Affect: Genealogy and Critique* (Chicago: University of Chicago Press, 2017): 307–9.

13. Paul Ceruzzi, *Computing: A Concise History* (Cambridge, MA: MIT Press, 2012), xv.

14. Jack Pemment, "Psychopathy versus Sociopathy: Why the Distinction Has Become Crucial," *Aggression and Violent Behavior* 18 (2013): 458–61.

15. M. Woodworth and S. Porter, "In Cold Blood: Characteristics of Criminal Homicides as a Function of Psychopathy," *Journal of Abnormal Psychology* 111 (2002): 436–45, quoted in Robert D. Hare, "Psychopathy: A Clinical and Forensic Overview," *Psychiatric Clinics of North America* 29 (2006): 714, 713, 715–716; Jessica Gurley, "A History of Changes to the Criminal Personality in the DSM," *History of Psychology* 12, no. 4 (2009): 293, 298.

16. This inquiry into interpersonal successes bears on notions of the "prosocial psychopath," "everyday psychopath," and "successful psychopath." See Adrianne John R. Galang, "The Prosocial Psychopath: Explaining the Paradoxes of the Creative Personality," *Neuroscience and Biobehavioral Reviews* 34 (2010): 1241–48. Galang equates psychopathy with the antisocial personality

and sociopathy and argues that lowered inhibition lends itself to greater creativity, as seen in a study of risk in card playing by persons with damage to the prefrontal cortex who displayed "flattened affective tone and expression" (1244, 1243). The "successful psychopath" has been described as an "everyday psychopath" who excels in business or other pursuits through behavior that is ruthless because it demonstrates willingness to ignore human costs. See Xanthé Mallet, "The God Complex: How to Spot an Everyday Psychopath," *Newsweek*, March 9, 2018, https://www.newsweek.com/psychopath-god-complex-812746. Cleckley had a different notion of successful psychopaths, by which he meant not people who achieved spectacular success through ruthlessness but people who contained their psychopathy enough to live fairly normal lives, punctuated by odd episodes of more clearly psychopathic behavior. He described these cases as suggestive of but incompletely manifesting full psychopathy. See Cleckley, *Mask of Sanity*, part 2.

17. Joseph H. Manson, Matthew M. Gervais, Daniel M. T. Fessler, and Michelle A. Kline, "Subclinical Primary Psychopathy, but Not Physical Formidability or Attractiveness, Predicts Conversational Dominance in a Zero-Acquaintance Situation," *PLoS ONE* 9, no. 11 (November 2014): e113135, 1, 9; Hare, "Psychopathy," 714.

18. Robert D. Hare, "Psychopathy: A Clinical Construct Whose Time Has Come," *Criminal Justice and Behavior* 23, no. 1 (March 1996): 27–28.

19. Javala Jarkko, Stephanie Griffiths, and Michael Mauran, *The Myth of the Born Criminal: Psychopathy, Neurobiology, and the Creation of the Modern Degenerate* (Toronto: University of Toronto Press, 2015) 128, quoting Cleckley, *Mask of Sanity* (1941), 259.

20. Herve Cleckley, *The Mask of Sanity*, 5th ed. (St. Louis: Mosby, 1976), 383.

21. Hare, *Manual for the Hare Psychopathy Checklist—Revised*, 2nd ed. (North Tonawanda, NY: Multi-Health Systems, 2003); Robert D. Hare and C. N. Newman, "The PCL-R Assessment of Psychopathy: Development, Structural Properties, and New Directions," in *Handbook of Psychopathy*, ed. C. Patrick (New York: Guilford, 2006), 58–88.

22. Jarkko, Griffiths, and Mauran, *Myth of the Born Criminal*, 224.

23. Paul H. Sandifer described a similar suite of behaviors in a 1948 address to the British Medico-Legal Society: "failure voluntarily to contribute to the welfare of the community"; "[a] tendency to disregard future consequences of present action, the calls of duty and the voice of conscience"; and "gross and widespread and global defect of conscience." See Sandifer, "Social and Medico-legal Problems of the Psychopath," *Medico-Legal Journal* 60, no. 4 (1992): 232, 242.

24. See G. E. Berrios, *The History of Mental Symptoms: Descriptive Psychopathology since the Nineteenth Century* (Cambridge: Cambridge University Press, 1996), 428; Hare, "Psychopathy: A Clinical Construct," 27.

25. Donald W. Black, "The Natural History of Antisocial Personality Disorder," *Canadian Journal of Psychiatry* 60 (2015): 309–14.

26. Cleckley, *Mask of Sanity*, 5th ed., 102–20.

27. Cleckley, *Mask of Sanity*, 5th ed., 77–93. Pete's case was not extreme, Cleckley noted, but he believed it belonged within the clinical description he was developing.

28. Patric Gagne, "He Married a Sociopath: Me," *New York Times*, October 16, 2020.

29. *Sociopath: A Memoir*, by Patric Gagne, Simon & Schuster, accessed October 14, 2023, https://www.simonandschuster.com/books/Sociopath/Patric-Gagne/9781668003183.

30. Marga Reimer, "Psychopathy and Personal Identity: Implications for Medicalization," in *Psychopathy*, 282.

31. Mat Pawluczuk, "[VR Transcripts] Mark Zuckerberg Interview in the Metaverse (Lex Friedman Podcast #398)," *Medium*, October 10, 2023, https://matpaw.medium.com/vr-transcripts-mark-zuckerberg-interview-in-the-metaverse-lex-friedman-podcast-398-b38663104a46.

32. Steven Johnson, *Interface Culture: How New Technology Transforms the Way We Create and Communicate* (New York: HarperCollins, 1997), 14. Interesting is the comment by graphical user interface creator Doug Engelbart, who compared it to "a very primitive pidgin English. It's like grunting and pointing, instead of speaking." Engelbart quoted in Nathan Rheingold, *Tools for Thought* (Cambridge, MA: MIT Press, 2000), 328.

33. For further discussion of and references to the work of Kay, Ihde, Suchman, Emerson, and Galloway, see Elizabeth Petrick, "A Historiography of Human-Computer Interaction," *IEEE Annals of the History of Computing* 42, no. 4 (October–December 2020): 8–23.

34. Ahmer Arif, Leo G. Stewart, and Kate Starbird, "Acting the Part: Examining Information Operations within #BlackLivesMatter Discourse," *Proceedings of the Association for Computing Machinery on Human-Computer Interaction* 2, no. CSCW 20 (November 2018): 26.

35. Johnson, *Interface Culture*, 14; Furio Hounsell, "Wherefore Art Thou . . . Semantics of Computation," paper presented at 3rd International Conference on the History and Philosophy of Computing, Pisa, Italy, October 8–11, 2016, in *History and Philosophy of Computing*, ed. Fabio Gadducci and Mirko Tavisanos (Cham, Switzerland: Springer 2016), 21, 11.

36. R. W. Rieber and Harold Vetter, "The Language of the Psychopath," *Journal of Psycholinguistic Research* 23 (1994): 1–28. A study of nonverbal indicators in psychopathic subjects yielded the surprising result that psychopathic offenders exhibited the same nonverbal behaviors when lying as did nonoffenders but were able to deceive successfully through other personal and nonverbal behaviors. One interpretation is that criteria for indicators of lying are not rigorous enough. See Jessica R. Klaver, Zina Lee, and Stephen D. Hart, "Psychopathy and Nonverbal Indicators of Deception in Offenders," *Law and Human Behavior* 31 (2007): 337–51.

37. Liam Magee, Vanicka Arora, and Luke Munn, "Structured like a Language Model: Analysing AI as an Automated Subject," *Big Data and Society* 10, no. 2 (2023), 7.

38. Meryl Alper, *Kids across the Spectrums: Growing Up Autistic in the Digital Age* (Cambridge, MA: MIT Press, 2023), 200; M. C. Coeckelbergh, C. Pop, R. Simut, A. Peca, S. Pintea, D. David, and B. Vanderborght, "A Survey of Expectations about the Role of Robots in Robot-Assisted Therapy for Children with ASD: Ethical Acceptability, Trust, Sociability, Appearance, and Attachment," *Science and Engineering Ethics* 22, no. 1 (2016): 47–65; Brian Scassalleti, "How Social Robots Will Help Us to Diagnose, Treat, and Understand Autism," *Robotics Research* 28 (2007): 552–63; Kathleen Richardson, *Challenging Sociality: An Anthropology of Robots, Autism, and Attachment* (Cham, Switzerland: Springer, 2018), 114–115, 132.

39. See Katherine Runswick-Cole, Rebecca Mallet, and Sammi Timimi, eds., *Rethinking Autism: Diagnosis, Identity, and Equality* (London: Jessica Kingsley, 2016); and Larry Arnold's review of the volume in *Good Autism Practice* 17, no. 2 (2016): 100–101. See also Richard Woods, Damian Milton, Larry Arnold, and Steve Graby, "Redefininig Critical Autism Studies: A More Inclusive Interpretation," *Disability and Society* 33. no. 6 (2018): 974–79.

40. Becky Dowley, "The Development and Importance of the Autistic Voice in Understanding Autism and Enhancing Services," *Good Autism Practice* 17, no. 1 (2016): 48–53. See also Ian Hacking, "Autistic Autobiography," *Philosophical Transactions of the Royal Society* 364 (2009): 1467–73.

41. Alper, *Kids across the Spectrums*; Bruno Bettelheim, "Joey: A Mechanical Boy," *Scientific American* 200 (1959): 116–30. "Remote control" is Bettelheim's phrase.

42. Sharon Richardson, "Affective Computing in the Modern Workplace," *Business Information Review* 37, no. 2 (2020).

43. Mark Andrejevich, "The Droning of Experience," *Fiberculture Journal* 25 (2015): 202–17, 206. Anderson's piece is "The End of Theory: The Data Deluge Makes the Scientific Method Obsolete," *Wired*, June 23, 2008.

44. Ronald C. Arkin and Lilia Moshkina, "Affect in Human-Robot Interaction," in *Oxford Handbook of Affective Computing*, ed. Rafael Calvao, Sidney D'Mello, Jonathan Gratch, and Arvid Kappos (Oxford: Oxford University Press, 2018), 2.

45. Caroline Bassett, "The Computational Therapeutic: Exploring Weizenbaum's ELIZA as a History of the Present," *AI and Society* 34 (2019): 803–12. Bassett's article led me to Andrejevich.

46. Brooks quoted in Wilson, *Affect and Artificial Intelligence*, 71–72, 71.

47. Uttama Lahiri, *A Computational View of Autism: Using Virtual Reality Technologies in Autism Intervention* (Cham, Switzerland: Springer, 2020), 169.

48. Brian Scassellati, "Theory of Mind for a Humanoid Robot," *Autonomous Robots* 12 (2002): 17, 23. On "gaze-related measures," see U. Lahiri, E. Bekele, E. Dohrmann, and Warren N. Sarkar, "Design of a Virtual Reality Based Adaptive Response Technology for Children with Autism," *IEEE Transactions on Neural Systems and Rehabilitation* 21, no. 1 (2013): 55–64.

49. Rana el Kaliouty, Rosalind Picard, and Simon Baron-Cohen, "Affective Computing and Autism," *Annals of the New York Academy of Sciences* 1093 (2006): 228, 230.

50. Bonnie Evans, *The Metamorphosis of Autism: A History of Child Development in Britain* (Manchester: Manchester University Press, 2017), 187–88.

51. Richardson, *Challenging Sociality*, 114; Gal A. Kaminka, "Curing Robot Autism: A Challenge," *Proceedings of the 2013 International Conference on Autonomous Agents and Multi-agent Systems* (International Foundation for Autonomous Agents and Multiagent Systems) (2013): 801, 802.

52. Simon Baron-Cohen, *Mindblindedness: An Essay on Autism and Theory of Mind* (Cambridge, MA: MIT Press, 1995), 82.

53. Richardson, *Challenging Sociality*, 129; Paola Pennisi, Alessandro Tonacci, Gennaro Tartarisco, Lucia Billeci, Liliana Ruta, Sebastiano Gangemi, and Giovanni Piogga, "Autism and Social Robotics: A Systematic Review," *Autism Research* 9 (2016): 175; Luthffi Idzhar Ismail, Thibault Verhoeven, Joni Dambre, and Francis Wyffels, "Leveraging Robotics Research for Children with Autism: A Review," *International Journal of Social Robotics* 11 (2019): 389–410; Richardson, *Challenging Sociality*, 129; Michael Abrams, "Teacher for Children with Autism," *Mechanical Engineering* (December 2018): 34–35; Robokind, accessed December 28, 2024, https://www.robokind.com; Pennisi et al., "Autism and Social Robotics," 166, 180–81; Richardson, *Challenging Sociality*, chapter 3.

54. Richardson, *Challenging Sociality*, 114–15, 125, 128, 131; Alper, *Kids across the Spectrums*, 94. Francisco Melo and 22 coauthors identified autism spectrum disorder as "persistent deficits" in social communication and social interaction; as therapy they proposed a networked robot system and a hardwired T-shirt providing physiological data more reliable, in their view, than

data not apparent in observing the child. See Francisco S. Melo, Alberto Sardinha, David Belo, Marta Couto, Miguel Faria, Anabela Farias, Hugo Gambôa, et al., "Project INSIDE: Towards Autonomous Semi-unstructured Human-Robot Social Interaction in Autism Therapy," *Artificial Intelligence in Medicine* 96 (2019): 198, 213; R. Picard, "Affective Computing: Challenges," *International Journal of Human-Computer Studies* 59 (2003): 59–64; and LouAnne E. Boyd, Jazette Johnson, Franceli Cibrian, Deanna Hughes, Eliza Delpizzo-Cheng, Karen Lotich, Sara Jones, et al., "Global Filter: Augmenting Images to Support Seeing the 'Big Picture' for People with Local Interference," *ACM Transactions on Computer-Human Interaction* 30, no. 3 (art. 45) (June 2023).

55. Emily M. Bender, Timnit Gebru, Angelina McMillan-Major, and Shmargaret Shmitchell, "On the Dangers of Stochastic Parrots: Can Language Models Be Too Big?" *Proceedings of the Fairness, Accountability, and Transparency Conference 2021* (Association for Computing Machinery) (March 2021): 617, 619.

56. Ben Tarnoff, "Weizenbaum's Nightmares: How the Inventor of the First Chatbot Turned against AI," *The Guardian*, July 25, 2023.

57. Alejo José Sison, Marco Tulio Daza, Roberto Gozalo-Brizuela, and Eduardo C. Garrido-Merchán, "ChatGPT: More than a 'Weapon of Mass Deception'; Ethical Challenges and Responses from the Human-Centered Artificial Intelligence (HCAI) Perspective," *International Journal of Human-Computer Interaction* 40, no. 17 (2023): 3, 13; https//doi.org/10.1080/10447318.2023.2225931.

58. Magee, Arora, and Munn, "Structured like a Language Model," 9, 7.

59. Natale, *Deceitful Media*, 132.

60. Fraser quoted in Tarnoff, "Weizenbaum's Nightmares."

61. Joseph Weizenbaum, *Computer Power and Human Reason: From Judgment to Calculation* (San Francisco: W. H. Freeman, 1976), 3.

62. Weizenbaum quoted in Rheingold, *Tools for Thought*, 153.

63. Joseph Weizenbaum, "ELIZA—A Computer Program for the Study of Natural Language Communication between Man and Machine," *Communications of the ACM* 9, no. 1 (January 1966): 42, 43.

64. Simone Natale, "If Software Is Narrative: Joseph Weizenbaum, Artificial Intelligence, and the Biographies of ELIZA," *New Media and Society* 21, no. 3 (2018): 712–28; Natale, *Deceitful Media*, 55.

65. Blaise Agüera y Arcas, "Do Large Language Models Understand Us?" *Daedalus* 151 (Spring 2022): 194; the sentence continues "and with many other animals."

66. Crawford, *Atlas of AI*, 157.

67. Xiaochang Li, "'There's No Data like More Data': Automatic Speech Recognition and the Making of Algorithmic Culture," *Osiris* 38 (2023): 165, 171–72.

68. Bender et al., "On the Dangers of Stochastic Parrots," 616.

69. Magee, Arora, and Munn, "Structured like a Language Model," 16, 17.

70. Yann LeCun, Yoshua Benigo, and Geoffrey Hinton, "Deep Learning," *Nature* 521 (May 28, 2015): 432.

16

Cryptography Goes Public

Contesting the Meaning of a New Field in the 1970s United States

Gili Vidan

It is a rare case that a historian, in writing a dense, millennia-spanning history of a practice, is credited by practitioners as giving rise to a field and setting a new research agenda. But David Kahn's 1967 book *The Code-breakers* has received just that kind of distinction. In the foreword to an influential textbook on the subject, Whitfield Diffie, one of the founders of the new field of "public cryptography," described the postwar literature on the field as "barren" prior to the arrival of Kahn's work and credited *The Codebreakers* with making "tens of thousands of people, who had never given the matter a moment's thought, aware of cryptography," inspiring a "trickle of new cryptographic papers" in the following decade.[1]

And it wasn't just soon-to-be professional cryptographers who took note of Kahn. During the 1970s, Kahn found himself in the position of a public intellectual on the subject of secrets: a keen cryptogram solver and military historian, he soon weighed in on the appropriate balance of power between state and citizens and the applicability of constitutional rights to scientific publishing. Throughout the 1970s, Kahn served as a spokesperson for the field of cryptography. It is hard to find writing on the subject of cryptography from this period (whether a technical report, conference paper, or news item) that does not pay some tribute to Kahn's work, if only as a citation. Full of sample codes and historical crypto-grams, *The Codebreakers* then formed its own kind of a password—if you knew to reference it, you were in the know. Kahn was also busy periodiz-ing the contemporary developments in the field (which he later wrote about in a second edition of *The Codebreakers*).[2] From historian to prophet, or at least cofounder, Kahn took an active role in shaping the public life of the new era of "public cryptography."

As awareness of the variety of data banks that average citizens were linked to spread, concerns over privacy and its demise dominated public discourse in this period.[3] But conversations about the broad concept of data privacy also focused attention on the changing material reality of communication and information. The increasing role that computers occupied in commercial enterprises and governmental bureaus became a matter of contestation alongside the forms of electronic communications individual citizens might one day conduct. The long-standing question of liberalism—the tension between the autonomy and freedom of the private individual and the demands of life in a shared broad community of fellow citizens—was being put to the test in the design of electronic communications and data-processing technologies. Computers both posed a threat to the status quo of privacy and vulnerability and offered the possibility of rethinking the ways the private relates to the public in the liberal order. By the end of the 1970s, cryptography, a mathematical field that until then had been primarily associated with spies and secretive military operations, occupied a prominent place in this debate about democracy. As Sarah Myers West argues, the decade saw a convergence of academia, industry, and policy institutions all searching for technical solutions to address privacy and security, a mode of materially enacting information control.[4] Encryption sat at the very boundary between the private and the public, regulating the means by which a connected society would incorporate its individual members.

This prominence, the public status of "public cryptography," was itself highly contested. In this chapter I explore how the nascent academic field of public cryptography emerged in the 1970s as a result of two highly contested dynamics: first, the applied content of the research as a way to secure trust outside a closed managerial system in a connected world and, second, the limits to the field's own ability to make its work publicly known and accessible.

Shaky, Paranoid Systems

In January 1972, a few months before an arrest at the Watergate hotel would turn wiretapping into a national obsession and bring down a president, *Spectrum*, the IEEE (Institute of Electrical and Electronics Engineers) magazine, published a feature article by Charles Beardsley surveying the emerging landscape of computer security. The litany of possible threats, ranging from intentional embezzlement to accidental use of magnetic adhesives (which could cause magnetically stored data to get erased), was "enough to make any system shaky and paranoid." Surveying a variety of potential remedies for the distraught data-processing manager—more scrupulous managerial procedures, physical storage locks, and cryptographic hardware solutions (including a requisite recommendation to read *The Codebreakers*)—Beardsley concluded by veering into social commentary. "The recent public outcry over privacy and data banks," he speculated, "may well indicate an anxiety caused by ignorance of and lack of participation in the decision-making processes that make use of computer-stored data." Computer security, he added, is simply the latest manifestation of a fundamental problem: "You can build locks, but who gets the key?"[5] In the following decade, the technical problem of cryptographic key management became increasingly entwined with the political question of who got to design the locks.

Cryptography most visibly entered the public record in May 1973. That year, the National Bureau of Standards (NBS) posted in the *Federal Register* a "solicitation of proposals" for a "cryptographic algorithm for protection of computer data during transmission and dormant storage." The rapid growth in data being processed, stored, and transmitted digitally across government and industry, the announcement explained, necessitated the development of such a standard. The flow of information was not only voluminous but also valuable, including the transfer of large sums of money, law enforcement communication, and medical information. "It is recognized that encryption," the solicitation continued, "represents the only means of protecting such data."[6] It may at first seem counterintuitive that the procedure for keeping such a vast array of sensitive (if unclassified) information secret was being solicited from the public through the very mechanism designed to create governmental transparency. But the purpose of the NBS standard-making process in this area was to solicit a design that would remain secure even if the algorithm itself was widely known. Dennis Branstad, a former National Security Administration (NSA) employee who moved to the NBS to lead its computer security initiative, described this feature as one of the requirements of the federal standard-to-be. Security derived from such a system would have to be "based not on the secrecy of the algorithm itself but only on the secrecy of the key."[7]

The distinction between the publicness of the algorithm, the procedure through which an encryption transformation takes place, and the crucial secrecy of the key, the input that is used to encrypt or decrypt, was widely understood in the field of cryptography. One reason was that an openly published algorithm would be subject to expert scrutiny and could have its security assessed by "mock attacks," resulting in a more accurate assessment of its potential vulnerabilities. This argument was later raised by academic researchers who argued that the NBS was not providing enough justification to support some of the design choices made in the eventual standard.[8] A second reason concerned the cost of making changes to a widely adopted algorithm in case its secrecy had been compromised, as compared with the relative ease of generating new keys in case those have been compromised, under the assumption that the algorithm being known does not in itself compromise the encryption process. It would mean that if a key was indeed compromised, there was no need to replace the entire authentication mechanism for a given institution, which could involve installing costly new hardware or developing new software; rather, one need only change the given key.

Another reason in support of a design consisting of a public algorithm whose security depended solely on the secrecy of the key was the perceived need to make it publicly available to US businesses and government agencies. By the late 1960s the NBS was undergoing major structural changes and rebranding that positioned it as an institution responsible for addressing the nation's everyday technical woes. An internal history of the institution describes the NBS during this period as stepping into the role of responding to a "nation in distress."[9] The NBS's involvement in developing an encryption standard was triggered by the perceived urgent need to develop a standard that could be employed by a wide variety of data-processing operations, within the federal government and in industry. Although numerous encoding devices were being developed and sold by private producers, NBS sprang into action on the belief that no system available on the market was actually secure and security had not been made enough of a priority.[10] NBS therefore felt responsible for highlighting the need for encryption and making a secure option available for general use.

Despite the talk of individual privacy and the potential general public use of an encryption standard, computer security experts at the NBS and in other professional circles considered a more specific set of threats. The propagation of computerized data-processing technology, time-sharing systems, and remote transmission of information left the systems vulnerable to exploitation by skilled, undetectable individuals.[11] Computer systems were "shaky and paranoid" because their deployment relied on trust in a new class of technical workers in a position to act against their employers and likely go undetected. By the decade's end, Branstad, together with psychologist Susan Reed, published their "Executive Guide for Computer Security," in which the authors argued that "without question, the 'trusted insider' is the greater threat to any computer system" and that the most common form of security breach is manipulating data in a system for personal gain.[12] Concern about embezzlement in businesses such as banks was an oft-cited incentive for urgent action. The opportunistic eavesdropper and the trusted system administrator presented the same threat. The aim of encryption, then, shifted from a focus on secret communication to the domain of limiting access to data and authenticating users, and the target audience became the business executive.

In her study of the development of computer security at the Association for Computing Machinery (ACM), Rebecca Slayton has demonstrated how different working groups conceived of security in quite different, at times contradictory, terms. Protection of data from unauthorized access ran against the grain of fault tolerance measures, which emphasized duplicating information across different systems to avoid its loss, thus creating redundancy, rather than limiting access.[13] For those who considered the business itself to be the most likely victim of computer crime, security most often meant the need to control individual employees' use of data. Donn B. Parker, a consultant at the Stanford Research Institute, emerged as one of the most vocal publicists of the risk of computer crime. Colleagues often commented on his proclivity for media appearances throughout the 1980s, which even resulted in a profile in *People* magazine.[14] Parker argued that with the introduction of computing into areas such as electronic funds transfer, the "nature, needs, and methods of trust" in business were changing in ways that allowed individuals to pose greater threats to the business in which they worked.[15]

Commentators on the new, public era of cryptography, from journalists like Kahn to technologists at the NBS, shared the conviction that this present context of computer security differed substantively from the preexisting security procedures of the military. Unlike Paul Edwards's account of the "closed world" that characterized the Cold War era of computerized military command-and-control centers, public cryptography seemed to have more in common with other civilian preoccupations of the 1970s. Coverage of cryptography's possible capacity to solve computer security concerns appeared on the pages of the NBS magazine, *Dimensions*, alongside reports on attempts to mitigate the risks of other precarious systems, including efforts to reduce domestic dependence on oil and the spread of house fires. In a world of increasingly interdependent, precarious systems, standardizing encryption was seen as mitigating some of the hazards of living in an interconnected society. The NBS's computer security efforts, though discussed under the heightened legislative and executive focus on increasing citizens' privacy and governmental transparency, seemed to go against the grain of privacy protections, which provided citizens with additional means of requesting information from institutions. Legislation such as the 1966 Freedom of Information Act, 1970 Fair Credit Reporting Act, and 1974 Privacy Act sought to redress the growing loss of control and power asymmetry felt by individuals at the hands of large data-collecting organizations. Public cryptography, as it developed at the NBS and with the needs of commercial enterprise in mind, sought to stabilize the control wielded by such organizations.

A focus on business needs reframed an old problem in cryptography—that of key distribution and management. If the security of the encryption standard hinged so strongly on the key, who should generate and distribute them? As NBS and other computer security experts framed the issue, this was a managerial problem that required each business to figure out its own access control structure and procedure for key issuance. On an individual user level, however, knowledge of the key would be shared between the authorized individual user and the system. Cryptography first went public as a practice of authentication, making the individual user legible to the system manager, rather than as a tool of hiding individual users' actions and private communications.

Concerns over the shakiness of computer systems in the commercial and civilian sector echoed a broader growing discourse about the power of the United States on the global stage and the exposure of citizens at home to developments happening far away from them. Interdependence was the watchword in international relations in the 1970s. As international relations historian Daniel Sargent has documented, "interdependence" was the term that framed how statesmen, economists, futurists, and standard-setting bureaucrats alike came to see the position of the United States in the world order. Political scientists such as Joseph Nye and Zbigniew Brzezinski (soon to be Jimmy Carter's national security adviser) announced the end of the postwar order and the rise of "complex interdependence," characterized by a perceived waning of the nation-state itself as the key structuring institution of global power.[16] Electronic communication was more than simply a site where interdependence was playing out. Brzezinski treated it as a defining feature of the interdependent era. In his 1970 book, *Between Two Ages*, Brzezinski described "the onset of the technetronic era"—a period marked by the social and political shaping power of technology and electronics, "particularly in the area of computers and communications." Brzezinski contrasted his neologism with Daniel Bell's description of the "post-industrial" society as a way to center the significance of electronics. The increasing salience of the electronic medium, however, was not leading toward greater global convergence or harmony. Rather, it collapsed the layers on which world events and daily life unfolded. Interdependence, Brzezinski argued, was "better characterized by interaction than by intimacy." The individual was ever more conscious of national and global events, altering what constituted the perimeter of local experiences. Yet this exposure to and awareness of systemic breakdowns was "all the more unsettling, precisely because the mutual confidence and reciprocally reinforcing stability that are characteristic of village intimacy will be absent." The result was "a nervous, agitated, tense, and fragmented web of interdependent relations."[17] Electronic communications did not simply fall into the category of "systems" that were now revealing a shakier foundation than previously thought. They were the very means through which a new form of connection between individuals was being forged, a new sense of societal ties: more interactions, but less trust in them.

The interdependence discourse, Sargent argues, soon found its way outside strictly academic or governmental circles and permeated the ways in which Americans made sense of the turbulent 1970s—from thinking about the globe's ecological instability to concerns over the individual psyche captured in Alvin Toffler's *Future Shock*.[18] Bookended by two oil shocks and the drawn-out end of the American War in Vietnam, the 1970s brought precarity to the fore of American public consciousness. On the bicentennial of the American Revolution, July 4, 1976, the *New York Times* ran a column titled "Interdependence Day." "Americans are ever more aware," the column asserted, "that interdependence with the other industrial nations in economics, and with the Soviet Union in nuclear stability, is more and more a two-way street."[19] Connectivity, through electronic communications, energy supply chains, and nuclear brinksmanship, required shoring up certain protections of the individual, the corporation, and the nation's boundaries in face of this transformation.

First Rule of Security

In March 1975 the NBS returned to the *Federal Register* to propose, and request comments on, the Data Encryption Standard (DES). The proposed standard was the result of a submission received from IBM's data security group and of continued collaboration between IBM, the NBS, and the NSA.[20] The NSA's involvement raised considerable suspicion among critics of the DES, especially about the decision to make the length of the key only 56 bits.[21] Perhaps the most vocal critics were Stanford associate professor Martin Hellman and his graduate student Whitfield Diffie. Diffie and Hellman both cited Kahn's *The Codebreakers* as the inspiration that led them to the field of cryptography, but they began working together thanks in part to the IBM team that had developed the DES. In the summer of 1974, Diffie visited the IBM team. IBM's Alan Konheim, who refused to answer any substantive questions about IBM's work, suggested Diffie look up Hellman. Konheim and Hellman had worked together during the latter's short stint at IBM, before Hellman became a professor at MIT and later Stanford. "I think," Diffie remarked in court testimony in 2013, "Alan Konheim wishes he'd never introduced us. We went on to oppose him on some points on the DES."[22] This four-decades-old recollection is somewhat of an understatement.

Soon after the publication of the proposed DES, Diffie published an exhaustive analysis of what he perceived as its weaknesses, and Hellman began a letter-writing campaign to fellow cryptographers, heads of the computer industry and professional societies, the NBS, members of the IBM team, members of Congress, and journalists, including David Kahn. Diffie and Hellman's critiques hardly passed unnoticed. Both were invited to attend the two workshops about the DES that NBS held in 1976, and they received various letters from colleagues and NBS staff that attempted to address their concerns. Yet Diffie and Hellman felt these responses sidestepped the issues they had raised. Chief among these was the length of the key. A shorter key, Diffie and Hellman argued, would make the system vulnerable to an exhaustive search attack, one in which an opponent simply tries all possible key combinations. While the processing power limits of the day might initially make such a search computationally infeasible, the cost of such an attack would rapidly decrease over time. In a 1975 letter to Willis Ware, a computer scientist at RAND and member of the National Privacy Commission, Hellman argued, "Since it is possible to greatly increase the size of the keyspace, at almost no additional cost, we do not understand why such a small size was chosen." NBS's response to such criticism, Hellman continued, was tantamount to saying, "Trust us, it's good." For Hellman, the involvement of the intelligence community in developing the standard presented a conflict of interest. "I believe that one of the first rules of security is to trust no one and nothing without evidence that the trust is well placed," Hellman concluded.[23]

Despite these reservations, the DES was adopted as a federal standard, implemented widely, updated routinely with its key size expanded, and eventually withdrawn and replaced with a new standard. Years afterward, Hellman and Diffie received a vindication of sorts regarding their suspicions over the NSA's key size recommendation. An internal history of Cold War cryptography at NSA, written in 1998 and partially declassified under the Freedom of Information Act in 2008 and 2010, argued that the decision to collaborate with the DES efforts was not unanimous among NSA staff. According to that account, NSA's support of the DES was partially an attempt to narrow the field of public cryptography to a single known algorithm that would make it easier for the agency to attempt and develop mechanisms to subvert it. The 56-bit key size was a compromise between a proposal of 64 bits and those pushing for an even lower, 48 bits. On a different contentious point probed by Diffie and Hellman, that of deriving the substitution tables called S boxes used in the transformation of data, however, the later consensus was that NSA intervened in order to strengthen IBM's design.[24]

The controversy over the DES and NSA's involvement in developing the standard highlights an important aspect of cryptography as a technology of trust. Although it is used in a context where trust is scarce, its successful deployment still heavily relies on trust in actors who are themselves removed from relevant transactions and communications. In his study of verifiable computer systems, Donald MacKenzie has argued that the context of security, one in which "trust is problematic," throws into relief the foundational role that trust commonly serves outside that context, where it often passes unnoticed. If the presumption is that actors are malicious, the steps taken to establish a secure environment can reveal the necessary conditions for trusting the system as a whole.[25] The DES was just such an attempt to mitigate the threat of distrusted individuals. But trust was constructed as problematic in a particular direction—as the employer or system administrator's distrust of the employee or individual. Diffie and

Hellman's distrust of the NSA, and by association also the NBS and IBM, highlights how trust in a particular institution is itself a precondition for trusting the technologies of trust that this institution puts forth.[26]

Today, Diffie and Hellman are best known for their own groundbreaking work on *public-key cryptography*, an architectural innovation that has since been held up as a solution to the need to rely on trusted intermediaries or on institutions. Its technical features and their affinity with anti-institutional politics have been explored effectively in various recent accounts of cryptography.[27] In the 1970s, debates centered on the institutional aspects of trust that shaped how public cryptography research was done as much as on the architectural features of Diffie and Hellman's public-key cryptography work.[28] Diffie and Hellman were two among several cryptographers in the nascent field constituting *public cryptography* as an avenue of open research.

Going Public

In news coverage of the 1970s debates over cryptography, the field was elevated above other forms of scientific knowledge as a matter of public concern greater than the abstract notion of free scientific enterprise. On March 20, 1980, the Government Information and Individual Rights Subcommittee in the US House of Representatives convened a panel of legal, scientific, and national security experts to address "the confrontation between public cryptography and national security interests." The panel appeared before the subcommittee just after Floyd Abrams, prominent lawyer for the press in the Pentagon Papers case, testified on the constitutional status of "born classified" scientific research.[29] The Nixon administration's legacy of secrecy and surveillance haunted the subcommittee's work.[30] Committee chairman Richardson Preyer, a Democrat from North Carolina, called on Kahn to open the panel. In his prepared testimony, Kahn took issue with the term *"public* cryptography" in the panel's title. The actual subject, in his view, was *nongovernmental* cryptography. The word "public," he explained, was too close to the burgeoning field of public-*key* cryptography, which might cause confusion with the more limited consideration of a particular architectural principle in the design of cryptographic systems. Furthermore, the word public "may be construed as governmental, as in 'public corporation,' when actually the opposite—private individuals—is meant."[31] Public cryptography, he pointed out, was defined as the deployment of cryptographic tools by citizens and commercial businesses, as opposed to *and indeed in opposition to* the government.

Kahn's frustration with the malleability and misuse of "public cryptography" during the panel presentation came with no acknowledgment of his own complicity. In 1979, he published a column in *Foreign Affairs* in which he inaugurated the era of "public cryptography." Titled "Cryptography Goes Public," Kahn's article presented the burgeoning field of public cryptography as a challenge to the state monopoly of the technology, both in its areas of application and use (i.e., outside the military sphere) and as a challenge to government censorship over knowledge production. Concluding his article, Kahn pointed to a further problem with public cryptography: the lack of public venues in which to debate and develop policies for the field. "So cryptology, in 1945 a nation's most closely held secret, has gone public," Kahn summarized. "But not even procedures or forums for coming to grips with the new problems have been settled on. Their evolving substance will be harder still to resolve."[32]

The panel that appeared before the House Subcommittee on Government Information and Individual Rights on March 20, 1980, reflected an attempt to develop some of these procedures. Up to that point, oversight of nongovernmental cryptographic research was limited to the ad hoc invocation and application of secrecy orders and munitions export regulation. In addition to Kahn, the panel also heard testimonies from NSA director Bobby Inman and University of Wisconsin computer scientist George Davida. Bobby Inman had been appointed director of the NSA in July 1977 and oversaw a period characterized, according to NSA historian Thomas Johnson, by a contentious relationship with the Carter administration and Democratic House members.[33] In his 1980 testimony, Inman argued that the uptick in public cryptography research posed a real threat to the agency's ability to carry out its intelligence operations, and thus a threat to national security. But he also recognized that the NSA needed to standardize the ways it was monitoring and intervening in public cryptography research through demands on limited NSF funding or patent secrecy orders. He cited a recent case in which his copanelist George Davida had his patent application referred to the NSA and placed under a secrecy order until public outcry resulted in the order being removed.[34]

When George Davida took his turn to testify, he argued that the very possibility of NSA having any sort of say about how he went about his research was unheard of. Not only was it a violation of the First Amendment, but by slowing down the development of public cryptography, Inman was actively undermining national security by leaving commercial and private data of US citizens and businesses vulnerable. The discussion then turned from debating the meaning of cryptography to arguing who got to define "national security interests." Was it not more urgent to make sure that troves of personal information of US citizens and commercial secrets of US businesses were protected through strong encryption than to protect some abstract, ill-defined national security interest?[35]

The Cryptographic Domain

Inman and Davida took their debate to the American Council on Education's (ACE) Public Cryptography Study Group, convened by NSF and

ACE in the spring of 1980. The group comprised university administrators, representatives from professional societies like the ACM and IEEE, academic cryptographers such as Davida, and NSA general counsel Daniel Schwartz. Hellman was included among the observers and retained an active consulting role throughout these meetings.[36] The group's initial purpose was to discuss the possibility of prepublication government evaluation of academic research in cryptography. The "public" of public cryptography once again shifted, from describing an intended domain of application to describing the mode through which knowledge about cryptography itself might be shared. The debate thus touched both on cryptography as a unique domain of knowledge—unique either for playing a too valuable role in connecting the electronic public or for the military's need for control over it to maintain national security—and on the nature of scientific knowledge and open publication in a free society.

These intersecting concerns drew attention from journalists and public commentators. Gina Kolata, a writer for *Science* and another authorized observer, covered the discussion in great detail, often suggesting the goal was to establish "prior constraints" on any form of cryptographic work that would have to first go through an NSA screening process. (Inman repeatedly protested Kolata's coverage, arguing that it did not accord him and his representatives the good faith with which they had entered the conversation and that, he believed, other members of the committee had.)[37] Kahn, though not attending the meetings himself, kept close tabs and published a couple of articles and opinion pieces protesting the notion that prior constraints could be applied to such a foundational field as cryptography.[38]

Although the group was primarily concerned with how academic research ought to be carried out and published, and although it often discussed the constitutional mandate against censoring such publication, the potential commercial need for cryptographic tools was a constant reference point. A direct link was drawn between the possibility of publishing research in the open and developing stronger and better means of computer security for US businesses. Keeping businesses safe and secure required good and strong cryptography, and good cryptography was the result of free scientific enterprise, whose cornerstone was publishing research in the open. This link echoed a pervasive vision at the time about the workings of science as an inherently public practice: that the goal of scientific production was to add to a collective pool of knowledge and that the very act of making knowledge public contributed to its vetting and thus secured its public authority.[39] One of the most vocal advocates of this view was John Ziman, a British physicist and science policy adviser who in the 1970s authored a series of popular spins on the work of sociologist Robert K. Merton, highlighting the social dimensions of the scientific enterprise and the role science itself played in constructing a public. To be considered scientific in the first place, Ziman argued, knowledge had to be "available in the public domain, rather than the memories and thoughts of each person." But to achieve that status, knowledge should not only be "poured out into the archives" but had to achieve consensus through "strong interactions between the human actors in the drama."[40]

In contemporary discourses about trust and academic knowledge, the act of publishing has come to stand in for an imagined process of contestation, vetting of ideas, and formation of a broad consensus. Even more interesting is the fact that, as the individuals involved in researching and publishing have become ever more specialized, the vision of a broad public audience for these publications has not only been sustained through the publications' own self-fashioning but has been the very mechanism through which the expertise of the specialized researchers is secured. In the 1970s the research community for cryptography was so small and intimate that indeed most of the concerned researchers could be included in the Study Group's in-person conversations, either as formal representatives or, as with Hellman and Kolata, as invited observers. A field concerned with building a world of trust among strangers was in fact represented by a rather small group of colleagues. Yet the same appeals to publishing in the open persisted, and cryptography emerged as a viable academic field, with all the markers of a scientific discipline.[41]

Appeals to procedures of public legitimacy continue to be a fundamental mode through which scientific expertise maintains its authority. The twentieth-century notion of the "information age" drew on an idealized form of an open and accessible scientific literature.[42] The argument in favor of the publicness of public cryptography thus fits within a long line of practices through which the institutions of science became authoritative. But it also relied on a vision of the communications environment that cryptographic knowledge would make possible. Ideas of how knowledge about cryptography ought to be made and circulated and how this knowledge might itself be used in constructing the network society were thus linked through the slippages between the many meanings of "public cryptography."

The debates within the ACE Public Cryptography Study Group about the status of cryptographic knowledge tended to focus on its unique characteristics rather than on questions of scientific freedom writ large. This focus allowed the group to at once narrow its scope, sidestepping the politics of Cold War knowledge and nuclear secrecy, and broaden its scope by claiming expertise over a critical field for the future of the electronic society. The uniqueness of cryptographic publication was premised not on the function of scientific articles but on a shared understanding that the networked society to come would urgently require such an infrastructure. Cryptography was itself a means for constituting the public. It therefore had to be considered as lying beyond the scope of standard scientific knowledge.

A year after its first meeting, the Study Group published a report establishing a voluntary prepublication NSA screening process for any researchers concerned about their work.[43] The review process was thus presented as a service, assisting concerned researchers who found themselves unsure of the national security impact of their work.

Davida submitted a "minority report" in which he argued that the computerization of society had created a large number of "electronic windows" and that cryptography needed to serve as a curtain. It would be paradoxical, then, to submit cryptographic research to NSA screening when its very purpose was to act as a balancing measure against this power asymmetry.[44] Public cryptography, in the eyes of some of its early practitioners and advocates, was a demarcation technology, creating the very boundaries between the private and the public. It wasn't only in the public interest; it was the means through which there could be a meaningful electronic public.

Notes

1. Whitfield Diffie, foreword, in Bruce Schneier, *Applied Cryptography: Protocols, Algorithms, and Source Code in C*, 2nd ed. (New York: J. Wiley & Sons, 1996), xv.

2. David Kahn, *The Codebreakers: The Story of Secret Writing*, revised ed. (New York: Scribner, 1996), 84.

3. On the growing concern with data banks at the time, see Sarah Elizabeth Igo, "The Record Prison," chapter 6 in *The Known Citizen: A History of Privacy in Modern America* (Cambridge, MA: Harvard University Press, 2018).

4. Sarah Myers West, "Cryptography as Information Control," *Social Studies of Science* 52, no. 3 (2022): 353–75, https://doi.org/10.1177/03063127221078314.

5. Charles W. Beardsley, "Is Your Computer Insecure?," *IEEE Spectrum* 9, no. 1 (1972): 67, 78.

6. 38 Fed. Reg. 12763 (May 15, 1973), online https://tile.loc.gov/storage -services/service/ll/fedreg/fr038/fr038093/fr038093.pdf.

7. Dennis K. Branstad, "Data Protection through Cryptography," *NBS (National Bureau of Standards) Dimensions* 59, no. 9 (September 1975): 196.

8. W. Diffie and M. E. Hellman, "Privacy and Authentication: An Introduction to Cryptography," *Proceedings of the IEEE* 67, no. 3 (1979): 421, https://doi .org/10.1109/PROC.1979.11256.

9. James F. Schooley, *Responding to National Needs: The National Bureau of Standards Becomes the National Institute of Standards and Technology, 1969– 1993*, NIST Special Publication 955 (Gaithersburg, MD: National Institute of Standards and Technology, 2000).

10. "Privacy and Security: Twin Challenges to Computer Technology," *NBS Dimensions* 58, no. 7 (July 1974): 147.

11. Donald A. MacKenzie, *Mechanizing Proof: Computing, Risk, and Trust*, Inside Technology (Cambridge, MA: MIT Press, 2001), 170–72.

12. Susan K. Reed and Dennis K. Branstad, "Executive Guide for Computer Security," *Computers and Security* 1, no. 3 (1982): 234–35.

13. Rebecca Slayton, "Framing Computer Security and Privacy, 1967–1992," in *Communities of Computing: Computer Science and Society in the ACM*, ed. Thomas J. Misa (New York: ACM Books; San Rafael, CA: Morgan & Claypool, 2017), 301–3.

14. "Literature about Donn Parker," n.d., Box 1, Folder 2, Donn B. Parker Papers, CBI 166, Charles Babbage Institute, University of Minnesota.

15. Quoted in Slayton, "Framing Computer Security and Privacy," 298–99. Parker, however, was quite skeptical of the possibility that technical fixes could protect businesses from such risk and instead believed that broader legislation was needed against computer crime, along with the development of a code of ethics for computer professionals that would emphasize the grave violation involved in betraying the trust of an employer. An early draft of this code of ethics contained language pertaining to the fair compensation of employees, but this provision was struck out in the final version.

16. Daniel J. Sargent, *A Superpower Transformed: The Remaking of American Foreign Relations in the 1970s* (Oxford: Oxford University Press, 2015), 7, 166.

17. Zbigniew Brzezinski, *Between Two Ages: America's Role in the Technetronic Era* (New York: Penguin Books, 1976), 9, 19.

18. Sargent, *A Superpower Transformed*, 168.

19. "Interdependence Day," *New York Times*, July 4, 1976, sec. Archives, https://www.nytimes.com/1976/07/04/archives/interdependence-day.html.

20. 40 Fed. Reg. 122134 (March 17, 1975), http://cdn.loc.gov/service/ll/fed reg/fr040/fr040052/fr040052.pdf. For more on IBM's data security group and its early work on the DES, see "The Standard," chapter 2 in Steven Levy, *Crypto: How the Code Rebels Beat the Government—Saving Privacy in the Digital Age* (New York: Viking, 2001).

21. While the official language from NBS declared that the key was 64 bits in length, 8 of these bits were used for a check function, making the length of the key that had to be guessed correctly in order to decrypt the ciphertext only 56 bits long.

22. TQP Development LLC v. Newegg, Inc., Shelly Holmes (E.D. Tex. 2013).

23. Martin Hellman to Willis H. Ware, December 30, 1975, Box 1, Folder 1, Martin Hellman Papers, University Archives, Stanford University, Stanford, CA.

24. Thomas R. Johnson, "Book III: Retrenchment and Reform, 1972–1980," in *American Cryptology during the Cold War, 1945–1989* (Washington, DC: Center for Cryptologic History, National Security Agency, 1998), 232, http://nsarchive2.gwu.edu/NSAEBB/NSAEBB260/nsa-6.pdf.

25. MacKenzie, *Mechanizing Proof*, 316.

26. On the relationship between trust in expert institutions and the production of mechanical objectivity, see Theodore M. Porter, *Trust in Numbers: The Pursuit of Objectivity in Science and Public Life* (Princeton, NJ: Princeton University Press, 1995).

27. In *Crypto*, Steven Levy frames the entire history that began with Diffie and Hellman as an antigovernment project. Some more recent works from media and STS scholars highlights how this understanding of the politics of public-key cryptography was itself formed over the past five decades through the taking up of cryptographic lore by activists and in particular the cypherpunks of the 1990s. See Sarah Myers West, "Survival of the Cryptic," *Limn*, no. 8 (February 2017), https://limn.it/survival-of-the-cryptic; Finn Brunton, *Digital Cash: The Unknown History of the Anarchists, Utopians, and Technologists Who Built Cryptocurrency* (Princeton, NJ: Princeton University Press, 2019); Jean-François Blanchette, *Burdens of Proof: Cryptographic Culture and*

Evidence Law in the Age of Electronic Documents (Cambridge, MA: MIT Press, 2012).

28. In 2015, Diffie and Hellman jointly received the Turing Award, the ACM's highest honor, for this work. For an assessment of the applied impact of their work on commercial computer security over the past five decades, see Paul van C. Oorschot, "Public Cryptography's Impact on Society: How Diffie and Hellman Changed the World," in *Democratizing Cryptography: The Work of Whitfield Diffie and Martin Hellman,* ed. Rebecca Slayton (New York: Association for Computing Machinery Books, 2022): 19–56.

29. On scientific secrecy and "born classified" knowledge in physics research, see Peter Galison, "Secrecy in Three Acts, Part IV: Mechanisms of Limiting Knowledge," *Social Research* 77, no. 3 (2010): 941–74; and Alex Wellerstein, *Restricted Data: The History of Nuclear Secrecy in the United States* (Chicago: University of Chicago Press, 2021).

30. See Johnson, "Book III," 193.

31. *The Government's Classification of Private Ideas: Hearings Before a Subcommittee of the Committee on Government Operations, House of Representatives, February 28, March 20, and August 21, 1980,* 96th Cong., 2nd sess., 144 (1981), http://hdl.handle.net/2027/mdp.39015082027817.

32. David Kahn, "Cryptology Goes Public," *Foreign Affairs* 58, no. 1 (Fall 1979): 141–59, https://doi.org/10.2307/20040343.

33. Johnson, "Book III," 190–93.

34. *Government's Classification of Private Ideas,* 426.

35. *Government's Classification of Private Ideas,* 417.

36. David Brandin to ACM Executive Committee via S. Weinstein, memo re "PCSG Meeting, October 6, 1980," October 10, 1980, Box 4, Folder "ACM Correspondence Public Cryptography Study Group," Daniel McCracken Papers, Charles Babbage Institute, University of Minnesota. Hellman's involvement in these conversations was motivated less by his objections to the DES than by his feeling, a few years prior, that his own work was coming under government censorship, in what has now come to be known as the Meyer Affair. In July 1977, Joseph A. Meyer, an IEEE member and NSA employee, sent a letter to the IEEE board raising concern over the publication of cryptography articles that might constitute a violation of arms export regulation. Meyer to Gannet, July 7, 1977, Box 1, Folder 9, Martin Hellman Papers, University Archives, Stanford University, Stanford, CA; Martin Hellman, "Autobiography (Uncompleted Draft)," 2013, Martin Hellman Papers, Stanford Digital Repository, https://purl.stanford.edu/kq639bj2341.

37. G. B. Kolata, "Policy on Cryptography Proposals," *Science* 210, no. 4469 (1980): 511, https://doi.org/10.1126/science.210.4469.511.b.

38. See, for example, David Kahn, "The Public's Secrets," *Cryptologia* 5, no. 1 (1981): 20–26, https://doi.org/10.1080/0161-118191855742.

39. At the time, the practice of "peer review" as an act of legitimizing scientific knowledge was not such a core component of scientific publication as we have come to view it today. The term "peer review" itself was often meant literally when referring to grant application reviews but was meant more figuratively when used in relation to published articles, often referring to the abstract possibility of one's research being open to the scrutiny of the scientific community. In a small research community like the nascent field of public cryptography, the peers were indeed rather clearly delineated, but the fear of censorship was argued along the metaphorical notion of making knowledge publicly available. For a timeline of when some key scientific journals formally instituted the

peer review mechanism for published articles, see Alex Csiszar, "Troubled from the Start," *Nature* 532, no. 7599 (2016): 306–8.

40. John M. Ziman, *Reliable Knowledge: An Exploration of the Grounds for Belief in Science* (Cambridge: Cambridge University Press, 1978), 6, 7.

41. Historians of scientific journals in 18th- and 19th-century Europe have argued that both trust in the institution of scientific expertise and the notion of authoritative and coherent knowledge were tied to a performance of publicness— procedures of publishing that enmeshed it more, not less, in commercial and political realities. See, for example, Thomas Broman, "The Habermasian Public Sphere and 'Science in the Enlightenment,'" *History of Science* 36, no. 2 (1998): 123; Alex Csiszar, *The Scientific Journal: Authorship and the Politics of Knowledge in the Nineteenth Century* (Chicago: University of Chicago Press, 2018), 16.

42. Csiszar, "Troubled from the Start," 286.

43. Hellman and Inman developed a close relationship, to the point that Hellman offered to be the first cryptographer to voluntarily submit his own research for review. Later on he also reflected on his initial criticism of the DES as perhaps not granting it sufficient credit for what it did offer. Hellman, "Rethinking the DES," *Datamation* (October 1985).

44. Susan Landau, "Primes, Codes, and the National Security Agency," *Notices of the American Mathematical Society* 30, no. 1 (1983): 7–10; "Report of the Public Cryptography Study Group," *Cryptologia* 5, no. 3 (1981): 130–42, https://doi.org/10.1080/0161-118191855931.

17

Nodes and Codes

Iterating with the State in México

Héctor Beltrán

At the 2015 "Hack CDMX" event in Mexico City, Hugo and Chavita show off their code to others who could appreciate it. "Mira todos estos imports" (Look at all these import files), Chavita told Hugo, as he pointed to the dozens of "import" statements in his Python file. (An import statement tells the current file to look in other files that contain previously written code that you can reuse for the task at hand.) "No tengo más de cuarenta líneas en cada *class*" (I don't have more than forty lines in each class), Chavita proudly explained. Hugo, who hadn't slept in the previous 30 hours, managed to follow Chavita's demonstration with his bloodshot eyes and confirm Chavita's accomplishments with an enthusiastic "Eres un chingón" (You're a badass).[1]

Indeed, the principles of reuse, simplicity, consistency, and efficiency and the ability to shuttle between different levels of abstraction are core tenets of computer science and metrics used to identify a talented computer programmer. Young people at the event use the time and space to share code from projects they have been working on, sometimes from their professional jobs, in which results-oriented managers fail to recognize the complexity and beauty of their creations (figure 17.1). Like the hackers with whom Gabriella Coleman conducted her research, they value cleverness, ingenuity, and wit and transfer this ethos to the process of making technology and writing smart pieces of code. The hackers, Coleman found, "revel in directing their faculty for critical thought toward creating better technology or more sublime, beautiful code."[2] If one can dissect, manipulate, reassemble, and solve the problem within the given constraints and tools at hand, one can create elegant, "original" code.

While some newcomers have learned the ways of the code worlds at the hackathon, others have carved out more permanent spaces to carefully cultivate their code work, along with corresponding attitudes and

Figure 17.1. Young people hack away at Hack CDMX in 2015. Photo by the author

values. "We loved to go to hackathons, so we made one that would extend more in time. We wanted to live the hackathon every day," Kike told me. He is one of the founders of Dev.F, the first "hacker school" in Latin America, created in 2014. Its founders envisioned the school as providing time and space beyond the usual ritual experience of the hackathon where participants are given 72 hours to network and create teams, use their coding skills to present technical solutions to identified social problems, and pitch their ideas to a panel of experts identified by an organizing entity. While thousands of prototypes might be started at these events, many hackathons have similar endings, where participants just shake hands and say goodbye and where much of what "gets built" never really gets built at all.[3] As Kike and fellow founders developed the hacker school, they also made explicit in their popular blog the 10 principles of the "hacker ethics" a Dev.F hacker must follow:[4]

1. *Give before you get*
2. No pedir permiso (Don't ask for permission)
3. Hacer > Hablar (Doing > Talking)
4. No existen excusas (No Excuses)
5. Resolver problemas (Solve problems)
6.1. Sigue tu curiosidad (Follow your curiosity)
6.2. Fracasar = Crecer (Failing = Growing)
7. Conoce tus herramientas y comunidades (Know your tools and communities)

8. Siempre aprender (Always be learning)
9. Involucrarse (Get involved)
10. Divertirse en el proceso (Have fun)

For example, rules number three and number four ("hacer > hablar" and "No existen excusas") call for productive, noncomplaining citizens. The hacker culture thus becomes a product to be sold by experts to hackers-in-the-making, and the language used in the ethic clearly aligns with neoliberal rhetoric that calls on young people to take matters into their own hands.

But Kike told me more about how the school helps burgeoning hackers connect with potential employers. Since the school is nomadic, he told me, each batch of students takes part in the 12-week program in a different part of Mexico City, many times holding their trainings and related events in coworking spaces and sometimes in tech companies and government offices. When they work in one of these facilities, they are not involved in the operations of the organizational entity, but they do promote Dev.F participants for advertised jobs with the company or government office. The idea is that a few participants might transition into professional roles in the organization that do not compromise the hacker ethic they have so carefully nurtured in the Dev.F program. "If there's no match, the hacker comes back to the next batch and we try again. It's like catching an exception," Kike told me.

The metaphors that participants at Dev.F used to describe their practices proved to be fundamental to their code work across various domains. A "batch," for example, is simply a group of items. The term can also be used to refer to a set of instructions or processes that run before or after a user interacts with the computer, as in "batch processing" or a "batch job." In the case of Dev.F, each batch of students developed their coding skills and hacker persona before they were presented to the potential "user," in this case a company that would potentially find the coding skills of the hacker to be of value to their business. If there was no "match," Kike explained, or if during the time that Dev.F worked in the organization's space, the students deemed that their coding skills would not be sufficiently valued or that the entity might compromise their ethics, then they would try again, or in computing terms, would "catch the exception." An exception occurs when there is an anomaly or unusual occurrence during the normal flow of execution of a computer program.

A defensive programmer plans for these exceptions by anticipating anything that can go wrong during a program's execution, and prepares special cases in the code to handle the exceptions. As each of their batches progresses through the program and the school organizes events within tech companies or state-sponsored spaces, they interview workers to get a feel for the workplace dynamics (e.g., retention rates, coder satisfaction, nepotism) and the political moves within. Instead of simply responding to job postings and allowing companies to exploit its programming skills, Dev.F envisions that its careful analyses of the company's inner workings will allow participants to create an overall better system.

In the same way, the hackathons and hacker school serve as spaces where individuals negotiate not only their belonging to communities but their position within broader political and economic processes. Hugo, for example, held a strong conviction about what a hacker should or should not be. He voiced some of his concerns with the hacker school: "Their website even uses the templates all the Silicon Valley start-ups use. What's the difference?" he asked. In disagreeing with the "authentic" hacker ethos being compromised, Hugo touched on one of the fundamental tensions I heard hacker entrepreneurs debate: the anticorporate hacker ethos versus hackers' embeddedness and dependence on tech companies. Other heated debates took place around the following common points of contention:

> The focus on autonomy and DIY ethos versus the promise of state protection or care
>
> Mexican hackers' desire for association with the Silicon Valley tech sector versus the reality of being exploited by it
>
> The diverse personal and professional aspirations of hackers versus the limited opportunities available in México
>
> The differing positions on migration and mobility versus staying and working in and for your country

In this chapter I demonstrate how a heterogeneous cast of characters brought different motivations and experiences to the world of hacking during a period of political transition in México.[5] Building on the work of scholars who have focused on hacker communities in the Global South,[6] I explore how the shifting meanings of hacking point to significant technical and political change.[7] By putting in the code work alongside a diverse and shifting group of programmers inside and outside the hackathon, across different hacker spaces, and by spending time with them in their daily lives, my ethnography highlights the ways young people position themselves in relation to nation-building projects and to narratives that promote the promise of technology and infrastructures.[8] The term *code work* refers to both the ways research participants use the underlying logics of coding and design principles as they develop software systems and how these logics become "good to think with" about the institutions and systems that function as elements in state-driven infrastructures that spatialize unequal opportunities.

To show how this code work became fundamental for my research participants across varied sites, I draw on data from interviews and participant observation I conducted between 2013 and 2017 in México and the San Francisco Bay Area. I attended more than 20 hackathons, spent time in coworking and hacking spaces, and attended related industry events. In México, I conducted more than 50 open-ended formal and informal interviews with hacker entrepreneurs in Mexico City and Xalapa. Most of them were college graduates in their 20s with degrees in computer science and engineering or related disciplines, though people from all ages (some were in their late 30s and early 40s) and professional backgrounds (journalists and artists) were attracted to hacker spaces. Although the sites I investigate in this article are situated in México, my ethnography is transnational in that I traveled frequently between México and the San Francisco Bay Area, sometimes accompanied by my research participants and sometimes running into them at various tech-related events and spaces.

One of the ways I point to the shifting tenor of hacking communities in México is by using the term "hacker entrepreneur" to identify my research participants. Although a minority of them tightly control the hacker identity, this hybrid term points to the way many often shift between the labels "hacker" and "entrepreneur" or in some cases see no difference between the labels. A popular conception of the hacker is someone who loves to program computers in the spirit of playfulness and exploration, and a popular conception of an entrepreneur is someone immersed in capitalistic or technocratic projects. A smaller subset of hacker projects, such as radical hacker collectives or hacktivists, were directly antagonistic to capital and sought to pursue social justice politics, but hacker entrepreneurs didn't conform to these strict demarcations. The term points to this fluidity between identities but also to the ways techno-entrepreneurial Silicon Valley–esque cultures have come to dominate hacking cultures in México and Latin America more broadly. Some collectives might have originated as free software development and advocacy projects in the late 20th century,[9] but they are now (whether they like it or not) tightly coupled with the technologies and infrastructures that have emanated from techno-entrepreneurial cultures. Still, my research participants proudly wore the hacker badge across diverse spaces and collectives, usually to reference the fact that they were able to put in the code work, able to immerse themselves in the code worlds and program computers.

Thus, instead of presenting simplified versions of coders or hackers as either duped neoliberal subjects or empowered coding heroes, I explore ethnographically how young people in México consolidate processes of self-making and being made in relation to hacking. Especially during a time of political transition, how do they come together across markers of difference to fill hacker spaces with meaning, hope, and critique?

Nodes

During the early to mid-2010s in México, the tech start-up scene surfaced in parallel to hype from economic analysts who projected that México was set to emerge as the "Aztec Tiger" economy. Popular media outlets announced that México was undergoing rapid economic growth that would raise the standard of living. My fieldwork overlapped with Enrique Peña Nieto's presidency. During this time, his administration's "reforms" followed a developmentalist logic aimed to move México beyond low-wage factory jobs and toward an entrepreneurial economy. Popular discourse claimed that México was producing graduates in engineering and technology at rates that challenged its international rivals, including the United States, also its main trading partner. University enrollment had tripled in thirty years to almost 3 million students, who were eager to join México's growing middle class. Thus, analysts concluded that while México was becoming a top producer of raw engineering "talent," it lagged far behind in basic measures of "innovation," such as numbers of patents secured, scientific papers published, and research and development investments made.[10]

Following these capitalistic and developmentalist narratives, hackathons and hackerspaces fit into the larger Mexican political-economic landscape as spaces in which to keep these recent graduates busy, as potential generators of companies that would create jobs for them and their colleagues, and as the type of infrastructure that could help México emerge on the global innovation stage. These spaces indexed the government's interest in rapid economic development defined by technical expertise and information technologies.

The government representatives responsible for the construction of iLab, one of the coworking spaces where I conducted research in Xalapa, were quite explicit about what they hoped to accomplish with the facility. Jorge Antonio Mansur, federal delegate of the Ministry of Economy in Veracruz, told me that more than "developing developers," the state was interested in creating a fun, innovative culture that young people would want to be a part of. To foment this new culture, iLab needed to be inviting and modern looking, "not like the government offices people want to run out of immediately." This is why so much money was invested in iLab's design; ultimately the government wanted to create a brand that could be subsequently propagated across the country via similar *nodes*. The nodes, modeled on coworking spaces, with rooms to learn, be creative, and code, would fulfill the promise of "Todos con el mismo chip" (Everyone with the same chip), the slogan behind many of Peña Nieto's programs designed to create a new culture of innovation and corresponding opportunities for young people (figure 17.2). Mansur even detailed how the sanitized, "modern" space was part of a ploy to attract other politicians to invest in software development and entrepreneurship. "A politician makes roads because you can see them. You can't see the development of developers for a long time. But the advantage of iLab is that we've made it visible," he proudly told me.

Figure 17.2. "Nodes" modeled on iLab have been constructed across México. This one is in Oaxaca City. Photo by the author

The use and display of technological infrastructures by state entities to advance underlying political agendas is not new and has been theorized in other contexts. "'Techno-politics" refers to a mode of politics that functions through invisibility.[11] Grounded in liberalism, governmental bodies seemingly leave citizens be, allowing them to go about their everyday affairs without government intervening. Actually, governments seek proxies in technological regimes, building sewers, computing networks, phone lines, and other infrastructure that is seen as technical and outside political processes. Achille Mbembe described the simulacral language that was part of an aesthetics of power in African postcolonial dictatorships.[12] The state used this simulacral language and citizens accepted it, but it was devoid of referential meaning. This system worked not because it generated legitimacy but because it provided specific events that dramatized state power and therefore reinforced it. Brian Larkin has found a similar pattern in Nigeria: technology is used as part of political rule, and state-sponsored projects—roads, bridges, radio, any "new" technologies—are linked to events, spectacular rituals meant to produce particular affective responses.[13] Not only are state officials always present in the mediated representations of these projects, before and after they are built (and even if they never are!), but the repetition of this pageantry across different media is meant to produce and address a modern subject and train them on how to react to these awe-inspiring projects.

Not surprisingly, then, state government offices could be found on the

first floor of iLab, as politicians frequently went up to the top floors to hear the latest start-up pitches and, of course, to take pictures with these hackers diligently working (figure 17.3). As I conducted my interview with Mansur, one of his aides interrupted us to let him know that a representative of a local television network was on their way. "Tell him I talked to the governor and he [the governor] won't be able to make it today," Mansur told him. The statement worked to show his direct political connections and influential reach, but also as a reminder that the industrious programmers above were in demand by publicity outlets. These code workers were effectively confirming México's "coming of age" while governments across Latin America raced to move their nations away from categorization as "developing countries."[14] More important, the celebration of *entrepreneurial* hackers working within these spaces helped to promote a political agenda wherein young people were asked to employ neoliberal discourse about taking initiative, being self-satisfied, not waiting on government action, and being "socially conscious."

These emerging model knowledge workers thus come to represent developmental narratives of progress across the globe, especially in the Global South. In Kenya, for example, techno-entrepreneurs are not only imagined to epitomize the "Africa rising" narrative but also become "over-researched" subjects, the focus of researchers and related audiences.[15] In Brazil, maker and hacker spaces fall in line with a "Pink Tide" of new, center-left governments that invest in projects of social innovation aimed

Figure 17.3. Politicians, complete with suits and crossed arms, frequently visit iLab for photo sessions. Photo from iLab's public Facebook page

at capacitating communities with technologies to tackle social problems themselves, but working-class women challenge some of these monolithic "creative" infrastructures by making the spaces intergenerational and carefully developing arrangements that are both productive and caring.[16] Indeed, techno-entrepreneurs from the Global South don't always neatly align with the saturated tropes of Silicon Valley personas—"'hackers' as rebels; 'software developers' as rock stars, 'tech entrepreneurs' as disruptive actors, and so on."[17] As Lin Zhang explored during a period of rapid economic and social change in China, people from all walks of life, "middle-class and working-class urbanites, peasants and peasant workers, and young, transnationally mobile middle-class women," must all grapple with remaking a Chinese self but also with the call of entrepreneurial reinvention.[18] As Zhang argues, the urge to become a "digital entrepreneur" promises a "silver bullet for a whole host of individual and collective ills," a seduction to join the "worldwide cult of entrepreneurialism" that pushes "individuals and nations alike to become more enterprising by reinventing themselves in profitable new ways."[19]

The model entrepreneurial hacker thus emerged as a valuable subject in the Mexican political-economic landscape, where the majority of young people were disconnected from institutional support. That is, only a minority of young people in México were connected to institutional circuits that allowed them to make decisions about their livelihoods regarding their health, work, education, and security.[20] A 2010 study showed that about 7 million young people aged 14 to 29 in México were either looking for employment, not enrolled in school, or fell under the broad category of "not economically active."[21] This label is accompanied by a more colloquial term, *ninis*, short for "ni estudian, ni trabajan" (they don't study, they don't work), often used to disparage young people who, the argument goes, have become a burden to the society or economy. This labeling has its counterparts in other parts of the world. Vered Amit and Noel Dyck highlight the bureaucratic category "NEET" (not in employment, education, or training), used to describe some British youth.[22] The authors point to a specific category of youth all over the world who have been affected by economic restructuring. These young people are just as likely to be framed as agents of social change and progress as they are to index social breakdown and a problem for society. Young people in México, particularly those from working-class backgrounds, express skepticism that university degrees will lead to gainful employment.[23] In interviews with youth in México and Spain, researchers found widespread identification with a "generation of disenchantment," as when research participants stated, "They fooled us, we did what they told us and in the end things aren't the way they told us they would be."[24] Alberto "Chung," for example, expressed to me his cynicism about the government sponsorship of the upcoming Xalapa hackathon:

> Don't ask how I got a draft of the call for the "Xalapa Hackathon,"
> but if you want to participate, you can start getting ready. We have
> to develop technological solutions to resolve mobility problems and
> municipal services . . . according to this [announcement] whether
> you win or not you have to submit your project, code, and docu-
> mentation, and, well, the prize is that you will appear in Xalapa's
> newspaper, you get a scholarship to iLab, and you will get a pat
> on the back from the municipal president . . . very tempting,
> right?

Chung's comments mirrored the comments from other "disenchanted" youth across the world that other scholars have pointed to and, more generally, the expressions of loss of hope for the future and loss of faith in the neoliberal project that other anthropologists have found in contemporary contexts.[25]

Indeed, this collective of "disenchanted youth" has proved important to México's political history. They were central to the election of left-leaning Andrés Manuel López Obrador (AMLO) in 2018. Running under the banner of the newly minted party, Movimiento de Regeneración Nacional (MORENA), AMLO won the most votes at the presidential level in Mexican history, an incredible 30 points more than his closest contender, and triumphed in 31 out of 32 states and in 80% of the country's municipalities. AMLO's political discourse seldom specified names or detailed his plans for change. An antiestablishment stance against the "mafia del poder" (mafia with power) and critique of the "system" was enough to win over young voters who held skeptical hopes for something new. Surprisingly, AMLO's victory and promise of renewal came with familiar faces. Old adversaries had now become MORENA members, and AMLO accepted everyone willing to join the party with little gatekeeping. This approach wasn't entirely surprising, as AMLO himself started his career as part of the Partido Revolucionario Institucional (PRI) (Institutional Revolutionary Party). To that extent, these transitions and shifting alliances represented politics as usual. From 1929 to 2000, the PRI held power while it constructed itself around the ideals of the Mexican Revolution and continually reinvented itself until it became the representative of what many scholars call "institutionalized revolution."[26]

Larissa Adler Lomnitz, Claudio Lomnitz, and Ilya Adler highlight how presidential campaigns in México are carried out as political rituals in order to preserve the same structure and organization under the guise of different campaigns.[27] Even before the party's presidential candidates are chosen, "hombres del presidente" (men of the president) and "hombres del sistema" (men of the system) collaborate in tactical maneuvers and strategies that ultimately determine which subjects emerge as the "politicians" (those who will maintain close relationships to the future president) and

Figure 17.4. iLab as transformed into a political billboard in 2021. Photo by the author

which will emerge as the "technicians" (those who will assure the system continues).

State-sponsored hackathons, hacker spaces, and coworking spaces endured from Peña Nieto's tenure to AMLO's, as "men of the system" and "men of the president" reorganized themselves around new projects and so-called reforms. Figure 17.4 provides a stark representation of these political moves. In a return trip to Xalapa in 2021, I found that iLab had been transformed from coworking space to political billboard. The logos that were part of the "branding" Mansur envisioned remained, but now the space was plastered with political propaganda from the current party vying for representation in state elections. Much grimmer than this transformation were the allegations that Mansur had helped Veracruz governor Javier Duarte launder money to acquire properties in Texas worth US$6.7 million. Duarte was formally charged by the Mexican attorney general and resigned before fleeing to Guatemala in 2016. Mansur also resigned. This was particularly disturbing news for me since, when I sat with him for the interview, he emphatically told me that the only way to fight corruption was to set clear expectations and rules and to let burgeoning hacker entrepreneurs know that the government would stand behind them and lend resources for any projects they developed. It wasn't that I was accustomed to believing every politician's promises, but these striking allegations and abrupt transitions left me with a bitter taste for

the predicaments many of my research participants frequently found themselves in.

Thus, as politicians and technicians shifted their tactics and messages,
with new proposals for infrastructures and systems that promised to change
the status quo in México, hacker entrepreneurs used the government-
provided spaces to carry out their own social analysis and critique of particular systems. The nodes were only one more initiative that promised
to make them the affluent knowledge workers of tomorrow. Those who
gravitated toward the code continued to hack away amid politics as usual,
which meant unreliable government sponsorships, shameless politicians,
and empty promises. A closer look at this code work across my research
spaces in México reveals how the underlying logics of software design provided heuristics for analyzing their precarious relationships to political
and economic entities.

Codes

Back at the hacker school, stories abound of unfortunate experiences with
government programs. Javo tells me about the time he applied for funding through the INADEM (Instituto Nacional del Emprendedor), established by Peña Nieto's administration. His team of developers proposed
a platform that would prevent vote fraud in elections. Although he was
technically awarded the funds, they didn't come in time before the next
administration took office, and so he never received any actual money. He
claims that it wasn't that big of a deal, since the money could have been
used only for market research and lawyer fees, not to pay for developer
salaries, which he asserts is the most important way a start-up can use its
seed funds. "A start-up is 99% team and 1% product," he tells me.

As this last statement reveals, Javo is quite conversant in both tech
start-up and hacker speak. This is revealing for three reasons. First, it
shows that Javo, like other developers I talked to, were concerned about
not being viewed as *just* developers. Several referenced India as an example of a country that had produced talented developers who lacked skills
as innovators or even leaders of potential companies.[28] Here they were
echoing perspectives by transnational IT companies that often rely on programmers for the lower-end, commodity side of the computer services
industry.[29] Framing employees in these "body shops" as interchangeable
with any other IT worker is in the best interest of companies that outsource labor, especially to countries in the Global South, in order to consistently grow their revenue. Curiously, Javo and fellow hackers in this
case agreed with Mansur and the state's rhetoric. By creating start-ups,
they could prove that they were innovative workers of tomorrow and not
"just" programmers who could easily be replaced.[30] (This is increasingly
relevant as analysts and consulting companies continue to frame México
as a place where a surplus of "talent" is waiting to be hired for 2–3 times

cheaper than in the United States, with specific geographical advantages over India.)[31] Second, Javo reminds us that the government is one of the primary sources of funding for technological start-up projects. Javo went on to tell me about the trials he had faced attempting to access private venture capital. In México there were very limited resources, and US investors were hesitant to fund Mexican start-ups. A fluent English speaker, trained in the prestigious Instituto Tecnológico de Monterrey ("México's MIT"), he had even gone as far as arranging to have a physical address in San Francisco make it appear to investors that his start-up and workers were physically in the United States. Finally, Javo's statement is important because it alludes to the type of ethnography necessary to understand the decisions of young people learning to work within and across the hacker and entrepreneurship worlds.

Indeed, the agility with which hacker-entrepreneurs maneuvered between seemingly contradictory spaces was quite revealing during my research. Early in my fieldwork, I visited multiple hacker spaces to get a feel for the differing political alignments and ideologies across hacker communities. One collective, more closely aligned with the images of hacker spaces that the general public might have, featured DIY workshops using open-source technologies. Its alignment with radical leftist politics was evidenced not only by participants' discourse but also by Zapatista posters and a communal community disdain for any member who might pull out an iPhone. This collective contrasted sharply with some of the corporate-sponsored hackathon spaces and events, complete with company-sponsored infrastructures to support software development (and sometimes gourmet catering), a reverence for Silicon Valley figures and discourses, and participants who compared the specs of their latest smartphones. To my surprise, I wasn't the only diligent "ethnographer" navigating these contrasting communities; across these spaces, I frequently ran into my research participants, plugging into and out of each collective with ease.

Hugo was one of the research participants I frequently encountered traversing what appeared to be conflicting spaces. I took advantage of a break at Hack CDMX to conduct an informal interview with him. Explaining why he was so tired, Hugo said he was working for a tech consulting firm in Mexico City and was spending 10 to 12 hours a day programming. He often had to work weekends with no extra pay because he was being paid by the project. His best friend, Memo, who came from a similar socioeconomic background, had recently landed a job doing back-end coding for a major bank in México. Memo now has a guaranteed salary perhaps worthy of his UNAM degree;[32] the 17,000 pesos per month (about US$850 in 2014) solidified his position as a lower-middle-class citizen who no longer had to worry about making ends meet. A monthly salary of 16,000–25,000 pesos (US$800–US$1,250) was the norm for a recent software engineering graduate from a reputable program; for someone with two to five years' experience, the monthly salary jumped to 30,000–50,000 pesos (US\$1,500–US\$2,500); and for a veteran programmer with ten or more years of experience, the monthly salary could reach close to 100,000 pesos (US\$5,000). These potential earnings and job security made Memo happy and his family proud of him.

Hugo, on the other hand, wasn't impressed with Memo's salary, much less with his "selling out." Although Hugo lacked job security and made less money working on projects for the consulting firm, he'd rather have worked on the more challenging projects the firm hands him, and, he said, even if the work was similar, they couldn't pay him enough to wear a suit and join the *godinez* (office workers) at the bank.[33] He is aware of the precarity of his situation, but instead of framing it negatively, he refers to his work arrangement as "loose coupling."

"Loose coupling" is a computing term that refers to a robust way to write code wherein data structures (or other components) can use other components in an interconnected system without needing to know the full details of their implementation. In this way, each component becomes more autonomous and can be used for different purposes by different components; elements become "coupled" and interdepend with very little (or no) direct knowledge of each other. Hugo goes on to recommend manuals and tutorials that further explain this software design so that I can appreciate its value. The term "loose coupling" Hugo uses to refer to his flexible work arrangement references at once his autonomy and his replaceability. Like many of the young people in attendance, Hugo contracted out his programming skills to diverse companies or start-ups. In his case, the consulting firm helped make these connections, especially with US-based companies looking for programmers who work for less than US-based software programmers.

Hugo then changed gears in our conversation to elaborate on the hackathon dynamics: "In the past, the politicians would arrive to distribute blenders and take pictures when the basket[ball] court was completed, if it was ever completed. Now, they arrive to distribute hackathon stickers and take pictures with the winning teams." The stickers Hugo refers to are primarily used as marketing material: they show the logos of tech companies, operating systems, development tools, and hackathon events, and participants like to decorate their laptops with them to create colorful, creative displays. Although Hugo criticized the practice of sticker distribution, associating it with politicians' "old" method of securing votes by gifting household electronic appliances such as blenders, he still proudly displayed his stickers on his laptop. Moreover, the varied events, companies, and technological platforms showed the conflicting and fleeting allegiances that currently make up his hacker world. Like the "loose coupling" approach he took to writing code, his sticker arrangement pointed to his flexible (and legible) networking capabilities.

Hugo talks about the politicians (as representatives of the state) in the same way as he does companies, as elements that can be both represented visually and reconfigured on his computer. Instead of thinking of these sticker arrangements as pointing to the seemingly incompatible collectives to which the hacker entrepreneurs belong, we can think of these shifting compositions as a canvas on which (fleeting) relationships are made explicit and negotiated. Loose coupling gives him a way to think through these relationships. In the same way, the hackathon serves as a space where individuals negotiate not only their belonging to communities but their position within broader political and economic processes. With a hesitancy similar to Hugo's regarding the hacker school "selling out" by adopting start-up cultural forms, my research participants used the hackathon as a space to debate common points of contention (outlined in this chapter's introduction): the desire of being associated with Silicon Valley infrastructures versus the reality of being exploited by them; the focus on autonomy and a DIY ethos central to the hacker world versus the promise of state care; the aspirations of hackers to benefit from their involvement with tech companies versus the practice of being exploited by them.

These shifting relationships and negotiations also point to the neoliberal knowledge economy and underlying processes of transnational capitalism that ask young people to work by the project and on their own time. Hacker entrepreneurs who wish to rise must learn to respond quickly and with agility to volatile market trajectories and frequently cross career, role, and political boundaries to perform their flexible or "latitudinal citizenship."[34] Market volatility becomes a way of life in which flexibility, instability, liquidity, and risk taking are interpreted as desirable challenges that the modern subject can manage by making calculated decisions.[35]

But unlike other entrepreneurs who embody the rhythms of neoliberal life (in México or elsewhere), Hugo and other hacker entrepreneurs stay close to the code. Concepts such as loose coupling serve to make sense of the shifting relationships, even when these relationships are with the institutions that promise to help them unpack these relationships (e.g., the hacker school). Their lived reality calls for young people to be flexible, but not too flexible; autonomous, but not too autonomous. As Ilana Gershon reminds us in her study of Silicon Valley job seekers and job hoppers, enacting a neoliberal self is tricky. Workers come to see themselves as projects that must steer through various possible obstacles and alliances, moment after moment, with each instance creating a possible contradictory dilemma.[36] Neoliberal workers develop different strategies of making do in precarious labor markets across different national and cultural contexts. Young people in South Korea, for example, learn to "think with play" when they immerse themselves in the digital gaming worlds and develop dispositions for thinking and acting in quickly shifting, unsettled circumstances; strategies for "productive slowness" allow gamers to take ownership over their time and activities in irregular labor arrangements.[37] In México, my research participants might not necessarily "escape" neoliberal work conditions, but the code work allows them to slow things down in order to take momentary snapshots of the way their economic and political connections are drawn together. The code work offers heuristics that might lead to transformative openings in volatile conditions as well as metaphors that provide the promise of autonomy as young people assume their positions as "neoliberal workers."

Techno-politics employs technological proxies that appear to be outside of politics in order to distract citizens and continue business as usual. In México, the hackathon and the hacker space serve as rituals for performing modernity and the promise of technology. But the hacker entrepreneurs construct new forms of mobility by traversing the bottom layers of these "new" technologies and infrastructures, where they can observe how different elements are related, how things "really work." Software's fundamental appeal is that it has the power to illuminate the unknown; the separation of software from hardware, interface from infrastructure, provides a powerful metaphor for the way the system works.[38] Young people adept at constructing and traversing these technical infrastructures feel less threatened or "trapped" by them and more inspired to "recapture them and turn them to new ends in the service of new worlds."[39] Between the code worlds and sociopolitical worlds, the code work gives hacker entrepreneurs a tool kit for social critique; it gives them the tools to think with the system, whether that system is the latest software infrastructure, socioeconomic program, or political reform.

Conclusion: Iterating with the State

I began my fieldwork as the technical instructor for a six-week incubator-style course designed to teach entrepreneurship and mobile internet technology skills to Mexican students. The workshops took place at UNAM in 2013 and were sponsored by MIT's Global Startup Labs.[40] At our first meeting, Memo showed up in a business suit and tie. His colleagues, who showed up in much more casual wear, ridiculed him. Perhaps Memo was confused by the call to create entrepreneurs *and* software developers and wanted to make sure we quickly associated him with the former; perhaps he got his hacker and entrepreneur agendas crossed up that day; perhaps he had read Anne Saxenian's book in which she attributes Silicon Valley's economic success largely to the way it developed a "laid back" California attitude and culture, complete with corresponding wardrobe, in direct contrast to the "buttoned up" style of the East Coast, and Memo associated MIT geographically with the latter.[41] But Memo's faux pas was an indicator of things to come. As I have argued, in order to iterate with the state, young people in México had to become adept at moving across social domains that appeared contradictory.

By exploring my research participants' multiple, overlapping, and contrasting relationships to the hacker entrepreneur cultural worlds, I highlight critiques that emerge about neoliberal work life from these *other(ed)* code workers themselves.[42] By shifting attention away from the results of making and hacking, I demonstrate how hacking functions as a mode of intervening in and positioning oneself in relation to existing economic and political structures.[43] Hackers in México shuttle between the different layers of abstraction that make up "the code" as they build technical systems using underlying software and design principles. They immerse themselves in these code worlds, navigating the politics of the knowledge economy while they reinterpret coding logics such as "loose coupling" and fundamental software concepts such as "batches" and "exceptions," in order to think about the ways they might reconfigure their relationships with state and economic entities that produce value from their hacking.

Hackathons, coworking spaces, entrepreneurial initiatives, and neoliberal "reforms" are seldom differentiated by politicians. But as the state plans to propagate nodes that promote the promise of technology but also threaten to simply respatialize unjust practices of corruption and inequality, young people use these spaces to observe these practices, rethink their relationship to them, and plan their course of action. If the "state" inserts itself on the bottom floor of iLab to monitor the work of the hacker entrepreneurs and use them as props for photo opportunities, the hacker school uses the coding metaphors to attempt to slow things down and reverse the relationship. That is, they attach themselves to entities to get a feel for the institution, to "interview" them while they figure out if it's an institution they want to commit to. Their work within these spaces might effectively help them dodge the problematic categorization of "ninis"; they inevitably perform for various publics the role of knowledge workers enthusiastically connected to their computers. But the real work, the code work, falls under the radar.

As each iteration of hacker entrepreneurs pushes forward their hopes and desires, they re-mediate pessimism and guarded optimism alongside each "new" version of modernity staged by corresponding political parties. Guided by their intimate knowledge of Mexican institutions, the hackers use coding logic to provide a foundation for honing their ability to manage themselves and their practices in order to scrutinize their relationships with the state, private companies, and their valued hacker communities. For these communities, fraught relationships with political and economic entities might be the reality of their existence, but coding logics provide the metaphors and (they hope) tools to think of these entities as elements in the current configuration, and to realign their own practices accordingly. The code work provides heuristics for analyzing the organization of entities and relationship between them, whether it is an element in a coding environment or an actor in a political-economic environment.

Around the same time that hacker entrepreneurs were honing their coding skills during my research, multitudes of citizens across México collectively protested the impunity, corruption, and violence that had come to characterize state practices, whereby *narcofosas* (drug-trade graves) with hundreds of unclaimed bodies frequently appeared in clandestine locations and dozens of protesting students went "missing" at the hands of state officials. In a particularly dramatic case well covered by the media, 43 students were forcibly disappeared from the teacher's college Ayotzinapa, in the state of Guerrero, on September 26, 2014. Mariana Mora argues that because these students were constructed as coming from a space of social backwardness, isolated from modernist nation-building efforts, they were treated as collateral damage; they were simply preventing the rest of the country from moving forward on a developmental scale.[44]

Although many of my research participants came from similar socio-economic backgrounds as the disappeared students, their activities in the hacker spaces helped them dodge a similar fate. At one level, their demonstrated activity in these spaces served to prevent their being labeled ninis, in that their activity might be qualified as "studying." Moreover, their studying fell in a "respectable" category, in that they weren't learning how to farm or becoming versed in transformative pedagogies, as the Ayotzinapa students were doing. The hacker entrepreneurs were performing their roles as affluent knowledge-workers-in-training within modern, sanitized coworking spaces. If they happened to come from marginalized backgrounds, they were demonstrating they were capable of overcoming intergenerational cultural and economic "backwardness." The code work thus offered a younger generation in México the conceptual tool kit to think about political economy on the ground in the context of changing state power, shifting meanings of entrepreneurialism, and a challenging era of political violence.

My research participants appropriated the discourses of flexibility and self-management at one level, while they allowed their code work to guide them at another. When faced with the developmentalist "reforms" of neoliberal economic change, middle-class youth may respond by exploiting their advantages vis-à-vis the poor, or by building solidarities to protest the status quo. Among Mexican hacker entrepreneurs, I found a heterogeneous cast of characters, motivations, and experiences; many of these workers were not just driven by misguided interest in exhibiting the entrepreneurial spirit in order to perform middle-class-ness. I've suggested that youth from diverse socioeconomic backgrounds gravitated toward code during a time of transition in México, when a newly formed leftist political party gained power after a century of rule by a party characterized by unmet promises, a party that continuously switched around its "men of the president" and "men of the system" to give an appearance of change. In this context, "hacking" emerged as a way for young people to make sense of their futures in a precarious state and economy, as a way to let the "code work" intervene in narratives that had delivered only false hopes, and as a way to think alongside the system responsible for reinstating unequal opportunities, iteration after iteration.

Notes

1. *Chingón* comes from the Spanish slang verb *chingar*, which translates roughly as "to screw." In his book *Dancing with the Devil: Society and Cultural Poetics in Mexican-American South Texas* (Madison: University of Wisconsin Press, 1994), José Limón conducted ethnography with South Texas *carnales* who exchanged aggressive idioms of sexual violation with one another, especially epithets using the word *chingar*. Limón argued that what could be read as homosexuality-in-play may also be interpreted as a moment for reversing the sociosexual idiom of *chingar* as practiced by *los chingones* (the screwers or men-in-charge), who continually violate the well-being and dignity of working-class men. For further discussion of performances of masculinity and construction of gender in Mexico hackathons, see my "The First Latina Hackathon: Recoding Infrastructures from México," *Catalyst: Feminism, Theory, and Technoscience* 6, no. 2 (2020).

2. Gabriella E. Coleman, *Coding Freedom: The Ethics and Aesthetics of Hacking* (Princeton, NJ: Princeton University Press, 2013), 118.

3. Lilly Irani, "Hackathons and the Making of Entrepreneurial Citizenship," *Science, Technology, and Human Values* 40, no. 5 (2015): 804. Lukas McIntosh and Caroline D. Hardin conducted an empirical study of nearly 12,000 project code repositories related to the popular Major League Hacking events between 2018 and 2019. They conclude that very few show patterns of consistent development, with only 7% of projects showing any activity after six months. See McIntosh and Hardin, "Do Hackathon Projects Change the World? An Empirical Analysis of GitHub Repositories," in *SIGCSE 2021: Proceedings of the 52nd Technical Symposium on Computer Science Education* (New York: Association for Computing Machinery, 2021).

4. Manuel Morato, "El Decálogo de la Cultura Hacker," *Medium*, January 31, 2015, https://medium.com/@ememorato/el-decalogo-de-la-cultura-hacker-95a6a45b5d1b; Manuel Morato, "The Ten Commandments of Hacker Culture," *Medium*, February 1, 2015, https://medium.com/@ememorato/the-ten-commandments-of-hacker-culture-4e183d570eb6.

5. This chapter draws in part from my article "Code Work: Thinking with the System in México," *American Anthropologist* 122, no. 3 (2020): 487–500.

6. Anita Say Chan, *Networking Peripheries: Technological Futures and the Myth of Digital Universalism* (Cambridge, MA: MIT Press, 2013); Yuri Takhteyev, *Coding Places: Software Practices in a South American City* (Cambridge, MA: MIT Press, 2012).

7. Gabriella E. Coleman and Christopher Kelty, "Preface: Hacks, Leaks, and Breaches," *Limn*, no. 8 (2017), https://limn.press/article/preface-hacks-leaks-and-breaches/.

8. Nikhil Anand, Akhil Gupta, and Hannah Appel, *The Promise of Infrastructure* (Durham, NC: Duke University Press, 2018); Shalini Shankar, *Desi Land: Teen Culture, Class, and Success in Silicon Valley* (Durham, NC: Duke University Press, 2008).

9. Christopher M. Kelty, *Two Bits: The Cultural Significance of Free Software* (Durham, NC: Duke University Press, 2008).

10. William Booth, "Mexico Is Now a Top Producer of Engineers, but Where Are Jobs?" *Washington Post*, October 28, 2012, https://www.washingtonpost.com /world/the_americas/mexico-is-now-a-top-producer-of-engineers-but-where -are-jobs/2012/10/28/902db93a-1e47-11e2-8817-41b9a7aaabc7_story.html.

11. Patrick Joyce, *The Rule of Freedom: Liberalism and the Modern City* (London: Verso, 2003); Timothy Mitchell, *Rule of Experts: Egypt, Techno-politics, Modernity* (Berkeley: University of California Press, 2002).

12. Achile Mbembe, *On the Postcolony* (Berkeley: University of California Press, 2001).

13. Brian Larkin, *Signal and Noise: Media, Infrastructure, and Urban Culture in Nigeria* (Durham, NC: Duke University Press, 2008).

14. Arlene Dávila, *El Mall: The Spatial and Class Politics of Shopping Malls in Latin America* (Oakland: University of California Press, 2016).

15. Angela Okune, "Open Ethnographic Archiving as Feminist, Decolonizing Practice," *Catalyst: Feminism, Theory, and Technoscience* 6, no. 2 (2020).

16. Liliana Gil, "A Fablab at the Periphery: Decentering Innovation from São Paulo," *American Anthropologist* 124, no. 4 (2022).

17. Sareeta Amrute and Luis Felipe Murillo, "Introduction: Computing in/from the South." *Catalyst: Feminism, Theory, and Technoscience* 6, no. 2 (2020): 3.

18. Lin Zhang, *The Labor of Reinvention: Entrepreneurship in the New Chinese Digital Economy* (New York: Columbia University Press, 2023), xvi.

19. Zhang, *Labor of Reinvention*, x.

20. Rosanna Reguillo, "La condición juvenil en el México contemporáneo: Biografías, incertidumbres y lugares" (The state of youth in contemporary Mexico: Biographies, uncertainties, and places), in *Los jóvenes en México*, ed. Rosanna Reguillo (México: FCE/Conaculta, 2010); Maritza Urteaga Castro Pozo, *La construcción juvenil de la realidad: Jóvenes contemporáneos* (México: Juan Pablos Eds., UAM—Iztapalapa, 2011); Mónica Valdez, "Jóvenes y datos: Panorama de la desigualdad," *Juventudes, culturas, identidades y tribus juveniles en el México contemporáneo: Suplemento diario de campo*, no. 56 (2009): 37–39.

21. Instituto Mexicano de la Juventud, *Encuesta Nacional de Juventud 2010* (México: INJUVE, 2010), http://www.imjuventud.gob.mx/imgs/uploads /Presentacion_ENJ_2010_Dr_Tuiran_V4am.pdf.

22. Vered Amit and Noel Dyck, "Pursuing Respectable Adulthood: Social Reproduction in Time of Uncertainty," in *Young Men in Uncertain Times*, ed. Vered Amit and Noel Dyck (New York: Berghahn Books, 2012).

23. Jayne Howell, "Getting Out to Get Ahead? Perspectives on Schooling and Social and Geographic Mobility in Southern Mexico," *Journal of Latin American and Caribbean Anthropology* 23, no. 2 (2017): 301–19.

24. Néstor García Canclini and Francisco Cruces, "Conversación a modo de prólogo," in *Jóvenes, Culturas Urbanas y Redes Digitales: Prácticas emergentes en las Artes, las Editoriales y la música*, ed. Néstor García Canclini, Francisco Cruces, and Maritza Urteaga Castro Pozo (Madrid: Fundación Telefónica, 2012), viii.

25. Lucia Cantero, "Sociocultural Anthropology in 2016: In Dark Times; Hauntologies and Other Ghosts of Production," *American Anthropologist* 119, no. 2 (2017): 308–18; Annelise Riles, "Market Collaboration: Finance, Culture, and Ethnography after Neoliberalism," *American Anthropologist* 115, no. 4 (2013): 555–69.

26. The party was founded in 1929 as the Partido Nacional Revolucionario (PNR), in 1938 it was dissolved and renamed Partido de la Revolución Mexicana (PRM), and in 1949 the PRM was dissolved and reborn as PRI.

27. Larissa Adler Lomnitz, Claudio Lomnitz Adler, and Ilya Adler, "El fondo de la forma: La campaña presidencial del PRI en 1988" (The function of the form: The PRI's presidential campaign in 1988), *Nueva Antropología* 11, no. 38 (1990): 45–82.

28. In examining the politics of entrepreneurial citizenship in India, Lilly Irani reveals that projects in the name of innovation implicitly reproduce long-standing divisions between "those who develop and those who must be developed." See Irani, *Chasing Innovation: Making Entrepreneurial Citizens in Modern India* (Princeton, NJ: Princeton University Press, 2019), 8.

29. Jeffrey R. Yost, *Making IT Work: A History of the Computer Services Industry*, (Cambridge, MA: MIT Press, 2017).

30. Sareeta Amrute has shown how Indian programmers, well aware of their precarious situation as talented but replaceable workers, deployed tactics such as bad code commenting to obscure parts of their cognitive labor and fight the logics of labor replaceability. See Amrute, *Encoding Race, Encoding Class: Indian IT Workers in Berlin* (Durham, NC: Duke University Press, 2016).

31. See, for example, Gustavo Parés, "In the Race for Tech Talent, the US Should Look to Mexico," *TechCrunch*, May 20, 2021, https://techcrunch.com /2021/05/20/in-the-race-for-tech-talent-the-us-should-look-to-mexico/; and the 2019 report by CodersLink, an organization dedicated specifically to connecting global companies with "world-class talent, no matter where they are located": "Tech Salaries Report 2019: A Quick Guide for Staying Competitive in the Tech Hiring Market," CodersLink, accessed January 1, 2015, https://coderslink.com /company/tech-salaries-report-2019/.

32. The Universidad Nacional Autónoma de México (National Autonomous University of Mexico) is the largest university in Latin America and consistently ranked as one of the world's top universities.

33. *Godinez* refers to a culture of urban office workers in Mexico who are usually salaried, nine-to-five employees consistently trying to outsmart their boss and find workarounds to make their routine jobs easier. The origins of the term are contested, and it can also be considered derogatory.

34. Aihwa Ong, *Flexible Citizenship: The Cultural Logics of Transnationality* (Durham, NC: Duke University Press, 2019).

35. Karen Ho, *Liquidated: An Ethnography of Wall Street* (Durham, NC: Duke University Press, 2009); Hirokazu Miyazaki, "The Temporalities of the Market," *American Anthropologist* 105, no. 2 (2003): 255–65.

36. Ilana Gershon, "Employing the CEO of Me, Inc.: US Corporate Hiring in a Neoliberal Age," *American Ethnologist* 45, no. 2 (2018): 175.

37. Stephen C. Rea, "Calibrating Play: Sociotemporality in South Korean Digital Gaming Culture," *American Anthropologist* 120, no. 3 (2018): 509.

38. Wendy Hui Kyong Chun, *Programmed Visions: Software and Memory* (Cambridge, MA: MIT Press, 2013).

39. Nick Seaver, "Captivating Algorithms: Recommender Systems as Traps," *Journal of Material Culture* 24, no. 4 (2019): 433.

40. The program was part of MIT's International Science and Technology Initiative, which "promotes development in emerging regions by cultivating young technology entrepreneurs." MIT GSL: Global Startup Labs, accessed February 16, 2025, https://gsl.mit.edu.

41. AnnaLee Saxenian, *Regional Advantage: Culture and Competition in Silicon Valley and Route 128* (Cambridge, MA: Harvard University Press, 1996).

42. Amrute, *Encoding Race, Encoding Class*; Irani, *Chasing Innovation*.

43. Morgan G. Ames, Silvia Lindtner, Shaowen Bardzell, Jeffrey Bardzell, Lilly Nguyen, Syed Ishtiaque Ahmed, Nusrat Jahan, Steven J. Jackson, and Paul Dourish, "Making or Making Do? Challenging the Mythologies of Making and Hacking," *Journal of Peer Production* 12 (2018): 1–21.

44. Mariana Mora, "Ayotzinapa and the Criminalization of Racialized Poverty in La Montaña, Guerrero, Mexico." *PoLAR: Political and Legal Anthropology Review* 40, no. 1 (2017): 67–85.

Epilogue

Artificial Intelligence—Braiding Irony, Paradox, and Possibility

Jeffrey R. Yost and Gerardo Con Díaz

Our two-day "Just Code" symposium at the Charles Babbage Institute in 2020 and the creation of this book occurred during a tumultuous half decade. The period included a global pandemic; the murder of George Floyd by police and the subsequent riots and peaceful protests; war and attacks to subvert democratic sovereignty (Russia's invasion of Ukraine and a US president's spurring insurrection at the US Capitol); Hamas's terrorism in Israel and the Israeli government's genocidal attacks, killing predominantly civilians in Gaza; and the reality and hype of rapidly deployed generative "artificial intelligence" (AI). Through the media and other means, the code (software) and codes (cultural and legal) of AI systems intersected all these occurrences and many others in areas of science, health, education, government, capitalism, and social politics.

The COVID-19 pandemic pushed our "Just Code" event online, a change that expanded its reach. More than 360 people from more than two dozen countries—including graduate students and scholars—virtually attended the two-day event in October 2020. All the chapters included in this book skillfully inform on the political economy of code and power imbalances, and they all lend weight to our analytic of representation, communication, and interaction code. Yet only one chapter is squarely on artificial intelligence—Stephanie Dick's stellar history of mathematically mapping the "standard" human head for computational-based facial recognition in the 1960s. Given the release and widespread use of OpenAI and Microsoft's ChatGPT-4 (Generalized Pre-trained Transformer) in late 2022, and the importance of historical contexts to these developments, this epilogue reflects on generative AI's prehistory and history. It highlights ironies, paradoxes, and appropriations of imagined futures or possibilities. We hope that this historical reflection can serve as a departing point for future work investigating the code analytic and the three major questions that organize this book, including and especially in the domain of AI.

Just over forty years ago, *Time* magazine announced that its 1982 "Man of the Year" was in fact the "Machine of the Year," the personal computer, with a cover headline "The Computer Moves In."[1] The image was of an almost ghostlike and perplexed white silhouette of a seemingly older man in his home, his hands resting on his legs, seated in a yellow chair, at a red table and with a white personal computer with monitor in front of him. There was speculation that AI might be the *Time* person or technology of the year for 2023, but the magazine chose to fuel another hype instead: that of icon, and new addition to the billionaire class, music superstar Taylor Swift. Sadly, in early 2024, only weeks later, Swift, like many other women (famous and not), suffered from the spread of deepfake AI pornography depicting her. In the Swift case, the increasingly toxic and misogynistic social platform X allowed it to go viral. One fake pornographic image had 47 million views before the image was taken down, compounding the harm.[2]

Back in 1982, despite *Time* magazine's accolades, personal computers (PCs) were less of a sensation than AI is today. For one thing, they were expensive. Depending on its configuration, an IBM PC was $1,565 to $3,000, or $5,000 to $9,000 in inflation-adjusted 2024 dollars. Fewer than 3 million personal computers of all makes and models were sold in 1982, and as IBM intended for its PC, many of its sales were for offices, not homes. By the end of 1984, the year the famous two-minute Orwellian Superbowl commercial introduced the iconic Macintosh, computers had "moved in" to only 8.2% of US households. In contrast, ChatGPT-4 had a million users within five days of its launch. This time to the million mark was 15, 60, and 250 times faster than Instagram, Facebook, and Netflix, respectively. Within two months, ChatGPT-4 had more than 100 million users, a feat nearly all other platforms took years to achieve.[3]

Even the memory of the early World Wide Web's spectacular bubble did not prepare us for AI's explosion in popularity. I had used "Braiding Irony, Paradox, and Possibility" as a subtitle once before, at the advent of the web, and did not expect to reuse it. This was for a chapter I (Yost) wrote back in late 1999, titled "Computers and the Internet" (and published in 2005).[4] There were many ironies and paradoxes and wide-ranging imagined futures in the early days of the World Wide Web, which was subsequently labeled Web 1.0. And here we are again, with the same kind of braid in a context that can be traced back directly to these early years of online engagement. Internet hype, the dot.com bubble, and our current generative AI situation have striking parallels: technological utopianism, technological solutionism, a Silicon Valley–based venture capital, a "California Ideology" of right-leaning libertarianism, "irrational exuberance," and the realities and fears of expanding inequalities. In that prior publication, the topics of race and gender discrimination figured prominently. They do in this epilogue as well.

As we move on to consider AI's place in our world, it is important to remember some of the lessons on the politics and economics of technological change that we learned during the dot.com bubble. John Chambers, Bill Gates, and Andy Grove, the CEOs of Cisco Systems, Microsoft, and Intel, respectively, rode the late-1990s-to-early-2000 bubble as heads of the three largest companies in information technology. They, of course, all suffered after the 2000–2002 tech bubble burst. The technology-laden NASDAQ market fell 78% from its prior peak and took more than a dozen years to recover in nominal terms. All three corporations were survivors of this so-called dot-com bust, which decimated yet-to-be-profitable companies, especially smaller enterprises that followed Amazon.com into the e-commerce arena.[5] Amazon and Microsoft thrived subsequently, reaching, in 2007 and 2015, respectively, their prior peaks and later greatly exceeding them. To this day, Intel and Cisco Systems have remained below their early–year 2000 capitalization, and far below such levels in real terms. Nvidia, based on AI graphical processing unit (GPU) demand and its 80%-plus GPU market share, is 12 times the size of Intel. With AI stocks, the question of whether the market is on the verge of crashing, or is more like in 1998 and a couple of years away, is uncertain. But sometime soon, a major correction is likely to bring to-the-moon dreams back to earth.

Technological change at the time was also subject to the industrial constraints that accompanied strong antitrust prosecution. In 1998 the Department of Justice (DOJ) sued Microsoft for violating the Sherman Antitrust Act. In late 1999, federal district judge Thomas Penfield Jackson ruled against Microsoft. Sentencing, in June 2000, ordered the breakup of Microsoft into separate companies, similar in form to the 1982 consent decree that sharded AT&T into the "Baby Bells."

Microsoft appealed the district court's decision, claiming prejudice on the part of Judge Jackson. A US appeals court upheld many findings of fact against Microsoft, but also found that Judge Jackson's communication with media "seriously tainted the proceedings" of the trial.[6] Microsoft, with a negotiating position strengthened by the appeals court opinion on the original ruling, reached a settlement with DOJ prosecutors whereby the software giant suffered only mild remedies far short of breakup. It set the stage for nonexistent or anemic efforts by the DOJ and Federal Trade Commission (FTC) in the past two decades to regulate information technology. The lone exception was FTC chair Lina Kahn's impressive leadership and aggressive regulatory efforts between 2021 and 2024, but it was an uphill battle against entrenched trillion-dollar corporations. The general inaction and lack of success with federal regulation have facilitated Microsoft's growth into one of several giants in both the cloud and AI.[7]

OpenAI was a nonprofit formed in December 2015. The pedigrees of its founders (Elon Musk of Tesla, Sam Altman of Loopt, Peter Thiel of PayPal, and others) help to explain why it has been part of the tech and financial elite from its origin. Undoubtedly other start-ups will be successful over time in AI, but the dominance of past highly capitalized information technology giants in this field stands out. This success extends in major part from the DOJ's and FTC's lack of regulation and ineffectiveness for decades. Today's top AI companies come from leadership in several key information technology products and services—smartphones, search engines, the cloud, social networking, e-commerce, and GPUs—and are the only US companies with trillion-dollar capitalizations: Apple, Microsoft, Alphabet, Amazon, Meta, and Nvidia. The technological, political, and social history of AI dates back long before the founding of these companies and centers on the desire to present and view machines with human-like qualities.

Making and Marketing Perceptions and Perceptrons

Since the advent of digital computing, media representations and public understandings of computers have centered on relations to and metaphors for humans and human brains. Late 1940s and early 1950s articles likened mainframe computers to "electronic brains." The War Department invested heavily in the technology during World War II, and when it publicly unveiled the earliest impactful digital computer in the United States, the Electronic Numerical Integrator and Computer (ENIAC), it emphasized its ability to "rebuild scientific affairs on new foundations."[8] Edmond Berkeley, an actuarial scientist and cofounder of the Association for Computing Machinery (ACM), added impetus to this metaphor in his classic 1949 book *Giant Brains; or, Machines That Think.*[9]

This metaphorical thinking was especially important during the early years of AI. In organizing a 1956 summer workshop of top scientists exploring ideas about information processing and automation at Dartmouth College, John McCarthy coined the term "artificial intelligence." In doing so, he expected it would help raise military funding for this area of research, and he was right. Meanwhile, after the summer, McCarthy left Dartmouth to join MIT, the birthplace of Whirlwind, SAGE, and the fast-emerging poster university for the military-industrial-academic complex. MIT's Jay Forester had used grandiose rhetoric to sell Whirlwind, and real-time processing, to the military.[10] Following the suit, McCarthy and other participants who backed his idea were doing the same with "artificial intelligence," making rhetorical appeals to military funders. It was marketing and presenting AI as significant science that could aid with the challenges of the escalating Cold War. Thus began habitual cycles of over-promising a future of human-like and superhuman machine intelligence, a future designed to appeal to the Department of Defense and to resonate with other funding sources.

In a July 1958 article, "Electronic 'Brain' Teaches Itself," the *New York Times* reported on a demonstration by Cornell University Aeronautical Laboratory Research psychologist Frank Rosenblatt's "embryo of an electronic computer named The Perceptron that [according to Rosenblatt] 'would be the first to perceive, recognize, and identify its surroundings without human training or control'" Rosenblatt and his group used the US Weather Bureau's $2 million IBM 704 computer to demonstrate that this "mechanical brain" could identify whether a randomly placed square was on the right or left. It was successful 97% of the time. In overly optimistic rhetoric for the nearer term that proved prescient for the more distant future, Rosenblatt predicted, "Later Perceptrons will be able to recognize people and call out their names. Printed pages, longhand letters and even speech commands are within its reach." He also emphasized that the machine needed no "priming," hinting at a generative capability from a generalized base. Rosenblatt's demonstration laid a core foundation for and added momentum to a neural network specialization within AI, one that had both biological–human brain and machine–computing research strands.[11]

Rosenblatt's demonstration built on and added to existing metaphorical thought and framing borrowed from human neurology. His development of the "perceptron" followed on and practically added to the pathbreaking theoretical work Warren McCulloch and Walter Pitts published in 1943 on "nervous nets," the "symbolic treatment of known nets," and a model for a machine neuron.[12] It also drew on and complemented past scholarship by Donald Hebb, in particular the models of neurons and brain cell interaction that were part of Hebbian theory, distilled in his seminal 1949 book, *The Organization of Behavior.*[13]

Playing Games and Processing Language

Computer power and speed have always been critical to the artificial intelligence field. In the late 1950s and 1960s no organizations had more combined computer power worldwide than IBM and the still quite secretive National Security Agency (where insiders joked that NSA stood for "No Such Agency"). NSA engaged in information gathering, analysis, cryptography, and computer security. IBM pioneered in mainframes, an industry it dominated globally with more than 60% market share. Fundamental to this dominance was its leadership in software, especially operating systems, programming languages (such as FORTRAN), and database management systems (Information Management Store). IBM bundled its applications and services to boost hardware sales and leases prior to the 1970s.

IBM played a major part in nurturing AI's role within computer science. Like General Electric and Bell Labs, IBM Research (headquartered at Thomas J. Watson Laboratories in Yorktown Heights, New York, and, over time, evident at its other prominent corporate research facilities also) was centrally concerned with scientific and technological development. Hardware was the overwhelming profit center focus at IBM, but the company was also deeply involved with the young discipline of computer science. Within this broad field, machine learning, a field drawing from precomputing pattern recognition and statistical analysis and lumped under the spreading umbrella of "artificial intelligence," had a special place.[14]

IBM's Arthur Samuel, who popularized the term "machine learning," was a pioneer in drawing from pattern recognition and statistics and applying IBM's computer resources, or compute, in the 1950s and 1960s to games. In 1959 Samuel published a landmark paper on the Samuel Checkers Playing Program. It was one of the first self-learning software programs, a computer program in which he coded the rules and objectives of the game as well as strategies. Through playing, gaining ever more data, the system learned. Samuel argued that through playing many games, the system could learn to beat its programmer.[15]

Checkers is a simple game with identical pieces. The major exception, of course, is reaching the end row and being crowned king (stacking two pieces on each other), thereby gaining additional powers to move and to capture. Given checkers' relative lack of sophistication, and the fact that IBM, Control Data Corporation (CDC), and other computer firms came out with increasingly powerful computers in the 1960s—the first supercomputers such as IBM Stretch and the CDC 6600—AI game playing quickly graduated.

Another square-grid, combat-oriented, and heavily gendered game—one that is extremely complex—would captivate AI scientists for decades: chess. In 1948, the pioneer who invented information theory, Claude Shannon researched whether there were more atoms in the universe or more different moves or variations in a game of chess. He reasoned that there were 10 to the 78th to 82nd power atoms in the universe, but there are more moves and variations in chess games. This hypothetical, finite, and undiscernible exact number became known as the Shannon Number, 10 to the 120th power.[16]

A focus on chess carried its own social and cultural baggage. For centuries, chess has been heavily gendered and played principally by boys and men. While not as restrictive as it once was, gender constructions and limiting diversity in chess continue, nationally and globally, and in competitive play as well as for leisure.[17] Of the players in International Chess Federation (FIDE) tournaments, only 10% are women. There are tournaments for both men and women, but most are gender segregated, and the federation has discriminated against women and transgender people. In 2023 FIDE issued a statement "FIDE Regulations on Transgender Chess Players." Its biased formal policy implies cisgendered women cannot be competitive with cisgendered men. It stated it "will recognize an individual's gender identity that is consistent with the identity they maintain in their non chess life and that has been confirmed by national authorities based on a due legal and formal process of change. FIDE recognizes that this is an evolving issue for chess . . . and may need to be evolved in the future in line with research evidence."[18]

More generally, the gamification of AI carried with it implicit gendering. McCarthy coined "artificial intelligence" to appeal to another gender-segregated space and institution, the US military. Key events in the 1960s intensified the Cold War, including the Bay of Pigs Invasion, Cuban Missile Crisis, the Gulf of Tonkin "advisors," and the Vietnam War. Computer applications to war gaming at the Pentagon and the Air Force–sponsored RAND Corporation, and computer models of input-output economics by C-E-I-R Inc. for the air force, escalated.[19] Against this backdrop, both McCarthy's marketing term and embracing studies of a male-gendered, warrior game of strategy and logic in chess seemed an excellent fit.

This Cold War construction of chess was true of the other superpower as well. Historian Nathan Ensmenger recounted how Russian mathematician and AI researcher Alexander Kronrod justified the inflated cost of computer time at the Soviet Institute of Theoretical and Experimental Physics by claiming that "chess was the Drosophila of AI." Herbert Simon—the Nobel laureate in economics, AI scientist (a key participant at Dartmouth in the summer of 1956), and codesigner of the Logic Theorist—published this analogy in 1973, which the AI research community picked up in succeeding years. And as Ensmenger insightfully argued, fruit fly research became an increasingly common metaphor for the disciplines of artificial intelligence and cognitive science.[20]

Chess remained a challenge and continuing focus of scores of AI researchers for decades. A mid-1980s AI chess project at Carnegie Mellon University moved to IBM by the end of that decade. After IBM took over, it changed the project's name to Deep Blue, drawing on the company's iconic nickname Big Blue. With fanfare Deep Blue took on world champion chess "grandmaster" Garry Kasparov in 1996 in a half-dozen matches, losing four of the six and winning twice. In a six-game rematch the following year, Deep Blue beat Kasparov with two wins and three draws.[21] Given the complexity of chess and its long-term centrality in AI research, it was an important moment in the AI community and a demonstration that became central to AI in popular culture. It symbolized what supercomputers and their programmers could achieve with the vast resources of a corporate giant in the 1990s, but it did not represent a thinking or intelligent machine, just one capable of processing copious amounts of data. It also drew unequaled attention to the world of chess, and the highly gendered space in which Kasparov achieved and added to his fame.

It therefore comes as no surprise that the history of AI is infused with the historical exclusions based on gender and race that characterize the history of computing more broadly. Although women programmed ENIAC, they were a stark minority in computer science, less than 15% in the 1950s and 1960s, and still less than 20% in the 1970s. Women's participation grew and peaked at over 35% in the mid-1980s, but then fell to below 20% in the 2000s. The proportion has been around 20 to 25% over the past decade.[22] Likely, gendered cultures, military funding, and a focus on chess all contributed to women's lower participation in AI than in computer science broadly. A Deloitte study reported 18% of the US AI workforce were women in 2010, while the National Science Foundation placed the computer–mathematical science workforce that year at 25.1%.[23] The gender imbalance may have been worse than AI overall in the increasingly prominent subspecialty of neural networks.

The Neural Information Processing Systems Conference long went by the acronym NIPS, with some male attendees making offensive jokes and wearing sexist T-shirts reading "my NIPS are NP-hard." A major women-led petition effort persuaded conference organizers to change the acronym to NeurIPs and to enact a new code of conduct starting in 2018. Threats of violence were made against some of the women pushing for this acronym change.[24]

The broader cultural and technological context of the Cold War also played a pivotal role in AI's development by generating demand and funding for specific kinds of AI systems. During the Cold War, there was heightened pressure to develop not only the most advanced weapons, command and control, logistics, and war gaming, but also the capacity to process vast amounts of textual and spoken communication. The pioneering work of Noam Chomsky and others in computational linguistics, as well as emergence of the AI subfield of natural language processing (NLP), was fundamental to both science and the military. In the late 1950s and 1960s symbolic NLP dominated with heuristics, or rules, to emulate, to advance human and machine communication. In 1960 MIT's John McCarthy designed LISP (LISt Processing), the start of evolving iterations of general-purpose programming languages in part optimized for symbolic NLP and AI. Inspired by McCarthy, it was Steve Russell (soon of pioneering Spacewar! computer game fame, in 1962) and other colleagues on the McCarthy-led team who developed it.

A desire and need to define and create "intelligence" accompanied this technological pressure. Back in the 1950s, Alan Turing had published his famed paper "The Imitation Game," or what later was referred to as a Turing test, in which human subjects would judge whether they were communicating with another human or a computer, based on teletype text.[25] This challenge was influential in the field, but fooling a subject does not mean a computer is legitimately an electronic brain, or is intelligent.

Increasing computational possibilities also generated the urge to create more interactive systems. The AI researcher most famous for a conversation software system in the mid-1960s was MIT computer scientist Joseph Weizenbaum. His ELIZA was a text system that used a psychoanalytical clinic style of Rogerian (after Carl Rogers) or person-centered psychotherapy, in which keywords are identified within answers and these are flipped into and incorporated systematically in subsequent questions on a computer screen. Participants who interacted with the system, and those who read about it, ascribed more to it—a level of understanding or intelligence—than it deserved.

Weizenbaum, himself, was surprised by this inflation to "intelligence" and was quick to caution against making this system into more than it was—a programmed set of rules to respond in a limited way and form without any broader contextual framework behind it. Whereas AI scientists, such as former MIT colleagues Marvin Minsky and Stanford's John McCarthy (who had moved from MIT in 1962), espoused grander visions of thinking machines, Weizenbaum emphasized differences between mere information processing and human thought or reason—ideas philosopher John Searle would reinforce and extend a half decade later.[26]

This mix of technological, political, and commercial pressures set the stage for an interesting paradox: *metaphors of AI have always been about humans and the human brain, but actual research and development in this field runs counter to these metaphors by backing away from complexity and finding shortcuts to show "progress."*

Communication scholar Xiaochang Li's research shows that IBM's Continuous Speech Recognition (CSR) group, which began at Watson Research Center in 1972, adopted a statistical approach whereby, as the group's director Fred Jelenik put it, "rather than exhaustively studying how people listen to and understand speech, we wanted to find the natural way for the machine to do it." This "brute force pattern recognition," as Li insightfully states, pried "speech recognition away from the study of human sensory and reasoning processes and toward the production of massive quantities of computer-readable speech and data." This approach foreshadowed what occurs with large language models (LLMs), a gearing of data, processes, and processing for false claims of mission accomplishment and progress toward human-like, artificial general intelligence, putative milestones achieved through ill-suited means that shove aside complex human and situational contexts and nuances. In IBM's CSR case, the outputs focused on individual words, not the perception and interpretation of speech.[27]

In 1969 Marvin Minsky and Seymour Papert published their influential *Perceptrons: An Introduction to Computational Geometry.*[28] The book presented a series of proofs highlighting challenges with multilayered perceptrons and was commonly interpreted as displaying the limits of the neural network model of AI, a critique that seemingly lent weight to a shift to symbolic AI in the 1970s and beyond (including expert systems and, more recently, the semantic web). Sociologist Mikel Olazaran usefully presented a more nuanced story of the perceptron debate of Rosenblatt, Minsky, and Papert and others. Olazaran's research highlights the fact that the traditional displacement of neural net research with symbolic AI was a construct of organizational influences and efforts in the symbolic

AI research community to better secure standing and funding, and that much within the history of the perceptron / neural net–versus–symbolic AI controversy is subject to "interpretive flexibility."[29] Also open to wide interpretation is the idea of and timings with AI seasonality.

Seasons of AI?

The years 2023 and 2024 saw great hype around and investment in AI. Journalists, computer scientists, and others are speculating on whether this investment will continue and how. One of the key ongoing questions is whether there might be an upcoming "AI Winter"—a period when frustration about overpromises and underdeliveries leads to a sharp decline in funding. Or perhaps AI has reached a maturity of perpetual warming, if not outright heat, when AI winters might be things of the past.

Using metaphorical seasons to discuss AI's history is, at best, problematic. It seems to have made sense to the elite group of scientists at a handful of top AI institutions (Carnegie Mellon, MIT, Stanford) working on particular areas of AI research, but not more broadly. It is also exemplary of how power dictates in the formation of narratives, and how those narratives endure and become the lore of a field. Several historians have commented usefully on this phenomenon, especially Aaron Mendon-Plasek, in emphasizing research more broadly in pattern recognition and machine learning, and highlighting a more gender, racial, and institutionally diverse set of actors, during supposed AI winters.[30] If there was such a thing as an AI winter, research did not slow or come to a halt. In fact, as historian Thomas Haigh has documented, during the first so-called AI winter, which began in the mid-1970s, researchers performed especially important organizational work and expansion.[31]

What does appear to wax and wane is federal, particularly military funding of highly resourced universities and labs. Since the Dartmouth AI summer workshop, top computer scientists at elite institutions have been hyperfocused on opportunities to use Department of Defense (DOD) research dollars. This was especially true in AI, where computer power and resources played an important real and perceived role.

The Eisenhower administration founded the DOD Advanced Research Projects Agency (ARPA) in 1958 in response to the launch of the Russian Sputnik satellite into orbit. From the start, ARPA's mission was to support big science and technology and to conduct research that would expand the frontiers of knowledge and strengthen the United States. In 1962, utilizing powerful air defense command-and-control backup mainframes no longer in service, the DOD launched the Information Processing Techniques Office (IPTO) to take advantage of these resources, and fortuitously selected psychologist and computer scientist J. C. R. Licklider to be its founding director. Licklider's work was critical in setting the early research agenda for computer science by principally funding top universities and creating centers of excellence in the areas of AI, time-sharing, graphics, and networking. IPTO also occasionally funded top scientists and created centers of excellence anew at prior, non-elite universities.[32]

It is difficult, if not impossible, to establish a direct causal link between overpromising and underdelivery on one hand and funding for AI research on the other. For example, Marvin Minsky, John McCarthy, and others overpromised and underdelivered with AI, but it would be a stretch to say that that this significantly impacted funding for AI. DARPA's (it added "Defense" to the name) disappointment with speech recognition progress at Carnegie Mellon University in the 1970s existed, but that is one area of AI research at one school. Of greater relevance in the context of the Nixon administration and the escalating conflict or war in Vietnam, the 1970–71 Mansfield amendment required a direct military purpose for DOD research dollars. Haigh usefully shows how the Association for Computing Machinery membership in the Special Interest Group on Artificial Intelligence (SIGART) grew by 50% between 1973 and 1978, to 3,500 people. This half-decade span represents the core years of the supposed AI winter. AI was a much larger field than a speech recognition project at Carnegie Mellon or machine translation work at several other elite schools (which the 1966 Automatic Language Processing Advisory Committee Report, admittedly, did negatively impact).[33]

Noticeable reductions in AI research funding have periodically unfolded around natural language processing, as it required tremendous computational power, and an (overly) optimistic appetite for what it might spawn has always been prevalent in the NLP and neural networks spaces. In the 1990s n-gram models and recurrent neural networks (RNNs) advanced NLP, and the rapid growth of the World Wide Web boosted the perceived need and opportunities for digital text and voice recognition applications. Increasingly, many of the larger projects developed at leading information technology companies' research and development labs, rather than at universities. In 2010 Apple released NLP virtual assistant Siri, followed by Amazon's Alexa a half decade later.

Still, one of the reasons why a more general direct causal connection between overpromising and research funding may not be established is that *much AI research has occurred outside the confines of any one field of technoscientific inquiry.* At elite and nonelite schools, new areas of AI were taking off and expanding. Most notably, at Stanford, Edward Feigenbaum, Joshua Lederberg, and Bruce Buchanon pioneered expert systems in organic chemistry compound prediction with DENDRAL in the mid-1960s, and with MYCIN, led by Edward Shortliffe and Feigenbaum, in medical diagnosis and antibiotic regimens in the 1970s. The National Institutes of Health funded this research.[34]

Successes and the announced Japanese Fifth Generation AI project did lead to increased fear, hype, and AI funding in the mid-1980s, but boosts as occasional responses to black swan times or events are different than the metaphorical summer growing season and winter dormancy. Further expert systems have been thriving at universities and at companies, where limiting domains has led to useful codification of knowledge and applications in medicine, logistics, enterprise resource planning, finance, and other areas.

Expert systems' effectiveness varies by domain, expert participation, cost, the quality of the heuristics, the constitution of the inference engine, and maintenance. The greatest challenge almost always is the knowledge acquisition (and representation) bottleneck. As such, the personnel (computer scientists, social scientists, and domain experts), management, safeguards, and maintenance of projects make all the difference. Like LLMs, expert systems also have at times produced social harms, such as with credit ratings and home loan evaluation systems amplifying biases.[35] Going beyond the limited domains of expert systems to LLMs, the risks greatly escalate. In general, LLMs and systems extending from them are a dangerous shortcut for efficiency that produce far less than expert outputs.

Matters of How Metaphors Matter

The year 2018 was a momentous one for LLMs and derivative bots. OpenAI created ChatGPT-1, and Google built BERT. Microsoft's investments and partnership with OpenAI began the following year. In late 2022 and 2023 OpenAI-Microsoft deployed ChatGPT-4, and soon thereafter, Google released Bard. These became fast media sensations. Such systems are trained on large volumes of online and print texts.

LLMs of the past decade surveille and scrape the web, publications, and other data indiscriminately and out of context. These systems are opaque and complex, and quality metaphors can help to better understand them, while lesser metaphors can distort. Georgetown's Helen Toner, who served on OpenAI's board of directors from 2021 to 2023, has argued for an "acting improv" metaphor for AI chatbots.[36] That metaphor is problematic as it highlights machines having human-like intelligence and creativity, but they do not. Her metaphor reflects more than a half century of anthropomorphizing machines and the human brain but is less useful than other types of metaphors given how LLM bots work.

In contrast, computational linguist Emily Bender and colleagues' "stochastic parrots" metaphor for generative AI is both compelling and illustrative of how these LLMs and chatbots are designed and operate, and their severe limitations and considerable risks.[37] Parrots, of course, are not human, but they can put together sentences, ones that are wholly blind to larger contexts. Rules and statistics often keep the responses plausible and reduce perceived hallucinations, but the very process of training on vast amounts of texts means output is acontextual, hallucinations are inevitable, and generated text is not trustworthy. When generated responses appear more on target, often the foundation is human labor, or annotators.

The Labor of Automation

Venkey Harinarayan earned his doctorate in computer science at Stanford, working under Jeffrey Ullman, in 1997, after earning an undergraduate computer science degree at Indian Institute of Technology, Madras, and a master's at University of California Los Angeles.[38] His influential dissertation and early publications focused on query processing in data warehousing and optimizing data cubes. He and four other engineers cofounded Junglee, a pioneering comparative-shopping e-commerce company, in 1996. Two years later, Amazon.com purchased Junglee for $250 million in Amazon stock. Becoming an Amazon executive, Harinarayan worked closely with Jeff Bezos to help flesh out the concept of Amazon Marketplace, a platform for third-party sellers that has become highly lucrative for Amazon over time.

In the early 2000s, Harinarayan developed the concept of microwork as part of Amazon Marketplace, an idea that evolved into Amazon Mechanical Turk or MTurk, which allows people to outsource work. This system is named after Hungarian inventor Wolfgang Ritter von Kempelen's automaton chess player, "the Turk," a 1770s hoax with a small man hiding inside a chess-playing "machine."

Amazon unveiled MTurk to the public at an MIT event, along with Amazon Web Services and cloud computing, in 2006. Harinarayan was lead engineer behind a US patent for this system, titled "Hybrid machine/human computing arrangement," filed in October 2001 and issued on March 27, 2007.[39] MTurk allows outsourcing of types of data-processing work—from transcription and image labeling and categorization to basic information research tasks. Clients list work tasks and the compensation. The latter ranges from $1 to over $10 an hour.

In both theory and practice, MTurk has centered on work that is difficult or more expensive to automate. However, automation (a term more descriptive than and used interchangeably with AI) rarely produces long-lived working systems. Such systems continually need to be maintained, improved, and updated. MTurk and its imitators exemplify something Amazon knew early on, namely that for automated systems, there is a tremendous and continuous need for microwork. Microwork reflects what media and gender studies scholar Lilly Irani has aptly characterized as "humans-as-a-service."

Irani argues that as crowdsourcing entrepreneurs keep menial tasks invisible, they and their clients can imagine workers are in "a better place," one with freedom in choosing tasks. She shows that, in fact, with MTurk "uneven rights, compensation, and safety are not aberrations, but rather constitutive of the roles and ideologies of high technology work—differences that matter in cultures of new media."[40] Early on at Amazon, as *The Economist* noted at the time, MTurk was euphemistically referred to as "artificial artificial intelligence."[41] A better way of putting it, as communication scholar Kate Crawford has, is that AI is neither artificial nor intelligent.[42] Humans, doing much dehumanizing work, help systems along and are central to what these systems appear to achieve, and the resulting automated elements are in no respect generalized intelligence.

Automation has been both an ideal and a fear for three-quarters of a century. Early on, it concerned factory production, and over time it came to encompass data processing and offices as well. In both settings. major design engineering and maintenance tasks are part of the automation systems, which involve human labor. In seeking to produce an "intelligent" system that can engage in conversations or self-drive an automobile, a tremendous labor task of annotators and trainers is integral to the system, and invisible to it. Nondisclosure agreements are the norm, and labor is usually low cost and done outside the United States.

The supply chain sometimes is headed by the principal AI corporation in charge of the project, and other times by individuals at companies or organizations listing on MTurk or the like, but often outsourcing companies are used, such as Scale, which works with Microsoft-OpenAI, Meta, Nvidia, and other giant AI companies, as well as the US Air Force and other entities in the public sector. The work is tedious and the pay is low. Trillion-dollar companies or contractors such as Scale (itself already a multibillion-dollar operation) outsource tedious labeling tasks to Kenya, where workers might make only $1 to $3 an hour after their extensive unpaid training for a particular assignment. All the work is task oriented, and the context of why they are doing what they are doing and who it is for is guarded information, which along with the levels of tedious concentration, adds to the unfulfilling nature of this task work.[43]

Augmentation Dreams and Automation Nightmares

J. C. R. Licklider, with his background as a psychologist, supported an early 1960s vision of developing "man-machine symbiosis." As the founding director of the Information Processing Techniques Office, Licklider set this project as a central concern. Licklider and his immediate successors as director in the mid- to late 1960s believed there were certain tasks machines did well (such as processing vast amounts of data quickly), and others that humans excelled at (such as creativity, decision making, judgment, and generalized intelligence). In that half decade, IPTO funded pathbreaking research in time-sharing, including MIT's Multics and Berkeley's Project Genie; created anew a center of excellence in graphics at the University of Utah; and supported AI pioneers such as Marvin Minsky, Herbert Simon, Allen Newell, and John McCarthy. It also launched the

ARPANET, which would grow from its first four nodes in 1969 to dozens in the 1970s, with, ironically, email becoming its first killer app.[44] Most IPTO funds went to universities in the 1960s, but the group also funded Douglas Engelbart and his Augmentation Research Center (ARC) at the Stanford Research Institute (SRI), which did early projects inventing and applying hyperlinks, the computer mouse, and other technologies that would become fundamental and ubiquitous to the internet and graphical user interfaces (GUI).

The changing federal funding landscape forced a shift in the institutional arrangements for research and development. The Mansfield Amendment of the 1970s forever changed IPTO by restricting the range of projects it could support. Quite remarkably, Xerox PARC, launched in the 1970s, seamlessly, without missing a step, took the baton in the relay race of basic research, human-machine symbiosis, and human-computer interaction. Its goal was to create the office of the future. With SRI and its ARC only five miles away in Palo Alto, Xerox PARC hired many scientists and engineers away, as well as others whom IPTO had funded earlier.

Historians and journalists have addressed the work of PARC's Charlie Thatcher and Butler Lampson on GUI and the Xerox ALTO, as well as the commercialized Star. They have also examined Apple's appropriation of this GUI technology on the Lisa and Macintosh. However, along the way, the work of an equally distinguished team consisting of Allen Newell, Stuart Card, and Thomas Moran has been significantly underappreciated. Newell spent summers and a portion of his weekends at PARC in the 1970s, but it was the top PhD student in his career, Stuart Card, who soon took the lead of PARC's User Interface Research Group, completing influential research on information retrieval, optimizing computer mouse performance, and human-computer interaction (HCI) models.

Newell and his HCI students were having more impact on systems and tools than those in the area of AI. Card was the lead author (along with Newell and Moran) on the much-used textbook *Psychology of Human-Computer Interaction,* published in 1983. This was only a year after University of Maryland's Ben Shneiderman and IIT's Bill Curtis founded an event in Gaithersburg, Maryland, with an unexpected 900 attendees that launched HCI's first professional conference, landing it a home within ACM the following year as SIGCHI, or SIG for Computer-Human Interaction (the sequence making it a pronounceable word).[45]

While AI winters, and especially the first, seem suspect, at least in common presentation (wide, and based on overpromising and underdelivering), elite schools did see AI research funds decline. HCI scientist Jonathan Grudin provocatively has posited that HCI thrives during AI winters and suffers during AI growing seasons. This is almost a law of computer science dynamics of the conservation of energy, interest, and funding.[46] Newell's work shows an overlapping interest among some top academics in the idea of an HCI model for AI. Interest has grown since a leading book, *Human-Centered AI*, by HCI cofounder Ben Shneiderman, appeared in 2022 and since the formation, a half-dozen years ago, of Stanford University's Institute for Human-Centered Artificial Intelligence (IHCAI), directed by distinguished AI scientist and computer vision specialist Fei-Fei Li.[47] Stanford's IHCAI, however, is geographically and institutionally challenged for truly human-centered AI as the region's "California Ideology" and the wealth and influence of giant tech and AI corporations and venture capital firms tend to skew research more toward maximizing profits, not user experience.[48]

It is debatable whether Douglas Engelbart's friend and associate the sociologist and philosopher Ted Nelson was selling snake oil with his 1974 book, *Computer Lib / Dream Machines*, and his ideas of "intertwingularity" (the symbiosis of networking, text, and interactive computing).[49] Steward Brand and many associated with such ideas in the 1970s increasingly showed their true colors—a right-leaning focus on property and libertarianism in the vein of Richard Barbook and Andy Cameron's "California Ideology," as communication scholar Frederick Turner has insightfully examined in *From Counterculture to Cyberculture*.[50] Outside the valley, but heavily influenced by it, the Clinton administration privatized the internet, the backbone of networks that the National Science Foundation (many of the regional networks had been funded by NSF) had been administering since the mid-1980s, a key span in extending neoliberal directions in technology.

One particularly original and insightful work of scholarship on hype in the era of the early privatized internet is Morgan Ames's masterful *The Charisma Machine: The Life, Death, and Legacy of One Laptop per Child*. The book illustrates how harmful utopianism is not just an exploitive tool of giant corporations but also infiltrates nonprofits, universities, and social policy. Examining One Laptop per Child in Paraguay, Ames shows that both well-meaning and self-serving efforts rested on false hopes, deception, and an MIT brand of technological arrogance and elitism. Of particular note, at its heart was a failed technology, an idea for powering machines by hand crank that "never materialized." Ames's research highlights how hype can victimize people and often harms people with the least economic resources, especially those in the Global South.[51]

The phrase "Data is the new oil," coined by British data scientist Clive Humby in 2006, was not just appropriate in characterizing the emergence of Web 2.0 and a culture of sharing but is especially apt as surveillance, as data analytics, and profiling have become ever more prevalent in business models and practices. The modeling, investment, extraction, and exploitation implied in the metaphor make it particularly fitting. Extractive industries and nations, whether with oil or data, exemplify and accelerate societal inequalities. And, of course, the intersectionality of energy and data extraction are considerable from AI modeling in drilling to natural gas, coal, nuclear energy, and other resources that power and cool (using water resources) vast data farms. The connection is evident widely and is fundamental to the "proof-of-work" energy-based mining consensus model launched in 2009 for the cryptocurrency Bitcoin.[52]

Paradoxically, apologists for AI—whether industry leaders speaking at Senator Charles Schumer's 2023 bipartisan AI Forum (including CEOs of Tesla, Palantir, Microsoft, Alphabet, OpenAI, and others), strategy consultants, or journalists—are profiteering from a delusion, a false claim and hope that AI can help people in their work and personal lives. Perhaps bridging the two are AI assistant bots that extend from LLMs. These will be additional tools encouraging and compelling users to share data used to surveille, analyze, profile, and exploit them.

(Not) Open AI

It is ironic and deceptive that OpenAI chose the name that it did and has stuck with it. The company went through a major organizational transformation midway through its eight-year span and continues to evolve in ways that allow its investors to profit from its technologies. It started as a nonprofit and used "open" to signify an allegiance to the open-source software movement, but it is something different.[53] There is in fact no accepted definition of what "open" in AI is, and as such it can be and is appropriated both for marketing and to deceive, shape, or defy regulatory efforts. Regardless of the chosen definition, OpenAI is quite far from any reasonable understanding of what constitutes "open" to most in the open-source software movement. This meaning includes ideals related to transparency and useability—including open-source code, open data sets, open model weights, open science (reproducibility), open or free use, openness to opt out of data collection, and openness to auditing to ensure fairness, accountability, and safety.

For some people, a name is just a name, but metaphors do matter, and "open" is an ill-chosen metaphor for OpenAI, for five reasons. First, OpenAI started as a nonprofit to produce artificial general intelligence that was to "benefit humanity as a whole, unconstrained by a need to generate a financial return." It kept that nonprofit umbrella while spinning out a limited profit corporation where millionaire and billionaire investors can secure a 10,000% return on their investment. Second, the open-source software movement was and is about sharing code, which is counter to OpenAI's practice. Third, it partnered with software, cloud, and AI giant Microsoft, which is using open AI to help jumpstart its anemic performance with search and information retrieval platforms. Google has long dominated search with a vast majority market share, and Microsoft, with its Bing search engine, has always been under 5%. Microsoft has long battled against the open-source software movement and thoroughly embraces the model of proprietary software. Fourth, Sam Altman, the high-profile CEO, emphasized that there was existential risk at the same time that he deployed a closed system on the world through Microsoft, a system that more than 100 million people soon used. This is anything but cautious. Fifth, there is a false competitive open ripple effect of growing: OpenAI's pre-emptive move led Google to release Bard, which is also a closed, proprietary system.

Closed systems are becoming the norm in LLMs. OpenAI (ChatGPT-4) and Google (Bard) have their own, and Meta is now seeking to differentiate itself by presenting a truly open alternative with its LLM Llama2. With restrictions and caveats, it is open for research and commercial use but is not open. It is not open-source code. Further, the reason for opening up with regard to allowing use is to develop momentum toward creating a standard, an ever-growing ecosystem that further empowers Meta, which is drawing back to a degree from the metaverse to invest heavily in AI.

Existential Power Plays

Despite all the rhetoric of the near-term possibility of artificial intelligence freeing workers from drudgery, making work higher-level and more rewarding, and boosting leisure time, artificial intelligence adds a great amount of low-paid unrewarding labor to train. It shifts labor, and in so doing, it reinforces and extends power imbalances in global society and amplifies social inequality. Industry CEOs like Sam Altman, Satya Nadella (Microsoft), Sundar Pichai (Google/Alphabet), and Mark Zuckerberg (Meta) sometimes acknowledge potential harm and a need for guardrails and ethical vigilance. Nonetheless they see existential risks of machines taking over—a view similar to or fully in concert with the fantasies of effective altruists.

Privacy and security of data constitute another major risk. Either bias or data breach risk has led major corporations to restrict and often fully ban ChatGPT and generative AI chatbots, including Citi, Northrup Grumman, iHeartRadio, Amazon, Apple, Deutsche Bank, Spotify, Wells Fargo, and Verizon. OpenAI and Microsoft irresponsibly have pushed to profiteer, and Google, Meta, and others are following suit. Ironically, corporate users with major past ethical blemishes, like Wells Fargo and Citi, will be important to setting and enforcing restrictive policies on internal generative AI chatbot use by their workforces. They will be fundamental to reducing harm, as will federal and international regulations.

Important ideas, such as transparency of data sets, weightings, nutrition-like labels and warnings, have been proposed by leading data scientist ethicists in the AI community and by thought leaders and critical voices such as Alondra Nelson, Timnit Gebru, Meredeth Whittaker, Rumman Chowdhury, and Joy Buolamwini. It is not surprising that the top ethical voices are women data scientists and social scientists, and mostly women of color, individuals who have seen the lack of diversity and the harmful proliferation of bias firsthand. Apart from these standout ethicists and activist data scientists combatting inequality, less knowledgeable and less well-intentioned ethical officers and consultants will pose a significant risk by continuing their rubber stamping. In computer security, community and organization-based education and credentialing have been fundamental, and it will need to be in AI ethics as well. Stanford has the largest human-centered AI institute, and while individuals do good work within it, it seems the wrong school, wrong set of donors, and wrong Silicon Valley setting to lead the way with ethical research and policy, and there is the need for many institutions and organizations learning from and checking one another.

Artificial intelligence has always centered on media presentations of future possibilities and, paradoxically, is headed in directions where outputs are less trustworthy, not more. Expert systems that target domains where greater attention is paid to studying datasets have been around for more than a half century. Training on smaller well-defined data sets produces greater reliability in AI systems. The new-wave LLMs and bots, trained on giant bodies of texts and images without context, ironically do not bring systems closer to a general artificial intelligence. Instead, they imitate in ways that serve to amplify historical inequalities and bring unprecedented levels of social biases and risks.

Notes

1. "The Computer Moves In," *Time* magazine, January 3, 1983

2. Kate Conger and John Yoon, "Explicit Deepfake Images of Taylor Swift Elude Safeguards and Swamp Social Media," *New York Times*, January 26, 2024; Emine Saner, "Inside the Taylor Swift Deepfake Scandal: 'It is Men Telling Women a Powerful Woman to Get Back in Her Box,'" *The Guardian*, January 31, 2024.

3. "ChatGPT Reaches 100 Million Users Two Months after Launch," *The Guardian*, February 2, 2023; "ChatGPT Reached 1 Million Users 250x Faster than Netflix," Blockchain Centre, March 29, 2023; @sama, Sam Altman, "ChatGPT Launched on Wednesday . . . ," Twitter, December 5, 2022.

4. Jeffrey R. Yost, "Computers and the Internet: Braiding Irony, Paradox, and Possibility," in *A Companion to American Technology*, ed. Carroll Pursell (New York: Wiley & Sons / Blackwell, 2005), 340–60.

5. Brent Goldfarb and David A. Kirsch, *Bubbles and Crashes: The Boom and Bust of Technological Innovation* (Stanford, CA: Stanford University Press, 2019).

6. Michael Brick, "US Appeals Court Overturns Microsoft Anti-trust Ruling," *New York Times*, June 28, 2001.

7. Jeffrey R. Yost. *Making IT Work: A History of the Computer Services Industry* (Cambridge, MA: MIT Press, 2017).

8. T. R. Kennedy Jr., "Electronic Computer Flashes Answers, May Speed Engineering." *New York Times*, February 15, 1946.

9. Edmund Berkeley, *Giant Brains; or, Machines That Think* (New York: Wiley & Sons, 1949).

10. Atsushi Akera, *Calculating a Natural World: Scientists, Engineers, and Computers during the Rise of U.S. Cold War Research* (Cambridge, MA: MIT Press, 2006).

11. "Electronic 'Brain' Teaches Itself," *New York Times*, July 13, 1958, E9. Rosenblatt initially completed The Perceptron in 1957.

12. Warren S. McCulloch and Walter Pitts, "A Logical Calculus of the Ideas Immanent in Nervous Activity," *The Bulletin of Mathematical Biophysics* 5 (1943): 115, 132.

13. Donald O. Hebb, *The Organization of Behavior: A Neuropsychological Theory* (New York: Wiley & Sons, 1949).

14. Jeffrey R. Yost, *The IBM Century: Creating the IT Revolution* (Los Alamitos, CA: IEEE Computer Society Press, 2011).

15. Arthur L. Samuel, "Some Studies in Machine Learning Using the Game of Checkers," *IBM Journal of Research and Development* 3, no. 3 (1950): 210–29.

16. Claude E. Shannon, "Programming a Computer for Playing Chess," *Philosophical Magazine*, 7th ser., 41, no. 314 (March 1950).

17. We wish to acknowledge commentary in an American Historical Association Congressional Briefing panel that I (Yost) appeared on with Janet Abbate, professor in the Department of Science, Technology, and Society at Virginia Tech. In wide-ranging comments on AI history, Dr. Abbate, briefly and very usefully, discussed the gendering of chess.

18. Juliana Kim, "World Chess Just Placed Restrictions on Both Trans Women and Trans Men," *NPR*, August 18, 2023.

19. Jeffrey R. Yost. *Making IT Work: A History of the Computer Services Industry* (Cambridge, MA: MIT Press, 2017).

20. Nathan Ensmenger, "Is Chess the Drosophila of Artificial Intelligence? A Social History of an Algorithm," *Social Studies of Science* 42, no. 1 (2012): 5–30.

21. Hsu Feng-Hsiung, *Behind Deep Blue: Building the Computer That Defeated the World Chess Champion* (Princeton, NJ: Princeton University Press, 2022).

22. Caroline Clark Hayes, "The Incredible Shrinking Woman," in *Gender Codes: Why Women Are Leaving Computing*, ed. Thomas J. Misa (New York: Wiley IEEE Press, 2010), 30; Thomas J. Misa. "Dynamics of Gender Bias in Computing." *Communications of the ACM* 64, no. 6 (June 2021): 78.

23. Deloitte US "The State of Women in AI Today." (Deloitte US); National Science Foundation. "STEM Education Data and Trends." (nsf.gov)

24. Holly Else, "AI Conference Widely Known as 'NIPS' Changes Its Controversial Acronym," *Nature News*, November 19, 2018.

25. Alan M. Turing. "Computing Machinery and Intelligence." *Mind* 43 (October 1950), 433–60.

26. Joseph Weizenbaum, *Computer Power and Human Reason: From Judgment to Calculation* (San Francisco: W. H. Freeman, 1976).

27. Xiaochang Li, "The Measure of Meaning: Automatic Speech Recognition and the Human-Computer Imagination," in *Abstractions and Embodiments: New Histories of Computing and Society*, ed. Janet Abbate and Stephanie Dick (Baltimore: Johns Hopkins University Press 2022), 352.

28. Marvin Minsky and Seymour A. Papert, *Perceptrons: An Introduction to Computational Geometry* (Cambridge, MA: MIT Press, 1969; expanded ed., 1988).

29. Mikel Olazaran, "A Sociological Study of the Official History of the

Perceptron Controversy," *Social Studies of Science* 26, no. 3 (August 1996): 611–59.

30. Aaron Mendon-Plasek, "Mechanized Significance and Machine Learning: Why It Became Thinkable and Preferable to Teach Machines to Judge the World," in *The Cultural Life of Machine Learning*, ed. J. Roberge and M. Castelle. (Cham, Switzerland: Palgrave Macmillan, 2021).

31. Thomas Haigh, "There Was No 'First AI Winter,'" *Communications of the ACM* 66, no. 12 (December 2023): 35–39.

32. Arthur L. Norberg, *Transforming Computer Technology: Information Processing for the Pentagon* (Baltimore: Johns Hopkins University Press, 1996).

33. Haigh, "There Was No 'First AI Winter.'"

34. Andrew S. Lea, *Digitizing Diagnosis: Medicine, Minds, and Machines in Twentieth-Century America* (Baltimore: Johns Hopkins University Press, 2023), 130, 147–55.

35. Joshua Lauer, *Creditworthy: A History of Consumer Surveillance and Financial Identity in America* (New York: Columbia University Press, 2017).

36. Helen Toner, "AI Chatbots Are Doing Something a Lot Like Improv," *Time* magazine, May 18, 2023.

37. Emily Bender, Timnit Gebru, Angelina McMillan-Major Shmargaret Shmitchell [Margaret Mitchell], "On the Dangers of Stochastic Parrots: Can Language Models Be Too Big?," *FAccT 2021 Proceedings* (Fairness, Accountability and Transparency, Association for Computing Machinery) (March 2021): 610–23.

38. Venkatesh, Harinarayan, "Query Processing in Data-Warehousing Environments," PhD dissertation, Department of Computer Science, Stanford University, 1997.

39. Harinarayan, V., et al. (2007), US Patent 7,197,459 "Hybrid machine/human computing arrangement." (filed October 2001, granted and published March 27, 2007).

40. Lilly Irani, "The Culture of Microwork," *New Media and Society* 17, no. 5 (May 2015): 731, 735.

41. "Artificial Artificial Intelligence," *The Economist*, June 10, 2006.

42. Kate Crawford, *Atlas of AI: Power, Politics, and the Planetary Costs of Artificial Intelligence* (New Haven, CT: Yale University Press, 2021).

43. Josh Dzieza, "AI Is a Lot of Work," *The Verge*, July 23, 2023.

44. Janet Abbate, *Inventing the Internet* (Cambridge, MA: MIT Press, 2000).

45. Jeffrey R. Yost, "Of Mice and Mentalité: PARC Ways to Exploring HCI, AI, Augmentation and Symbiosis, and Categorization and Control," *Interfaces: Essays and Reviews in Computing and Culture* 2 (March 2021): 12–26.

46. Jonathan Grudin *From Tool to Partner: The Evolution of Human-Computer Interaction* (New York: Springer, 2017).

47. Ben Shneiderman *Human Center AI* (New York: Oxford University Press, 2022).

48. Richard Barbrook and Andy Cameron, "The California Ideology" *Mute* (September 1995).

49. Ted Nelson, *Computer Lib / Dream Machines* (1974; reprint, n.p.: Microsoft Press, 1987).

50. Frederick Turner, *From Counterculture to Cyberculture: Stewart Brand, the Whole Earth Network, and the Rise of Digital Utopianism* (Chicago: University of Chicago Press, 2006).

51. Morgan Ames, *The Charisma Machine: The Life, Death, and Legacy of One Laptop per Child* (Cambridge, MA: MIT Press, 2019).

52. Jeffrey R. Yost, "Bitcoin's Baptism in the Heavenly City of 21st Century Financiers," *Blockchain and Society* (January 2024).

53. David Widder Gray, Sarah West, and Meredith Whittaker, "Open (for Business): Big Tech, Concentrated Power, and the Political Economy of Open AI," Social Science Research Network, August 17, 2023.

Contributors

Jennifer Alexander is an Associate Professor in Mechanical Engineering at the University of Minnesota, where she is the Director of Graduate Studies for the Program in the History of Science, Technology, and Medicine.

Héctor Beltrán is an Associate Professor of Anthropology at the Massachusetts Institute of Technology. He is the author of *Code Work: Hacking across the US/México Techno-Borderlands* (Princeton University Press, 2023).

Shun-Ling Chen is an Associate Research Professor and the Codirector of the Information Law Center at Institutum Iurisprudentiae, Academia Sinica.

Gerardo Con Díaz is a Professor of Science and Technology Studies at the University of California, Davis. He is the author of *Everyone Breaks These Laws: How Copyrights Made the Online World* (Yale University Press, 2025) and *Software Rights: How Patent Law Transformed Software Development in America* (Yale University Press, 2019).

Stephanie Dick is an Assistant Professor in the School of Communication at Simon Fraser University. Her work explores the history of mathematics, computing, and the mind in the postwar United States.

Hamid R. Ekbia is a University Professor and the Director of the Autonomous Systems Policy Institute at the Maxwell School of Citizenship and Public Affairs, Syracuse University.

Megan Finn is an Associate Professor in the School of Communication at American University and an Affiliate Associate Professor in the Information School at the University of Washington. She is the author of *Documenting Aftermath: Information Infrastructures in the Wake of Disasters* (MIT Press, 2018).

Mar Hicks is a historian of technology and an Associate Professor of Data Science at the University of Virginia, Charlottesville. Hicks is the author of *Programmed Inequality* (MIT Press, 2017) and a coeditor of *Your Computer Is on Fire* (MIT Press, 2021).

Ya-Wen Lei is a Professor in the Department of Sociology at Harvard University. She is also affiliated with the Fairbank Center for Chinese Studies and the Weatherhead Center for International Affairs at Harvard.

Mariel García Llorens is an anthropologist and science and technology studies scholar who researches digital money in Latin America. She holds a PhD from the University of California, Davis.

A. P. Janani was an MSc (Digital Society) student at the International Institute of Information Technology Bangalore at the time that the research for chapter 3 was conducted.

Meg Leta Jones is a Provost's Distinguished Associate Professor for the Master's in Communication, Culture & Technology Program at Georgetown University, where she researches rules and technological change with a focus on computers and privacy.

Shreeharsh Kelkar is a Continuing Lecturer in Interdisciplinary Studies at the University of California, Berkeley, and is currently working on a manuscript about the practices and politics of data science.

Dylan Mulvin is an Associate Professor of Media and Communications at the London School of Economics and Political Science.

Nico Wen-Li Ong is a research intern and editorial assistant for Gerardo Con Díaz at the University of California, Davis, and helped the co-editors prepare this book for production.

Elisa Oreglia is a Reader in Global Digital Cultures at King's College London.

Justin Petelka is a PhD candidate at the University of Washington's Information School and a Cybersecurity Fellow of the International Policy Institute at the University of Washington's Henry M. Jackson School of International Studies.

Elizabeth Petrick is an Associate Professor of History at Rice University and the author of *Making Computers Accessible: Disability Rights and Digital Technology* (Johns Hopkins University Press, 2015).

Elizabeth Semler is a historian of science, medicine, and technology at the University of California, Berkeley's Oral History Center.

Janaki Srinivasan is an Associate Professor of Digital South Asian Studies at the University of Oxford. At the time that research for chapter 3 was conducted, she was on the faculty at the International Institute of Information Technology Bangalore.

Gili Vidan is an Assistant Professor in the Department of Information Science of the Cornell Bowers College of Computing and Information Science and a field member in Cornell University's Department of Science and Technology Studies and the Media Studies Program.

Jeffrey R. Yost is the Director of the Charles Babbage Institute and Research Professor of the History of Science, Technology, and Medicine at the University of Minnesota. He has published eight books, including *Making IT Work: A History of the Computer Services Industry* (MIT Press, 2017).

Index

Page numbers in *italic* refer to figures; page numbers in **bold** refer to tables.

Abbate, Janet, 261
Abraham, Spencer, 281
Abrams, Floyd, 380
A/B testing, 306, 307, 315
access, rectification, cancellation, and objection (ARCO) rights, 89, 90, 91, 93, 94
ACE Public Cryptography Study Group, 381–82, 383, 384
Adler, Ilya, 397
Advanced Research Projects Agency (ARPA), 420
affective computing, 347, 357, 358, 359
Affero General Public License (AGPL), 130, 134, 135
Agarwal, Anant, 314
Agüera y Arcas, Blaise, 363, 364
Alaska Airlines, 95, **96,** 99, 100, 108n49
Alexa, 16, 99, 421
Alexander, Jennifer, 19, 209n12
algorithmic audit, 112, 113, 121
algorithmic collusion, 116, 122–23
algorithmic system, 112, 113, 114, 121
Alibaba, 3, 39
All-China Federation of Trade Unions (ACFTU), 24
Allen, Fran, 261
Alper, Meryl, 357, 361
Altman, Sam, 413, 428
Amazon: AI chatbots, 428; algorithmic pricing systems, 14, 109–10, 121; book publishers and, 148, 149, 153, 159–60, 162, 163–64, 165; business model of, 114, 272; competition between Apple and, 153, 160, 161; e-book pricing strategy, 148–49, 152–53, 159, 160, 164–65; growth of, 119, 413; Marketplace, 423; Mechanical Turk (MTurk), 276, 423–24;

purchase of Junglee, 423; responses to DSAR requests, 94, **96,** 98–99, 100; Web Services, 134; withdrawal from SSPL, 136–37
American Competitiveness and Workforce Improvement Act, 281
American Disabled for Attendant Programs Today (ADAPT), 198
American Society of Mechanical Engineers, 361
Americans with Disabilities Act (ADA), 198
Ames, Morgan, 226, 229; *The Charisma Machine,* 426
Amit, Vered, 396
Amrute, Sareeta, 339, 408n30
Ananny, Mike, 113
Anderson, Chris, 359
Anderson, John, 311
Anderson, Sonya, 247
Andreesen, Marc, 70
Andrejevich, Mark, 359
Anti-Capitalist Software License (ACSL), 139, 140, 141
Anti-996 License, 138, 139–40, 141
antitrust law, 111, 114, 117–18, 120, 123, 270, 413
Apache open-source licenses, 134–35, 139
APIs (application programming interfaces), 305, 307, 321n39
Apple: AI chatbots, 428; book publishers and, 14, 148–50, 153–54, 155–57, 159, 160–61, 164–65; business model, 116, 272; competition between Amazon and, 153, 160, 161; e-book pricing, 148–49, 151, 155, 157, 158, 165; Facebook ad campaign against, 80; GUI technology, 425; internal research records, 149–50, 151, 156; iOS update, 64, 67, 76, 80; lawsuit

against, 149–50; portable devices, 150; privacy regulations, 64–65, 74, 79, 81; smartphone operating system, 119
Apple's Worldwide Podcasting, 155
Applied Data Research, Inc. (ADR), 260
App Store, 150
App Tracking Transparency policy, 64, 80
Arif, Ahmer, 356
Arkin, Ronald, 359
ARPANET, 289, 425
Article 29 Data Protection Working Party (A29WP), 73, 75, 76, 77
artificial intelligence (AI): anthropology of, 222; automated problem-solving, 175, 189n6; chess project, 417; cloud-based, 145n26; codes of, 1; copyright and, 12; criticism of, 419–20, 428–29; development of, 411, 419–21, 425; gamification of, 416, 417; IBM and, 415; importance of speed for, 415; invention of the term, 414; metaphors for, 419, 422–23; notion of "responsible," 143; as open source, 132; outputs of, 429; privacy and security of data, 428; reliability of, 429; restrictions on chatbots, 428; *Time* magazine on, 412; top companies, 413–14; training of, 429
assemblage, 296
Association for Computing Machinery, 266
asymmetric contracts, 29, 30, 57
AT&T, 114, 119, 120, 413
Augmentation Research Center (ARC), 425
Ausloos, Jef, 102
autism: computational view of, 359–60; computer therapies for,

autism (*cont.*)
357–58, 361–62; deception deficit, 360–61; definition of, 344, 357, 359–60, 370n54; inability to understand social interactions, 358, 359; masking, 357–58, 360; obstruction test, 358
Automatic Data Processing (ADP), 260
Automatic Language Processing Advisory Committee Report, 421
Azure Cloud Switch, 133

Ballmer, Steve, 132, 133
Banerjee, Uptal K., 333, 334
banks: digital payment systems and, 212, 217, 223, 226–27, 228; popular distrust in, 227, 228, 234n51, 235n58
Barbook, Richard, 426
Barksdale, Jim, 70
Barnes & Noble, 152, 158, 160
Baron-Cohen, Simon, 360
Barratt, Tom, 24
Bassett, Caroline, 359, 362
Bassett, Ross, 326, 333, 334, 337, 339; *The Technological Indian*, 326
"batch job," 390–91
Beaman, Jean, 197
Beardsley, Charles, 373
Beer, Anthony Stafford, 326
behavioral codes, 2, 6, 7
Behlendorf, Brian, 68–69
Bell, Daniel, 377
Beltrán, Héctor, 19
Bender, Emily, 346, 362, 363, 422
Benigo, Yoshua, 364
Benjamin, Ruha, 195
Bergmanis, Maris, 241
Berkeley, Edmond: *Giant Brains; or, Machines That Think*, 414
Berners-Lee, Tim, 68, 69
Bertillon, Alphonse, 174, 185
Bettelheim, Bruno, 358
Better than Cash Alliance (BTCA), *216*, 217, 226, 227
Bewell, Alan, 275
Bezilla, Robert, 204
Bezos, Jeff, 423
Big Data, 18, 295, 297, 308, 309
Bim platform: development of, 213, 214, 221–22, 223, 227, 231; infrastructure of, 225, 234n46; official launch of, 220; promotion of, 234n51; technology of, 233n33; testing of, 213, 215, 217–19, 224–26, *225*, 229–30, 233n21
Bitcoin, 427

Bledsoe, Woody: career of, 175; facial recognition algorithm of, 172, 174, 176–83, 184–86, 187; interest in pattern recognition, 175–76, 190n15
Bloor, David, 173
Boeing, corporate culture, 241–42
book publishing, 14, 148–51, 153–54, 155–57, 159, 160–61, 164–65
Bork, Richard, 117
Bouk, Dan, 188
Bowen, William, 312, 313
Bradford, Anu, 92
Brand, Steward, 426
Branstad, Dennis, 374, 375
British Tabulating Machines Corporation (BTM), 336
Brooke Group Ltd. v. Brown & Williamson Tobacco Corp., 118
Brooks, Rodney, 359
Brown Shoe Co. v. United States, 117
Brussels Effect, 87, 92
Brynjolfsson, Erik, 295, 308, 309
Brzezinski, Zbigniew, 377
Buber, Martin, 360
Buolamwini, Joy, 429
Burnett, D. Graham: *Trying Leviathan*, 92
Burton, Antoinette, 328
Bush, George W., 139
Business Software Alliance, 132

Cader, Michael, 159
California privacy acts, 76, 77–78, 80, 87, 91, 94, 95
Cameron, Andy, 426
Campbell-Kelly, M., 268
Card, Stuart, 425
Carey, Allison, 198
Carnegie Mellon University (CMU), 310, 312, 417
Carter, Jimmy, 270, 377, 381
cash economy: challenges of, 227, 232n14; popularity of, 212, 213, 215, 227–28, 231n5
CBS Corporation, 153
Central Intelligence Agency (CIA), 175, 184
Ceruzzi, Paul, 347
Chambers, John, 412
ChatGPT: code of, 1; creation of, 422; release of, 411; restrictions on, 428; simulation of humanness, 362, 364; training of, 363–64; users of, 412
Chen, Shun-Ling, 14
chess, 416, 417
Chile: banking system, 326; face recognition in identification of the dead in, 174, 175, 190n11

China: Big Tech corporations in, 123; civil society, 57; digital entrepreneurs in, 396; employment classification, 56; food delivery platform economy, 24, 25, 32–34, 36; "ideology work," 38; labor law regime, 26, 30, 34, 35, 44, 56; manufacturing sector, 35; overworked developers in, 138; US embargo on selling computers to, 326, 336
choice architectures, 66
Chomsky, Noam, 418
Chowdhury, Rumman, 429
Christensen, Clay: *The Innovator's Dilemma*, 322n44
Chrome browser, 76, 132
Cisco Systems, 412–13
citizenship: cultural, 197, 199, 207–8; digital, 193, 194, 199, 202; disability and, 194, 195–96, 197–99, 207; economic, 197, 202; theory of, 194, 196–97
Clayton Antitrust Act (1914), 117
Cleckley, Herve: clinical observation of psychopaths, 351–53, 367n5; definition of psychopathic personality, 344–45, 347–48, 350, 354, 366; *Mask of Sanity*, 345, 350
Clever, Iris, 178, 184
clickwrap agreements, 31, 57
Clinton, Bill, 279
COBOL (Common Business Oriented Language), 256, 266, 274, 276, 277–78, 283–84, 288
code analytic, 4, 7, 9, 10, 11, 12, 13, 14, 15, 18
code-breaking, 336
codes: audits, 112; bias in, 2, 16; creation of, 1–2, 17; "decentering," 12; definition of, 6, 113; governance function of, 12–14; as law, 3, 65; as object, 12; political economy of, 3, 4, 5, 6, 411; registers of, 7–8, 10; separation between users and makers of, 17–19; studies of, 2, 3, 4, 12; types of, 1, 2
cognitive task analysis, 313, 323n59
Cold War: key events of, 417; technological development during, 175, 185, 326, 334, 335, 336–37, 414, 417–18
Coleman, Gabriella, 388
collective contention, 27–28, 30, 32
collusion, 110–11, 115
Colombo Plan for Cooperative Economic and Social Development in Asia and the Pacific, 330
Commercial Privacy Bill of Rights Act, 76

Committee for Skeptical Inquiry, 172
Commons Clause, 134, 135, 136, 140
Communications Supplementary
    Activities–Washington (CSAW),
    237
computational linguistics, 418
computational "solutions," 172–73,
    174, 187–88, 189n6
Computer Fraud and Abuse Act, 71
computer industry: "benching" prac-
    tices, 282, 290n32; "body shop-
    ping" practices, 280, 282–83, 286,
    287, 288; British and American
    leadership in, 237, 256, 325–26,
    330; gendered labor, 327–29,
    417–18; globalization of, 275;
    in India, 18, 326, 327, 330; labor
    politics in, 274, 276, 278, 286–87;
    offshoring practices, 275, 280, 283;
    outsourcing practices, 274–76,
    279–80, 283, 283, 286; politics
    and, 326, 336–37; transfer of
    technology to less industrialized
    nations, 325–27, 330
computerization: colonial relations
    and, 330–33, 338; global reach of,
    325, 327, 340
computer languages, 283, 284, 415
computer-mediated communication:
    benefits of, 193–94, 201–3, 204,
    205–6; vs. face-to-face communi-
    cation, 200–201, 204–5, 208; idea
    of hidden identity in, 201, 204; as
    leveling technology, 196, 199, 203,
    208; people with disabilities and,
    194–95, 199–207; research on,
    199–200, 202–3, 207, 209n12;
    shortcomings of, 195, 196, 201;
    trials of, 201, 203–4
Computer Professionals for Social
    Responsibility, 70
computer programmers, 276, 289,
    320n23
computers: history of, 414–16; opacity
    of, 355; proliferation of, 279, 300,
    303
computer security, 373–74, 375
computer specialists, 301–2
computing labor force: flexibility
    of, 402, 405; H1-B visa holders,
    280–82, 283, 285, 287, 289; immi-
    grant, 399–401, 404; neoliberal
    market and, 395, 397, 405; precar-
    ious situation of, 402–3, 408n30;
    women in, 241, 242, 245–46,
    247–49, 251n23, 252n26, 253n53,
    254n54
consent, 65–68, 71, 81
"consent dilemma," 67

Consumer Privacy Protection Act, 76
Consumer Project on Technology, 70
content platforms, 304
Continuous Speech Recognition
    (CSR), 419
contracts, 29, 30, 31, 56–57, 66
Control Data Corporation (CDC):
    benefits materials, 243–44, 244,
    253n41; corporate culture of,
    236–37, 239, 240–44, 246, 249–50,
    251n22, 254n55; discriminatory
    policies of, 237, 247; employment
    practices, 236, 241, 242–43, **247,**
    249; gendered segregation of labor,
    243–44, 246–48, 252n30, 254n54,
    261; history of, 16, 236, 237–38,
    239, 242, 252n24; innovative proj-
    ects of, 239, 240, 241; job's grades,
    239–40, 245, 246–47, 251n17; man-
    agement of, 249; organizational
    structure of, 240, 240; pay struc-
    ture, 236, 239–40, 242, 245–47,
    251n17; people-first policy, 240;
    purchase of IBM Service Bureau
    Division, 258, 262; rural produc-
    tions sites, 246–47, **247**; systems
    services, 260; upward mobility,
    248, 253n51; women employees,
    241, 242, 245–46, 247–49, 251n23,
    252n26, 253n53, 254n54; Work
    Planning policy, 248, 249
Conway, Melvin, 256
Conway's Law, 256
Cook, Tim, 67, 81
cookies: consent code and, 67; in
    e-commerce, 68–69, 70, 72, 75–76;
    opt-out, 71, 73; regulation of,
    71–73. See also HTTP cookies
Cooter, Roger, 10
copyleft, 128, 137
copyright: licenses, 142, 143; software
    and, 131–32
copyright law, 11–12, 65, 129, 142, 143
Cortada, J., 261
Cote, Denise, 149
COVID-19: economic impact of, 80,
    232n14; video-conferencing dur-
    ing, 206
Crawford, Kate, 113, 424
Creative Commons licenses, 142
Crockford, Douglas, 139
cryptography: during Cold War, 379;
    commercial need for, 382; debate
    over meaning and status of,
    380–81, 382–83; key distribution
    and management, 376, 379, 380,
    385n21; military use of, 1; nongov-
    ernmental, 380; as practice of
    authentication, 374–75, 376–77;

during World War II, 237. See also
    public cryptography
Cue, Eddy: career of, 150; market
    research, 150–51; negotiations with
    Big Six publishers, 149, 153–56,
    158–59, 161, 164; pricing strategy,
    154–55, 157, 158
Cullinane/Cullinet, 260
cultures of solidarity, 27
Curtis, Bill, 425
cyberbullying, 346, 367n9
cybersecurity, 275, 285–86

D'Adderio, Luciana, 303
DALL-E transform creative works, 12
Dame-Griff, Avery, 195
data access rights, 88
Data Encryption Standard (DES),
    378–80
data privacy, 77, 87, 93, 373
Data Protection Directive (DPD95),
    71, 76, 89
data protection law, 66, 90–92, 93,
    102
data scientists, 308–9, 322n54
data storage, 302
data subject access requests (DSARs),
    101; analysis of, 13, 88, 95; com-
    panies' responses to, **96**; corporate
    interpretation of, 93–94, 97; in
    different legal regimes, assessment
    of, 94; transparency of, 90, 102
Davida, George, 381, 382, 384
"default effect," 66
defaults: categories of, 66, 67–68,
    71–72, 81n7; consent dilemma
    and, 66–68; natural, 81n7; privacy
    policy and, 66–67; user preferences
    and, 66
Department of Defense (DOD): In-
    formation Processing Techniques
    Office (IPTO), 420–21; research
    funding, 421
Department of Health, Education
    and Welfare 1973 report, 90
Department of Justice (DOJ): lawsuit
    against IBM, 256, 257, 262, 270,
    271; lawsuit against Microsoft, 413
Desai, Morarji, 333
Deutsche Bank, 428
Dev.F ("hacker school"), 389, 390–91,
    399
Dewitte, Pierre, 102
Diakopolous, Nicholas, 112
Díaz, Con, 14, 67, 93; Everyone Breaks
    These Laws, 11
Dick, Stephanie, 15, 123
Diffie, Whitfield, 372, 378, 379, 380,
    386n28

Digital Equipment Corporation (DEC), 270
digital gaming, 402–3
digital rights management, 65, 72
digital Taylorism, 23
digital technologies: as communication tool, 193–94, 206; inequality and, 195, 199, 226
Dijkstra, Edsger, 278
dispute tree, concept of, 28, 44, 50, 52
disruptive innovation, 322n44
Dohle, Markus, 161
domain experts, 309, 322n54
Do Not Call List, 73
Do Not Track (DNT), 67, 73, 74–76, 77–78, 79, 80
dot.com bubble, 412–13
Doty, Nicholas, 75
DoubleClick, 71, 73
Douglas, Mary, 173
Dowley, Becky, 358
Drake, Willis, 238, 251n12
DREAM (Development of Robot-Enhanced Therapy for Children with Autism), 360
dress code, 1, 6
Duarte, Javier, 398
Dunn, Ashley, 274
Dyck, Noel, 396
dynamic pricing, 113, 116

e-book prices: agency model of, 149, 155, 156, 158, 165; dispute over, 148–49; for new releases, 156–57; vs. physical book prices, 155, 156, 160; weighted-average increase of, 164; wholesale model of, 149, 152, 155, 157, 158, 162
e-books: alternative platforms, 152–53; projected revenue, 151, 152
Eckert Mauchly Computer Company, 251n12
e-commerce: tech giants' profits from, 64–65, 80; technical and legal defaults in, 72; use of cookies, 68–69, 70, 72, 75–76
Edelman, Lauren, 93
Edelstein, David N., 270
Edwards, Paul, 288, 376
edX: business model of, 310, 314, 315–16, 318; course-making process, 318; establishment of, 314–15; institutional partners, 315, 316, 318; mission of, 315, 317; number of courses, 317, 318; platform model of, 314, 316; software features, 316–17; study of, 323n69; technical and social labor on, 317; "verified certificate" of, 316

Ehmke, Caroline Ada, 139, 140
Eisen, Michael, 109
Eisenhower, Dwight D., 420
Ekbia, Hamid, 13, 14, 67
"Electronic 'Brain' Teaches Itself" (New York Times article), 414
Electronic Communications Privacy Act, 71
Electronic Data Systems (EDS), 260, 274
Electronic Information Exchange System, 200
Electronic Numerical Integrator and Computer (ENIAC), 414, 417
Electronic Privacy Information Center (EPIC), 70
Ele.me (platform company), 24, 25, 36, 37, 39, 40
ELIZA, 362–63, 418
Ellcessor, Elizabeth, 199
Ellis, Kate, 195
Elman, Julie Passanante, 198
Elyachar, Julia, 229
Emerson, Lori, 355
Engelbart, Douglas, 369n32, 425, 426
Engineering Research Associates (ERA), 237–39, 250, 250n5, 251n11, 251n12
Engstrom, Howard, 237
Ensmenger, Nathan, 241, 278, 320n23, 417
Equal Pay Act (1963), 245
Ericsson (IT company), 218, 221, 222
ethical licenses, 138–40, 142
Eubanks, Virginia, 9, 221
European Economic Community (EEC), 89, 334
European Union: antitrust law, 120; data protection and regulation, 89–90; ePrivacy Directive, 71–73, 75, 77, 80; privacy protection in, 71–72, 76–77, 99–100; transnational companies in, 92–93, 94–95
European Union's General Data Protection Regulation (GDPR), 13
"Executive Guide for Computer Security" (Reed and Branstad), 375
experimental governmentality, 230, 235n66
expertise: definition of, 297–98; vs. experts, 297
Eyal, Gil, 297

Facebook: advertising revenue of, 64, 80; antitrust lawsuit against, 120; competitors of, 80, 120; COVID-19 impact on, 80; "download my data"

function, 97; efforts to divest, 122; markets, 94; platform, 116; responses to DSAR requests, 96; tracking of users' activity, 64–65, 79, 80
facial recognition: accuracy of, 182, 188; Bledsoe's algorithm of, 172, 174, 176–83, 184, 185–86, 186, 187, 188; database of "known" faces, 176–77, 177, 182, 186; debates on, 189n4; eugenics and, 174, 175, 185; forensic applications, 174, 189–90n11; on Kaplan daguerreotype, 171–72; as policing tool, 174, 186–87; racial biases in, 172, 177, 184, 187, 189n4; RAND tablet, 178, 179, 180; redefinition of facial features in, 174, 177–80, 179; rotation, lean, and tilt corrections, 180–82, 181, 186; "standard head" use in, 180–83, 183, 184, 188, 190–91n22; study of, 15
Fair Credit Reporting Act of 1970, 90, 376
FairPlay rights system, 154
Fassett, Bruce, 274
Federal Trade Commission (FTC), 71, 73, 117, 271
Feigenbaum, Edward, 421
Ferenczi, Sánder, 345
fingerprints matching, 174, 185, 191n32
Finn, Megan, 65
Fleischer, Doris, 198
Flipkart, 95, 96, 100
food delivery couriers: career plans, 35; choice between service and gig platforms, 35–36; collective actions of, 33–34, 40; contractual relationships, 33, 37, 39; court decisions in cases related to, 34–35; deaths and injuries of, 35; demographics of, 33, 35; disciplinary actions against, 39; dispatch algorithms, 37; freedom of, 35; full-time, 36; identity confirmation, 39; management and control of, 37, 38, 39; meetings of, 38; migrant workers, 35; performance metrics, 37; prior work experiences, 35; remuneration of, 38–39; social insurance, 35; social media groups, 33; training of, 40. See also gig platform couriers (GPCs); service platform couriers (SPCs)
Forester, Jay, 414
Forsythe, Diana, 222
FORTRAN programming language, 255–56, 258–59, 276, 277, 415

Fowler, Donald, 284
Fraser, Colin, 362
Frechette, Claude, 171, 172
free and open source software (FOSS): business sector and, 128, 133; copyright regime and, 132, 137; developers of, 141; expansion of influence of, 132; licensing, 129, 133, 134, 142; moral dilemma of, 138; politics and, 128, 134–40; power structures, 128; principles of, 127, 141; self-regulation of, 14, 133
Free Software Foundation (FSF), 129, 130
Free Software Movement (FSM), 127
free spaces for collective contention, 28, 30, 32
Frehill, Lisa, 241
Future of Privacy Forum, 73

Gaithersburg, Maryland, 425
Galang, Adrianne John R., 367n16
Gallagher, Mary, 26
Gallati, Robert, 184
Galloway, Alexander, 355
Galton, Francis, 174, 175, 185
gamification, 24, 37
Gandhi, Indira, 333
García Llorens, Mariel, 195
Gartner Group, 274
Gates, Bill, 70, 131, 132, 412
Gates, Kelly, 188
Gebru, Timnit, 429
Geiger, Rose Mary, 251n23
General Data Protect Regulation (GDPR): collection of personal data, 89; compliance with, 92, 93–94, 101; data access rights, 88, 90; DNT standards, 80; efficacy of, 88; evidence of trickledown effect, 67, 95, 102; expectations from, 99–100; implementation of, 87; in multinational context, 77, 87, 91; passing of, 76; scope of, 87, 94
General Dynamics, 267
General Public License (GPL), 129, 130
generative artificial intelligence (GenAI), 142, 411, 412, 422, 428
Gershon, Ilana, 402
gig platform couriers (GPCs): bystanders, 53–54; collective actions of, 40–41, **42**, 50–55, 57; communication with platform representatives, 50; complaints of, 42, 47; demographic characteristics, 55, 57–58; disciplinary rules, 48; exploitation of, 50; freedom to decline dispatch assignments, 48; legal control of, 48–50; organizational control of, 49, 50–51; remuneration of, 47, 48; *vs.* service platform couriers, 25; social media groups, 50, 53; solidarity of, 56; technological control of, 26, 47–48; training sessions, 49
gig platforms: architecture of, 26–27, 38, **41**; construction of grievances under, 47; contractual power of, 57; control and management under, 53, 54, 56–57; dispatch algorithms, 38, 47; legal and technological power of, 49–50; management and supervision tasks, 38; organizational design of, 39–40; patterns of collective contention, **43**; *vs.* service platforms, 25–26
Gillespie, Tarleton, 65
GitHub, 133, 138, 145
Global Privacy Control (GPC), 67, 77–78
Gluck-Thaler, Aaron, 178
GNU ("GNU's not Unix") project, 129, 130
Goggin, Gerard, 194, 206
Goods, Caleb, 24
Google: advertising revenue, 64; business model, 132, 272; JSON License project, 139; market share, 427; platform, 116; privacy regulations, 64–65, 73, 76, 79; public relations, 11, 271; release of Bard, 422, 428; responses to DSAR requests, 94, **96,** 100; search engine, 2, 8, 15, 112, 221; smartphone operating system, 119; US copyright law and, 11–12
Google Analytics, 64
Google Takeout, 97
GPL version 3 (GPLv3), 130, 136
Grad, Burt, 269
Grandinetti, Russ, 160
graphical user interfaces (GUIs), 305, 307, 321n39, 347, 354–55, 425
grappling with code, 17–19
Greenspan, Alan, 117
Greenwood, Joy, 267
Griesbach, Kathleen, 29
Griffiths, Stephanie, 350
Grove, Andy, 412
Grubhub, 57

H1-B visas, 280–82, 283, 285, 287, 289, 343n50
Hachette, publisher, 151, 152, *153,* 154, 155, 158, 161, 163, *164*

hackathons, 19, 388–89, 391, 393, 396–97, 398, 400, 401–4, 406n1
Hack CDMX event, 388–89, *389,* 391–92, 400
Hacker, Jacob, 5
hackers: backgrounds of, 405–6; code work of, 388, 390, 391, 399, 403, 404–5; communities of, 391, 392; connection with potential employers, 390; demographics of, 392, 405; dependence on tech companies, 391; entrepreneurship of, 392, 395; ethics of, 389–90, 391; government programs for, 399; politics and, 399, 400, 401–2; popular conception of, 392; salaries, 400–401
Haigh, Thomas, 301, 420, 421
Halloween Documents, 131, 132
Halverson, Barbara, 241
Hamid, Sarah T., 187, 188
Hamraie, Aimi, 199
Hardin, Caroline D., 406n3
hardware and software products, 64–65, 256, 257–58, 268, 301, 304, 415
Hare, Robert, 345, 347–48, 349, 366, 367n5
Hare's Psychopathy Checklist–Revised, 348, 349, 350
Harinarayan, Venkey, 423
HarperCollins, publisher, 151, *153,* 154, 159, 160, 161–62, *164*
Harrington, Joseph, 110, 111
Harvard Business Review, 295, 308
Hayek, Friedrich, 116
Hebb, Donald: *The Organization of Behavior,* 415
HEC-2M computer, 336
Hellman, Martin, 378, 379, 380, 383, 385n28, 386n36, 387n43
Henke, Louise, 251n23
Hertel-Fernandez, Alexander, 5
Heyck, Hunter, 299
Hicks, Mar, 18, 241, 261, 288
Hildebrandt, Mireille, 81n7
Hill, Jack, 238
Hiltz, Starr Roxanne, 200, 204, 205; *Network Nation,* 201–2
Hindustan Computers Limited (HCL), 333
Hinton, Geoffrey, 364
"HiPPO" ("highest-paid person's opinion"), 308, 309
Hippocratic License, 139, 140
Hirschmann, Nancy, 199
Hollerith, Herman, 258
Holtman, Koen, 69
Hopper, Grace Murray, 261
HotWired.com, 68

House Subcommittee on Government Information and Individual Rights, 380, 381
HTTP cookies, 67, 68–72, 75–76
"HTTP State Management Mechanism," 69
human-computer interaction (HCI) models, 425–26
human-computer interfaces: as host of aggression, 356; human vulnerability in, 346, 361, 366; *vs.* interaction between person and machine, 354–55; psychopathy and, 344–46, 347, 354, 356, 364–65; semantic deficiency of, 356; as site of masking, 346, 355, 365; studies of, 354, 355; user-friendly forms, 355, 365
Humby, Clive, 426
Hyman, Richard, 23

IBM (International Business Machines): achievements of, 255–56, 258–59, 260; AI and, 415; antitrust lawsuits against, 256, 257, 259, 262, 270, 271; company towns, 260; competitors of, 261, 264, 265, 268; Continuous Speech Recognition (CSR) group, 419; corporate culture, 255, 256; customer technical representatives (CT reps), 266; data-processing services, 268; Deep Blue project, 417; development of Data Encryption Standard, 378; employment strategies, 262–63; field workers, 264, 265; gender disparity, 17, 257, *257*, 261, 262, 266, 267; hardware and software products, 257–58, 415; historiography of, 255, 256, 257–58, 259–60, 270; maintenance and systems application, 261; market share, 415; open-systems standards, 114, 115; resistance to selling latest technologies to India, 326, 333; sale representatives, 262–63, 267; services-oriented model, 257, 258, 259, 269, 271, 272n2; speech recognition group, 363; systems engineers (SEs), 267, 268–69; "THINK" motto of, 255; transition to computing, 265, 416; women employees, 256–57; workforce fluidity, 265–66. *See also* System Service Women's Corps (SSWC)
IBM 650, 265
IBM Fellows, 266
IBM Service Bureau Corporation (SBC), 258, 262, 266, 268, 269, 270

IBM Service Bureau Division (SBD): creation of, 258, 261–62, 263, 271; data-processing operation, 265; during Great Depression, 263–64; offices location, 263; photo of workers at Singapore office of, *262*; plans for, 264; sale to CDC, 258, 262, 269; "Siberia" moniker, 262, 266; women workforce of, 269–70
IBM Systems Engineering, 267
IBM Systems Journal, 267
iBookstore, 148–49, 155, 157, 158
Ihde, Don, 355
iHeartRadio, 428
iLab, 393–95, *394, 395*, 397, 398, *398*, 404
Improving Web Advertising Business Group, 78
IMS database, 255, 259
India: Administrative Staff College of, 331–32, 333–34, 335, 338, 340; adoption of Western technologies, 325–27, 330–31, 332–33, 334–37, 338–39; banking system in, 326; British exports to, 327, 329, 330, 343n38; class divisions in, 329–30, 338–39, 340; computerization in, 325, 327, 330, 333–34, 340; data protection and regulation, 87, 89, 91–92, 94; digital payment system in, 212, 218; IT companies, 275, 276, 279, 280, 282, 338; native computing industry, 18, 326, 327; training centers for technocrats, 330, 331–34, 340; transnational companies in, 94–95; US economic sanctions, 275
Indian Institute of Technology (IIT), 330, 337
Indian IT workers: expertise of, 275, 283, 334–35, 339; gendered labor of, 328–29; in global IT services, role of, 327, 334, 399; H1-B holders, 285, 339, 343n50; mobility of, 280, 339; precarious situation of, 408n30; training of, 330, 331–32; US software industry and, 275–76, 280
Indian Statistical Institute, 331, 336, 337
Indian Supreme Court's 2017 Puttaswamy judgment, 100
Indigenous populations, biases against, 9
Informatics Inc., 260
information age, notion of, 383
informational asymmetry, 9
Information Processing Techniques Office (IPTO), 420–21, 424, 425

information symmetry, 116
information technology (IT): *vs.* Big Data, 18; development of, 3; management and, 295, 300, 301, 302; programmers' role in, 300–301, 302, 308; regulation of, 65, 413; studies of, 4–5; top companies in, 414
Infrastructure Protection Digest (journal), 285
initial public offerings (IPOs), 118
injustice frames of collective action, 27, 30, 40
Inman, Bobby, 381, 387n43
innovation, 306, 307, 322n44
Instagram, 65
Institute for Human-Centered Artificial Intelligence (IHCAI), 426
InstructGPT, 362, 364
Intel, 412–13
intellectual property laws, 129
intelligent tutoring systems (ITSs), 310–11, 312
Interbank, 231, 235n67
interdependence discourse, 377–78
International Chess Federation (FIDE) tournaments, 416
International Computers and Tabulators (ICT), 331
International Computers Limited (ICL), 332, 333
internet: advertising on, 64, 72, 78–79, 80; privatization of, 426
Internet Advertising Bureau (IAB), 72, 75–76
Internet Engineering Task Force (IETF), 69, 74
interoperability: concept of, 120
iPad, 148, 149, 150–51, 154, 162
iPhone, 77, 80, 150
iPod Touch devices, 150
Irani, Lilly, 276, 408n28, 423
iTunes Store, 150, 151, 156, *158*, 162

Jackson, Thomas Penfield, 413
Jacobs, Meg, 197
Jacobson, Gloria, 251n23
Jarkko, Javla, 350
Jelenik, Fred, 419
Jobs, Steve, 148, 150–51, 153, 154, 155, 157–58, 161–63, 164
Johnson, Lyndon B., 185
Johnson, Ruth, 251n23
Johnson, Steven, 355, 356
Johnson, Thomas, 381
Jones, Meg Leta, 13, 90
JSON License, 139, 140

Judiciary Committee report "Investigation of Competition in Digital Markets," 119
Junglee, 423

Kafer, Alison, 194
Kahn, David: *The Codebreakers*, 372, 378; debates over cryptography, 380–81, 382
Kahn, Lina, 413
Kaminka, Gal, 360
Kaminski, Margot E., 90
Kaminsky, Dan, 73
Kaphan, Shel, 69
Kaplan, Albert, 171, 172
Kaplan daguerreotype, *170*, 171–72
Kar, Sohini, 227
Kasparov, Garry, 417
Kay, Alan, 355
Kelkar, Shreeharsh, 18
Kennedy, Anthony, 119
Kent, Mike, 195
Kenya: cash economy, 212; M-Pesa project in, 212, 215, 222, 223, 229; techno-entrepreneurship in, 395
Kerr, Elaine, 204, 205
Kerr, Ian, 66, 81n7
Kessler-Harris, Alice, 242
Keynesian industrial policy, 306
key performance indicators (KPIs), 36–37
Khan, Lina, 271
Khan Academy, 315
Kindle: e-book pricing, 148, 149, 153; introduction of, 152; *vs.* iPad, 150–51; search for alternatives to, 152–53; text-to-speech feature of, 155, 160
Kindle Digital Text Platform, 160
King, John Leslie, 288
Kolata, Gina, 382, 383
Kolko, Gabriel, 119
Konheim, Alan, 378
Kristol, Dave, 69
Kronrod, Alexander, 417
Kuhn, Bradley M., 136
Kyle, Jamie, 127, 128, 138, 139

labor code, 2
labor control, 23–24, 28, 29–30
Lagardere, Arnaud, 152, 153
Lahiri, Uttama, 359
Lakatos, Imre: *Proofs and Refutations*, 173
Lamoreaux, Naomi, 9, 119
Lampson, Butler, 425
large language models (LLMs), 362, 363, 419, 422, 428
Larkin, Brian, 394

Lawless, William, 266
Lawrence, Peter: *The Making of a Fly*, 109
learning-by-doing, 311
Leavitt, Harold, 295, 300
LeCun, Yunn, 364
legal code, 87, 88, 97, 100–101, 102
legal defaults, 65, 66, 71, 72, 81n7
Lei, Ya-Wen, 12, 13
Leibowitz, Jon, 73
Lerna project, 127, 128
Lessig, Lawrence, 3, 65
Levy, Karen, 65–66
Leys, Ruth, 347
Li, Fei-Fei, 426
Li, Xiaochang, 363, 419
license plate identification, 185, 191n32
licenses: alternative models, 139, 140, 141–42; ethical, 138–40, 142; open source model, 130–31, 134–35, 137
Licklider, J. C. R., 420, 424
Light, Jennifer, 241
Limón, José: *Dancing with the Devil*, 406n1
Lincoln, Abraham, *170*, 171
Linkenhoker, Virginia, 264
Linker, Beth, 199
Linux, 133
LISP (LISt Processing), 418
Little, Peter: *Toxic Town*, 260
Lizabeth Cohen: *A Consumer's Republic*, 197
Llorens, Mariel García, 16
Loftus, Margaret, 248, 253n51
Lomnitz, Claudio, 397
Lomnitz, Larissa Adler, 397
Longmore, Paul, 197
"loose coupling," 401, 402, 404
López Obrador, Andrés Manuel (AMLO), 397, 398
Lowenthal, Alan, 76
Lukita platform, 231
Lynda.com, 315

Maccarrone, Vincenzo, 24, 55
machine learning, 49, 112, 145n26, 190n15, 295–96, 415–16, 420
machine-like personality, 344–45, 346, 347–50, 351, 354, 364
Machine Platform Cloud (McAfee and Brynjolfsson), 309
machines, anthropomorphization of, 344–45
MacKenzie, Donald, 88; *An Engine, Not a Camera*, 101
Macmillan, publisher, 151, 158, 161, 163, *164*
Mactaggart, Alastair, 77

Magee, Liam, 364
Mahalanobis, Prasanta Chandra, 335, 336, 337
Major League Hacking events, 406n3
Making IT Work (Yost), 267, 327
Making of a Fly, The (Lawrence), 109, **110**
"management information system" (MIS), 301, 302, 321n28
Mansfield Amendment, 425
Mansur, Jorge Antonio, 393, 395, 398, 399
Martin, Philip, 281
masked personality, 345, 346, 351, 353, 355, 360–61, 366
Mask of Sanity (Cleckley), 345, 350
Massachusetts Institute of Technology (MIT), 310, 314, 403, 408n40, 424
massive open online courses (MOOCs), 298, 309, 310, 314, 317
Matsushita Electric Industrial Company v. Zenith Radio Corporation, 118
Mauran, Michael, 350
Maurer, Bill, 227
Maurice v. Judd, 92
Mayer, Jonathan, 74, 79
Maynard, Terrill, 285–86
Mbembe, Achille, 394
McAfee, Andrew, 295, 308, 309
McCarthy, John, 414, 418, 419, 421, 424
McCulloch, Warren, 415
McDonald, Aleecia, 74, 75, 76
McIntosh, Lukas, 406n3
McIntosh, Madeline, 156
McNeal, Ramona, 193
Mechanical Turk (MTurk), 276, 423–24
Medina, Eden, 12, 174, 189n11, 255
mega cloud services, 134–35
Mehta, Dewang, 286
Meituan (platform company), 24, 25, 36, 37, 39, 40, 52–53
Melo, Francisco, 370n54
Mendon-Plasek, Aaron, 420
Menon, Nikhil, 336
Merton, Robert K., 382
Meta, 424, 428
México: computer talents in, 388, 392, 395, 399–401, 403–4; crime, 405; economic growth, 392–93; hackathons, 19, 388–89, 391, 393, 396–97, 398, 400, 401–4, 406n1; Peña Nieto's presidency, 393; political development, 397–98, 405–6; social unrest, 405; university enrollment, 393; urban office

México (*cont.*)
workers in, 401, 408n33; youth unemployment, 396
Meyer, Joseph A., 386n36
Mhyrvold, Nathan, 70
microfinance, 223, 229
Microsoft: acquisition of GitHub, 133; business model, 272; DNT standards, 74, 75; dot.com bubble crisis and, 412–13; FOSS community and, 132–33; lawsuit against, 145, 271, 413; market share, 427–28; OpenAI, 424; software copyright policy, 131, 132, 133
Microsoft Internet Explorer, 74, 131
microwork, 276, 423
Migration Policy Institute (think tank), 282
Miller, Clair, 241, 253n51
Milo (robot), 361
Minich, Julie, 194
Minnesota computing industry, 237, 250n5
Minsky, Marvin, 419, 421, 424
mobile money, 214, 215, 218, 221, 222, 227, 229, 231
Modelo Perú (Peru Model): bank partners of, 212, 217, 223, 226–27, 228, 231, 235n67; development of, 212–13, 214–15; government support of, 226; implementation of, 225–26; promotional campaign, 216, 216–18, 219, 220, 228–31; stakeholders, 226; targeted users, 213–14, 216–17, 219–21, 222–23, 224–25, 230; telecom partners of, 222, 223; testing of, 233n21. *See also* Bim platform
Moerer, Keith, 155, 156, 157
MongoDB Inc., 135–37, 141, 146n36
Montulli, Lou, 68, 69
Monzo, 94, **96**, 97
Moonves, Leslie, 153, 154, 158
Mora, Mariana, 405
Moran, Thomas, 425
Moshkina, Lilia, 359
Mossberg, Walt, 162
Mossberger, Karen, 193
most favored nation (MFN) provision, 157, 158, 161, 163
Motorola (telecommunications company), 274
Mozilla Firefox, 74, 79, 131, 132, 137
M-Pesa project, 212, 215, 222, 223, 229, 233n33
mug shots, 174, 185, 186–87
Mullaney, Frank, 238, 239
Mulvin, Dylan, 17
Mundie, Craig, 132

Murdoch, James, 161
Murphy, Michelle, 230
Murray, Brian, 160
Musk, Elon, 413
Myers, Polly, 241

Nadella, Satya, 133, 428
Nader, Ralph, 70
Nair, Rukmini Bhaya, 279
Nakamura, Lisa, 195
Napack, Brian, 158
Narayan, Devika, 287
Narayanan, Arvind, 74, 79
NASA Mission Control, 266
National Association of Elementary School Principals, 70
National Association of Software and Service Companies (NASSCOM), 282
National Bureau of Standards (NBS), 374, 375, 376, 378, 380
National Infrastructure Protection Center (NIPC), 285
National Science Foundation, 184
National Security Agency (NSA), 374, 378, 379–80, 381–82, 384, 415
natural language processing (NLP), 418, 421
Nelson, Alondra, 429
Nelson, Ted: *Computer Lib / Dream Machines*, 426
Netflix, 307–8
Netscape Navigator, 68–69, 70, 131, 137
Netscape Public License (NPL), 137
Network Advertising Initiative (NAI), 71, 73
networked computing, 275, 319
Network Nation (Hiltz and Turoff), 201–2, 205
Neural Information Processing Systems Conference, 418
Newell, Allen, 175, 424, 425
Newell, Christopher, 206
New York State Identification and Intelligence System (NYSIIS), 123, 175, 184, 185–86, 188, 191n29
Nickell, Joe, 172
Nielsen, Kim, 198
Noble, Safiya, 8, 221; *Algorithms of Oppression*, 15
Norris, William, 236, 237, 240, 246, 247
Northrup Grumman, 428
Nourry, Arnaud, 163
NovaSoft Information Technology, 275
Nvidia, 272, 413, 414, 424
Nye, Joseph, 377

Office of Naval Research, 184
offshoring practices, 275, 280, 283
Olazaran, Mikel, 419
One Laptop Per Child development project, 226, 426
OneTrust privacy management service, 98
online ethnography, 33, 34
Online Store, 150
OpenAI, 271, 413, 422, 427–28
Open Courseware (OCW), 310
open education resources (OER) movement, 310
Open Innovation Network, 133
Open Learning Initiative (OLI), 298, 309, 311–13, 314, 318
Open Source Initiative (OSI), 130, 141
open source licensing, 130–31, 134–35, 137
open source software (OSS), 128, 130–31, 133
open-system standards, 114–15, 116
optical character recognition (OCR), 175
opt-in/opt-out consent, 64, 66, 68, 72, 73–74, 75, 76, 77
Organisation for Economic Co-operation and Development (OECD), 89
Organization for Ethical Source, 141
Outsourcer, The (Sharma), 338
outsourcing, 274–76, 279–80, 283, 286
Oxford Handbook of Affective Computing (Arkin and Moshkina), 359

Pagos Digitales Peruanos (Peruvian Digital Payments, or PDP), 218–20, 221, 225–26, 227, 234n46, 234n51
Palo Alto Research Center (PARC), 303, 355, 425
Pangrazio, Luci, 193
Papert, Seymour, 419
Parker, Donn B., 376
Parker, John, 251n11
Partneo software, 110
pattern recognition, 175–76, 190n15
Paytm, 95, **96**
PC (personal computer), 114–15, 129, 255, 271, 304, 411–12
Pearson, publisher, 157, 162
peer review practice, 386n39
Peña Nieto, Enrique, 393, 398, 399
Penguin, publisher, 151, 152, *153*, 157, 159, 161, 163, *164*
people with disabilities: citizenship of, 194, 195–96, 197–99, 207; civil rights of, 196, 202; computer-

mediated communication and, 194, 199–208; economic contribution of, 202; invisibility and exclusion of, 194–95, 199, 205; online identity, 206; social participation of, 205; study of, 206, 207

Perot, Ross, 260

personality disorders, 347–48

Peru: Andahuaylillas district, 215–16, 217, 218, 219, 220, 221, *225*, 229, 230; banks, 228, 231, 234n51, 235n58, 235n67; cash economy, 213, 215, 227–28, 231n5, 232n14; currency, 234n40; digital payment services, 16, 212, 213–14, 220, 227–28; informal jobs, 232n14. *See also* Modelo Perú (Peru Model)

Petelka, Justin, 13, 67

Petrick, Elizabeth, 354, 355

Philip, Kavita, 339

phrenology, 10, 182

Picasso, Pablo, 309

Pichai, Sundar, 428

Pierce, John, 363, 364

Pierson, Paul, 5

Pitts, Walter, 415

platform architecture: concept of, 13, 24, 55; legal and organizational design of, 30; multidimensionality of, 28–29; relationship to labor contention, 26–27, 29–30, 31–32, 54, 57

platform economy, 23–24, 29, 30, 32, 42, 55–56

platform expertise, 296, **296**, 297, 298, 302–9, 319

platforms: definition of, 305; implementation of, 306; modern monopolies and, 114; openness of, 115; power of, 116, 121; purpose of, 307; types of, 116, 304, 305; users of, 115

policing, use of technology in, 175, 185, 187, 191n32

political economy, 2–3, 5, 9–12

Pollock, Neil, 303

portable devices, 150

Porter, Beth, 315

Poskett, James, 10

Powell, Lewis, 118

power imbalances, 9, 17

Powers-Samas company, 330

Powers Tabulating Machine Company, 330

Preyer, Richardson, 380

pricing practices, 109–10, 111, 113, 114, 116, 117, 118, 149

Privacy Interest Group (PING), 74, 79

privacy policies: approaches to, 66–67, 92; business interests and, 77, 78–79; consent dilemma of, 66–68, 71–72, 80; legislative efforts, 76, 79–80

programming: construction of, 296–97; decision making and, 297, 298–300; definition of, 296, 319; elements of, 298–302; future of, 319; languages, 255–56; management and, 297; open online learning and, 309, 312; platform expertise and, **296,** 309; as styles of reasoning, 298; techniques, 295

Programming assemblage, 296, 298, 299, 302, 314

projective geometry, 190–91n22

Psychology of Human-Computer Interaction, 425

psychopaths: clinical observation of, 349–50, 351–52; deceptive and manipulative behaviors, 348–49, 350, 351, 361; definition of, 347–48, 349, 350, 366; machine-like personality of, 344–45, 346, 347–50, 351, 354; mask of, 351, 353, 366; nonverbal behaviors, 357, 369n36; practical limits of, 365; self-destructive behaviors, 349, 351, 352–54, 368n23; socially successful, 349, 353, 367–68n16

psychopathy: in context of criminality, 349, 353–54; as form of semantic disorder, 356–57; studies of, 345, 349, 351, 367n5

Psychopathy Checklist–Revised (PCL-R), 345

public cryptography, 19, 372, 373, 376, 380, 381, 384, 385n27

publishers: "Big Six" houses, 151, *153*, 154; e-book pricing strategy, 152; market value, 151; negotiations with Amazon, 148–49, 152–54, 163–64; negotiations with Apple, 148–54, 155–56, 160–61

Pugh, Emerson, 260, 261

punched cards, 256, 328–29, 330

Raff, Daniel, 9

Rajaraman, Vaidyeswaran, 338

RAND Corporation, 178, 300, 417

Random House, publisher, 151, *153*, 156, 161, 164, *164*

Ravneberg, Bodil, 206

Raymond, Eric, 130, 131

Reagan, Ronald, 270

Recoding Gender (Abbate), 261

Red Hat, 134

Redis Labs, 134, 135, 136, 146n35

Redmayne, Charlie, 160

Reed, Susan: "Executive Guide for Computer Security," 375

Rehabilitation Act Amendments (1973), 210n37

Reidenberg, Joel, 65

Reidy, Carolyn, 153, 154, 155, 158

Reko, Molly Lou, 249, 254n54

Reliance Jio, 95, **96**

Remington Rand Inc., 238–39, 251n12, 264

Resnick, Sandra, 248–49, 253n53

Richardson, Kathleen, 357, 360

rights consciousness, 102, 108n56

Robinson-Patman Act, 120

Robokind company, 361

robotics, 360, 361

Rogers, Carl, 418

Rometty, Virginia, 261

Rosaldo, Renato, 197

Rosenblat, Alex, 29

Rosenblatt, Frank, 414–15

Rosental, Claude, 226

Rosenthal, Caitlin, 9

Rosewell, James, 78, 79

Roy, Ananya, 229

Russell, Steve, 418

Russian Internet Research Agency, 356

Ryanair, 95, **96,** 98

Safari browser, 74

SAGE (Semi-Automatic Ground Environment), 300

Sammet, Jean, 266

Samuel, Arthur, 416

Samuel Checkers Playing Program, 416

Sargent, Daniel, 377

Sargent, John, 163

"satisficing," concept of, 189n6, 320n15

Saudi Arabia, 282

Saul, Kevin, 156–57

Savage, Ben, 79

Scardino, Marjorie, 162

Scassellati, Brian, 359

Schaefer, Ellen Kerksieck, 267

schizophrenia, 344, 351, 358

Schlombs, Corinna, 255

Schwartz, Paul M., 92

science and technology studies (STS), 221, 226, 230

Security and Exchange Commission (SEC), 118

Sefton-Green, Julian, 193

Semler, Elizabeth, 16, 261

Senate Committee on Commerce, Science, and Transportation, 73

Senate Special Committee on the
    Year 2000 Technology Problem,
    285
Server Side Public License (SSPL),
    135, 136–37, 140
service platform couriers (SPCs):
    collective actions of, 40–41, **42,**
    44, 46–47, 54–55; complaints of,
    42–43, 44; delivery tasks per day,
    43; demographic characteristics,
    55, 57–58; disciplinary rules,
    43–44; *vs.* gig platform couriers,
    25; legal control of, 42–45; orga-
    nizational control of, 45–46; rela-
    tionship with management, 43,
    45–46; remuneration of, 44, 45,
    46; solidarity of, 56; technological
    control of, 26, 42–43
service platforms: architecture of, 36,
    41; control and management in,
    25–26, 29, 30–32; dispatch algo-
    rithms, 43, 44, 45; labor dimen-
    sion of, 56; patterns of collective
    contention, 40–43, **43**
Shannon, Claude, 416
Sharma, Dinesh, 280, 282; *The Out-
    sourcer,* 338
Sherman Antitrust Act (1890), 111,
    117, 413
Shneiderman, Ben, 425; *Human-
    Centered AI,* 426
Shortliffe, Edward, 421
short message service (SMS), 233n33
Simon, Herbert, 175, 189n6, 300,
    320n15, 417, 424; *Administrative
    Behavior,* 299
Simon & Schuster, publisher, 151,
    153, *153,* 154, 158, 161, *164*
Sink, Eric, 69
Siri, 16, 421
Skafish, Peter, 8
Slayton, Rebecca, 375
small businesses, 64, 80
Sobel, R., 261
social media, manipulation of,
    355–56; platforms, 305
sociopathy, 344, 347–48, 349, 353,
    368n16
Söderström, Sylvia, 206
software: as commodity, 129, 260–61;
    copyright and, 129, 131–32; inter-
    pretation of concept of, 93; licens-
    ing, 138–40, 141–42; self-learning,
    416; studies of, 2, 5; systems, 113;
    tech corporations' control over,
    256. *See also* free and open source
    software (FOSS); open source soft-
    ware (OSS)

software code, 87, 88, 100–101
software designers, 141, 303–4, 307,
    308
Soghoian, Chris, 73
Soltani, Ashkan, 77, 79
source-available licenses, 134, 140,
    142
source code, 6, 7, 113, 127, 129–30,
    134, 135–36
Soviet Union, computing technology
    in, 326, 332, 334, 336, 337
Space Race, 336
Special Interest Group on Artificial
    Intelligence (SIGART), 421
speech recognition, 363
Sperry Rand Corporation, 239, 249,
    251n22
Spotify, 94, **96,** 97, 428
Srikrishna Committee report, 91,
    105n28
Stallman, Richard, 129
Stamm, Sid, 74
Stanford Research Institute (SRI),
    425
Stanford University, 57, 314
Starbird, Kate, 356
Stark, Luke, 29
Steinberg, Marc, 305
Stewart, Leo, 57, 356
Stockman, Daniel, 127
Strathern, Marilyn, 8
Subramanian, Ramesh, 200, 326, 333
Suchman, Lucy, 29, 355
Swift, Taylor, 412
Swisher, Kara, 67, 162
systems engineering, 256, 258, 260,
    261, 263, 264, 267
System Service Women's Corps
    (SSWC), *263,* 264–65, 266–67

Tabulating Machine Company, 258
Taj Technologies, 282–83
Tassinari, Arianna, 24, 55
Tata Consultancy Services, 282
Tata Iron and Steel Corporation, 329
tech companies, 121, 122, 123
techno-entrepreneurs, 392, 395–96
technological defaults, 66, 72, 81n7
Technological Indian, The (Bassett),
    326
technology: contracts and, 31, 65–66;
    in policing, 175, 185, 187, 191n32;
    religious critique of, 344; social
    impact of, 325–26, 340. *See also*
    digital technologies
techno-politics, 394, 403
Teen, Kathleen, 5
Temin, Peter, 9

Tencent, 3, 39
Thatcher, Charlie, 425
Thiel, Peter, 413
Thille, Candace, 312
Thornton, James, 239
Three Faces of Eve, The (film), 351
Time magazine: "Machine of the
    Year," 411
Title VII of the Civil Rights Act
    (1964), 245
Toffler, Alvin: *Future Shock,* 377
Tolbert, Caroline, 193
Toner, Helen, 422
Trachtenberg, Jeffrey: "Publisher in
    Talks with Apple over Tablet," 159
tracking of users' activity: consent
    issues, 66–67, 71, 80, 81; opt out
    from, 64, *65,* 67, 73–74; tech cor-
    porations' practices of, 64–65, 76,
    79
Tracking Protection Working Group
    (TPWG), 74–75
TRUSTe, 74
Tunki platform, 231
Turing, Alan: "The Imitation Game"
    paper, 418
Turner, Frederick: *From Countercul-
    ture to Cyberculture,* 426
Turoff, Murray, 200; *Network Nation,*
    201–2

Uber, 29, 39, 47, 48, 50, 57, 116, 307
Ullman, Jeffrey, 423
Umansky, Lauri, 197
UNAM (National Autonomous
    University of Mexico), 400, 403,
    408n32
United Kingdom: Competition and
    Markets Authority, 78; comput-
    ing industry in, 325, 326–27, 329,
    330; data protection legislation,
    103n10; Equality Act of 2010, 120;
    gendered labor, 327–28, 329; pri-
    vacy policies, 75–76; Right to In-
    formation Act (RTI), 90, 105n27;
    "Save the Cookies" campaign in,
    72; transfer of technology to India,
    325–26, 330–31, 332–33, 334–37,
    340, 343n38; youth unemploy-
    ment, 396
United States: census data, 258;
    computing industry in, 325–26;
    data protection and regulation, 89,
    90; economic sanctions on India,
    275; embargoes on selling comput-
    ers to China and the USSR, 326,
    336, 343n38; Freedom of Infor-
    mation Acts (FOIAs), 90, 376;

immigration labor codes, 276, 280–81, 287; import of computer talent, 339, 400–401; information privacy rights, 90–91; Privacy Act of 1974, 90, 376; privacy policies in, 75–76, 77; transnational companies in, 94–95
United States v. Apple, Inc., 149
United States v. Philadelphia National Bank, 117
unstructured supplementary service data (USSD), 233n33
users, 303, 304, 305, 306–8
US Immigration and Customs Enforcement (ICE), 127
Usselman, Steven, 255, 259, 268, 269, 272n2
U.S. v. AT&T, 114

Van Vechten, Anne, 265
Vatis, Michael, 285, 286
Veen, Alex, 24
Venmo, 95, **96**
Verizon, 428
VHS player, as platform, 304
Vidan, Gili, 19
Viljoen, Salomé, 67, 80

Virginia's Consumer Data Protection Act, 91
VisualBasic, 283

Waldman, Ari, 93
Walker, Lucille, 251n23
Wallet platform, 218, 221
Walsh, Taylor, 312
Ware, Willis, 379
Ware reports, 90, 104n17
war gaming, 417
Watson, Thomas J., Jr., 256, 266
Watson, Thomas J., Sr., 256, 264
Watson Research Center, 415, 419
Web 2.0, 426
Weizenbaum, Joseph, 362–63, 364, 418–19
Wells Fargo, 428
West, Sarah Myers, 373
Whisler, Thomas, 295, 300
Whittaker, Meredeth, 429
Wilcox, Maggie, 267
Williams, Al, 266
Williams, Robin, 303
Wilson, Elizabeth, 345
Wiretap Act, 71
Woodward, Joanne, 351

World Privacy Forum, 73
World Wide Web, 412
World Wide Web Consortium (W3C), 69, 74, 75, 76, 77, 78–79

Xalapa Hackathon, 396–97
Xerox PARC, 268, 425
Xiang, Biao, 280, 282

Y2K crisis: challenges of, 274, 278–79, 285; as labor issue, 274–76, 279–85, 286, 287–89; problem of missing digits, 277; repairs of, 275–76, 280, 289; roots of, 274–75, 277, 278; security concerns, 275, 285–86
Yape platform, 231
Yost, Jeffrey R., 17, 267; *Making IT Work,* 327
YouTube, 65, 304–5, 306, 315

Zaffiris, Robert, 160
Zames, Frieda, 198
Zboray, Michael, 274–75
Zeng, Ming, 116
Zhang, Lin, 396
Ziman, John, 382
Zuckerberg, Mark, 80, 428

Printed and bound by CPI Group (UK) Ltd, Croydon, CR0 4YY

09/07/2026

14916211-0001